PRINCIPLES OF ACTIVE NETWORK SYNTHESIS AND DESIGN

GOBIND DARYANANI

Bell Telephone Laboratories

JOHN WILEY & SONS, New York • Chichester • Brisbane • Toronto

Library of Congress Cataloging in Publication Data

Daryanani, Gobind.
Principles of active network synthesis and design.

Includes bibliographies and index.
1. Electric networks. 2. Electronic circuit design.
3. Electric filters. I. Title.

TK454.2.D27 621.319′2 76-20659
ISBN 0-471-19545-6

Printed in the United States of America

10 9 8 7 6 5

To Carol

PREFACE

Integrated circuit technology profoundly influences the design of networks for voice and data communication systems. Integration allows the realization of these networks with small-size and low-cost resistors, capacitors, and active elements; thereby eliminating the need for inductors, which are relatively bulky and expensive. Furthermore, active RC networks provide advantages their passive counterparts do not, such as standardization and modularity of design, switchability, and ease of manufacture. These features have revolutionized the design of modern voice and data communication systems. More and more, the engineer is being faced with the challenges and problems of active-RC network design. The purpose of this book is to provide the knowledge to meet these challenges.

The approach used in the book is to develop the fundamental principles of active and passive network synthesis in the light of practical design considerations. Active Network Synthesis is a particularly good vehicle for introducing many general design concepts, such as performance versus cost trade-offs, technological limitations, and computer aids. These ideas are presented in a simple way to allow assimilation by the undergraduate electrical engineer, and are closely related to the practical world of engineering.

The book is suitable for a basic course on network synthesis or an intermediate course on circuits. The first two chapters describe some simple analysis tools and basic properties of active and passive networks. In Chapter Three the student is introduced to the world of filters: active, passive, electromechanical, and digital. Examples from voice communication systems are used to illustrate the applications of the basic filter types. In Chapter Four, the filter approximation problem is discussed, with stress on the use of the standard approximation functions rather than on their theoretical development.

An important criterion in a practical design is the sensitivity of the resulting circuit to deviations in elements caused by manufacturing tolerances and environmental changes (i.e., temperature, humidity, and aging). In keeping with the practical orientation of the book, sensitivity is treated in Chapter Five, prior to discussing the synthesis of circuits. This permits the synthesis steps to be closely linked with this all-important figure of merit and also allows alternate circuit realizations to be compared on the basis of their sensitivities.

The synthesis of passive RLC networks is considered in Chapter Six with emphasis on the synthesis of the double-terminated ladder filter, a structure most often used in the design of passive filters. This structure also serves as a starting point for the synthesis of the coupled active filters described in Chapter Eleven.

Chapters Seven to Ten deal with operational amplifier realizations of the biquadratic function, which is the fundamental building block used in both cascaded and coupled active filters. The single amplifier circuit realizations are the subject of Chapters Eight and Nine, and the three amplifier realizations are covered in Chapter Ten. These circuits are compared on the basis of sensitivities to passive and active elements, spreads in component values, ease of tuning, and types of filter functions that can be realized. A brief introduction to the design of coupled filters and of gyrator and frequency-dependent-negative-resistor realizations is presented in Chapter Eleven.

In Chapter Twelve the nonideal properties of the operational amplifier and their effects on filter performance is explored in greater detail than in the preceding chapters. Finally, Chapter Thirteen describes the complete design sequence, emphasizing computer aids, cost minimization, and design optimization. The last part of this chapter briefly describes the discrete, thick-film, thin-film, and integrated circuit technologies used in manufacturing the filter, concentrating on the principles of design instead of on specific details, which are expected to change with technological advances.

Two computer programs are included in the text discussion (Appendix D) as aids in design. The MAG program computes the magnitude, phase, and delay of functions; and the CHEB program evaluates the Chebyshev approximation function for a given set of filter requirements. These programs are written in ANSI FORTRAN IV, which should be compatible with most computers. Copies of the program on cards can be obtained from me. Equations and sample tables (Chapter Four) can be used in lieu of the computer programs.

Although the book is primarily aimed at the undergraduate level, it can certainly be used for a first-year graduate course, and by engineers entering the field of active filters. The principal prerequisite is a basic circuits course. The book is designed to be covered in a one-semester course but, if need be, several of the sections in Chapters Six, Eleven, and Twelve may be omitted without loss of continuity. At a minimum, Chapters One to Five and Seven to Ten should be covered in the course. In a graduate course the technical publications referenced at the end of each chapter may be used as supplementary material.

The material for this book evolved from my six years of work in the Network Analysis and Synthesis Department at Bell Laboratories. This department has been deeply involved in the area of active filters since the inception of this field. The present form of the book originated from an undergraduate course I taught at Southern University (Baton Rouge), where I was a Visiting Professor on a program sponsored by Bell Laboratories and from a similar one-semester course given to Bell Laboratories engineers as part of the company's continuing education program.

Gobind Daryanani

ACKNOWLEDGMENTS

It is with great pleasure that I take this opportunity to thank my colleagues at Bell Laboratories for their help and encouragement. In particular I owe much to Paul Fleischer for his complete and thorough review of the entire book. He made significant improvements in the choice of material for the book, the manner of presentation, and clarification of many ideas.

I am very grateful to my colleagues Dan Hilberman, Joseph Friend, James Tow, Renato Gadenz, Douglas Marsh, Ta-Mu Chien, George Thomas, and George Szentirmai for their thorough reviews and many excellent suggestions on the sections related to their areas of expertise. I am especially grateful to my wife, Carol, for her help with the computer programs.

I thank Bell Laboratories for the support provided me in the writing of this book. Specifically, I thank my department head Carl Simone for his constant encouragement. My appreciation also goes to Darlene Kurotschkin and Patricia Cottman for the typing of the manuscript.

G. D.

CONTENTS

4. The Approximation Problem

5. Sensitivity

6. Passive Network Synthesis

7. Basics of Active Filter Synthesis

8. Positive Feedback Biquad Circuits

9. Negative Feedback Biquad Circuits

10. The Three Amplifier Biquad

11. Active Networks Based on Passive Ladder Structures

12. Effects of Real Operational Amplifiers on Active Filters

13. Design Optimization and Manufacture of Active Filters

APPENDIXES

1.

NETWORK ANALYSIS

In order to develop the design procedures for active and passive networks, it is first necessary to have good analysis techniques. While there are several different methods* for analyzing these networks, *nodal analysis* will be used in this book. This method of analysis is simple, quite general, and very suitable for active and passive filter circuits. Although it is assumed that the student is familiar with the principles and use of nodal analysis, a brief review is given in this chapter. In particular, the analysis of circuits containing *operational amplifiers, resistors, and capacitors*, which are the elements constituting most active filters, is covered in detail. The examples chosen not only review nodal analysis but also serve to introduce some elementary principles of synthesis. Computer aids that can be used for the analysis of networks are referenced at the end of the chapter.

1.1 RLC PASSIVE CIRCUITS

In this introductory section we review the s domain nodal analysis of passive networks by considering the following simple example of a circuit containing resistors, capacitors, and an independent current source.

Example 1.1

Find the s domain function $V_3(s)/I_1(s)$ for the circuit shown in Figure 1.1*a* (known as a bridged-T network).

Solution

The first step in the analysis is to express the admittance of each element in the s domain, as shown in Figure 1.1*b*. In this circuit the voltages at nodes 1, 2, and 3 with respect to ground are designated $V_1(s)$, $V_2(s)$, and $V_3(s)$, respectively.

The node equations are obtained by using Kirchhoff's current law at nodes 1, 2, and 3, as follows:†

node 1:

$$\frac{1}{R_1}(V_1 - V_2) + sC_1(V_1 - V_3) = I_1$$

or

$$\left(\frac{1}{R_1} + sC_1\right)V_1 - \frac{1}{R_1}V_2 - sC_1V_3 = I_1 \tag{1.1}$$

* Some other methods [4] for analyzing networks use mesh analysis, the indefinite matrix approach, signal-flow graph techniques, and the state-space approach.

† Hereafter, I and V are used to mean $I(s)$ and $V(s)$.

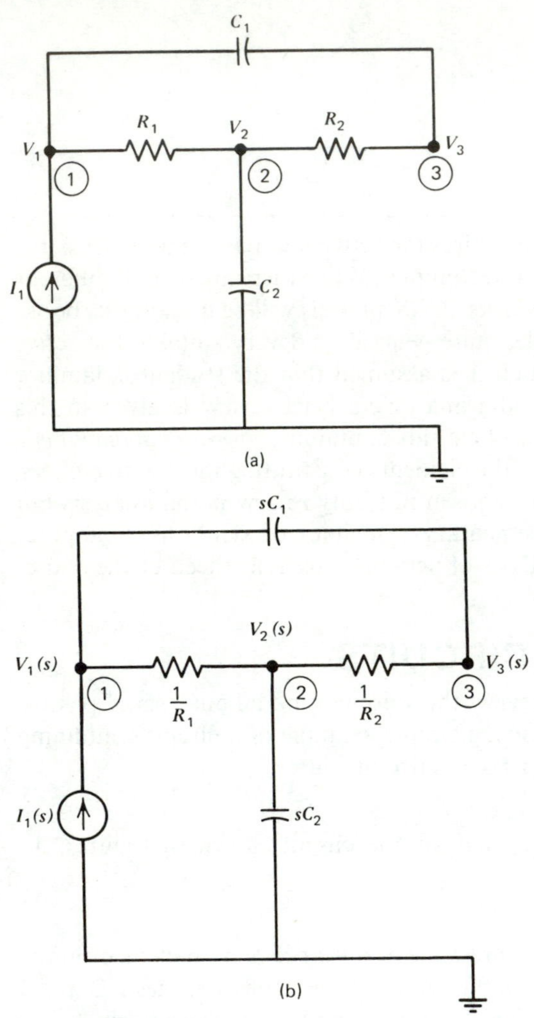

Figure 1.1 (*a*) Circuit for Example 1.1.
(*b*) Circuit showing admittances in *s* domain.

node 2:

$$\frac{1}{R_1}(V_2 - V_1) + \frac{1}{R_2}(V_2 - V_3) + sC_2V_2 = 0$$

or

$$-\frac{1}{R_1}V_1 + \left(\frac{1}{R_1} + \frac{1}{R_2} + sC_2\right)V_2 - \frac{1}{R_2}V_3 = 0 \qquad (1.2)$$

node 3:

$$\frac{1}{R_2}(V_3 - V_2) + sC_1(V_3 - V_1) = 0$$

or

$$-sC_1V_1 - \frac{1}{R_2}V_2 + \left(\frac{1}{R_2} + sC_1\right)V_3 = 0 \tag{1.3}$$

The nodal matrix representation of the above equations is

$$\begin{bmatrix} \frac{1}{R_1} + sC_1 & -\frac{1}{R_1} & -sC_1 \\ -\frac{1}{R_1} & \frac{1}{R_1} + \frac{1}{R_2} + sC_2 & -\frac{1}{R_2} \\ -sC_1 & -\frac{1}{R_2} & \frac{1}{R_2} + sC_1 \end{bmatrix} \begin{bmatrix} V_1 \\ V_2 \\ V_3 \end{bmatrix} = \begin{bmatrix} I_1 \\ 0 \\ 0 \end{bmatrix} \tag{1.4}$$

Using Cramer's rule [4], V_3/I_1 is given by

$$\frac{V_3}{I_1} = \frac{\begin{vmatrix} \frac{1}{R_1} + sC_1 & -\frac{1}{R_1} & 1 \\ -\frac{1}{R_1} & \frac{1}{R_1} + \frac{1}{R_2} + sC_2 & 0 \\ -sC_1 & -\frac{1}{R_2} & 0 \end{vmatrix}}{\begin{vmatrix} \frac{1}{R_1} + sC_1 & -\frac{1}{R_1} & -sC_1 \\ -\frac{1}{R_1} & \frac{1}{R_1} + \frac{1}{R_2} + sC_2 & -\frac{1}{R_2} \\ -sC_1 & -\frac{1}{R_2} & \frac{1}{R_2} + sC_1 \end{vmatrix}} \tag{1.5}$$

$$= \frac{\frac{1}{R_1R_2} + sC_1\left(\frac{1}{R_1} + \frac{1}{R_2} + sC_2\right)}{\left\{\begin{array}{l} \left(\frac{1}{R_1} + sC_1\right)\left[\left(\frac{1}{R_1} + \frac{1}{R_2} + sC_2\right)\left(\frac{1}{R_2} + sC_1\right) - \frac{1}{R_2^2}\right] \\ \qquad - \frac{1}{R_1}\left(\frac{sC_1}{R_2} + \frac{1}{R_1R_2} + \frac{sC_1}{R_1}\right) \\ \qquad - sC_1\left(\frac{1}{R_1R_2} + \frac{sC_1}{R_1} + \frac{sC_1}{R_2} + s^2C_1C_2\right) \end{array}\right\}} \tag{1.6}$$

which simplifies to

$$\frac{V_3}{I_1} = \frac{s^2 + s\left(\frac{1}{R_1 C_2} + \frac{1}{R_2 C_2}\right) + \frac{1}{R_1 R_2 C_1 C_2}}{s^2\left(\frac{1}{R_1} + \frac{1}{R_2}\right) + s\left(\frac{1}{R_1 R_2 C_1}\right)} \tag{1.7}$$

Observations

1. Since nodal analysis uses Kirchhoff's current law, it is most convenient if the independent sources are current sources. Most circuits, however, are driven by voltage sources. The formulation of the nodal equations of circuits containing voltage sources requires the use of Norton's theorem, which states that a voltage source V in series with an impedance Z can be replaced by a current source $I = V/Z$, in parallel with the impedance Z. Figure 1.2 illustrates this equivalence. Typically, Z would be the internal impedance of the voltage source.

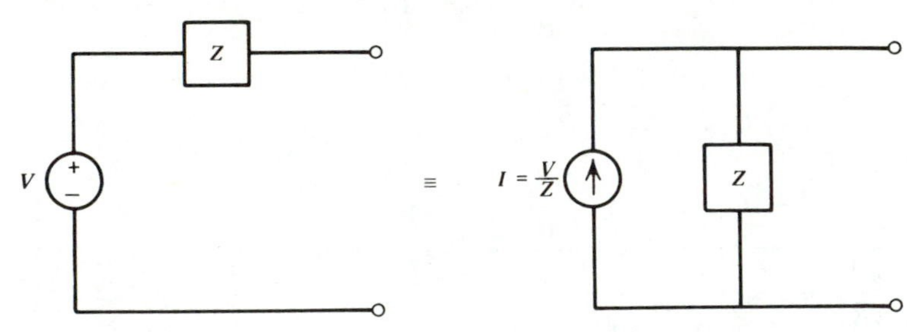

Figure 1.2 Norton's equivalent circuit of a voltage source in series with an impedance.

2. The nodal determinant, which is the denominator of Equation 1.5, is symmetrical about the diagonal. Thus the term in the ith row and jth column [i.e., the (i, j) term], is equal to the (j, i) term. This is a characteristic property of the nodal (and mesh) determinants of RLC networks.*
3. Suppose we had the problem of finding a network that will realize the function:

$$\frac{V_0}{I_{IN}} = \frac{s^2 + cs + d}{ns^2 + as} \tag{1.8}$$

where the coefficients c, d, n, and a are positive numbers. Since this function has the same form as Equation 1.7, it should be possible to realize it using

* All reciprocal networks have this property [4].

the circuit of Figure 1.1a. The element values of the desired circuit can be obtained by equating the coefficients of equal powers of s in (1.7) and (1.8):

$$n = \frac{1}{R_1} + \frac{1}{R_2} \tag{1.9}$$

$$a = \frac{1}{R_1 R_2 C_1} \tag{1.10}$$

$$c = \left(\frac{1}{R_1} + \frac{1}{R_2}\right)\frac{1}{C_2} = \frac{n}{C_2} \tag{1.11}$$

$$d = \frac{1}{R_1 R_2 C_1 C_2} = \frac{a}{C_2} \tag{1.12}$$

From (1.11) and (1.12) it is seen that the given coefficients must satisfy the relationship

$$\frac{n}{c} = \frac{a}{d} \tag{1.13}$$

and the capacitor C_2 is given by

$$C_2 = \frac{n}{c} = \frac{a}{d}$$

One choice of resistors that satisfies Equations 1.9 is

$$R_1 = \frac{2}{n} \qquad R_2 = \frac{2}{n}$$

Substituting in (1.10), the remaining unknown, C_1, is

$$C_1 = \frac{n^2}{4a}$$

Thus we see that if the given coefficients satisfy the relationship $n/c = a/d$, the transfer function of Equation 1.8 can be realized by the circuit of Figure 1.1a with

$$C_1 = \frac{n^2}{4a} \qquad C_2 = \frac{a}{d} \qquad R_1 = \frac{2}{n} \qquad R_2 = \frac{2}{n} \tag{1.14}$$

The above discussion indicates that whenever a circuit is **analyzed** the results of the analysis can be used for the **synthesis** of a related class of functions.

4. Let us next consider the synthesis of a slightly different function:

$$\frac{V_0}{I_{IN}} = \frac{s^2 + cs + d}{ns^2 + as + b} \tag{1.15}$$

A comparison with Equation 1.7 tells us that the circuit of Figure 1.1*a* will not work if *b* is nonzero. The questions the reader may now ask are:

(a) Can the function be realized with real (nonnegative) elements using a different circuit?

(b) Suppose the function is determined to be realizable; how does one proceed to find the required circuit and the element values for the circuit?

The first question relates to the **realizability** of functions—we will have more to say about this in Chapter 2. The answers to the second question constitute the synthesis of networks, and the major part of this text is devoted to this problem. ■

1.2 RLC CIRCUITS WITH ACTIVE ELEMENTS

Thus far we have discussed the analysis of *RLC* circuits with independent current and voltage sources. Next we will consider the analysis of *RLC* circuits containing active devices. The model of an active device will always include a voltage or current source whose value depends on a voltage across, or a current through, some other part of the circuit. Thus, to be able to analyze circuits containing active devices, it becomes necessary to study the nodal formulation of circuits with *dependent* current and voltage sources.

1.2.1 DEPENDENT CURRENT SOURCES

The two types of dependent current sources encountered in active networks are the voltage controlled current source (*VCCS*, Figure 1.3*a*) and the current controlled current source (*CCCS*, Figure 1.3*b*).

Examples of these types of sources are the *VCCS* model of the transistor (Figure 1.4*b*), and the *CCCS* model of the transistor (Figure 1.4*c*). The models shown are commonly referred to as hybrid-π models.

The first step in the nodal equation formulation of these circuits is to express

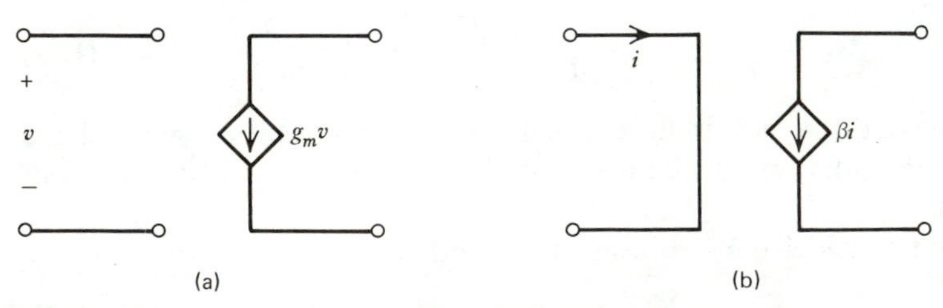

Figure 1.3 (*a*) Voltage controlled current source (*VCCS*). (*b*) Current controlled current source (*CCCS*).

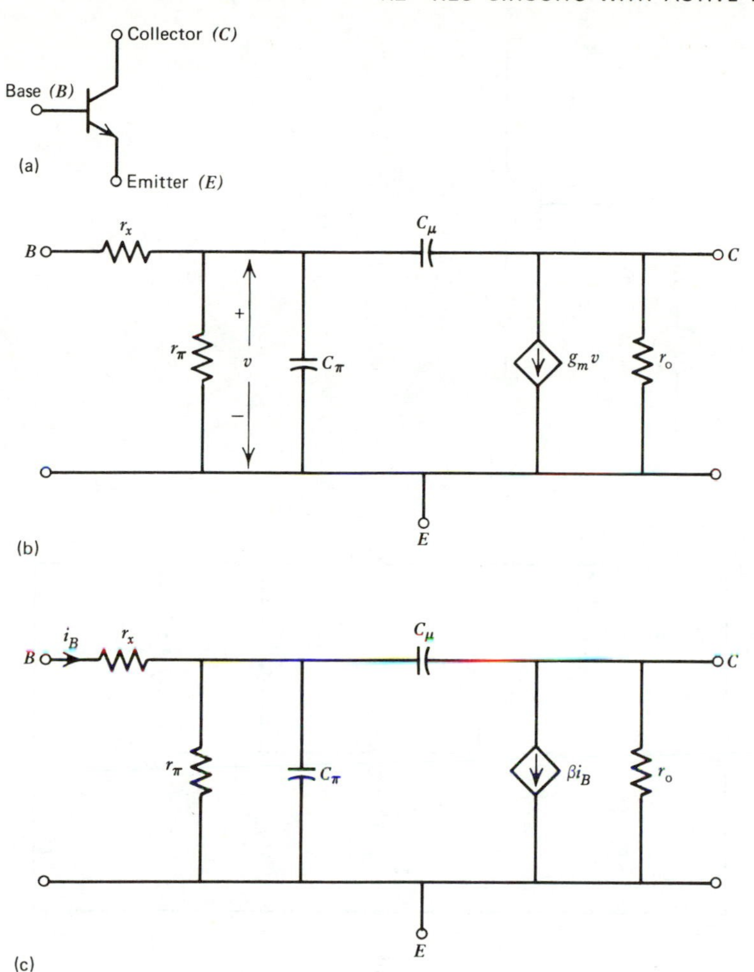

Figure 1.4 Hybrid-π models for a transistor: (*a*) Symbol; (*b*) *VCCS* model; (*c*) *CCCS* model.

the dependent current sources in terms of the circuit node voltages. The node equations can then be written as before, treating the dependent current sources as if they were independent sources. Finally, the equations are rearranged so that only the independent sources occur on the right hand side of the equality. The following example illustrates the procedure.

Example 1.2

Find the nodal matrix equation of the transistor circuit of Figure 1.5*a*, using the hybrid-π model of Figure 1.4*b*. Note that for the purpose of ac analysis, the collector of the transistor is effectively at ground potential.

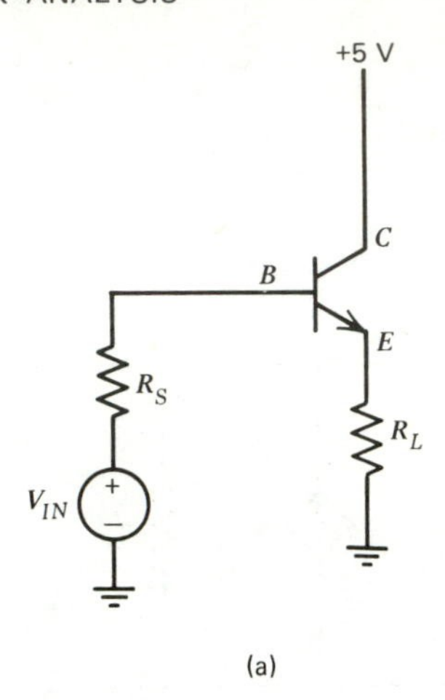

(a)

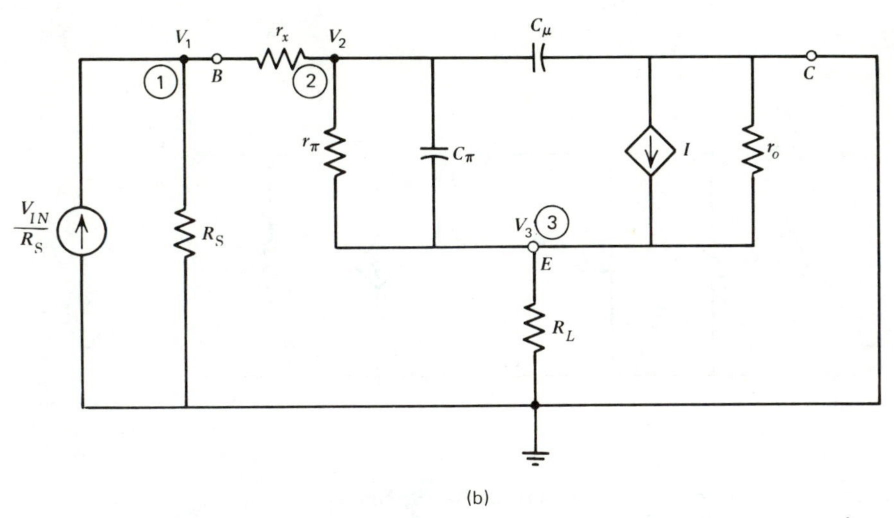

(b)

Figure 1.5 (*a*) Circuit for Example 1.2. (*b*) Equivalent circuit.

Solution

The circuit, with the transistor modeled, is shown in Figure 1.5*b*. In this figure the current source I is given by

$$I = g_m(V_2 - V_3)$$

The node equations are

node 1:

$$V_1\left(\frac{1}{R_S} + \frac{1}{r_x}\right) - V_2\left(\frac{1}{r_x}\right) = \frac{V_{IN}}{R_S} \tag{1.16}$$

node 2:

$$-V_1\left(\frac{1}{r_x}\right) + V_2\left(\frac{1}{r_x} + \frac{1}{r_\pi} + sC_\pi + sC_\mu\right) - V_3\left(\frac{1}{r_\pi} + sC_\pi\right) = 0 \quad (1.17)$$

node 3:

$$-V_2\left(\frac{1}{r_\pi} + sC_\pi\right) + V_3\left(\frac{1}{R_L} + \frac{1}{r_0} + \frac{1}{r_\pi} + sC_\pi\right) = g_m(V_2 - V_3) \quad (1.18)$$

The third equation is rearranged so that all of the node voltages are on the left side of the equality:

$$-V_2\left(\frac{1}{r_\pi} + sC_\pi + g_m\right) + V_3\left(\frac{1}{R_L} + \frac{1}{r_0} + \frac{1}{r_\pi} + sC_\pi + g_m\right) = 0$$

The nodal matrix equation is, therefore:

$$\begin{bmatrix} \frac{1}{R_S} + \frac{1}{r_x} & -\frac{1}{r_x} & 0 \\ -\frac{1}{r_x} & \frac{1}{r_x} + \frac{1}{r_\pi} + sC_\pi + sC_\mu & -\left(\frac{1}{r_\pi} + sC_\pi\right) \\ 0 & -\left(\frac{1}{r_\pi} + sC_\pi + g_m\right) & \frac{1}{R_L} + \frac{1}{r_0} + \frac{1}{r_\pi} + sC_\pi + g_m \end{bmatrix} \begin{bmatrix} V_1 \\ V_2 \\ V_3 \end{bmatrix} = \begin{bmatrix} \frac{V_{IN}}{R_S} \\ 0 \\ 0 \end{bmatrix} \quad (1.19)$$

which can be solved for V_1, V_2, and V_3 using Cramer's rule.

Observation

The presence of the dependent current source $g_m(V_2 - V_3)$ makes the nodal determinant nonsymmetrical [the (2, 3) term is not the same as the (3, 2) term]. Such asymmetry usually occurs in circuits containing dependent sources. ■

1.2.2 DEPENDENT VOLTAGE SOURCES

The two types of dependent voltage sources encountered in active networks are the voltage controlled voltage source (*VCVS*) and the current controlled voltage source (*CCVS*), shown in Figure 1.6*a* and 1.6*b*, respectively.

Examples of the *VCVS* are the triode (Figure 1.7) and the differential operational amplifier (Figure 1.8). One example of the use of a *CCVS* is in modeling the gyrator (see Problem 1.8).

A circuit containing dependent voltage sources can be analyzed by converting the voltage sources to current sources using Norton's theorem, as explained in Example 1.1. This circuit, with dependent current sources, can then be analyzed just as in the last section. The following example illustrates the procedure.

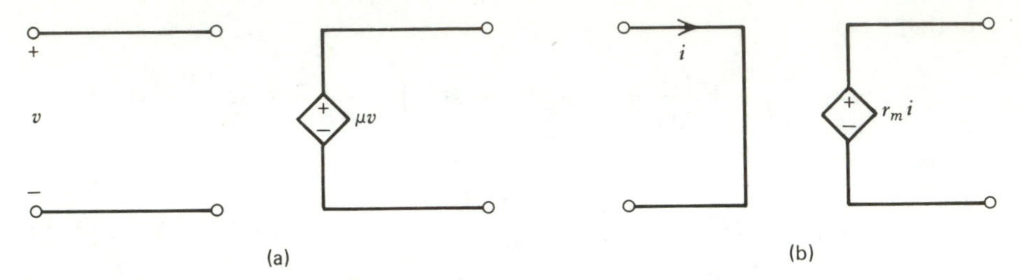

Figure 1.6 (*a*) Voltage controlled voltage source (*VCVS*). (*b*) Current controlled voltage source (*CCVS*).

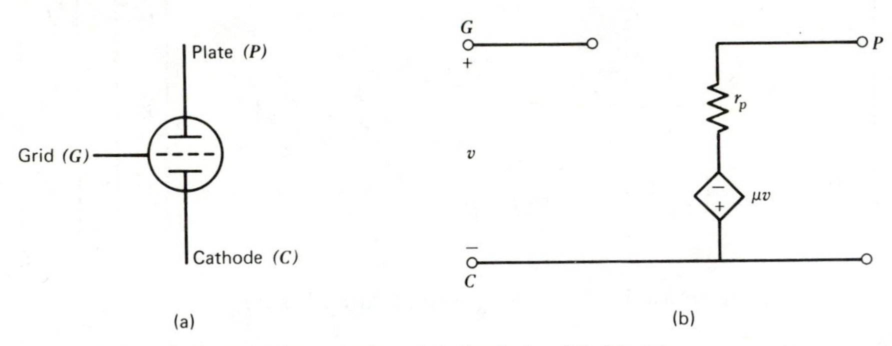

Figure 1.7 *VCVS* model for a triode: (*a*) Symbol. (*b*) Model.

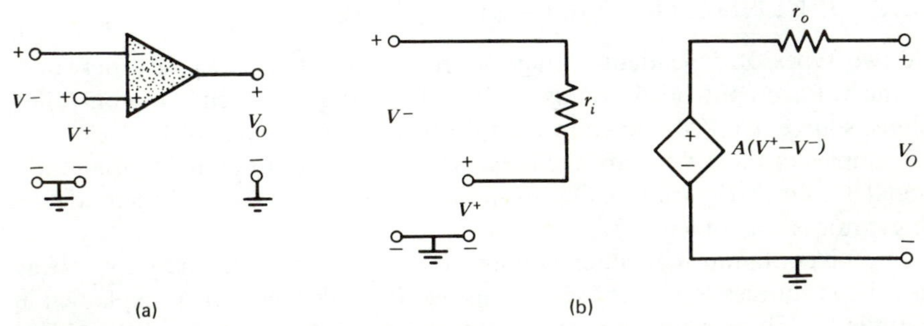

Figure 1.8 *VCVS* model for a differential operational amplifier: (*a*) Symbol. (*b*) Model.

Example 1.3

Find the function V_O/V_{IN} for the operational amplifier (hereafter abbreviated as op amp) circuit shown in Figure 1.9*a*.

Solution

The equivalent circuit, obtained by using the op amp model of Figure 1.8*b*, is shown in Figure 1.9*b*. The nodal equations for the current source equivalent of this circuit (Figure 1.9*c*), are:

node 1:

$$V_1\left(\frac{1}{R_S} + \frac{1}{r_i} + \frac{1}{R_F}\right) - V_O\left(\frac{1}{R_F}\right) = \frac{V_{IN}}{R_S} \tag{1.20}$$

node 2:

$$-V_1\left(\frac{1}{R_F}\right) + V_O\left(\frac{1}{R_F} + \frac{1}{r_O}\right) = -\frac{AV_1}{r_O} \tag{1.21}$$

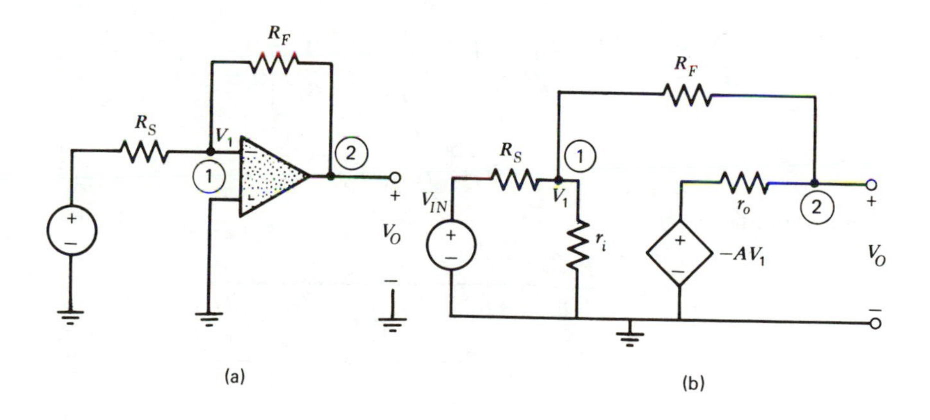

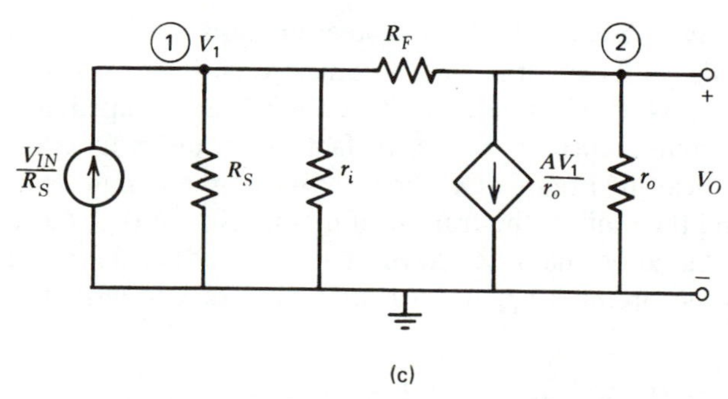

Figure 1.9 (*a*) Circuit for Example 1.3. (*b*) Equivalent circuit using voltage sources. (*c*) Equivalent circuit using current sources.

Rearranging the second equation so that the dependent voltages are on the left hand side of the equality, we get

$$-V_1\left(\frac{1}{R_F}-\frac{A}{r_O}\right)+V_O\left(\frac{1}{R_F}+\frac{1}{r_O}\right)=0$$

From the above, the nodal matrix equation is

$$\begin{bmatrix} \frac{1}{R_S}+\frac{1}{r_i}+\frac{1}{R_F} & -\frac{1}{R_F} \\ -\frac{1}{R_F}+\frac{A}{r_O} & \frac{1}{R_F}+\frac{1}{r_O} \end{bmatrix} \begin{bmatrix} V_1 \\ V_O \end{bmatrix} = \begin{bmatrix} \frac{V_{IN}}{R_S} \\ 0 \end{bmatrix} \tag{1.22}$$

Therefore, V_O is given by

$$V_O=\frac{\begin{vmatrix} \frac{1}{R_S}+\frac{1}{r_i}+\frac{1}{R_F} & \frac{V_{IN}}{R_S} \\ -\frac{1}{R_F}+\frac{A}{r_O} & 0 \end{vmatrix}}{\begin{vmatrix} \frac{1}{R_S}+\frac{1}{r_i}+\frac{1}{R_F} & -\frac{1}{R_F} \\ -\frac{1}{R_F}+\frac{A}{r_O} & \frac{1}{R_F}+\frac{1}{r_O} \end{vmatrix}}$$

which simplifies to

$$\frac{V_O}{V_{IN}}=-\frac{R_F}{R_S}\cdot\frac{1}{1+\dfrac{\left(1+\dfrac{r_O}{R_F}\right)\left(1+\dfrac{R_F}{R_S}+\dfrac{R_F}{r_i}\right)}{A\left(1-\dfrac{r_O}{AR_F}\right)}} \tag{1.23}$$

The alert reader will have observed that the above analysis could not be used if the output impedance of the op amp were assumed to be zero. This is so because the application of Norton's theorem on an ideal, zero-impedance voltage source yields an infinite current source. In fact, the transfer function for the circuit with $r_O = 0$ can be obtained by first analyzing the circuit as in the above, and then taking the limit of the transfer function as $r_O \to 0$. A more direct and simpler way is based on the observation that $V_O = -AV_1$. Thus the only unknown voltage in the circuit is V_1; and it is only necessary to write the one node equation:

$$V_1\left(\frac{1}{R_S}+\frac{1}{r_i}+\frac{1}{R_F}\right)+AV_1\left(\frac{1}{R_F}\right)=\frac{V_{IN}}{R_S} \tag{1.24}$$

from which

$$V_O = -AV_1 = -A\left(\frac{\dfrac{V_{IN}}{R_S}}{\dfrac{1}{R_S} + \dfrac{1}{r_i} + \dfrac{1}{R_F} + \dfrac{A}{R_F}}\right)$$

which immediately yields the desired function

$$\frac{V_O}{V_{IN}} = -\frac{R_F}{R_S} \cdot \frac{1}{1 + \dfrac{1}{A}\left(1 + \dfrac{R_F}{R_S} + \dfrac{R_F}{r_i}\right)} \tag{1.25}$$

This equation is the same as Equation 1.23, with $r_0 = 0$. ■

1.3 SIMPLIFIED ANALYSIS OF OPERATIONAL AMPLIFIER CIRCUITS

The analysis of circuits containing operational amplifiers is very much simplified if the operational amplifiers are assumed to be ideal. The ideal op amp is assumed to have the following properties:

(a) The gain is infinite.
(b) The input impedance is infinite and the output impedance is zero.

In practice, the gain of an op amp is a function of frequency. For a typical op amp (Figure 1.10) the gain at frequencies below 10 kHz, where active filters find the greatest application, is in excess of 10,000. The input impedance of real op amps is around 500 kΩ, and the output impedance is in the order of 300 Ω. In most circuits the ideal op amp is a good approximation to the real-world op amp; this is evidenced by comparing theoretical analyses assuming ideal op amps, with experimental results using the real op amps.

The output voltage of the ideal differential op amp is (Figure 1.8*b*):

$$V_0 = A(V^+ - V^-)$$

so

$$V^+ - V^- = \frac{V_0}{A} \tag{1.26}$$

Now in any useful circuit, all the voltages and currents must be finite. Therefore, for the output voltage V_O to be finite in a circuit containing an ideal operational amplifier ($A = \infty$), the potential difference between the input terminals of the amplifier must be zero. Moreover, since the input impedance is infinite, the input current to the ideal op amp is zero. These results are restated here for

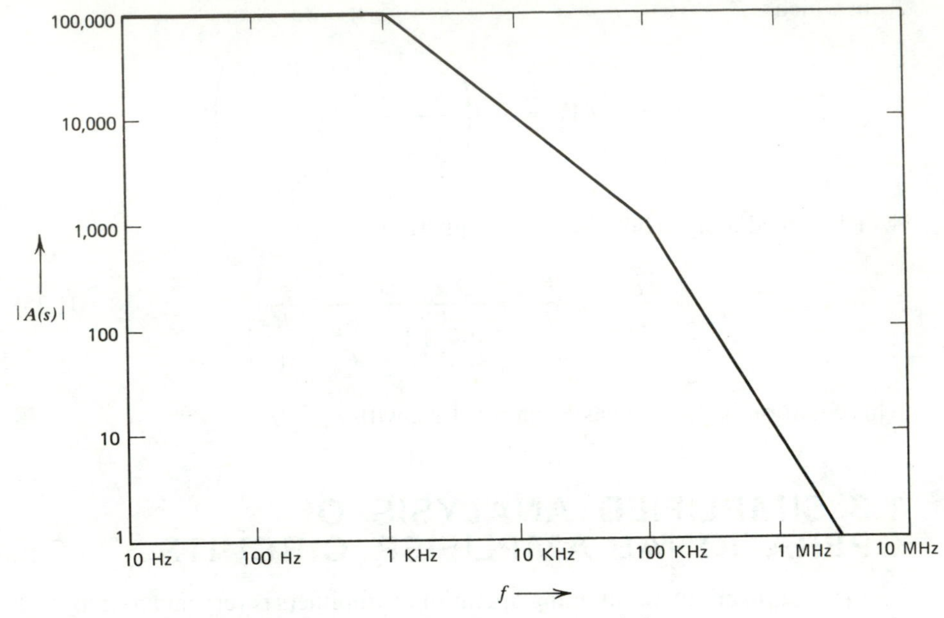

Figure 1.10 Typical op amp gain versus frequency characteristic.

emphasis. For an ideal op amp:

Rule 1: *The potential difference between the input terminals is zero.*
Rule 2: *The current into each input terminal is zero.*

The simplified analysis of op amp circuits using these two rules is illustrated in the following examples.

Example 1.4
Find the function V_O/V_{IN} for the circuit of Figure 1.9*a*, assuming an ideal op amp.

Solution
The positive terminal of the op amp is at ground potential, therefore, by Rule 1, the voltage at the negative terminal is

$$V_1 = 0 \tag{1.27}$$

From Rule 2, no current goes into the negative input terminal of the op amp; therefore, by summing the currents at node 1, we get

$$\frac{V_{IN} - V_1}{R_S} + \frac{V_O - V_1}{R_F} = 0 \tag{1.28}$$

From (1.27) and (1.28):

$$\frac{V_O}{V_{IN}} = -\frac{R_F}{R_S} \tag{1.29}$$

Observations

1. Equation 1.23 is seen to reduce to Equation 1.29 under the assumptions of an ideal op amp ($A = \infty$, $r_i = \infty$, and $r_0 = 0$).
2. The above op amp circuit inverts the input voltage and scales it by the factor R_F/R_S. It is commonly referred to as the **inverting amplifier** structure.
3. In this example the negative input terminal of the op amp is at ground potential at all times, so a *virtual ground* is said to exist at this terminal.

■

Example 1.5

Find the function V_O/V_{IN} for the circuit shown in Figure 1.11.

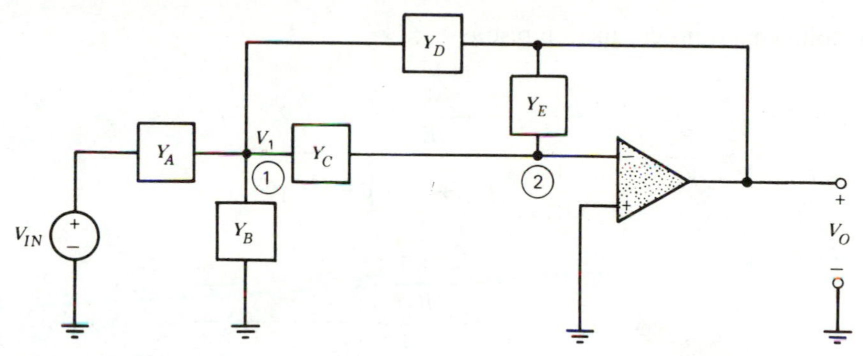

Figure 1.11 Circuit for Example 1.5.

Solution

Let V_1 be the voltage at node 1. Node 2 is seen to be at virtual ground. Since there are two unknown voltages, V_1 and V_0, we need two node equations. Choosing nodes 1 and 2 for the node equations* and using Rule 2, we get

node 1:

$$V_1(Y_A + Y_B + Y_C + Y_D) - V_0(Y_D) - V_{IN}(Y_A) - 0(Y_C) = 0$$

node 2:

$$0(Y_C + Y_E) - V_1(Y_C) - V_0(Y_E) = 0$$

Solving

$$-V_0 \frac{Y_E}{Y_C}(Y_A + Y_B + Y_C + Y_D) - V_0 Y_D = V_{IN} Y_A$$

$$\frac{V_0}{V_{IN}} = -\frac{Y_A Y_C}{Y_E(Y_A + Y_B + Y_C + Y_D) + Y_C Y_D} \tag{1.30}$$

* Observe that the node equation for node V_0 cannot be written, since the current from the output of the op amp is indeterminate.

Observation

The above result may be used to synthesize the function

$$\frac{V_O}{V_{IN}} = \frac{-s}{s^2 + 2s + 1} \tag{1.31}$$

In Equation 1.30, if we select the admittances as

$$Y_A = \frac{1}{R_1} \qquad Y_B = 0 \qquad Y_C = sC_1 \qquad Y_D = sC_2 \qquad Y_E = \frac{1}{R_2}$$

the following transfer function results

$$\frac{V_O}{V_{IN}} = \frac{-\dfrac{sC_1}{R_1}}{\dfrac{1}{R_2}\left(\dfrac{1}{R_1} + sC_1 + sC_2\right) + s^2C_1C_2}$$

$$= \frac{-\dfrac{1}{R_1C_2}s}{s^2 + s\left(\dfrac{1}{R_2C_2} + \dfrac{1}{R_2C_1}\right) + \dfrac{1}{R_1R_2C_1C_2}} \tag{1.32}$$

This expression has the form of Equation 1.31. A comparison of (1.31) and (1.32) yields the following three equations in four unknowns

$$\frac{1}{R_1C_2} = 1$$

$$\frac{1}{R_2C_2} + \frac{1}{R_2C_1} = 2$$

$$\frac{1}{R_1R_2C_1C_2} = 1$$

One solution to this set of equations is

$$R_1 = R_2 = 1 \qquad C_1 = C_2 = 1$$

The transfer function is therefore realized using the circuit shown in Figure 1.12. The technique used for the above synthesis is called the **coefficient matching technique**—details of which will be presented in Chapter 7. ■

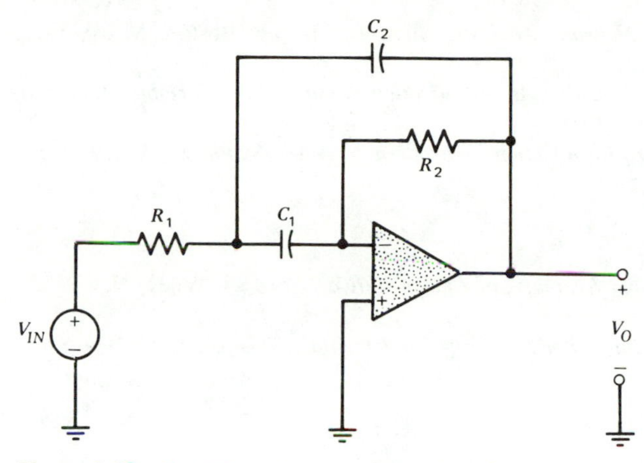

Figure 1.12 Realization for transfer function of Equation 1.31.

1.4 CONCLUDING REMARKS

Network analysis is used as a tool in several steps of the design procedure. In this book, the nodal analysis method described in this chapter will be used exclusively.

Several computer programs are available for the analysis of circuits. Reference 10 has a good summary of the presently existing programs. Among the programs that analyze circuits in the s domain are CORNAP [12], ASTAP [13], SLIC [11], and LISA [9].

In the discussion of Example 1.1, we mention that before attempting to synthesize a function it is necessary to determine whether or not the function can be realized using a given class of components. This question of realizability is the subject of the next chapter. The synthesis of passive networks is discussed in Chapter 6, while active network synthesis techniques are described in Chapters 7 to 11. The concluding chapters deal with practical problems encountered in the design procedure.

FURTHER READING

Analysis of passive networks

1. P. M. Chirlian, *Basic Network Theory*, McGraw-Hill, New York, 1969.
2. C. A. Desoer and E. S. Kuh, *Basic Circuit Theory*, McGraw-Hill, New York, 1969, Chapter 10.
3. T. S. Huang and R. R. Parker, *Network Theory: An Introductory Course*, Addison-Wesley, Reading, Mass., 1971.

4. S. Karni, *Intermediate Network Analysis*, Allyn and Bacon, Boston, Mass., 1971, Chapter 2.
5. R. A. Rohrer, *Circuit Theory: An Introduction to the State Variable Approach*, McGraw-Hill, New York, 1971.
6. M. E. Van Valkenburg, *Introduction to Modern Network Synthesis*, Wiley, New, York, 1960, Chapter 2.

Analysis of active networks

7. S. K. Mitra, *Analysis and Synthesis of Linear Active Networks*, Wiley, New York, 1969, Chapter 4.
8. G. S. Moschytz, *Linear Integrated Networks Fundamentals*, Van Nostrand, New York, 1974, Chapter 3.

Computer aids

9. K. L. Deckert and E. T. Johnson, "LISA 360-A program for linear systems analysis," IBM Program Inform. Dept., Hawthorne, N.Y.
10. J. Greenbaum, "A library of circuit analysis programs," *Circuits and Systems Newsletter*, **7**, No. 1, February 1974, pp. 4–10.
11. T. E. Idleman, F. S. Jenkins, W. J. McCalla, and D. O. Pederson, "SLIC—A simulator for linear integrated circuits," *IEEE J. Solid-State Circuits*, *SC-6*, August 1971, pp. 188–203.
12. F. F. Kuo and J. F. Kaiser, Eds., *Systems Analysis by Digital Computer*, Wiley, New York, 1965, Chapter 3.
13. W. T. Weeks, et al., "Algorithms for ASTAP—A Network analysis program," *IEEE Trans. Circuit Theory*, **CT-20**, No. 6, November 1973, pp. 628–634.

PROBLEMS

1.1 *RC ladder analysis.* Write nodal equations for the *RC* ladder network shown, and determine the function V_O/V_{IN}.

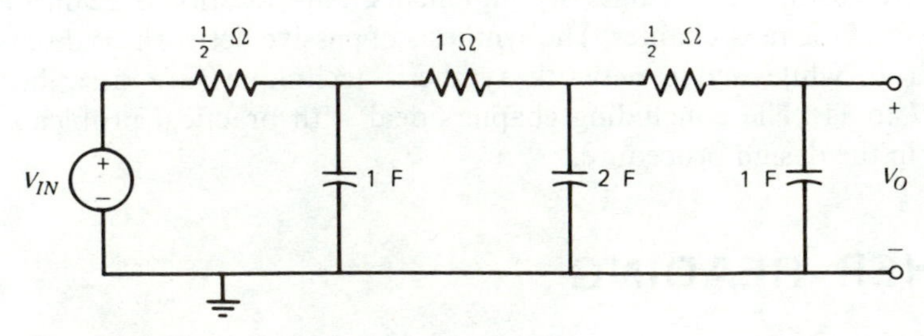

Figure P1.1

1.2 *LC ladder analysis.* Express V_O/V_{IN} for the *LC* ladder network of Figure P1.2, in the form:

$$K\frac{(s^2 + a)(s^2 + b)}{(s^2 + c)(s^2 + d)}$$

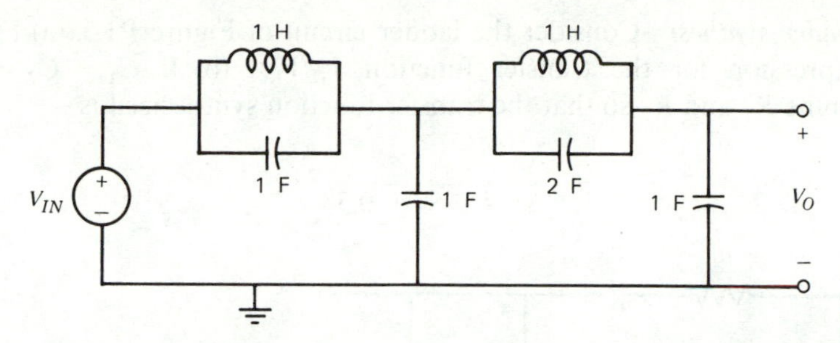

Figure P1.2

1.3 *Twin-T RC network analysis.* Show that the transfer function V_O/V_{IN} for the Twin-T *RC* network shown is

$$\frac{s^2 + \dfrac{1}{R^2C^2}}{s^2 + \dfrac{4}{RC}s + \dfrac{1}{R^2C^2}}$$

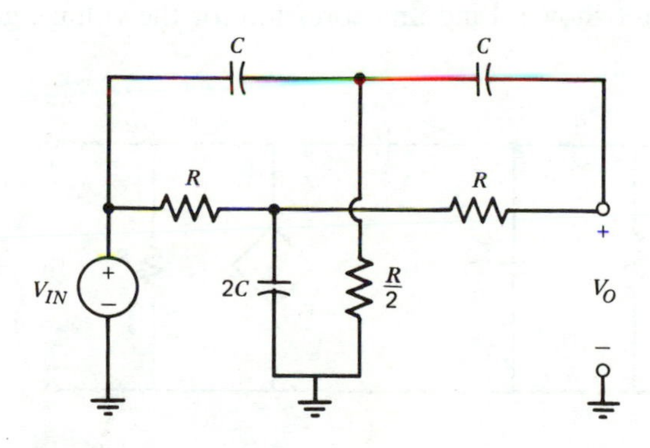

Figure P1.3

1.4 *Bridged-T network analysis.* Show that the transfer function V_O/V_{IN} of the bridged-T circuit shown reduces to $Z_2/(Z_1 + Z_2)$.

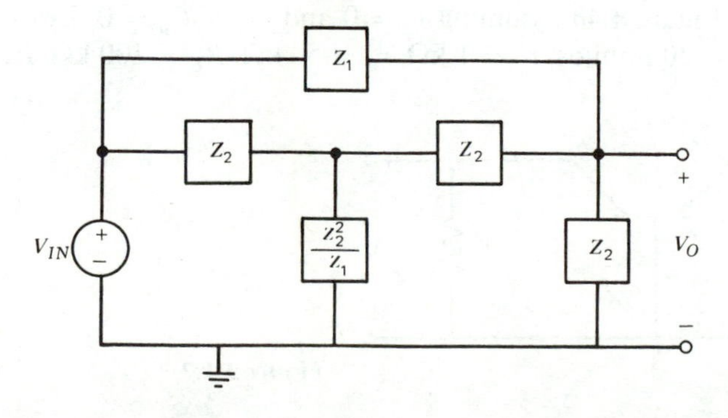

Figure P1.4

1.5 *RC ladder synthesis.* Consider the ladder circuit of Figure P1.5. (a) Find an expression for the transfer function V_O/V_{IN}. (b) If $C_1 = C_2 = 1$, determine R_1 and R_2 so that the transfer function synthesized is

$$\frac{s}{s^2 + 2.5s + 0.5}$$

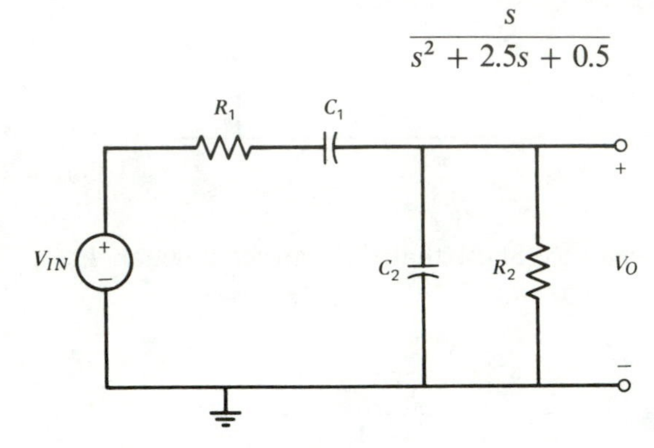

Figure P1.5

1.6 *Transistor circuit analysis.* A transistor amplifier has the dependent source equivalent circuit shown. Find an expression for the voltage gain V_O/V_{IN}.

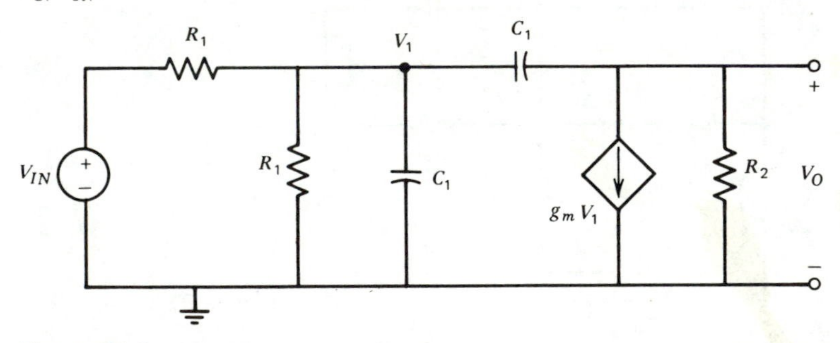

Figure P1.6

1.7 *Common-base transistor amplifier analysis.* Determine the gain function V_O/V_{IN} for the common-base amplifier shown, using the transistor equivalent circuit of Figure 1.4*b* assuming $r_x = 0$ and $C_\pi = C_\mu = 0$. Evaluate the gain for $g_m = 50$ mmhos, $r_\pi = 1$ kΩ, $r_O = 50$ kΩ, $R_L = 100$ kΩ, $R_S = 20$ kΩ.

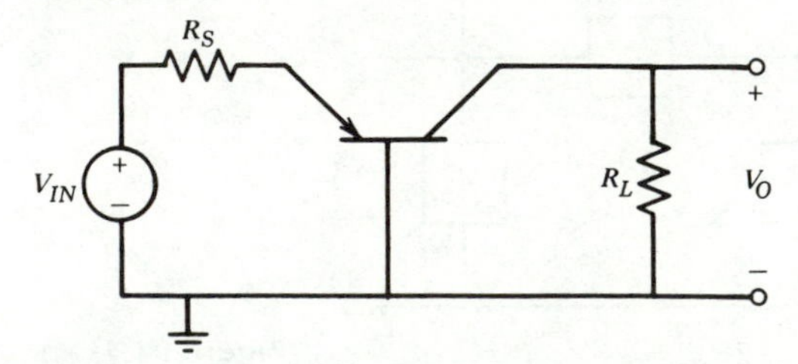

Figure P1.7

1.8 *Gyrator realization of inductor.* The equivalent circuit for a gyrator is shown in Figure P1.8. If port 2 of the gyrator is terminated by a capacitor C, show that the input impedance V_1/I_1, seen at port 1, is that of an inductor of value CR^2.

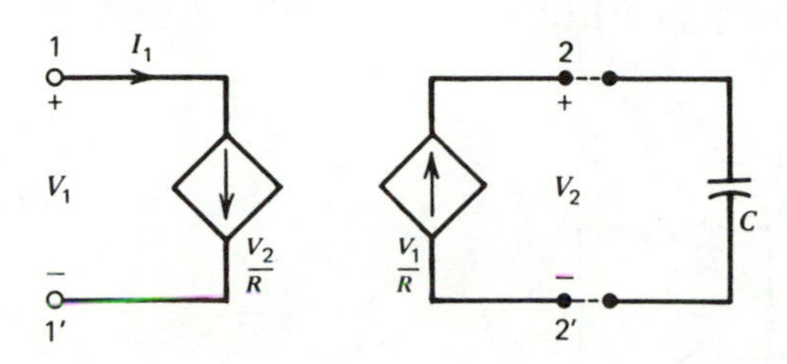

Figure P1.8

1.9 *Effects of finite gain, input resistance, and output resistance in op amp circuit.* Consider the noninverting amplifier of Figure P1.9.

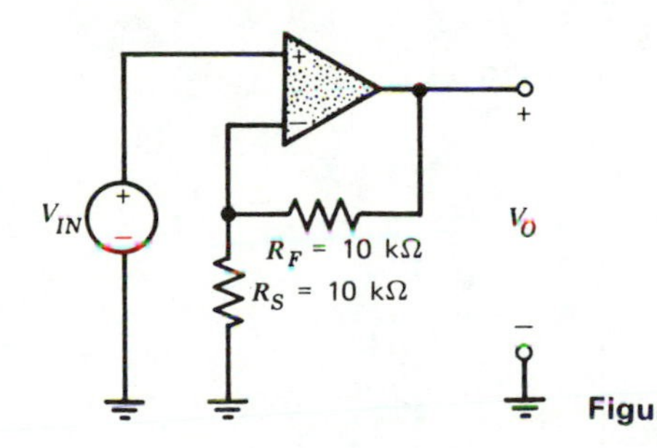

Figure P1.9

(a) Using the equivalent circuit of the op amp of Figure 1.8*b* show that the voltage gain is given by

$$\frac{V_O}{V_{IN}} = \frac{1 + \dfrac{R_F}{R_S} + \dfrac{r_O}{r_i}\dfrac{1}{A(s)}}{1 + \dfrac{1}{A(s)}\left[1 + \left(1 + \dfrac{r_O}{R_F}\right)\left(\dfrac{R_F}{R_S} + \dfrac{R_F}{r_i}\right)\right]}$$

(b) Determine the voltage gain if the op amp is assumed to be ideal.

(c) Assuming the op amp gain is modeled as

$$A(s) = \frac{2\pi 10^6}{s}$$

$r_i = \infty$ and $r_O = 0$, compute the magnitude of the voltage gain $|V_O/V_{IN}|$ at 1 kHz, 10 kHz, 100 kHz, 1 MHz, and 10 MHz. Sketch the gain versus frequency.

(d) Compute the magnitude of the gain at 10 kHz for a real op amp with $A(s)$ modeled as in (c) if $r_i = 500$ kΩ and $r_O = 300$ Ω.

In Problems 1.10 to 1.22, assume the op amps to be ideal

1.10 *Summer, integrator, voltage follower.* Determine the output voltage for (a) the inverting summer; (b) the inverting integrator; and (c) the voltage follower circuits shown in Figure P1.10.

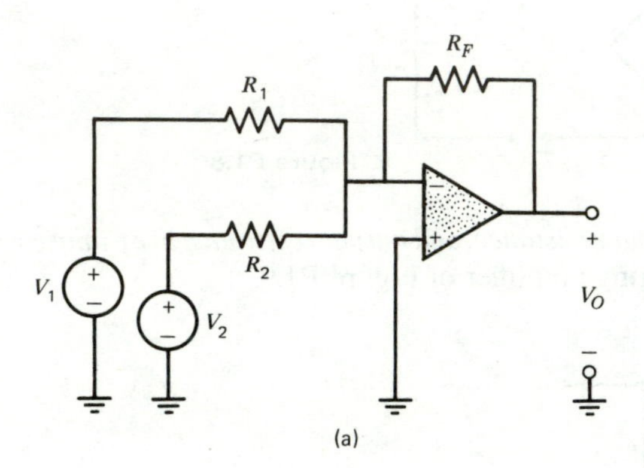

(a)

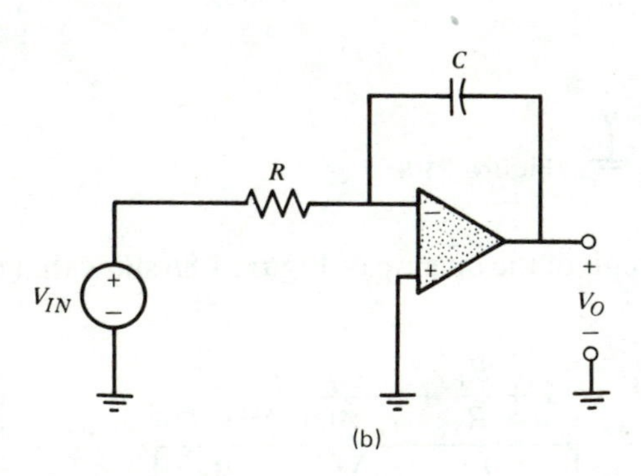

(b)

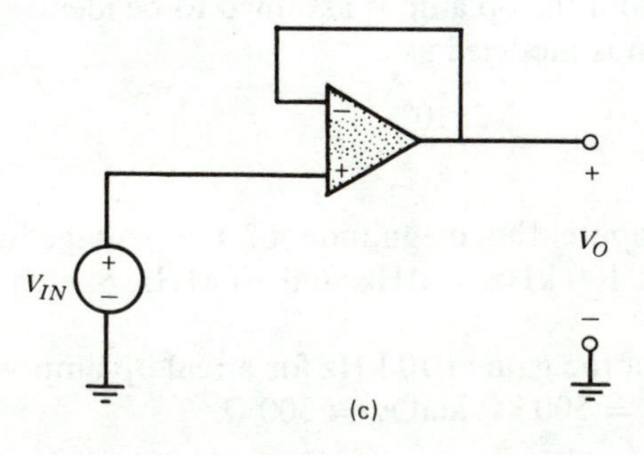

(c)

Figure P1.10

1.11 *Synthesis using op amps.* Use the results of Problem 1.10 to synthesize the following transfer functions:

(a) $\dfrac{V_O}{V_{IN}} = -400$

(b) $\dfrac{V_O}{V_{IN}} = \dfrac{1000}{s}$

(c) $\dfrac{V_O}{V_{IN}} = \dfrac{100}{s^2}$

1.12 *Leaky integrator.* Analyze the leaky integrator circuit shown to obtain the transfer function V_O/V_{IN}. Use the result to synthesize:

(a) $\dfrac{V_O}{V_{IN}} = \dfrac{-3}{s+4}$

(b) $\dfrac{V_O}{V_{IN}} = \dfrac{1000}{s+2000} + \dfrac{2000}{s+4000}$

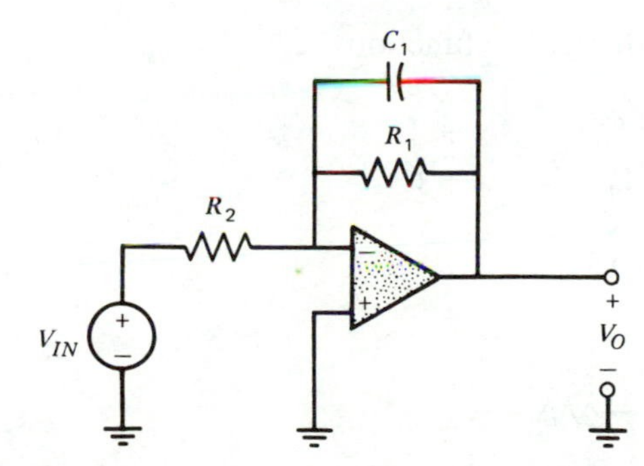

Figure P1.12

1.13 *Differential summer.* Find an expression for the output voltage for the differential summer shown. Show how the circuit can be used to obtain the weighted difference of the two voltages $V_O = -3V_1 + 2V_2$.

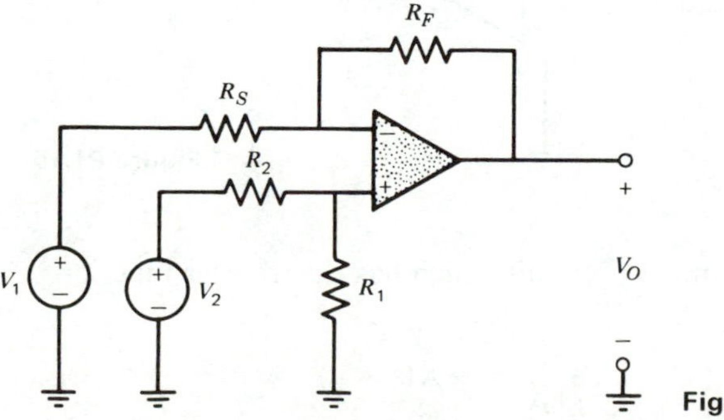

Figure P1.13

1.14 *Noninverting integrator.* Show that the circuit of Figure P1.14 realizes a noninverting integrator.

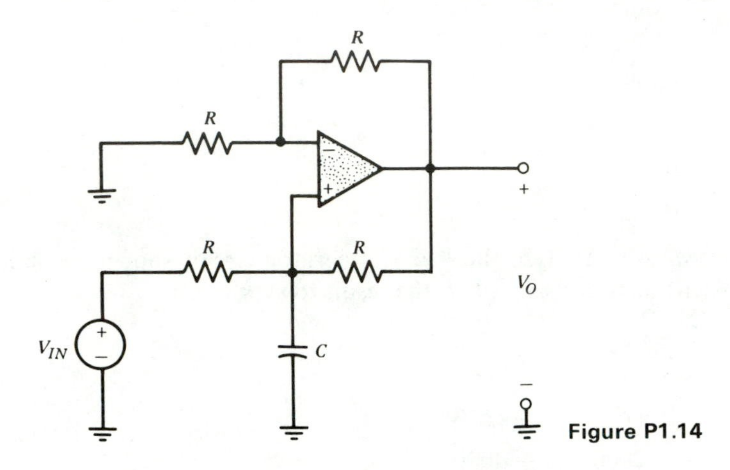

Figure P1.14

1.15 *Synthesis using op amps.* Synthesize the function

$$\frac{V_O}{V_{IN}} = -\frac{s+4}{s+6}$$

using the circuit of Figure P1.15.

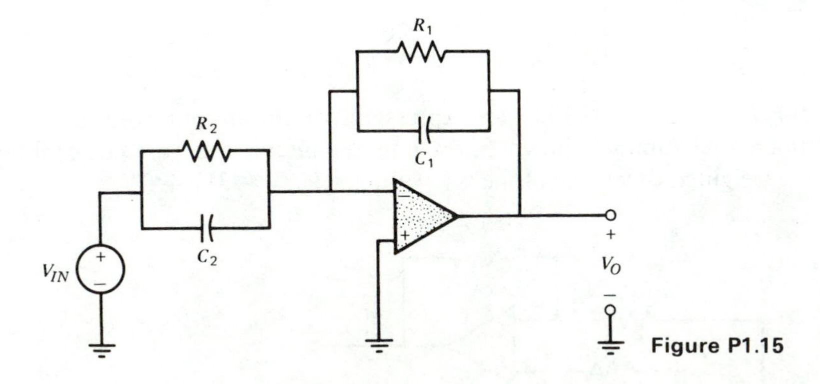

Figure P1.15

1.16 Find a single op amp *RC* circuit which has the transfer function

$$\frac{V_O}{V_{IN}} = -K(s+a)$$

1.17 *Op amp circuit analysis.* Analyze the circuit of Figure P1.17 to obtain the transfer function V_O/V_{IN} in the form

$$K \frac{s}{s^2 + as + b}$$

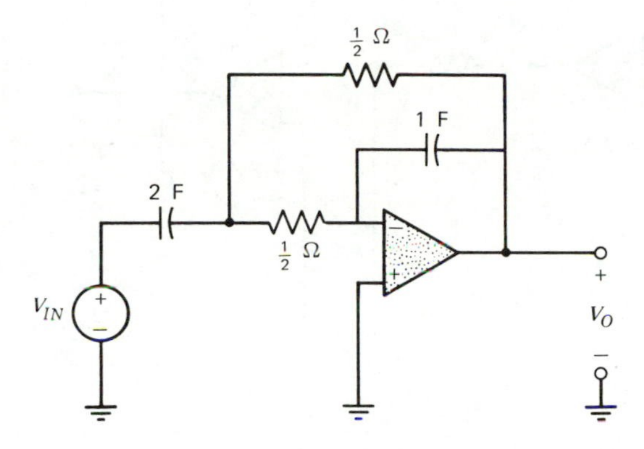

Figure P1.17

1.18 Show that the op amp RC circuit of Figure P1.18 has the transfer function

$$\frac{V_O}{V_{IN}} = \frac{-s}{s^2 + 0.5s + 1}$$

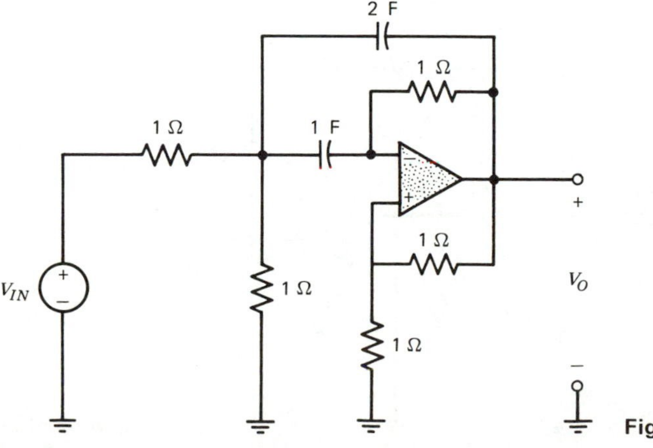

Figure P1.18

1.19 Determine V_O/V_{IN} for the two op amp RC circuit of Figure P1.19.

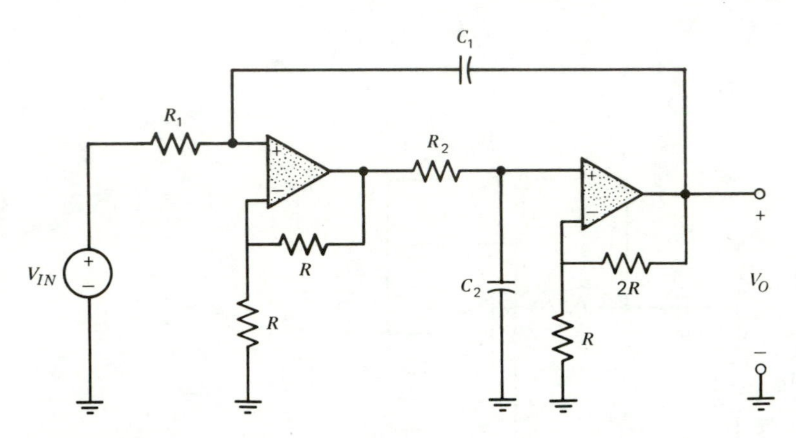

Figure P1.19

1.20 *Op amp circuit synthesis.* Find V_O/V_{IN} for the circuit shown. If $C_1 = C_2 = 1$, find a set of values for R_1, R_2, and R_3 to realize the function

$$\frac{V_O}{V_{IN}} = \frac{-2}{s^2 + 7s + 8}$$

Is the answer unique? Explain.

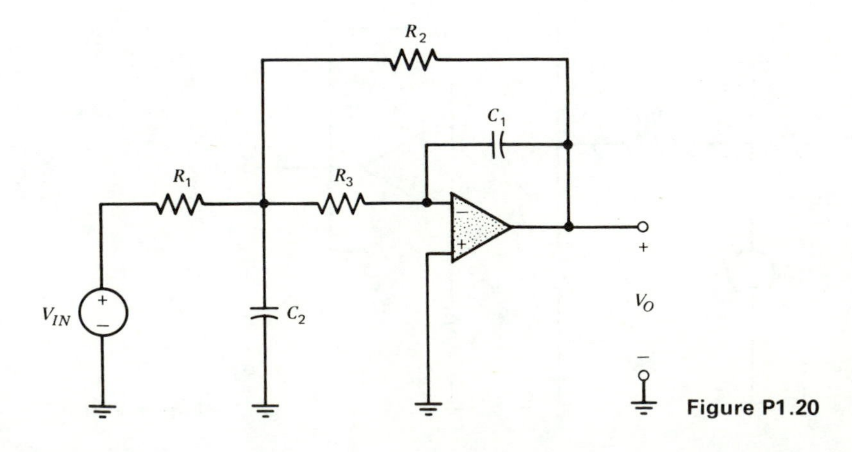

Figure P1.20

1.21 Synthesize the function

$$\frac{s^2}{s^2 + 6s + 8}$$

using the topology of Figure P1.21.

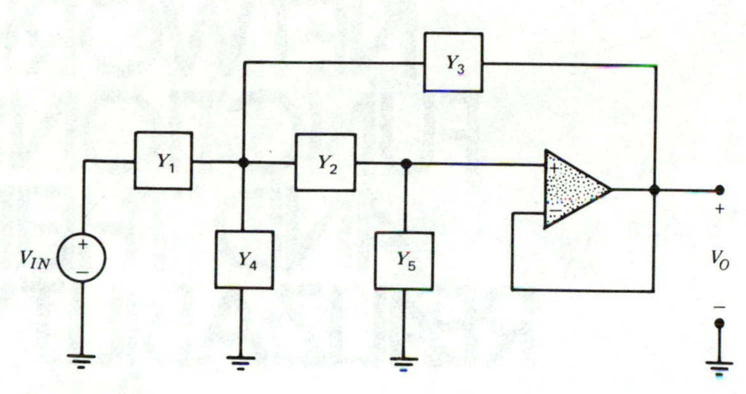

Figure P1.21

1.22 *Capacitance multiplier.* The circuit shown is a capacitance multiplier. Show that the input impedance V_1/I_1 is that of a capacitor of value $(1 + R_2/R_1)C$.

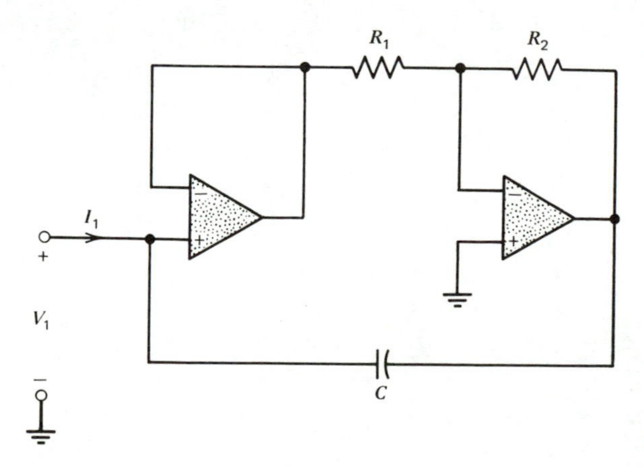

Figure P1.22

2.

NETWORK FUNCTIONS AND THEIR REALIZABILITY

In a typical synthesis problem the designer is provided with a loss requirement, such as that shown in Figure 2.1. Any function whose loss characteristic lies outside the shaded region is said to satisfy the filter requirements. Obviously there is an infinitude of functions that could be used. However, a restriction on the function is that it be realizable using a given set of passive and/or active components. It is important, therefore, that we be able to predict the kinds of functions that are realizable using a given group of components. To study the design of active filters it is necessary to consider the properties of *RC* networks, *RLC* networks, and active *RC* networks. This chapter develops the properties that network functions must have to be candidates for realization using these three component groups.

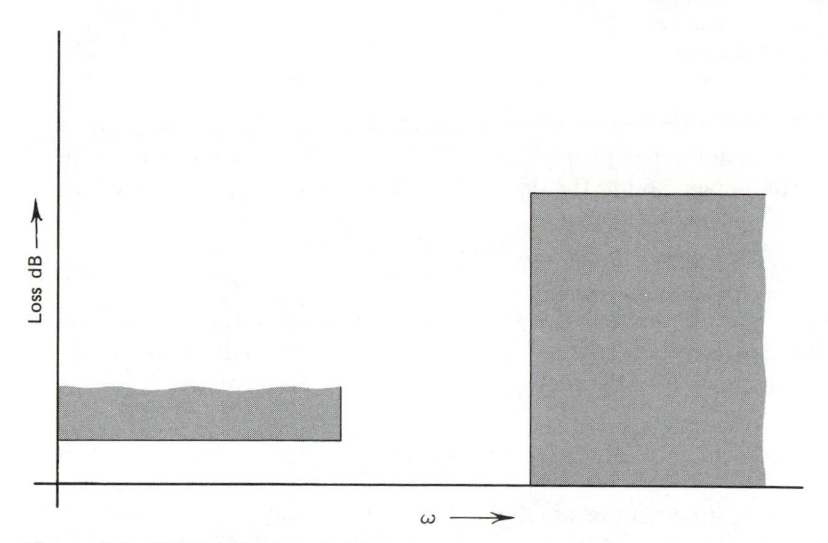

Figure 2.1 Typical filter requirements.

2.1 NETWORK FUNCTIONS

Consider the general two-port network shown in Figure 2.2*a*. The terminal voltages and currents of the two-port can be related by two classes of network functions, namely, the driving point (dp) functions and the transfer functions.

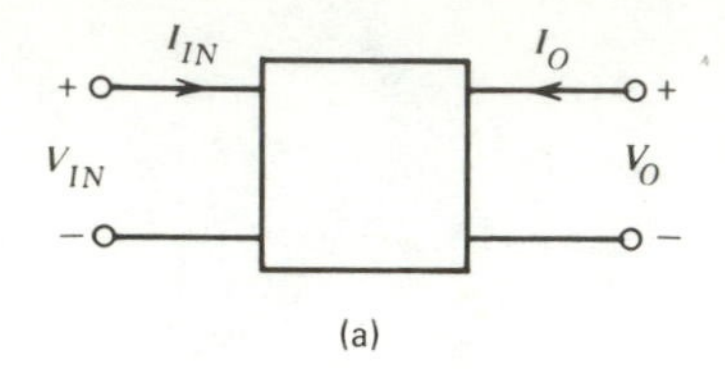

(a)

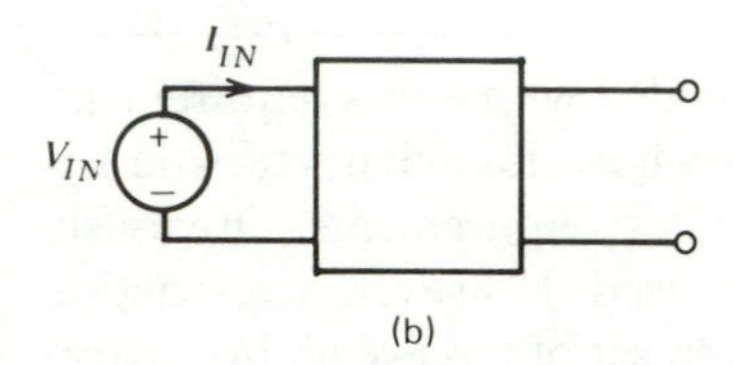

(b)

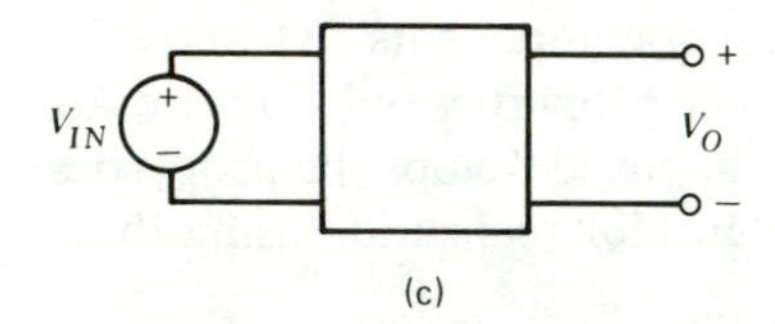

(c)

Figure 2.2 (*a*) A two port network.
(*b*) Measuring input impedance.
(*c*) Measuring voltage gain.

The dp functions relate the voltage at a port to the current at the same port. Thus, these functions are a property of a single port. For the input port (with the output port an open circuit) the dp impedance function $Z_{IN}(s)$ is defined as

$$Z_{IN}(s) = \frac{V_{IN}(s)}{I_{IN}(s)} \tag{2.1}$$

This function can be measured by observing the current I_{IN} when the input port is driven by a voltage source V_{IN} (Figure 2.2*b*). The dp admittance function $Y_{IN}(s)$ is the reciprocal of the impedance function, and is given by

$$Y_{IN}(s) = \frac{I_{IN}(s)}{V_{IN}(s)} \tag{2.2}$$

The output port dp functions are defined in a similar way.

The transfer functions of the two-port relate the voltage (or current) at one port to the voltage (or current) at the other port. The possible forms of transfer functions are:

1. The voltage transfer function, which is a ratio of one voltage to another voltage.
2. The current transfer function, which is a ratio of one current to another current.

3. The transfer impedance function, which is the ratio of a voltage to a current.
4. The transfer admittance function, which is the ratio of a current to a voltage.

The voltage transfer functions are defined with the output port an open circuit, as

$$\text{voltage gain} = \frac{V_O(s)}{V_{IN}(s)} \tag{2.3}$$

$$\text{voltage loss (attenuation)} = \frac{V_{IN}(s)}{V_O(s)} \tag{2.4}$$

To evaluate the voltage gain, for example, the output voltage V_O is measured with the input port driven by a voltage source V_{IN} (Figure 2.2*c*). The other three types of transfer functions can be defined in a similar manner. Of the four types of transfer functions, the voltage transfer function is the one most often specified in the design of filters.

The functions defined above, when realized using resistors, inductors, capacitors, and active devices, can be shown to be the ratios of polynomials in s with real coefficients. This is so because the network functions are obtained by solving simple algebraic node equations, which involve at most the terms R, sL, sC and their reciprocals. The active device, if one exists, merely introduces a constant in the nodal equations, as is shown in Chapter 1. While this term renders the nodal determinant asymmetrical, the solution still involves only the addition and multiplication of simple terms, which can *only* lead to a ratio of polynomials in s. In addition, all the coefficients of the numerator and denominator polynomials will be real. Thus, the general form of a network function is

$$H(s) = \frac{a_n s^n + a_{n-1}s^{n-1} + a_{n-2}s^{n-2} + \cdots + a_0}{b_m s^m + b_{m-1}s^{m-1} + b_{m-2}s^{m-2} + \cdots + b_0} \tag{2.5}$$

where

$$a_n \neq 0 \qquad b_m \neq 0$$

and all the coefficients a_i and b_i are real. If the numerator and denominator polynomials are factored, an alternate form of $H(s)$ is obtained:*

$$H(s) = \frac{a_n(s - z_1)(s - z_2)\cdots(s - z_n)}{b_m(s - p_1)(s - p_2)\cdots(s - p_m)} \tag{2.6}$$

In this expression $z_1, z_2, \ldots, z_n$ are called the *zeros* of $H(s)$, because $H(s) = 0$ when $s = z_i$. The roots of the denominator $p_1, p_2, \ldots, p_m$ are called the *poles* of $H(s)$. It can be seen that $H(s) = \infty$ at the poles, $s = p_i$.

* It is assumed that any common factors in the numerator and denominator have been canceled.

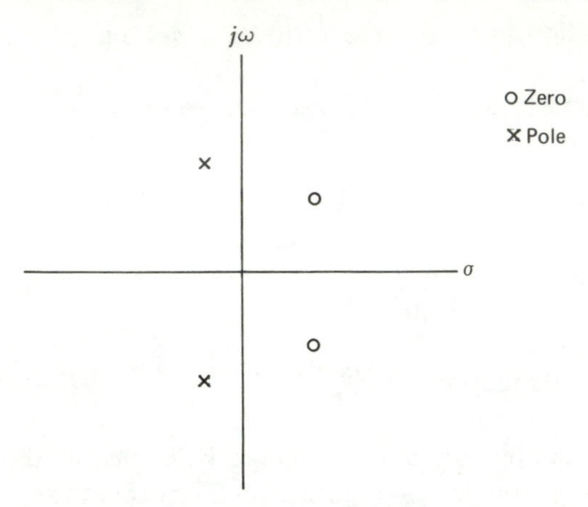

Figure 2.3 Poles and zeros plotted in the complex s plane.

The poles and zeros can be plotted on the complex s plane ($s = \sigma + j\omega$), which has the real part σ for the abscissa, and the imaginary part $j\omega$ for the ordinate (Figure 2.3).

2.2 PROPERTIES OF ALL NETWORK FUNCTIONS

We have already seen that network functions are ratios of polynomials in s with real coefficients. A consequence of this property is that complex poles (and zeros) must occur in conjugate pairs. To demonstrate this fact consider a complex root at $s = -a - jb$ which leads to the factor $(s + a + jb)$ in the network function. The jb term will make some of the coefficients complex in the polynomial, unless the conjugate of the complex root at $s = -a + jb$ is also present in the polynomial. The product of a complex factor and its conjugate is

$$(s + a + jb)(s + a - jb) = s^2 + 2as + a^2 + b^2 \tag{2.7}$$

which can be seen to have real coefficients.

Further important properties of network functions are obtained by restricting the networks to be *stable*, by which we mean that a bounded input excitation to the network must yield a bounded response. Put differently, the output of a stable network cannot be made to increase indefinitely by the application of a bounded input excitation. Passive networks are stable by their very nature, since they do not contain energy sources that might inject additional energy into the network. Active networks, however, do contain energy sources that could join forces with the input excitation to make the output increase indefinitely. Such unstable networks, however, have no use in the world of practical filters and are therefore precluded from all our future discussions.

A convenient way of determining the stability of the general network function $H(s)$ is by considering its response to an impulse function [7], which is obtained by taking the inverse Laplace transform of the partial fraction expansion of the function.

If the network function has a simple pole on the real axis, the impulse response due to it (for $t \geq 0$) will have the form:

$$h(t) = \mathscr{L}^{-1} \frac{K_1}{s - p_1} = K_1 e^{p_1 t} \tag{2.8}$$

For p_1 positive, the impulse response is seen to increase exponentially with time (Figure 2.4*a*), corresponding to an unstable circuit. Thus, $H(s)$ cannot have poles on the positive real axis.

Suppose $H(s)$ has a pair of complex conjugate poles at $s = a \pm jb$. The contribution to the impulse response due to this pair of poles is

$$\begin{aligned} h(t) &= \mathscr{L}^{-1}\left(\frac{K_1}{s - a - jb} + \frac{K_1}{s - a + jb}\right) = \mathscr{L}^{-1} \frac{2K_1(s - a)}{(s - a)^2 + b^2} \\ &= 2K_1 e^{at} \cos bt \end{aligned} \tag{2.9}$$

Now if a is positive, corresponding to poles in the right half s plane, the response is seen to be an exponentially increasing sinusoid (Figure 2.4*b*). Therefore, $H(s)$ cannot have poles in the right half s plane.

An additional restriction on the poles of $H(s)$ is that any poles on the imaginary axis must be simple. A double pole, for instance, will contain the following term in the impulse response:

$$\begin{aligned} h(t) &= \mathscr{L}^{-1}\left(\frac{K_1}{(s + jb)^2} + \frac{K_1}{(s - jb)^2}\right) = \mathscr{L}^{-1} \frac{2K_1(s^2 - b^2)}{(s^2 + b^2)^2} \\ &= 2K_1 t \cos bt \end{aligned} \tag{2.10}$$

which is a function that increases indefinitely with time (Figure 2.4*c*). Thus, double poles on the imaginary axis cannot be permitted in $H(s)$. Similarly, it can be shown that higher order poles on the $j\omega$ axis will also cause the network to be unstable.

From the above discussion we see that $H(s)$ has the following factored form

$$H(s) = \frac{N(s)}{\prod_i (s + a_i) \prod_k (s^2 + c_k s + d_k)} \tag{2.11}$$

where $N(s)$ is the numerator polynomial and the constants associated with the denominator a_i, c_k, and d_k are real and nonnegative. The $s + a_i$ terms represent poles on the negative real axis and the second order terms represent complex

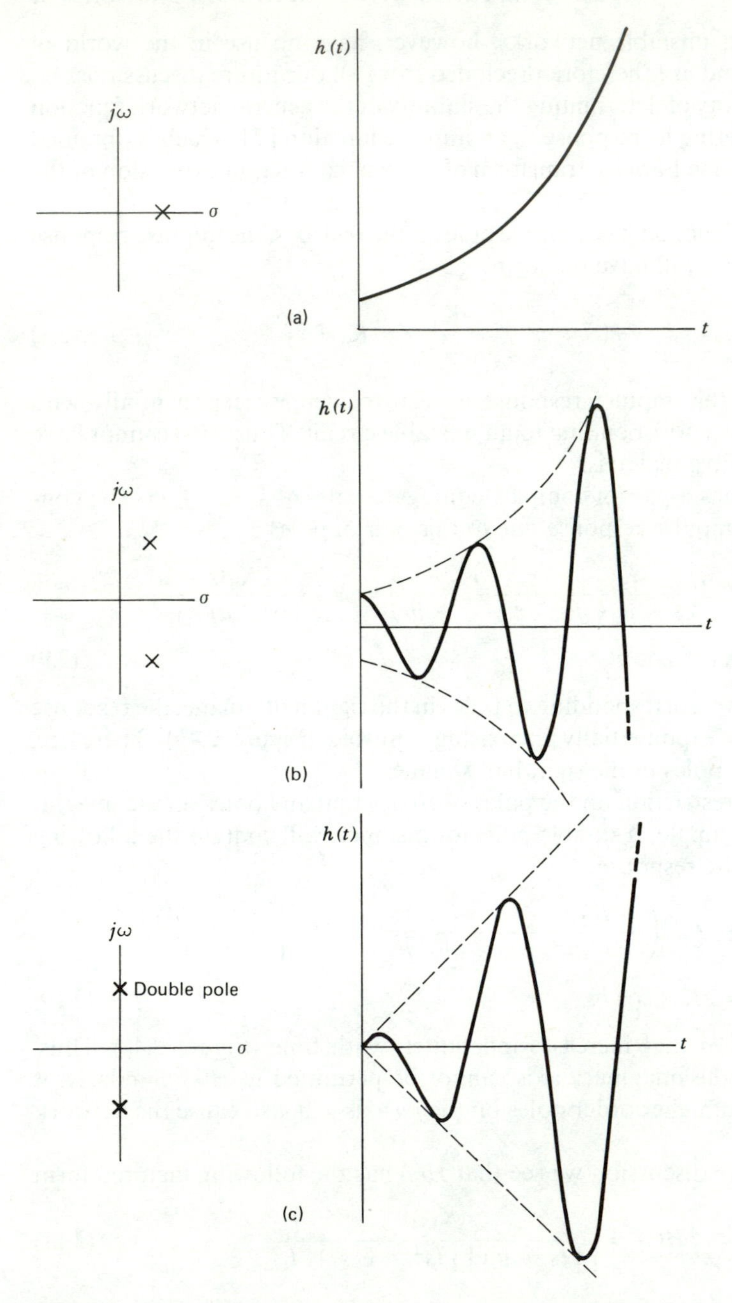

Figure 2.4 Impulse response due to: (*a*) A pole on the positive real axis. (*b*) Complex poles in the right half *s* plane. (*c*) Double poles on imaginary axis.

conjugate poles in the left half s plane. It is easy to see that the product of these factors can only lead to a polynomial, all of whose coefficients are real and positive; moreover, none of the coefficients may be zero unless all the even or all the odd terms are missing.

In summary, the network functions of all passive networks and all stable active networks

- Must be rational functions in s with real coefficients.
- May not have poles in the right half s plane.
- May not have multiple poles on the $j\omega$ axis.

Example 2.1
Check to see whether the following are stable network functions:

$$\text{(a)}\ \frac{s}{s^2 - 3s + 4} \qquad \text{(b)}\ \frac{s - 1}{s^2 + 4}$$

The first function cannot be realized by a stable network because one of the coefficients in the denominator polynomial is negative. It can easily be verified that the poles are in the right half s plane.

The second function is stable. The poles are on the $j\omega$ axis (at $s = \pm 2j$) and are simple. Note that the function has a zero in the right half s plane; however, this does not violate any of the requirements on network functions.

■

2.3 PROPERTIES OF DRIVING POINT FUNCTIONS

In this section we derive some simple properties that all dp functions must have. The properties will apply to impedance and admittance function realized by using some or all of the permissible components.

Driving point functions must satisfy the general properties of network functions mentioned in Section 2.2. In addition, since the reciprocal of a dp function is also a dp function, its zeros must satisfy the same constraints as its poles. Therefore, a driving point function:

- May not have poles or zeros in the right half s plane.
- May not have multiple poles or zeros on the $j\omega$ axis.

A corollary of the second property is that the degree of the numerator can differ from the degree of the denominator by no more than one. The reasoning is that at infinite frequency ($s = j\infty$), the terms associated with the highest power of s in Equation 2.5 dominate, so that

$$H(s)|_{s=j\omega} \cong \frac{a_n s^n}{b_m s^m}$$

Now if n differs from m by more than one, $H(s)$ will either have a multiple pole ($n > m$) or a multiple zero ($m > n$) on the imaginary axis, at $s = j\infty$.

2.3.1 PASSIVE RLC DRIVING POINT FUNCTIONS

Passive *RLC* networks contain no energy sources and as such they can only dissipate—but not deliver—energy. This dissipative nature of passive networks imposes a further restriction on the dp function. The restriction is that if the function is evaluated at any point on the $j\omega$ axis, the real part will be non-negative. Mathematically

$$\operatorname{Re} Z(j\omega) \geq 0 \text{ for all } \omega \tag{2.12}$$

A heuristic argument [7] to justify this statement follows. Suppose the network function has a negative real part at frequency $s = j\omega_1$. Then the function at $s = j\omega_1$ can be written as

$$Z(j\omega_1) = -R(\omega_1) + jX(\omega_1)$$

We can place a capacitance (or inductance) whose reactance is $-jX(\omega_1)$ in series with the network (Figure 2.5), to cancel the reactance term in the original circuit. This new network, which is also passive, has an impedance

$$Z_1(j\omega_1) = -R(\omega_1)$$

which is a negative resistance. If we applied a voltage source at this frequency to this network, the network would *deliver* current to the source, thereby violating the passive nature of the circuit. Thus the network cannot have a negative real part for any frequency on the $j\omega$ axis.

Another property of passive dp functions is that the residues of $j\omega$ axis poles (defined in Appendix A) must be real and positive. We refer the reader to [12] for the proof.

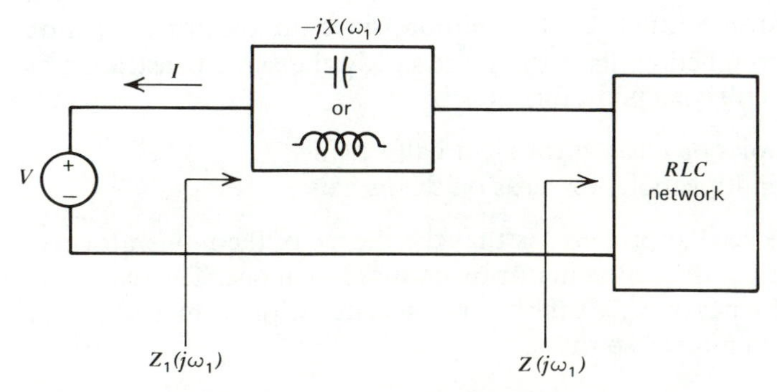

Figure 2.5 Cancellation of the reactance of the *RLC* dp impedance $Z(j\omega_1)$.

Summarizing, a passive driving point function:*

- Must be a rational function in s with real coefficients.
- May not have poles or zeros in the right half s plane.
- May not have multiple poles or zeros on the $j\omega$ axis.
- May not have the degrees of the numerator and denominator differing by more than one.
- Must have a nonnegative real part for all $s = j\omega$.
- Must have positive and real residues for poles on the $j\omega$ axis.

2.3.2 PASSIVE RC DRIVING POINT FUNCTIONS

In this section we consider networks consisting of resistors and capacitors. Such networks will be needed in the synthesis of active filters using operational amplifiers, resistors, and capacitors. It is expected that by restricting the class of usable components in a network, further limitations are imposed on the corresponding network function. *RC* driving point functions must necessarily satisfy the properties of passive dp functions. Additional properties that must be satisfied by *RC* networks are described in the following.

It was shown that the poles and zeros must be in the left half plane for all dp functions. The following is a heuristic argument to show that the poles and zeros for *RC* networks must lie on the negative real axis of the s plane. If any poles are off the axis, the corresponding term in the impulse is given by Equation 2.9,

$$h(t) = 2K_1 e^{at} \cos bt$$

which, for negative a, is an exponentially decaying sinusoid (Figure 2.6). However, such a sinusoidal response requires capacitors and inductors in the circuit to store and release energy in alternate half cycles of the response. A network consisting of only resistors and capacitors cannot produce such a response; therefore, the corresponding network function cannot have poles off the axis.

Since the reciprocal of a dp function is also a dp function, the zeros also have this same restriction.

Let us next consider the behavior of *RC impedance* functions at *dc* and at infinity. At *dc* all the capacitors become open circuits so the *RC* network will reduce to a resistor, or to an open circuit. If it is resistive, the function is a positive constant at *dc*; if it is an open circuit, the function has a pole at *dc*.

* The properties are equivalent to the *positive real* (p.r.) property of passive driving point functions which states that:

1. $H(s)$ is real for real s
2. $\text{Re}[H(s)] \geq 0 \qquad \text{for Re } s \geq 0$

The p.r. property is mentioned here for completeness; a detailed discussion of this topic can be found in any standard textbook on passive network synthesis [12].

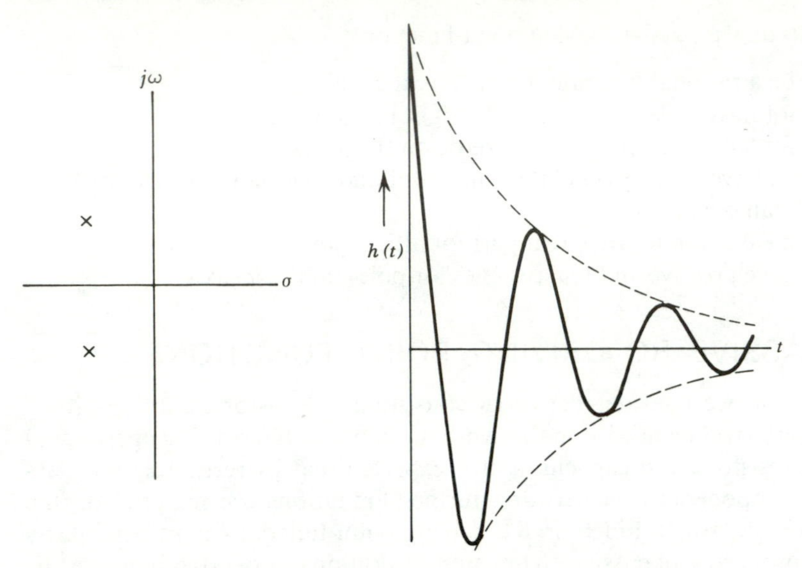

Figure 2.6 Impulse response due to complex poles in the left half s plane.

Again at infinite frequency the capacitors are short circuits so the *RC* network will reduce to a resistor, or to a short circuit. If it is resistive the function is a positive constant; if a short circuit the function has a zero at infinite frequency.

Using the above properties, the *RC* dp impedance function can be written in the form

$$Z_{RC}(s) = K \frac{(s + z_1)(s + z_2) \cdots (s + z_n)}{(s + p_1)(s + p_2) \cdots (s + p_m)} \tag{2.13}$$

where K, z_i, and p_i are real and positive. The partial fraction expansion of such a function with simple poles, from Appendix A, is

$$Z_{RC}(s) = \frac{K_0}{s} + K_\infty + \sum_i \frac{K_i}{s + p_i} \tag{2.14}$$

where the residues can be shown to be real and positive [12]. In this expansion, the K_0 term is present if the function has a pole at the origin, and the K_∞ term is present whenever the function is a constant at $s = \infty$.

Since the poles and zeros of $Z_{RC}(s)$ are all on the negative real axis, we can graphically characterize the *RC* dp function by plotting $Z(\sigma)$ versus σ, for $-\infty < \sigma < 0$. The slope of the function $Z(\sigma)$ can be obtained by summing the derivatives of the terms in Equation 2.14. Since

$$\frac{d}{d\sigma}\left(\frac{K_i}{\sigma + p_i}\right) = \frac{-K_i}{(\sigma + p_i)^2} \tag{2.15}$$

and the K_i's are positive, the slope $dZ(\sigma)/d\sigma$ must be negative. A consequence of the negative slope property is that the poles and zeros of *RC* dp functions must alternate, that is, the function cannot have adjacent poles or zeros [12]. Furthermore, the fact that $Z(\sigma)$ must be a positive constant or infinite at dc, combined with the negative slope property, leads to the conclusion that the root closest to the origin must be a pole. By a similar argument the root furthest from the origin is seen to be a zero.

As an example, consider the plot of the impedance function

$$Z_1(s) = \frac{3}{4}\frac{(s+2)(s+4)}{(s+1)(s+3)}$$

For real *s*, the corresponding impedance function is given by

$$Z_1(\sigma) = \frac{3}{4}\frac{(\sigma+2)(\sigma+4)}{(\sigma+1)(\sigma+3)}$$

This function is sketched in Figure 2.7. The *dc* value of the function is obtained by letting $s = \sigma = 0$

$$Z_1(\sigma)|_{\sigma=0} = 2$$

The infinite frequency value $Z_1(\sigma)$ is seen to be 3/4.

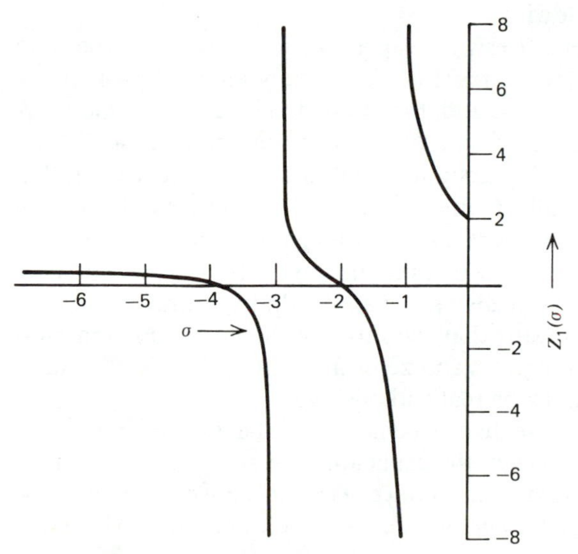

Figure 2.7 $Z(\sigma)$ versus σ plot for an *RC* dp impedance.

In summary, *RC* dp impedance functions satisfy all the properties associated with *RLC* dp functions and in addition:

- The poles and zeros lie on negative real axis and alternate.
- At *dc* the dp impedance function is either a positive constant or has a pole.
- At infinity the dp impedance function is either a positive constant or has a zero.
- The residues of $Z_{RC}(s)$ are real and positive.
- The slope of $dZ(\sigma)/d\sigma$ is negative.

Proceeding as in the above, *RC* dp *admittance* functions can be shown to have the following properties [12]:

- The poles and zero lie on the negative real axis and alternate.
- At *dc* the dp admittance function is a positive constant or has a zero.
- At infinity the dp admittance function is either a positive constant or has a pole.
- The residues of $Y_{RC}(s)$ are real and negative.*
- The slope $dY(\sigma)/d\sigma$ is positive.

2.3.3 PASSIVE LC DRIVING POINT FUNCTIONS

In this section the properties of *LC* dp immittance (i.e., impedance and admittance) functions are considered. Besides satisfying the properties of all dp functions, the *LC* immittance functions must also satisfy some other properties, which are described in the following.

Of the three passive elements (resistor, capacitor, and inductor), the only one that can *dissipate* energy is the resistor. Thus a fundamental property of *LC* networks is that they are lossless and, therefore, do not dissipate energy. A consequence of this characteristic is that all the poles and zeros of an *LC* dp function must be on the $j\omega$ axis. The argument leading to this conclusion is that if there were poles in the left half s plane, the impulse response would contain decaying factors; but an *LC* network cannot dissipate energy and hence it cannot have a decaying term. Thus, the poles cannot be off the $j\omega$ axis. Moreover the reciprocal of the given function, also being an *LC* dp function, cannot have poles off the $j\omega$ axis—which means that the zeros of the given function must also be on the $j\omega$ axis. Since the poles and zeros are on the $j\omega$ axis, they must be simple and their residues must be real and positive.

Let us consider the infinite frequency behavior of the *LC* dp impedance function $Z_{LC}(s)$. At infinite frequency the capacitors are short circuits and the inductors are open circuits. Thus the impedance of the *LC* network is either zero or infinite; in other words, $Z_{LC}(s)$ either has a zero or a pole at infinity. Moreover,

* The residues of $Y_{RC}(s)/s$ can be shown to be positive [12].

since the degree of the numerator cannot differ from that of the numerator by more than one, the pole or zero at infinity must be simple. By a similar argument, $Z_{LC}(s)$ can be seen to have a simple pole or a zero at *dc*. Therefore, the *LC* dp impedance function must have one of the following forms:

$$Z_{LC}(s) = K\frac{s(s^2+\omega_{z_1}^2)(s^2+\omega_{z_2}^2)\cdots(s^2+\omega_{z_n}^2)}{(s^2+\omega_{p_1}^2)(s^2+\omega_{p_2}^2)\cdots(s^2+\omega_{p_n}^2)} \tag{2.16a}$$

or

$$Z_{LC}(s) = K\frac{(s^2+\omega_{z_1}^2)(s^2+\omega_{z_2}^2)\cdots(s^2+\omega_{z_n}^2)}{s(s^2+\omega_{p_1}^2)(s^2+\omega_{p_2}^2)\cdots(s^2+\omega_{p_n}^2)} \tag{2.16b}$$

Thus, the *LC* dp function can be seen to be an odd function over an even function, or an even function over an odd function (i.e., an odd rational function).

The partial fraction expansion of such a function, from Appendix A, is

$$Z_{LC}(s) = \frac{K_0}{s} + K_\infty s + \sum_i \frac{2K_i s}{s^2+\omega_{p_i}^2} \tag{2.17}$$

where all the K's can be shown to be real and positive [12]. The term K_∞ is present in the expansion whenever $Z_{LC}(s)$ has a pole at $s = j\infty$; and K_0 is present whenever $Z_{LC}(s)$ has a pole at the origin.

Since all the poles and zeros are on the $j\omega$ axis, we can graphically describe the *LC* dp function by plotting $Z(j\omega)$ versus $j\omega$ or, equivalently, by plotting the so-called *reactance* function

$$X(\omega) = \frac{1}{j}Z(j\omega)$$

versus ω. The slope $dX(\omega)/d\omega$, obtained by differentiating the terms in the expansions of Equations 2.16, is seen to be positive. A consequence of the positive slope property is that the poles and zeros of *LC* dp functions must alternate [12].

For example, let us consider the plot of the following *LC* dp impedance function

$$Z_1(s) = \frac{s(s^2+4)(s^2+16)}{(s^2+1)(s^2+9)(s^2+25)}$$

The corresponding reactance function

$$X_1(\omega) = \frac{\omega(-\omega^2+4)(-\omega^2+16)}{(-\omega^2+1)(-\omega^2+9)(-\omega^2+25)}$$

is sketched in Figure 2.8.

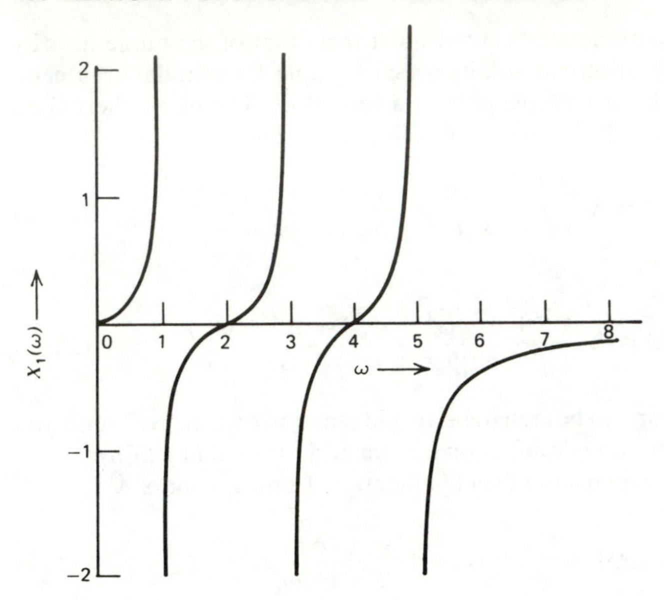

Figure 2.8 Plot of the reactance function $X(\omega)$ versus ω for an *LC* dp function.

In summary, an *LC* dp impedance function must satisfy the properties associated with *RLC* dp functions and in addition:

- The poles and zeros lie on the $j\omega$ axis, are simple, and alternate.
- The dp function must have a pole or a zero at *dc*.
- The dp function must have a pole or zero at infinity.
- The function must be odd/even or even/odd.
- The slope $dX(\omega)/d\omega$ is positive.

The symmetry of the above properties of $Z_{LC}(s)$ suggests that its reciprocal, $Y_{LC}(s)$, will have the same properties. This is indeed the case, as shown in [12].

2.4 PROPERTIES OF TRANSFER FUNCTIONS

Transfer functions realized by *RLC* passive circuits or by active *RC* circuits, must satisfy the properties of general network functions. We recall that these are:

- The function is rational in *s* with real coefficients.
- Complex poles and zeros occur as conjugate pairs.

Further, for gain transfer functions:

- The function does not have poles in the right half s plane.
- The poles on the $j\omega$ axis are simple.

There is no restriction on the location of complex zeros and, in general,

- The zeros can be anywhere in the s plane.*

2.5 MAGNITUDE AND PHASE PLOTS OF NETWORK FUNCTIONS

Much insight can be gained into the properties of a given network function by plotting its magnitude and phase versus frequency. In this section we will describe a simple way of obtaining such plots.

A general network function can be represented as

$$H(s) = \frac{N(s)}{D(s)} = K\frac{\prod_{i=1}^{n}(s - z_i)}{\prod_{i=1}^{m}(s - p_i)} \tag{2.18}$$

where the poles and zeros can be real or complex. The magnitude of $H(j\omega)$ in decibels (dB) is defined to be

$$\text{Magnitude} = 20\log_{10}|H(j\omega)| \tag{2.19}$$

$$= 20\log_{10}|K| + \sum_{i=1}^{n} 20\log_{10}|j\omega - z_i| - \sum_{i=1}^{m} 20\log_{10}|j\omega - p_i| \tag{2.20}$$

and the phase in degrees (or radians) is defined as

$$\text{Phase} = \tan^{-1}\frac{\text{Im } H(j\omega)}{\text{Re } H(j\omega)} \tag{2.21}$$

$$= \sum_{i=1}^{n}\tan^{-1}\left(\frac{\text{Im}(j\omega - z_i)}{\text{Re}(j\omega - z_i)}\right) - \sum_{i=1}^{m}\tan^{-1}\left(\frac{\text{Im}(j\omega - p_i)}{\text{Re}(j\omega - p_i)}\right) \tag{2.22}$$

The exact calculation of the magnitude and phase using the above expressions is a tedious process. However, for many purposes, an approximate sketch proves to be entirely adequate. Such approximate sketches of the magnitude and phase functions, called *Bode plots*, can be obtained with relative ease by

* For an *RLC* two-port in which the input and output ports each have a terminal connected to ground, some restrictions do exist on the location of the zeros. Fialkow and Gerst [3] showed that, for such a grounded three terminal *RLC* network, the zeros cannot lie on the positive real axis, nor can they lie in a wedge with vertex at the origin which is symmetrical about the positive real axis and forms an angle of $2\pi/n$, where n is the degree of the numerator.

using the techniques described in the remainder of this section.

In factored form, $N(s)$ and $D(s)$ are made up of four kinds of terms:

(a) A constant term, K.
(b) The factor s, representing a root at the origin.
(c) The factor $s + \alpha$, representing a real root.
(d) The factor $s^2 + as + b$, representing complex conjugate roots.

Since the magnitude (in dB) of a product of terms is equal to the sum of the magnitudes of the factors in the product, the problem of sketching the magnitude of $H(j\omega)$ reduces to that of sketching the four basic terms. This summation property also holds for the phase. Let us consider each of these terms separately.

(a) The constant term K.

The magnitude function $20 \log_{10}|K|$ is positive for $|K| > 1$, and negative for $|K| < 1$. The phase function $\tan^{-1}(0/K)$ is $0°$ for $K > 0$ and $180°$ for $K < 0$. These functions are plotted in Figure 2.9.

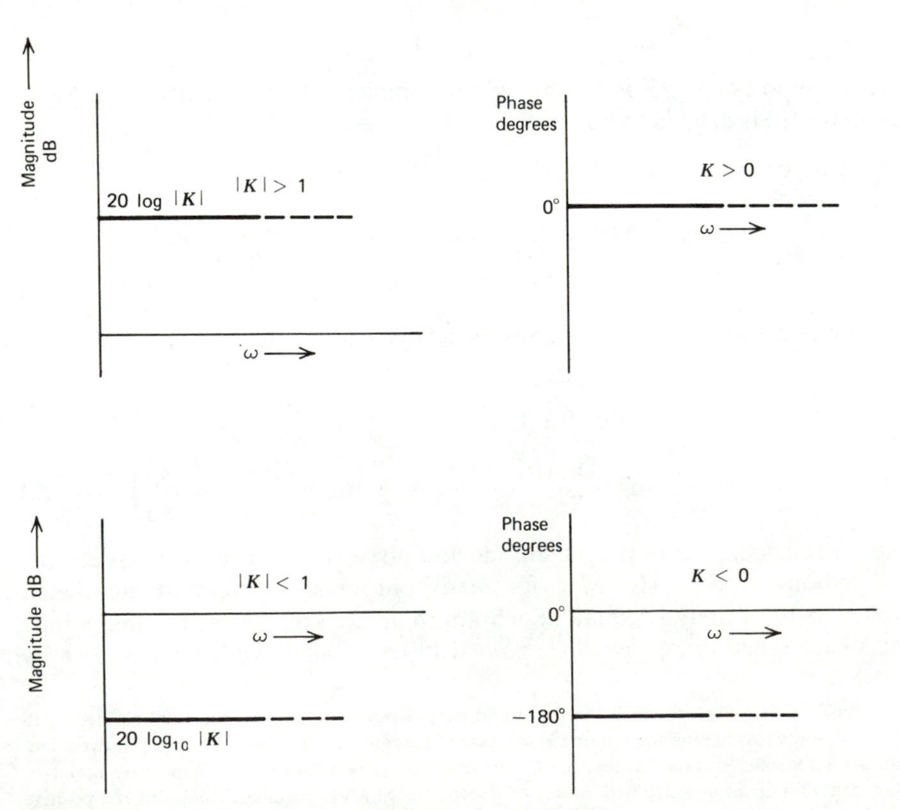

Figure 2.9 Magnitude and phase plots for a constant K.

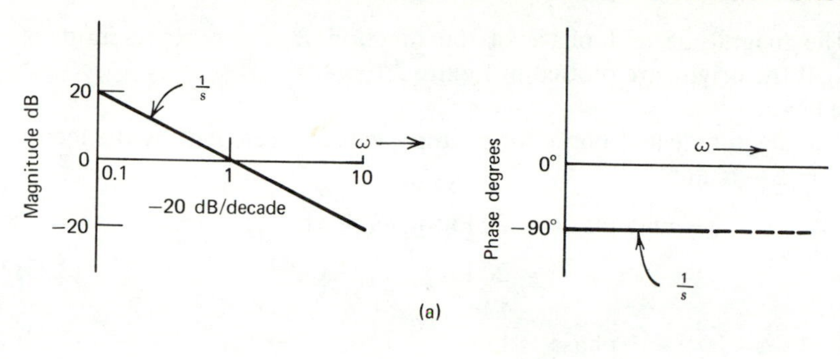

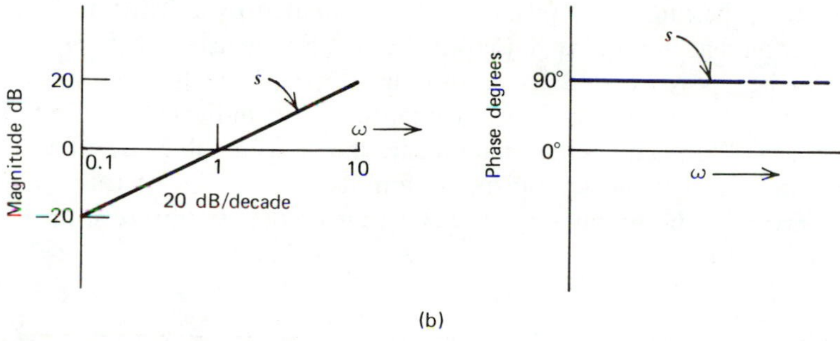

Figure 2.10 Magnitude and phase plots for (*a*) $1/s$ (*b*) s.

(b) The factor s.

Consider the case when the root at the origin is a pole. The magnitude and phase of a pole at the origin, represented by $H(s) = 1/s$, are

$$\text{magnitude} = 20 \log_{10}\left|\frac{1}{j\omega}\right| = -20 \log \omega$$

$$\text{phase} = \tan^{-1}(0) - \tan^{-1}\left(\frac{\omega}{0}\right)$$

$$= -90^\circ$$

These are plotted in Figure 2.10*a*. Observe that the magnitude decreases by 20 dB if the frequency is multiplied by 10. Equivalently, the magnitude decreases by 6 dB* if the frequency is doubled. Thus the slope of the magnitude plot is -20 dB/decade, which is the same as -6 dB/octave.†

* More exactly, the slope is $20 \log_{10} 2 = 6.02$ dB/octave.

† The number of decades separating two frequencies f_1 and f_2 is $\log_{10}(f_2/f_1)$; and the number of octaves is $\log_2(f_2/f_1) = [\log_{10}(f_2/f_1)] \div (\log_{10} 2) = 3.322 \log_{10}(f_2/f_1)$.

The magnitude and phase of the function $H(s) = s$, representing a zero at the origin, are plotted in Figure 2.10*b*.

(c) The factor $s + \alpha$.

The magnitude and phase for a simple zero, represented by the factor $H(s) = s + \alpha$, are

$$\text{magnitude} = 20 \log_{10}|j\omega + \alpha| = 20 \log_{10}(\omega^2 + \alpha^2)^{1/2} \tag{2.23}$$

$$\text{phase} = \tan^{-1}\left(\frac{\omega}{\alpha}\right) \tag{2.24}$$

These functions are plotted in Figure 2.11*a* and *b*. At low frequencies (i.e., $\omega \ll \alpha$), the function $H(s)$ can be approximated by α. Thus, the low frequency gain asymptote is $20 \log_{10}\alpha$. At high frequencies, where $\omega \gg \alpha$, $H(s)$ can be approximated by s, and the slope of the high frequency asymptote is 6 dB/octave. These asymptotes are indicated by dotted lines in Fig. 2.11*a*. The actual magnitude, shown as a solid line, departs most from the asymptotic approximation at $s = j\omega = j\alpha$. At this point the departure of the actual magnitude curve from the asymptote approximation is 3.01 dB.

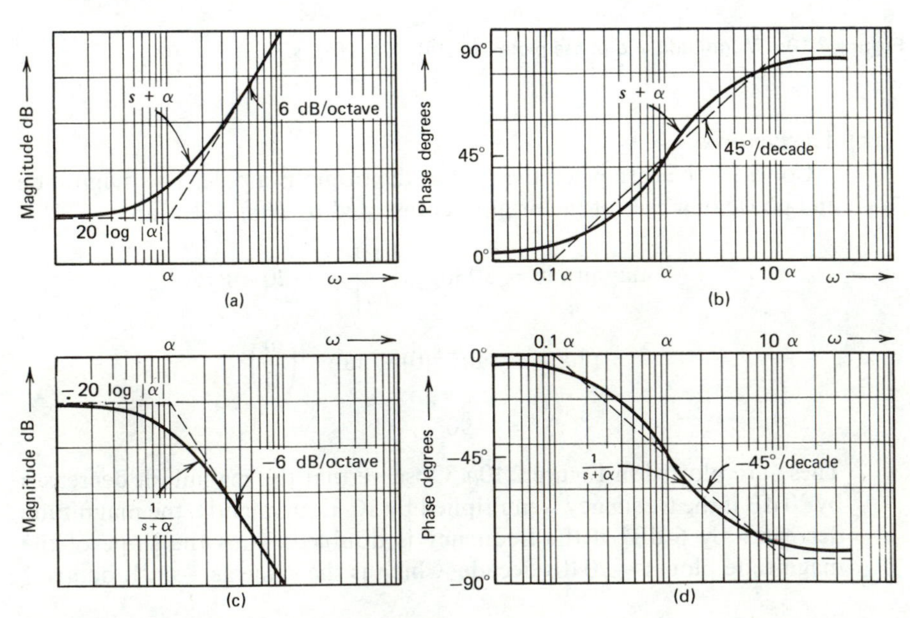

Figure 2.11 (*a*) Magnitude and (*b*) Phase plots for $s + \alpha$. (*c*) Magnitude and (*d*) Phase plots for $\dfrac{1}{s + \alpha}$.

From Equation 2.24, the phase asymptote at low frequencies is 0° and at high frequencies it is 90°. A straight-line approximation to the phase characteristic is obtained by joining these asymptotes by a line whose slope is 45° per decade, as shown in Figure 2.11*b*. Observe that the phase shift at the frequency $\omega = \alpha$ is 45°, and the error of approximation is zero at this frequency. The exact phase characteristic is drawn as a solid curve in the figure.

The magnitude and phase for a pole at $s = -\alpha$, represented by $1/(s + \alpha)$, are plotted in Figure 2.11*c* and *d*, respectively.

We will see that these straight-line segment approximations provide a quick and easy way of sketching the magnitude and phase plots of network functions.

(d) The factor $s^2 + as + b$.

Let us consider a pair of complex poles represented by

$$H(s) = \frac{1}{s^2 + as + b} \tag{2.25}$$

At *dc* the magnitude of this function is

$$20 \log_{10}\left(\frac{1}{b}\right)$$

and the phase is

$$-\tan^{-1}\left(\frac{0}{b}\right) = 0°$$

An infinite frequency, the function approaches $1/s^2$ so the magnitude decreases at -40 dB/decade, and the phase is $-180°$.

The frequency at which the magnitude achieves a maximum is obtained by equating the derivative of $|H(j\omega)|$ to zero:

$$\frac{d}{d\omega}\left|\frac{1}{-\omega^2 + aj\omega + b}\right| = 0$$

Solving this equation

$$\omega_{\max} = \sqrt{b}\sqrt{1 - \frac{a^2}{2b}} \qquad \text{for } \frac{a^2}{2b} < 1 \tag{2.26a}$$

$$= 0 \qquad \text{for } \frac{a^2}{2b} \geq 1 \tag{2.26b}$$

If $a^2/2b \ll 1$, then

$$\omega_{\max} \cong \sqrt{b} \tag{2.27}$$

This frequency is known as the **pole frequency**, ω_p. In terms of the given coefficients

$$\text{pole frequency } \omega_p = \sqrt{b} \tag{2.28}$$

At this pole frequency, the magnitude in dB is given by

$$20 \log_{10}\left|\frac{1}{(j\sqrt{b})^2 + aj\sqrt{b} + b}\right| = 20 \log_{10}\left(\frac{1}{a\sqrt{b}}\right) \tag{2.29}$$

and the phase is

$$-\tan^{-1}\left(\frac{a\sqrt{b}}{0}\right) = -90^\circ$$

Using the low frequency asymptote, the high frequency asymptote, and the value at ω_p, we can obtain an approximate sketch of the magnitude, as shown in Figure 2.12*a*.* The height of the bump at ω_p, relative to the low frequency asymptote, is seen to be

$$20 \log_{10}\left(\frac{1}{a\sqrt{b}}\right) - 20 \log_{10}\left(\frac{1}{b}\right) = 20 \log_{10}\left(\frac{\sqrt{b}}{a}\right) \tag{2.30}$$

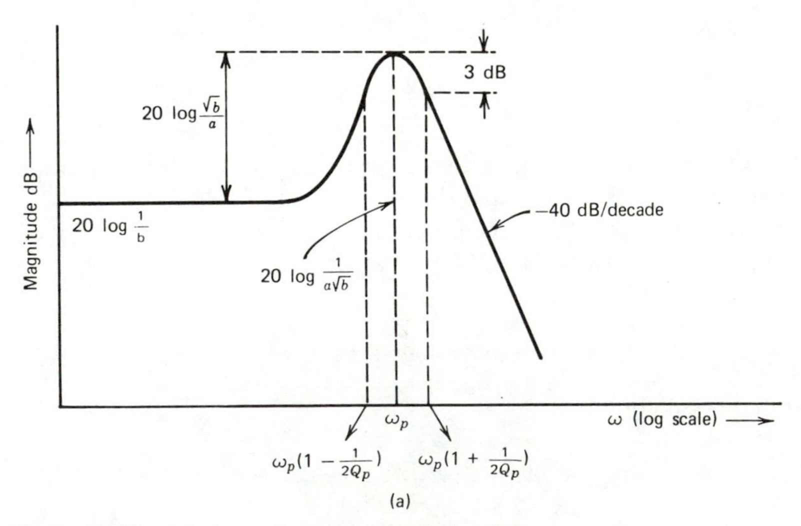

Figure 2.12 (*a*) Approximate magnitude sketch for

$$\frac{1}{s^2 + as + b} \quad \text{for} \quad Q_p = \frac{\sqrt{b}}{a} \gg 1.$$

* The approximation in the location of the maximum and subsequent relationships are reasonably good when $Q_p \geq 5$.

The parameter $\sqrt{b}/a$ determines the height of the bump at the pole frequency and is known as the **pole Q** (or Q_p):

$$\text{pole } Q = Q_p = \frac{\sqrt{b}}{a} \tag{2.31}$$

From (2.28) and (2.31), the network function representing a pair of complex poles can also be expressed in terms of ω_p and Q_p as

$$H(s) = \frac{1}{s^2 + \dfrac{\omega_p}{Q_p}s + \omega_p^2} \tag{2.32}$$

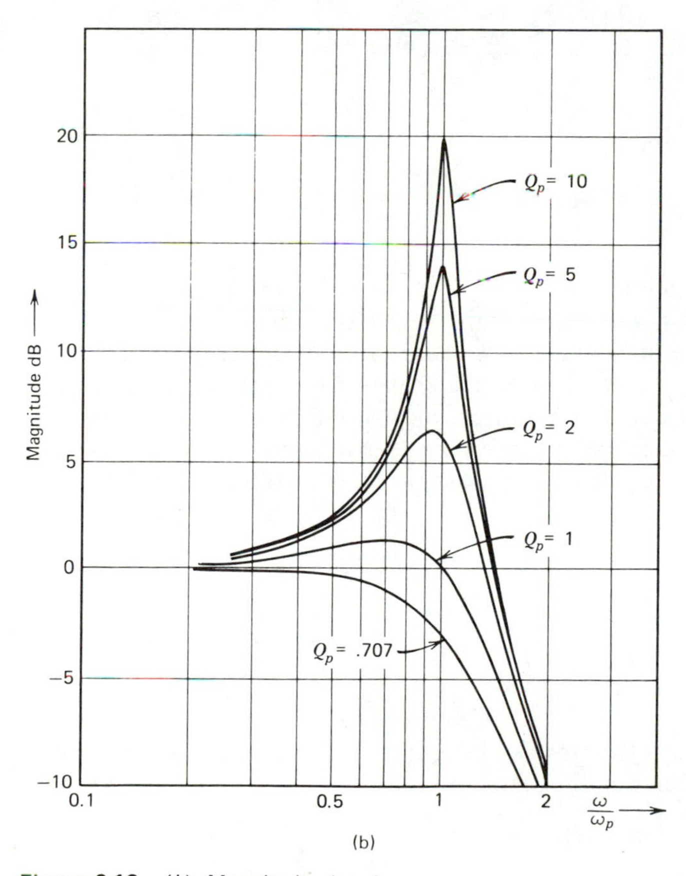

Figure 2.12 (*b*) Magnitude plots for

$$\frac{1}{s^2 + (\omega_p/Q_p)s + \omega_p^2} \quad \text{for different } Q_p.$$

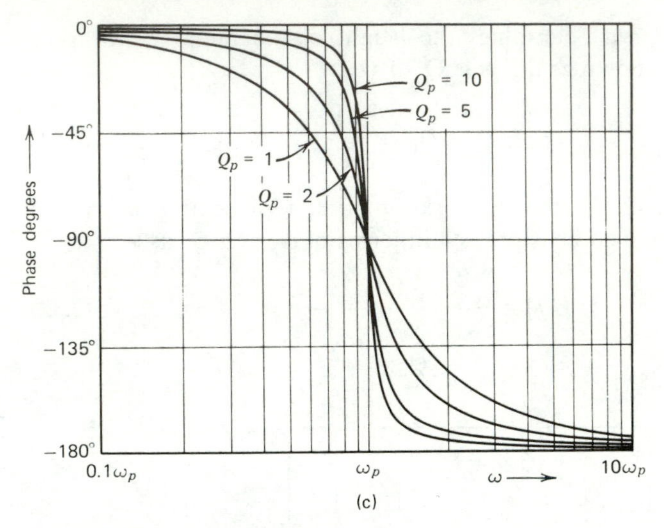

Figure 2.12 (*c*) Phase plots for

$$\frac{1}{s^2 + (\omega_p/Q_p)s + \omega_p^2} \quad \text{for different } Q_p.$$

Plots of the magnitude and phase of this function for different values of Q_p are shown in Figure 12.12*b* and *c*, respectively. It can be seen that the sharpness of the magnitude and phase plots near the pole frequency increases with Q_p. For pole Q's greater than 5 the maximum magnitude occurs essentially at the pole frequency; for Q_p's less than 0.707, the function does not exhibit a bump and the maximum is the dc value.

Let us next consider the magnitude of the function of Equation 2.32 at

$$s = j\omega_1 = j\left(\omega_p \pm \frac{\omega_p}{2Q_p}\right) \tag{2.33}$$

Substituting in (2.32),

$$H(j\omega_1) = \frac{1}{\left(-\omega_p^2 \mp \frac{\omega_p^2}{Q_p} - \frac{\omega_p^2}{4Q_p^2}\right) + j\frac{\omega_p}{Q_p}\left(\omega_p \pm \frac{\omega_p}{2Q_p}\right) + \omega_p^2}$$

If $Q_p \gg 1$, the ω_p^2/Q_p^2 terms can be ignored relative to the ω_p^2/Q_p terms. Then the magnitude in dB is given by

$$20 \log_{10}|H(j\omega_1)| \cong 20 \log_{10} \frac{1}{\left|\mp \frac{\omega_p^2}{Q_p} + j\frac{\omega_p^2}{Q_p}\right|}$$

$$= 20 \log_{10}\left(\frac{1}{\sqrt{2}\, a\sqrt{b}}\right) \tag{2.34}$$

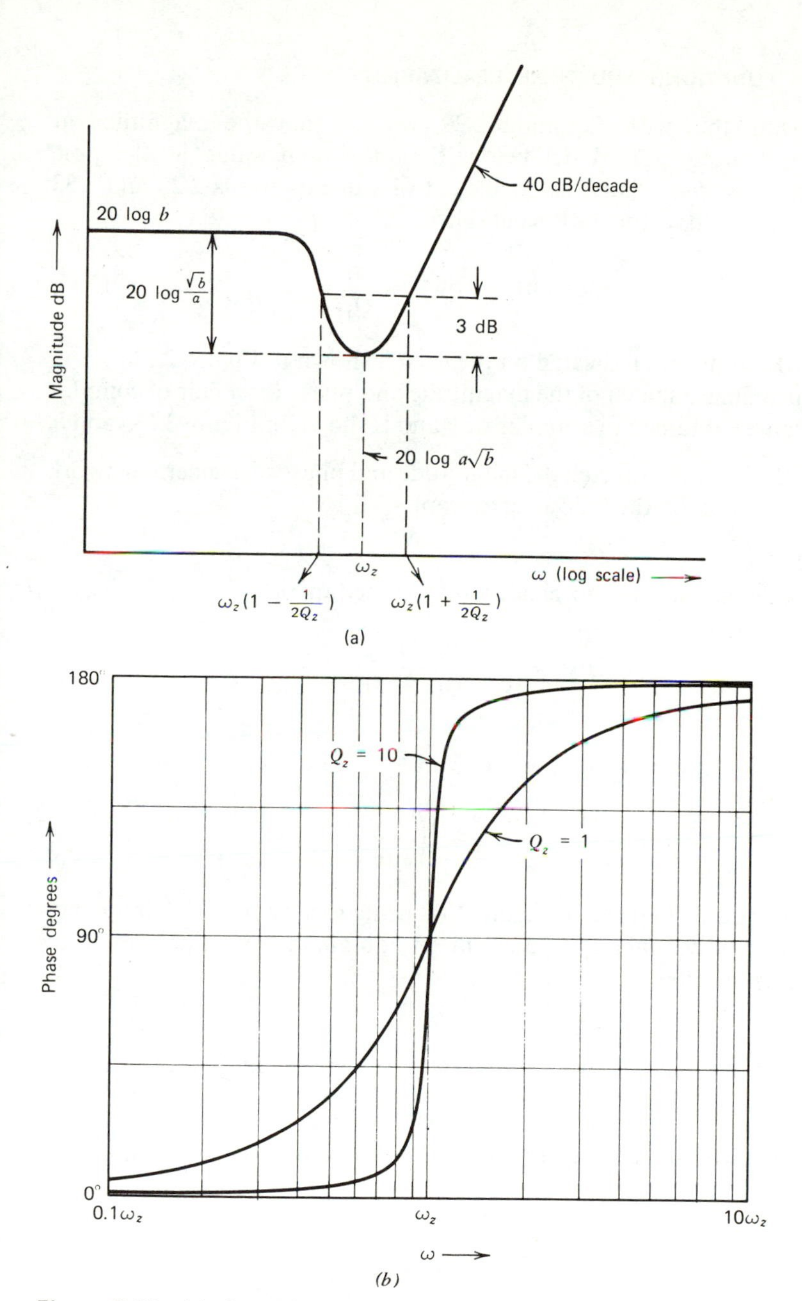

Figure 2.13 (*a*) Approximate magnitude sketch for

$$s^2 + as + b \quad \text{for} \quad Q_z = \frac{\sqrt{b}}{a} \gg 1.$$

(*b*) Phase plots for

$$s^2 + \frac{\omega_z}{Q_z} s + \omega_z^2 \quad \text{for different } Q_z.$$

Comparing this with Equation 2.29, we see that the magnitude at $s = j(\omega_p \pm \omega_p/2Q_p)$ is 3 dB below the maximum value at the pole frequency $s = j\omega_p$. Thus the coefficient of s in Equations 2.25 and 2.32 can be identified as the 3 dB **bandwidth**:

$$\text{bandwidth} = (\text{bw})_p = \frac{\omega_p}{Q_p} = a \tag{2.35}$$

The 3 dB points are indicated on the sketch shown in Figure 2.12*a*.
An approximate sketch of the magnitude and phase for a pair of complex zeros can be obtained in a similar way and is shown in Figure 2.13*a* and *b*.

The use of Bode plots to sketch the magnitude and phase of a general network function is illustrated by the following examples.

Example 2.2
Sketch the Bode magnitude and phase plots for the function

$$H(s) = \frac{4(s+2)}{(s+1)(s+4)}$$

Solution
The given function can be written in the form:

$$(2)\left(\frac{s+2}{2}\right)\left(\frac{4}{s+4}\right)\left(\frac{1}{s+1}\right)$$

Each of these terms is sketched in Figure 2.14, using the low and high frequency asymptotes. The individual plots are summed to obtain the solid line magnitude and phase plots for $H(s)$. ■

Example 2.3
Find a function that has the magnitude sketch shown in Figure 2.15.

Solution
Comparing the given sketch with Figure 2.12*a*, the form of the function is seen to be

$$H(s) = \frac{K}{s^2 + as + b}$$

Since the maximum occurs at $\omega = 2$, from Equation 2.27, $b \cong 4$. The constant K is evaluated by considering the gain at dc. From the figure this gain is 6 dB. Thus

$$\frac{K}{b} = 2 \qquad \text{so} \qquad K = 8$$

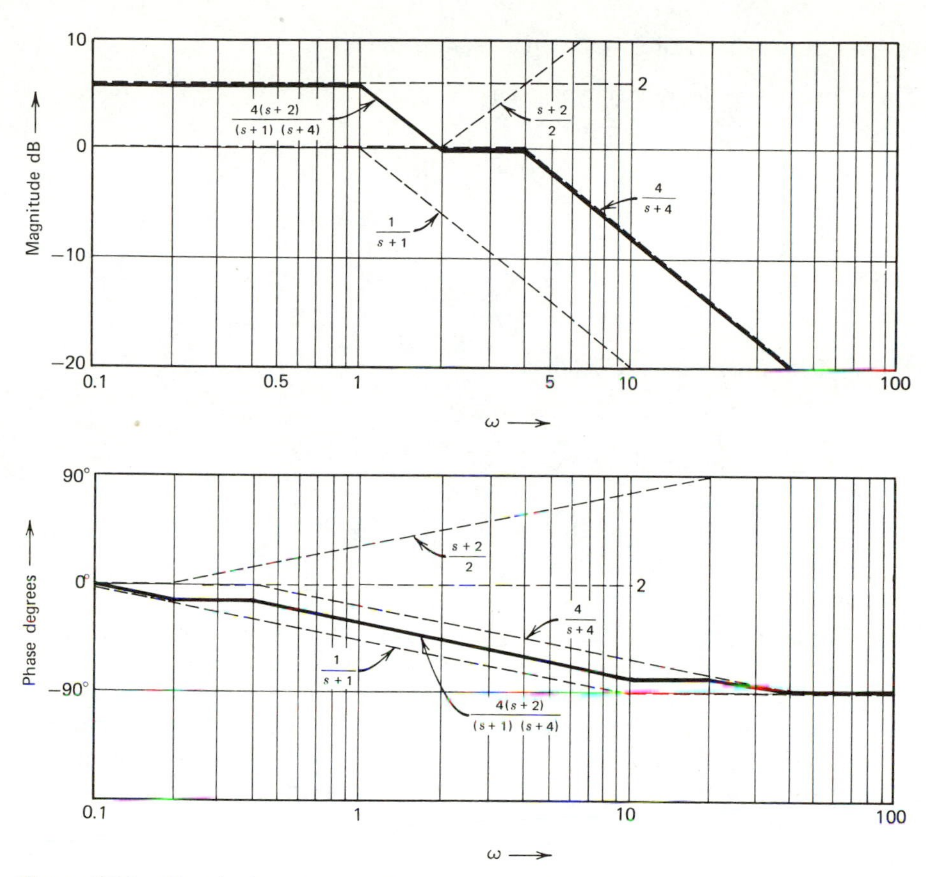

Figure 2.14 Magnitude and phase plots for Example 2.2.

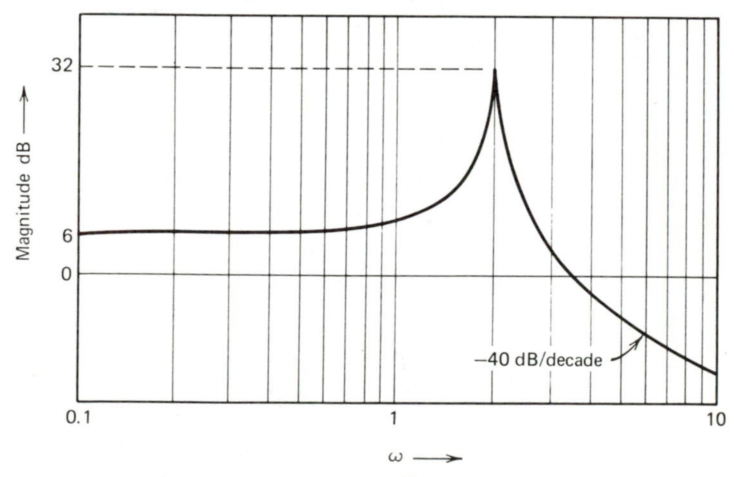

Figure 2.15 Magnitude plot for Example 2.3.

The gain at the pole frequency is given by

$$20 \log_{10}\left|\frac{K}{a\sqrt{b}}\right| = 32$$

Thus

$$\frac{K}{a\sqrt{b}} = 39.8$$

which yields the remaining coefficient

$$a = 0.1$$

The desired function is therefore

$$H(s) = \frac{8}{s^2 + 0.1s + 4}$$

Observations

1. The pole Q in this example is $Q_p = \sqrt{b}/a = 20$. For such a high pole Q, the approximations used in the computations are very good.
2. This example illustrates the problem of obtaining a function that approximates a given magnitude sketch. This step, known as the **approximation** problem, is an important part of the synthesis of filters. We devote an entire chapter (Chapter 4) to various ways of obtaining approximation functions.

2.6 THE BIQUADRATIC FUNCTION

In this section we consider the second-order function

$$H(s) = K\frac{(s + z_1)(s + z_2)}{(s + p_1)(s + p_2)} \tag{2.36}$$

This function, known as a **biquadratic**, is the basic building block used in the synthesis of a large class of active filters. The details of the synthesis will be covered in Chapters 7 to 10.

Equation 2.36 may be written as

$$H(s) = K\frac{s^2 + (z_1 + z_2)s + z_1z_2}{s^2 + (p_1 + p_2)s + p_1p_2} \tag{2.37}$$

For complex poles and zeros

$$z_1 = \text{Re } z_1 + j \text{ Im } z_1 \qquad z_2 = \text{Re } z_1 - j \text{ Im } z_1$$
$$p_1 = \text{Re } p_1 + j \text{ Im } p_1 \qquad p_2 = \text{Re } p_1 - j \text{ Im } p_1$$

Thus

$$H(s) = K \frac{s^2 + (2 \operatorname{Re} z_1)s + (\operatorname{Re} z_1)^2 + \operatorname{Im}(z_1)^2}{s^2 + (2 \operatorname{Re} p_1)s + (\operatorname{Re} p_1)^2 + \operatorname{Im}(p_1)^2} \tag{2.38}$$

As mentioned in Section 2.5 a pair of poles can be represented in terms of ω_p (the pole frequency) and Q_p (the pole Q). Similarly a pair of zeros can be described by ω_z and Q_z. In terms of these parameters

$$H(s) = K \frac{s^2 + \dfrac{\omega_z}{Q_z} s + \omega_z^2}{s^2 + \dfrac{\omega_p}{Q_p} s + \omega_p^2} \tag{2.39}$$

This form of the biquadratic is particularly useful in the sketching of the function, as shown in the following.

The *dc* gain is

$$20 \log_{10} \left| K \frac{\omega_z^2}{\omega_p^2} \right| \tag{2.40}$$

and the infinite frequency gain is

$$20 \log_{10} |K| \tag{2.41}$$

In the last section it was shown that the maximum value for the complex poles occurs approximately at the pole frequency ω_p. For biquadratics in which the zero is far removed from the pole (that is, $\omega_z/\omega_p \gg 1$ or $\omega_p/\omega_z \gg 1$), the location of this maximum is unaffected by the complex zeros. From (2.38) and (2.39), the pole frequency is related to the location of the pole in the s plane by*

$$\omega_p = \sqrt{(\operatorname{Re} p_1)^2 + (\operatorname{Im} p_1)^2} \tag{2.42}$$

which is the radial distance from the origin to the pole location. Similarly, the biquadratic will have its minimum value approximately when the numerator is at a minimum. This minimum occurs at

$$s = j\omega \cong j\omega_z \qquad \text{for} \qquad Q_z \gg 1$$

The zero frequency ω_z is related to the zero location by

$$\omega_z = \sqrt{(\operatorname{Re} z_1)^2 + (\operatorname{Im} z_1)^2} \tag{2.43}$$

* For real poles, from (2.37) and (2.39), $\omega_p = \sqrt{p_1 p_2}$.

The pole Q, which determines the sharpness of the bump at ω_p, is obtained from (2.38) and (2.39):*

$$Q_p = \frac{\omega_p}{(\text{bw})_p} = \frac{\sqrt{(\text{Re } p_1)^2 + (\text{Im } p_1)^2}}{2 \text{ Re } p_1} \tag{2.44}$$

and the zero Q is given by

$$Q_z = \frac{\omega_z}{(\text{bw})_z} = \frac{\sqrt{(\text{Re } z_1)^2 + (\text{Im } z_1)^2}}{2 \text{ Re } z_1} \tag{2.45}$$

Very often the zeros are on the $j\omega$ axis in which case $Q_z = \infty$.

Summarizing this section, for a biquadratic function:

- The maximum occurs approximately at ω_p.
- The minimum occurs approximately at ω_z.
- Q_p is a measure of the sharpness of the maximum.
- Q_z is a measure of the sharpness of the minimum.
- The *dc* magnitude is $20 \log_{10}|K\omega_z^2/\omega_p^2|$.
- The infinite frequency magnitude is $20 \log_{10}|K|$.

The approximations improve as the pole and zero Q's increase and as the pole and zero frequencies get further apart.

Example 2.4

Sketch the gain versus frequency for the voltage transfer function:

$$T(s) = 10 \frac{s^2 + 16}{s^2 + 2s + 100}$$

Solution

From Equation 2.39

$$K = 10 \qquad \omega_z = 4 \qquad Q_z = \infty \qquad \omega_p = 10 \qquad Q_p = 5$$

The *dc* gain is

$$20 \log_{10}\left[10\left(\frac{16}{100}\right)\right] = 4.08 \text{ dB}$$

The gain at the pole frequency, $\omega = 10$, is

$$20 \log_{10}\left|10 \frac{-100 + 16}{j20}\right| = 32.46 \text{ dB}$$

At the zero frequency the gain is

$$20 \log_{10}(0) = -\infty \text{ dB}$$

* For real poles $Q_p = \sqrt{p_1 p_2}/(p_1 + p_2)$, which has a maximum value of 1/2.

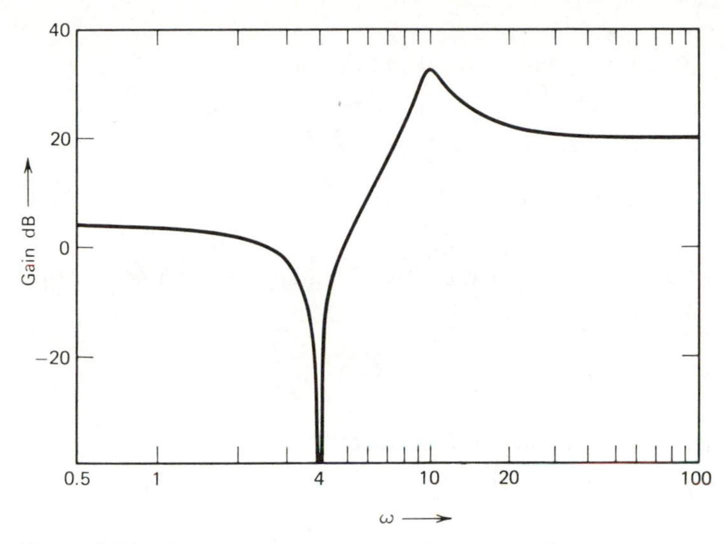

Figure 2.16 Gain versus frequency plot for Example 2.4.

and the gain at infinite frequency is

$$20 \log_{10}(10) = 20 \text{ dB}$$

Using these computed gains, the transfer function can be sketched as in Figure 2.16. ■

2.7 COMPUTER PROGRAM FOR MAGNITUDE AND PHASE

A computer program, MAG, for the exact evaluation of the magnitude and phase of rational functions is listed in Appendix D.* The network function is described in the following factored form:

$$H(s) = \prod_{i=1}^{N} \frac{m_i s^2 + c_i s + d_i}{n_i s^2 + a_i s + b_i} \tag{2.46}$$

The frequencies at which the function is to be evaluated are described in terms of a start frequency FS, a frequency increment FI, and final frequency FF, all in hertz. The magnitude and phase are calculated using Equations 2.19 to 2.22. The following example illustrates the use of the program.

* The program also computes delay, a quantity that will be defined in Chapter 3.

Example 2.5

Compute the magnitude and phase of the network function

$$H(s) = \frac{s(s+3)(s+7)}{(s+1)(s+5)(s^2+s+81)}$$

at frequencies 0.1 Hz to 2 Hz in steps of 0.1 Hz.

Solution

Two second-order functions are needed to describe the given function. Writing the given function as

$$H(s) = \frac{s^2+3s+0}{s^2+6s+5} \cdot \frac{0.s^2+s+7}{s^2+s+81}$$

the coefficients for the two sections are seen to be:

$$m_1 = 1 \quad c_1 = 3 \quad d_1 = 0 \quad n_1 = 1 \quad a_1 = 6 \quad b_1 = 5$$
$$m_2 = 0 \quad c_2 = 1 \quad d_2 = 7 \quad n_2 = 1 \quad a_2 = 1 \quad b_2 = 81$$

The desired frequencies are described by

$$FS = 0.1 \quad FI = 0.1 \quad FF = 2.$$

The computed magnitude and phase are listed in the following table.

Freq HZ	Gain dB	Phase Deg
0.1	−30.991	67.207
0.2	−27.091	56.403
0.3	−25.223	53.109
0.4	−23.785	52.788
0.5	−22.437	53.477
0.6	−21.072	54.403
0.7	−19.637	55.237
0.8	−18.082	55.798
0.9	−16.347	55.917
1.0	−14.343	55.338
1.1	−11.926	53.539
1.2	−8.836	49.196
1.3	−4.588	37.922
1.4	0.656	1.139
1.5	−1.076	−59.961
1.6	−5.891	−82.728
1.7	−9.269	−90.605
1.8	−11.707	−94.153
1.9	−13.585	−95.996
2.0	−15.101	−97.021

■

2.8 CONCLUDING REMARKS

In this chapter we have developed various properties of network functions. Since the network function must satisfy these, they are called **necessary** conditions. However, we have not guaranteed that if the conditions are satisfied we will be able to realize the function by a physical network. In other words, the properties developed were not shown to be **sufficient** conditions. Such sufficient conditions do indeed exist for different categories of networks—refer to [12] for further discussion. We may mention in passing that the positive real property for passive dp network functions (Section 2.3.1) is both necessary and sufficient. In other words, these conditions must be satisfied by a passive dp function; moreover, a p.r. function is always realizable using passive networks [12].

We have limited our discussions to *RLC*, *LC*, *RC*, and active-*RC* networks, because these are the networks that occur most frequently in the study of active filters. Similar properties can be developed for *RL* networks [12].

All our work in this text deals with lumped components, which lead to rational polynomial functions. If we included distributed elements, the network functions would be nonrational containing terms such as $\sqrt{s}$ and $\coth(s)$. The theory and realization of these functions is covered in texts on distributed networks [4].

FURTHER READING

1. C. M. Close, *The Analysis of Linear Circuits*, Harcourt Brace Jovanovich, New York, Chapter 6.
2. C. A. Desoer and E. S. Kuh, *Basic Circuit Theory*, McGraw-Hill, New York, 1969, Chapters 13 and 15.
3. A. Fialkow and I. Gerst, "The transfer function of general two-port RC networks," *Quart. Appl. Math.*, *10*, July 1952, pp. 113–127.
4. M. S. Ghausi and J. J. Kelly, *Introduction to Distributed-Parameter Networks*, Holt, New York, 1968.
5. G. S. Moschytz, *Linear Integrated Network Fundamentals*, Van Nostrand, New York, 1974, Chapter 1.
6. E. A. Guillemin, *Synthesis of Passive Networks*, Wiley, New York, 1957, Chapters 1 and 2.
7. F. F. Kuo, *Network Analysis and Synthesis*, Second Edition, Wiley, New York, 1966, Chapters 8 and 9.
8. H. Ruston and J. Bordogna, *Electric Networks: Functions, Filter Analysis*, McGraw-Hill, New York, 1966, Chapter 2.
9. J. L. Stewart, *Circuit Theory and Design*, Wiley, New York, 1956, Chapters 2 and 3.
10. H. H. Sun, *Synthesis of RC Networks*, Hayden, New York, 1967, Chapter 2.
11. M. E. Van Valkenburg, *Network Analysis*, Third Edition, Prentice-Hall, New York, 1974, Chapters 10 and 11.
12. M. E. Van Valkenburg, *Introduction to Modern Network Synthesis*, Wiley, New York, 1960.

PROBLEMS

2.1 *Pole-zero diagrams.* Sketch the s domain pole-zero diagrams for the following functions:

(a) $\dfrac{s^2 + 2s + 5}{s^2 + 4s + 5}$

(b) $\dfrac{5s(s + 1)}{(s + 1 - 3j)(s + 1 + 3j)}$

(c) $\dfrac{s(s^2 + 1)}{(s^2 + 2)}$

2.2 *Network functions, realizability.* Which of the following are stable network functions? Of the stable network functions, which can be realized as *dp* functions and/or transfer functions? Give reasons.

(a) $\dfrac{s + 1}{s - 1}$

(b) $\dfrac{s - 1}{s + 1}$

(c) $\dfrac{s + 4}{s^2 + 1}$

(d) $\dfrac{(s - 1 + j)(s - 1 - j)}{(s + 2 - j)(s + 2 + j)}$

(e) $\dfrac{(s + 2)^2}{(s + 1)(s + 3)}$

(f) $\dfrac{s + 1}{(s + 2)^2}$

(g) $\dfrac{1}{s^2 + 2s + 4}$

(h) $\dfrac{s^2 - s + 1}{s^2 + s + 1}$

(i) $\dfrac{s^3 - s + 1}{s^3 + s + 1}$

2.3 *Impedance functions from p-z patterns.* Determine the *dp* impedance functions corresponding to the pole-zero patterns shown in Figure P2.3.

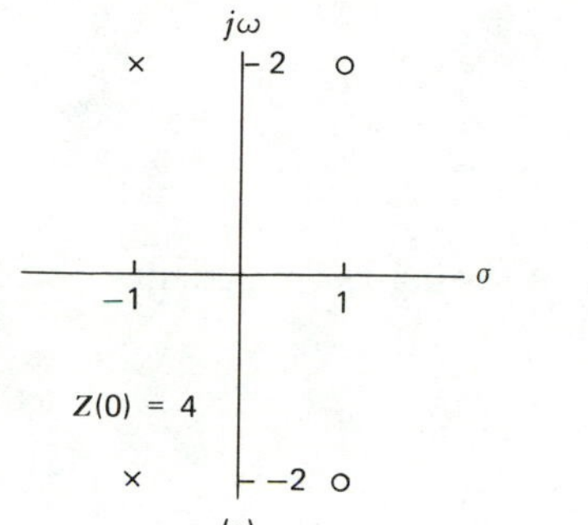

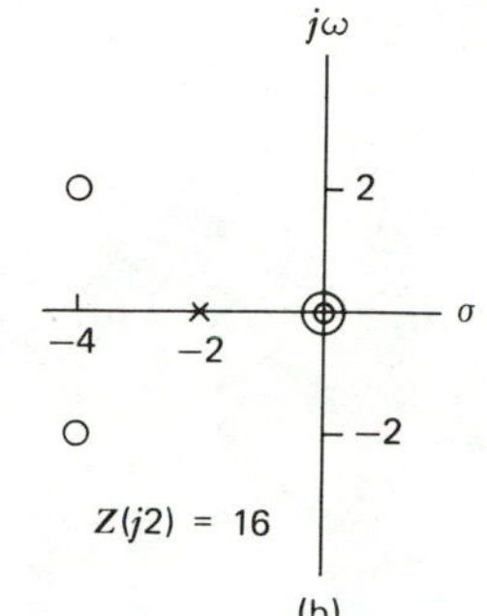

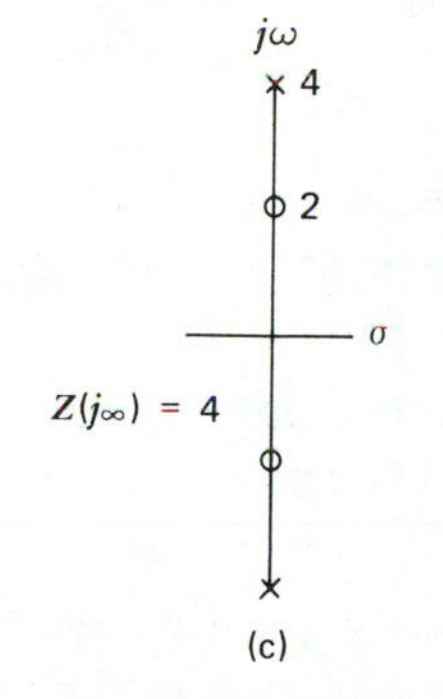

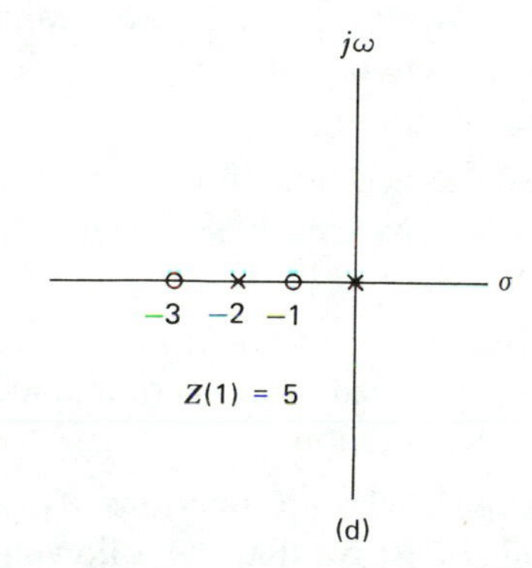

Figure P2.3

2.4 *RC impedance functions.* Which of the following are *RC* impedance functions? Give reasons.

(a) $\dfrac{(s+2)(s+6)}{(s+4)(s+8)}$

(b) $\dfrac{(s+3)(s+7)}{(s+2)(s+5)}$

(c) $\dfrac{s(s+1)(s+3)}{(s+2)(s+4)}$

2.5 Sketch $Z(\sigma)$ versus σ for the following *RC* impedance functions:

(a) $4\dfrac{(s+2)(s+6)}{(s+1)(s+4)}$

(b) $\dfrac{s+1}{s(s+2)}$

2.6 *LC impedance functions.* Which of the following are *LC* impedance functions? Give reasons:

(a) $\dfrac{(s^2+1)(s^2+3)}{(s^2+2)(s^2+4)}$

(b) $\dfrac{s(s^2+4)(s^2+6)}{(s^2+3)(s^2+8)}$

2.7 Sketch the reactance functions for the following *LC* impedance functions:

(a) $\dfrac{s(s^2+4)}{(s^2+1)(s^2+9)}$

(b) $\dfrac{(s^2+1)(s^2+4)}{s(s^2+2)}$

2.8 *RC dp function.* An *RC dp* function has infinite impedance at $\sigma = -2, -6$ and has zero impedance at $\sigma = -4, -8$. The impedance at *dc* is 5.33. Find the function.

2.9 An *RC dp* impedance function has poles at $\sigma = -1$ and $\sigma = -4$; and zeros at $\sigma = -2$ and $\sigma = -\sigma_0$. The impedance at infinite frequency asymptotes to 2 Ω, and the *dc* impedance is 6 Ω. Find σ_0.

2.10 *LC dp function.* An *LC* dp function has infinite impedance at 1000 Hz and 4000 Hz, and the impedance is zero at 2500 Hz. The impedance at 500 Hz is 1 kΩ. Find the function.

2.11 *RLC dp functions.* The *dp* functions Z_1, Z_2, and Z_3 can be realized as *RLC* impedances. Prove that the following functions can also be realized as *RLC* impedances:

(a) $\dfrac{1}{Z_1}$

(b) $\dfrac{Z_1 Z_2}{Z_1 + Z_2}$

(c) $Z_1 + \dfrac{1}{\dfrac{1}{Z_2} + \dfrac{1}{Z_3}}$

Give counterexamples to show that the following cannot always be realized as *RLC* impedances:

(d) $Z_1 Z_2$

(e) $Z_1 - Z_2$

(f) $\dfrac{Z_1}{Z_2}$

2.12 *Properties of RL dp functions.* Give reasons for the following properties of *RL* impedance functions:

(a) The root closest to the origin is a zero.

(b) The root furthest from the origin is a pole.

(c) The slope of $Z(\sigma)$ vs σ is positive.

(d) Poles and zeros must lie on the negative real axis.

(e) Poles and zeros must alternate.

2.13 *Transfer and dp functions.* Prove that if a function is realizable as a *dp* function it is also realizable as a transfer function. Give a counterexample to prove that the converse is not true.

2.14 *Sketches of dp functions.* Determine the *dp* functions corresponding to the sketches shown in Figure P2.14.

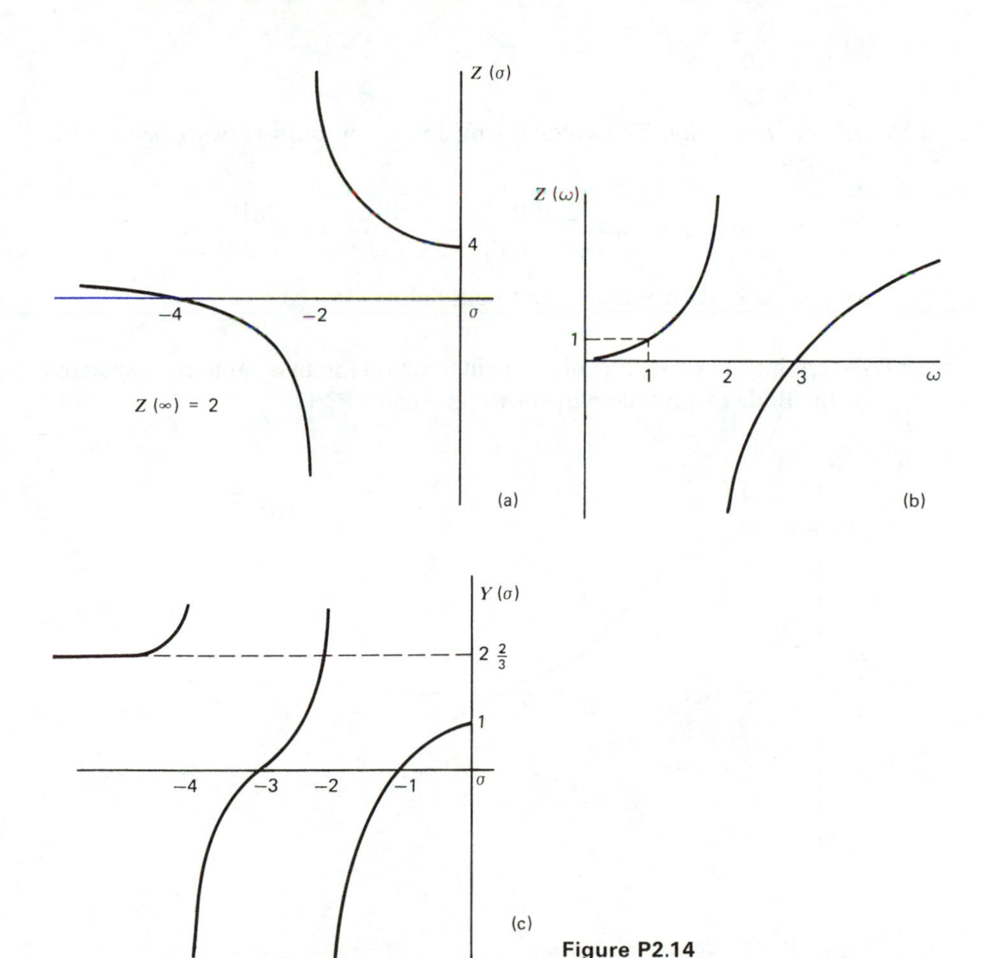

Figure P2.14

2.15 *Bode plots, real roots.* Sketch the Bode magnitude and phase plots for the following functions:

(a) $\dfrac{4}{s + 4}$

(b) $\dfrac{s}{s + 2}$

(c) $\dfrac{0.7s(s + 5.5)}{(s + 8.7)}$

(d) $\dfrac{2(s + 5)(s + 20)}{(s + 2)(s + 12)}$

(e) $\dfrac{s^2}{(s + 4)(s + 8)}$

2.16 *Op amp Bode plot.* The voltage gain for an op amp is described by the function:

$$A(s) = 10^5 \cdot \frac{2\pi 100}{(s + 2\pi 100)} \cdot \frac{2\pi 10^4}{(s + 2\pi 10^4)} \cdot \frac{2\pi 10^6}{(s + 2\pi 10^6)}$$

Sketch the Bode magnitude and phase plots for $A(s)$.

2.17 Determine the s domain voltage gain function for an op amp characterized by the Bode magnitude plot shown in Figure P2.17.

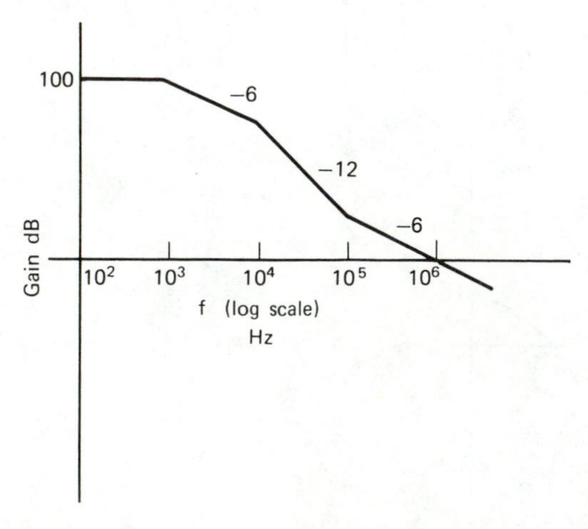

Figure P2.17 Slopes in dB/octave.

2.18 *Functions from Bode plots.* Determine the functions corresponding to the Bode magnitude plots shown in Figure P2.18.

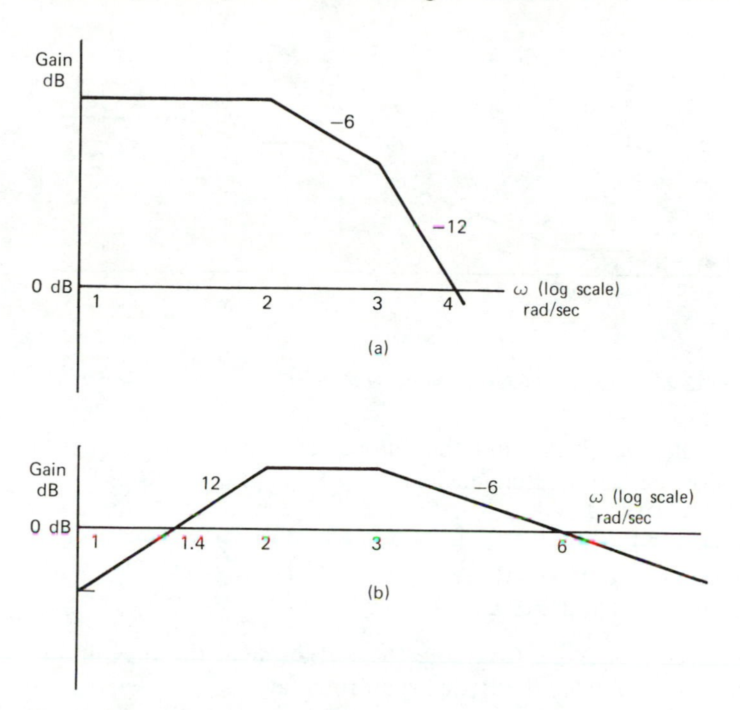

Figure P2.18 Slopes in dB/octave.

2.19 *Bode plots, phase.* Obtain the Bode phase plots from the magnitude plots given in Problem 2.18.

2.20 Determine the impedance function corresponding to the given Bode phase plot. The magnitude of the function at *dc* is 6 dB.

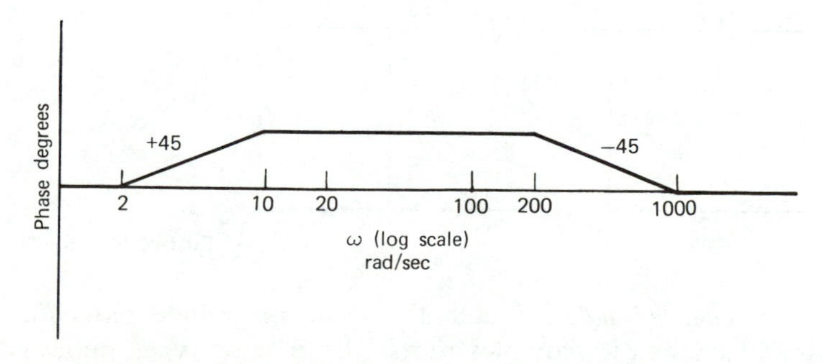

Figure P2.20 Slopes in degrees per decade

2.21 Repeat Problem 2.20 for the phase plot of Figure P2.21, given that the magnitude at *dc* is 0 dB.

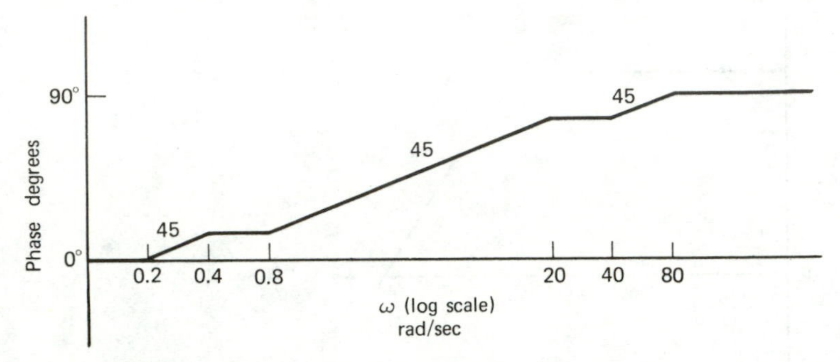

Figure P2.21 Slopes in degrees per decade

2.22 *Octaves, decades.* Determine the number of octaves and decades between the following pairs of frequencies:
(a) 10 Hz, 1000 Hz.
(b) 0.5 Hz, 20 Hz.
(c) 1700 rad/sec, 4200 rad/sec.
(d) 1 rad/sec, 16 rad/sec.

2.23 *Bode plots, real roots.* Compute the deviation of the magnitude of the function $(s + a)$ from the Bode asymptotes, at:
(a) The break frequency $\omega = a$.
(b) One octave above the break frequency.

2.24 Repeat Problem 2.23 for the phase of the function $(s + a)$.

2.25 *Bode plots, given network.* Sketch the Bode magnitude plot for the transfer function V_O/V_{IN} of the *RC* network shown.

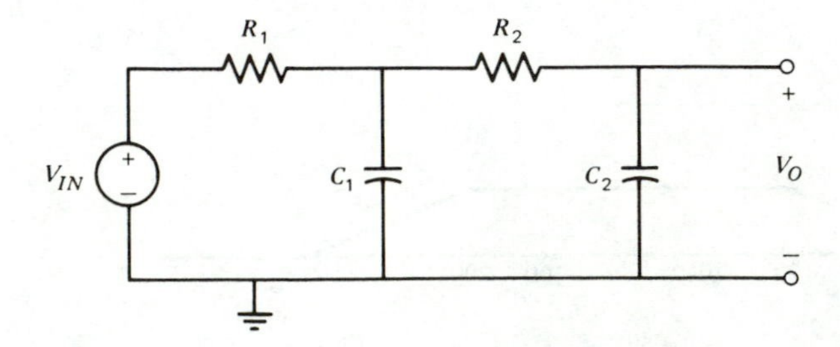

Figure P2.25

2.26 *Bode plots, complex poles.* Sketch the Bode magnitude plots for the following functions with complex roots. The function types, indicated in parenthesis, will be described in Chapter 3.

(a) $\dfrac{50}{s^2 + s + 25}$ (low-pass)

(b) $\dfrac{10s}{s^2 + 10s + 100}$ (band-pass)

(c) $\dfrac{10s^2}{s^2 + 2s + 36}$ (high-pass)

(d) $\dfrac{50}{s^2 + 5s + 25}$ (low Q low-pass)

2.27 *Bode plots, complex poles and zeros.* Sketch the Bode magnitude plots for

(a) $4\dfrac{s^2 + 25}{s^2 + 2.5s + 100}$ (high-pass notch)

(b) $4\dfrac{s^2 + s + 100}{s^2 + 2.5s + 100}$ (gain equalizer)

(c) $\dfrac{s^2 - 2.5s + 25}{s^2 + 2.5s + 25}$ (all-pass)

2.28 *Complex poles, peak magnitude.* Show that the maximum magnitude for the function of Equation 2.32, for $\omega_p = 1$ and $Q_p > 0.707$, is

$$20 \log_{10}\left(\frac{Q_p}{\sqrt{1 - \dfrac{1}{4Q_p^2}}}\right) \text{dB}$$

Verify this in the $Q_p = 2$ curve of Figure 2.12.

2.29 *Biquadratic parameters.* Identify the biquadratic parameters K, ω_z, ω_p, Q_p, Q_z in the following functions

(a) $3\dfrac{s^2 + 4s + 16}{s^2 + 2s + 25}$

(b) $\dfrac{4s^2 + 36}{2s^2 + 6s + 112}$

(c) $\dfrac{(s + 1 - 2j)(s + 1 + 2j)}{(s + 4 + 3j)(s + 4 - 3j)}$

2.30 Determine the biquadratic parameters for the functions having the pole-zero diagrams shown in Figure P2.30. Note that K cannot be obtained from pole-zero diagrams.

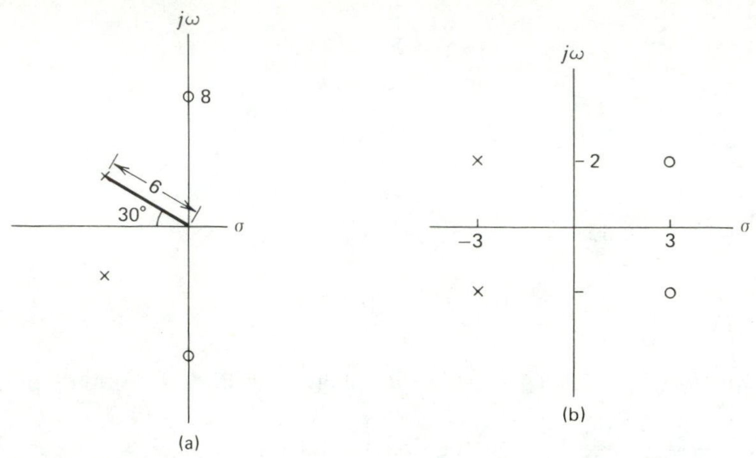

Figure P2.30

2.31 *Biquadratic functions from sketches.* Determine the biquadratic functions corresponding to the magnitude sketches of Figure P2.31.

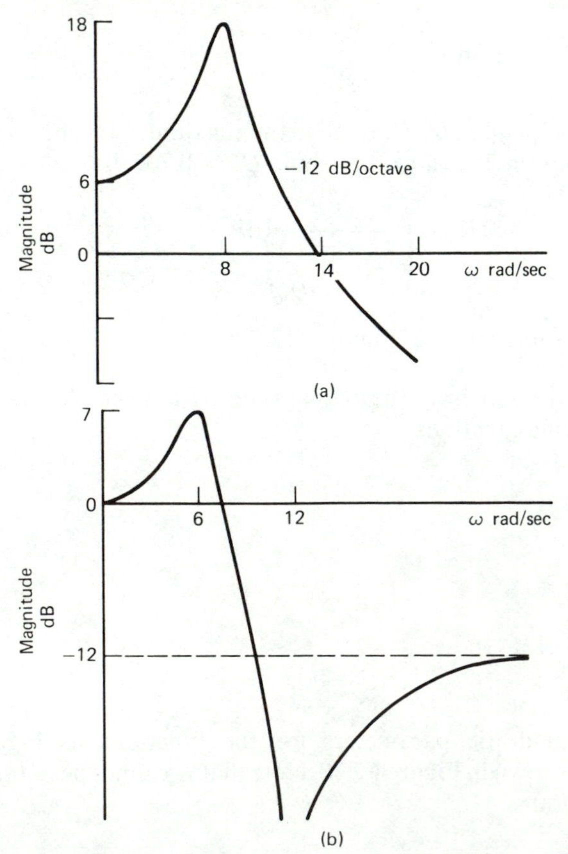

Figure P2.31

2.32 *Biquadratic function.* A biquadratic function has poles at $-10 \pm 1000j$, zeros at $\pm 1200j$ and the magnitude at infinite frequency is 0 dB. Determine:

(a) An expression for the function.
(b) The pole Q.
(c) The magnitude at 1000 and 1200 rad/sec.
Use the results to sketch the function.

2.33 A biquadratic function has a pole Q of 7 and its pole frequency ω_p is 100 rad/sec. The zero frequency ω_z is 200 rad/sec, the zeros being on the imaginary axis. It is also given that the *dc* gain is 10 dB. Find the function.

2.34 *Band-pass biquadratic.* Consider the biquadratic function

$$\frac{s}{s^2 + \dfrac{\omega_p}{Q_p} s + \omega_p^2}$$

which has a band-pass filter characteristic (filter functions will be discussed in Chapter 3).

(a) Show that the magnitude of the function is maximum at the pole frequency ω_p.
(b) Find exact expressions for the two frequencies ω_1 and ω_2 at which the magnitude is 3 dB below that at ω_p. Show that $\omega_1 \omega_2 = \omega_p^2$.
(c) Determine the phase at ω_p, ω_1, and ω_2.

2.35 *MAG computer program.* Use the MAG computer program to obtain the exact magnitude and phase plots for the following function:

$$\frac{0.088s^2 + 2.24(10)^6}{s^2 + 2844s + 1.355(10)^8} \cdot \frac{s^2 + 4.84(10)^8}{s^2 + 2935s + 2.411(10)^8}$$

Choose frequencies from 100 Hz to 4000 Hz in steps of 100 Hz for the computation.

2.36 Use the MAG program to plot the exact magnitude characteristic for the op amp gain characteristic of Problem 2.16.

3.

INTRODUCTORY FILTER CONCEPTS

A filter is a network used to shape the frequency spectrum of an electrical signal. These networks are an essential part of communication and control systems. Filters are classified according to the functions they perform, as low-pass, high-pass, band-pass, band-reject, amplitude equalizers, and delay equalizers. In this chapter we describe these filter types and illustrate their applications with examples taken from voice and data communications.

Before the 1960s, filters for voice and data communication systems were manufactured using passive *RLC* components. In recent years, the emergence of hybrid integrated circuit technology has opened up the large field of active *RC* filters. The main features and applications of active and passive filters will be discussed in Section 3.2. To show where active and passive filters fit in the world of filter applications, we also briefly describe other filtering techniques using electromechanical, digital, and microwave filters.

3.1 CATEGORIZATION OF FILTERS

In this section we describe each of the following filter types: low-pass, high-pass, band-pass, band-reject, amplitude equalizers, and delay equalizers, and illustrate their use with examples.

3.1.1 LOW-PASS

The basic function of a low-pass (*LP*) filter is to pass low frequencies with very little loss and to attenuate high frequencies. A typical low-pass requirement is shown in Figure 3.1. To meet this requirement a filter characteristic is sought that stays outside the shaded region. The *LP* filter is required to pass signals from *dc* up to the **cutoff frequency** ω_P, with at most A_{max} dB of attenuation. This band of frequencies, from *dc* to ω_P, is known as the **passband**. Frequencies above ω_S are required to have at least A_{min} dB of attenuation. The band of frequencies from ω_S to infinity is called the **stopband**, and ω_S is referred to as the **stopband edge frequency**. The frequency band from ω_P to ω_S is called the **transition** band. The four parameters ω_P, ω_S, A_{min}, and A_{max} completely describe the requirements of the *LP* filter shown in Figure 3.1.

A more general stopband characteristic could have different amounts of attenuation in sections of the stopband, as shown in Figure 3.2. Here the

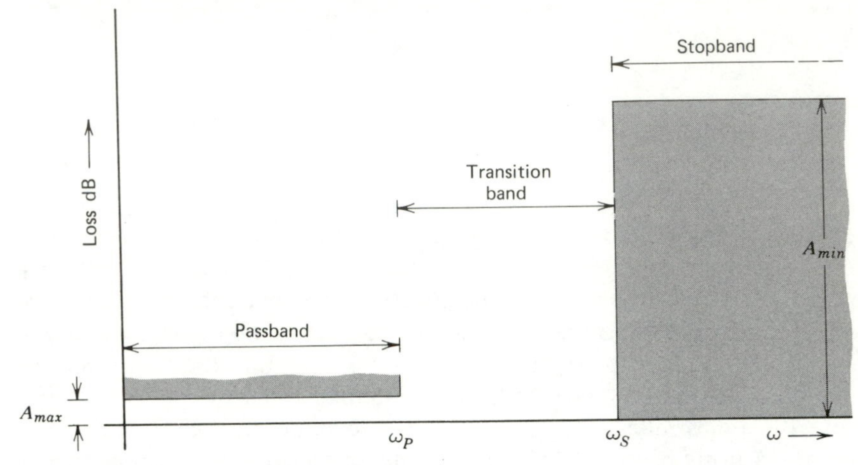

Figure 3.1 Low-pass filter requirements.

attenuation from ω_{S_1} to ω_{S_2} must be at least $A_{\min_1}$, while the attenuation from ω_{S_2} to infinity must be a minimum of $A_{\min_2}$. In most applications, however, the stopband requirement has just one level; therefore, future discussions will be restricted to this case.

A second-order gain function that realizes a low-pass characteristic is

$$\text{Gain} = \frac{V_O}{V_{IN}} = \frac{b}{s^2 + as + b} = \frac{\omega_p^2}{s^2 + \dfrac{\omega_p}{Q_p} s + \omega_p^2} \tag{3.1}$$

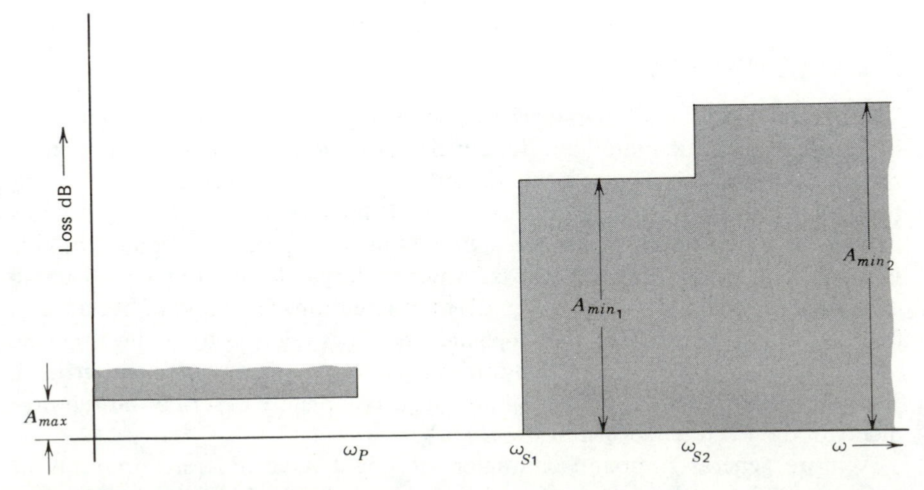

Figure 3.2 Non-flat stopband.

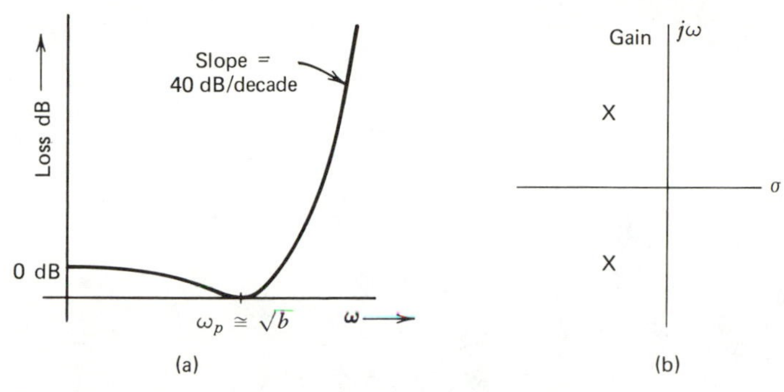

Figure 3.3 A second-order *LP* function: (*a*) Loss; (*b*) Pole-zero plot.

The corresponding loss function is given by*

$$\text{Loss} = \frac{V_{IN}}{V_O} = \frac{s^2 + as + b}{b} \tag{3.2}$$

As shown in the sketch of Figure 3.3*a*, the low frequency loss approaches unity (0 dB); while at high frequencies the loss increases as s^2, that is, at 40 dB/decade. The second-order low-pass gain function has a pair of complex poles, as illustrated in Figure 3.3*b*. The location of the poles of Equation 3.1 determines the shape of the filter response in the passband. As discussed in Section 2.5, for high Q poles the bump in the passband occurs at the pole frequency, ω_p; and the sharpness of the bump increases with the pole Q.

A familiar application of *LP* filters is in the tone control of some hi-fi (high fidelity) amplifiers. The treble control varies the cutoff frequency of a *LP* filter and is used to attenuate the high frequency record scratch-noise and amplifier hiss.

3.1.2 HIGH-PASS

A high-pass (*HP*) filter passes frequencies above a given frequency, called its cutoff frequency. A typical *HP* requirement is shown in Figure 3.4. The passband extends from ω_P to ∞, and the stopband from *dc* to ω_S. As in the *LP* case, the parameters ω_S, ω_P, $A_{\min}$, and $A_{\max}$ completely characterize the *HP* filter requirements.

* It is common practice in the active filter literature to express the transfer function as a gain function (V_O/V_{IN}). Also, pole-zero diagrams are usually drawn for the gain function. On the other hand, filter requirements and magnitude sketches are usually drawn for the loss function (V_{IN}/V_O).

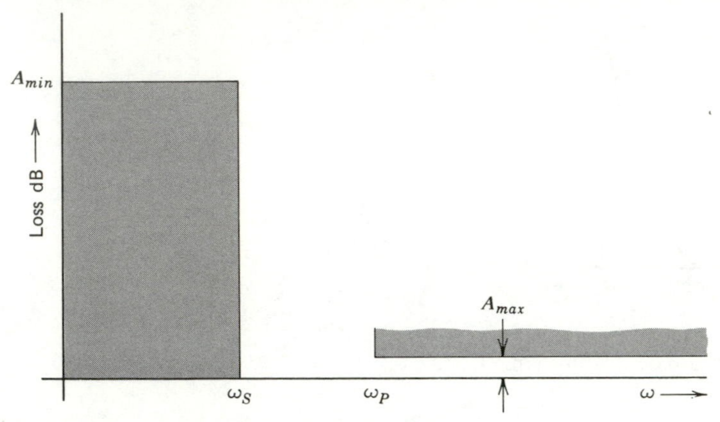

Figure 3.4 High-pass filter requirements.

A second-order gain function that has an *HP* characteristic is given by

$$\frac{V_O}{V_{IN}} = \frac{s^2}{s^2 + as + b} = \frac{s^2}{s^2 + \dfrac{\omega_p}{Q_p} s + \omega_p^2} \tag{3.3}$$

This gain function has a pair of complex poles and a double zero at the origin, as shown in Figure 3.5*a*. From the magnitude sketch of the loss function (Figure 3.5*b*), it is seen that the high frequency loss approaches unity, while the low frequency loss increases at 40 dB/decade.

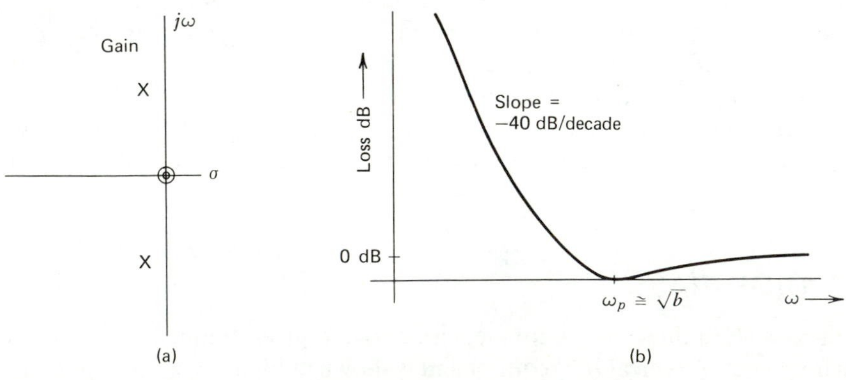

Figure 3.5 Second-order *HP* filter: (*a*) Pole-zero plot; (*b*) Loss.

3.1.3 BAND-PASS FILTERS

Band-pass filters pass frequencies in a band of frequencies with very little attenuation, while rejecting frequencies on either side of this band. Figure 3.6 shows a typical *BP* requirement. The passband from ω_1 to ω_2 has a maximum

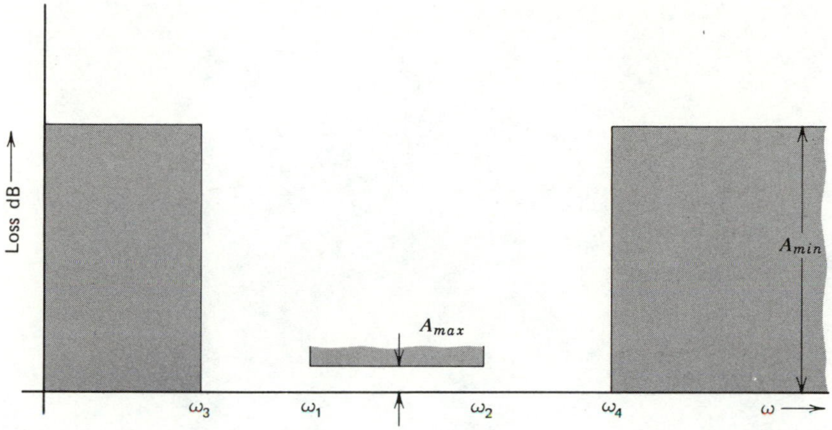

Figure 3.6 Typical band-pass filter requirements.

attenuation of $A_{\max}$ dB; the two stopbands from *dc* to ω_3, and ω_4 to ∞, have a minimum attenuation of $A_{\min}$ dB. A second-order transfer function that has a band-pass characteristic is

$$\frac{V_O}{V_{IN}} = \frac{as}{s^2 + as + b} = \frac{\dfrac{\omega_p}{Q_p} s}{s^2 + \dfrac{\omega_p}{Q_p} s + \omega_p^2} \tag{3.4}$$

This function has a pair of complex poles in the left half *s* plane and a zero at the origin (Figure 3.7*a*). At low frequencies, and at high frequencies, the loss increases as *s*, that is, at 20 dB/decade. At the pole frequency, $\omega_p = \sqrt{b}$, the loss is a constant equal to unity. A sketch of the loss function is shown in Figure 3.7*b*.

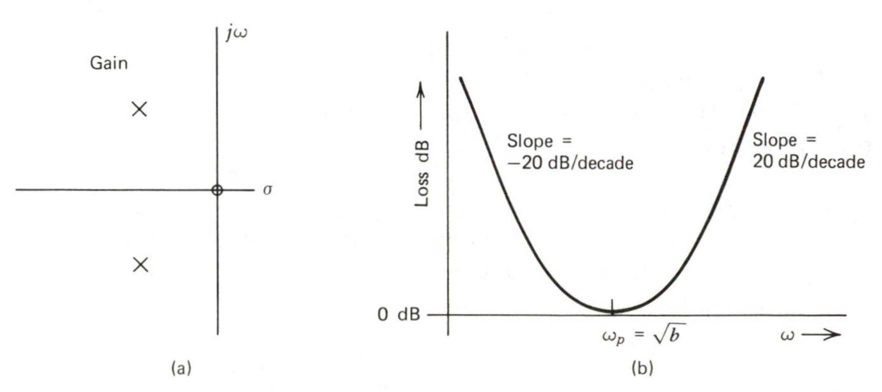

Figure 3.7 A second-order *BP* filter: (*a*) Pole-zero plot; (*b*) Loss.

Figure 3.8 TOUCH–TONE® telephone set.

An interesting example that illustrates the application of low-pass, high-pass, and band-pass filters is in the detection of signals generated by a telephone set with push buttons (Figure 3.8), as in TOUCH–TONE® dialing. As the telephone number is dialed a set of signals is transmitted to the telephone office. Here, these signals are converted to suitable *dc* signals that are used by the switching system to connect the caller to the party being called.

In TOUCH–TONE dialing the 10 decimal digits 0 to 9 and an additional six extra buttons (used for special purposes) need to be identified. The signaling code adopted provides 16 distinct signals by using 8 signal frequencies in the frequency band 697 Hz to 1633 Hz. The 8 frequencies or tones are divided into two groups, four low-band and four high-band, as illustrated in Figure 3.9. Pressing a push button generates a pair of tones, one from the high-band and one from the low-band. Each push button is therefore identified by a unique pair of signal frequencies.

At the telephone office, these tones are identified and converted to a suitable set of *dc* signals for the switching system, as illustrated by the block diagram of Figure 3.10. After amplification, the two tones are separated into their respective

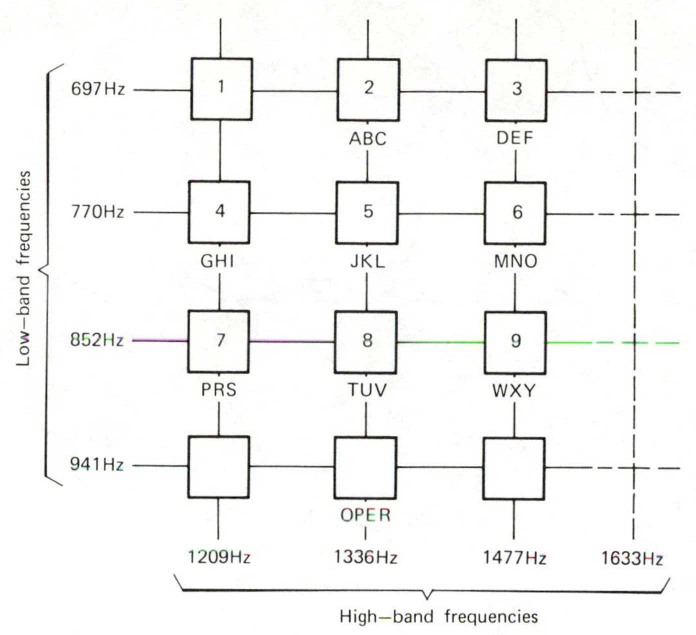

Figure 3.9 Frequency assignments for TOUCH-TONE® dialing.

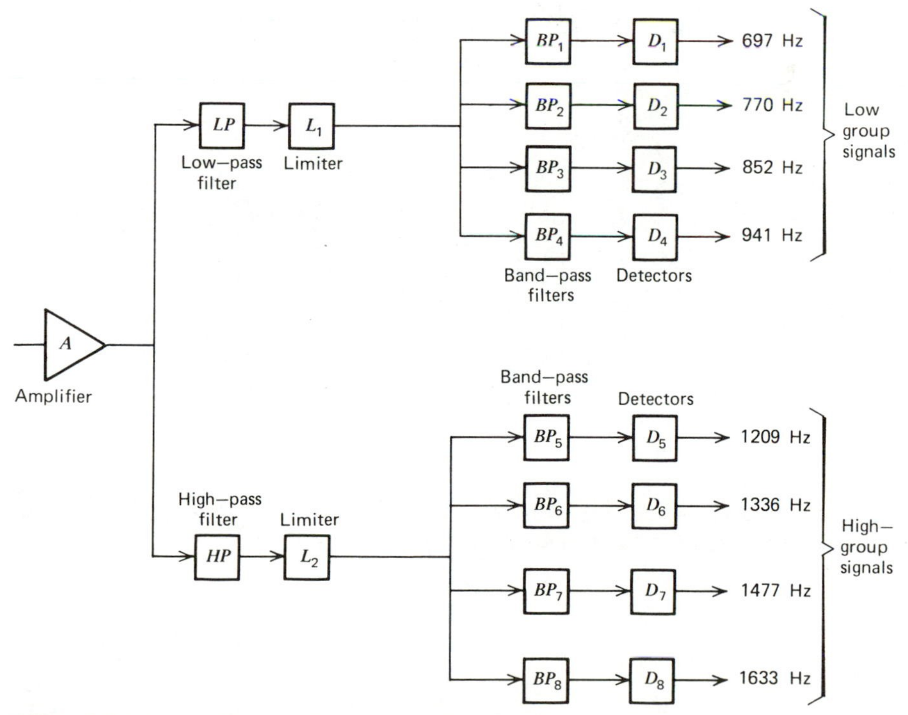

Figure 3.10 Block diagram of detection scheme.

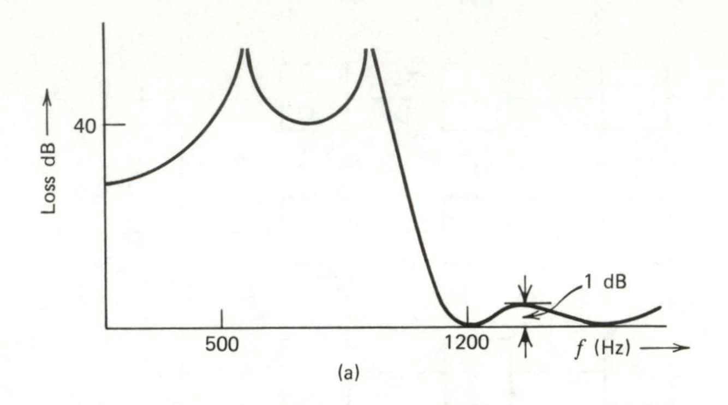

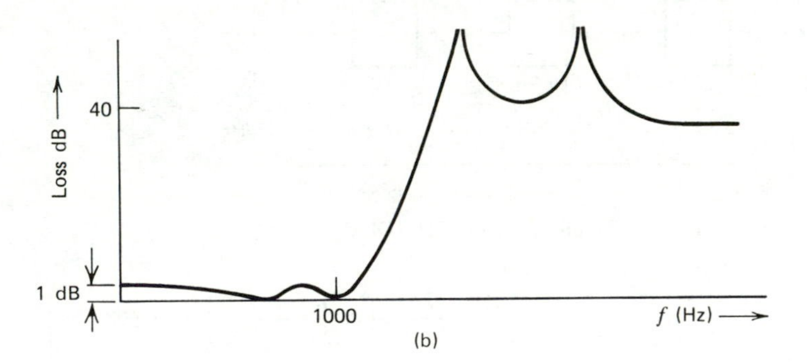

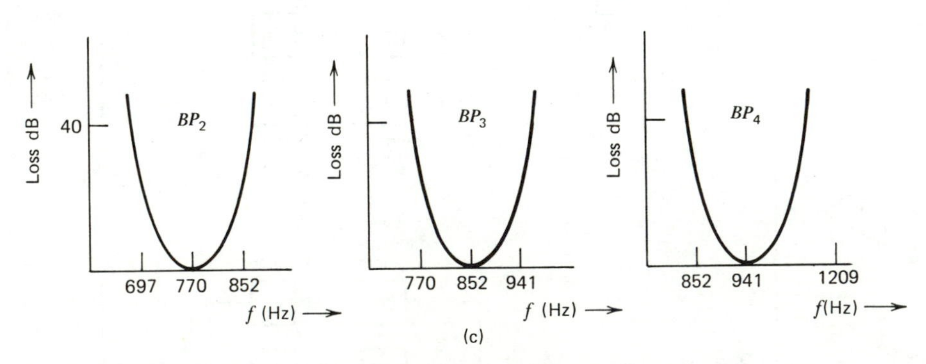

Figure 3.11 Filters for detection of tones: (*a*) High-pass filter; (*b*) Low-pass filter; (*c*) Some band-pass filters.

groups. The high-pass filter passes the high-group tones with very little attenuation while also attenuating the low-group tones, as illustrated by the *HP* filter characteristic of Figure 3.11*a*. The low-pass filter passes the low-group tones, while attenuating the high-group tones (Figure 3.11*b*). The separated tones are then converted to square waves of fixed amplitude using limiters. The next step in the detection scheme is to identify the individual tones in the respective groups. This is accomplished by the 8 band-pass filters shown in Figure 3.10. Each of these *BP* filters passes one tone, rejecting all the neighboring tones. Typical band-pass characteristics are indicated in Figure 3.11*c*. The band-pass filters are followed by detectors that are energized when their input voltage exceeds a certain threshold voltage, and the output of each detector provides the required dc switching signal.

3.1.4 BAND-REJECT FILTERS

Band-reject (*BR*) filters are used to reject a band of frequencies from a signal. A typical *BR* filter requirement is shown in Figure 3.12. The frequency band to be rejected is the stopband from ω_3 to ω_4. The passbands extend below ω_1, and above ω_2.

A second-order function with a band-reject characteristic is

$$\frac{V_O}{V_{IN}} = \frac{s^2 + d}{s^2 + as + d} = \frac{s^2 + \omega_z^2}{s^2 + \dfrac{\omega_p}{Q_p} s + \omega_p^2} \tag{3.5}$$

where $\omega_z = \omega_p$. This function has complex poles in the left half s plane and complex zeros on the $j\omega$ axis. Also the pole frequency is equal to the zero

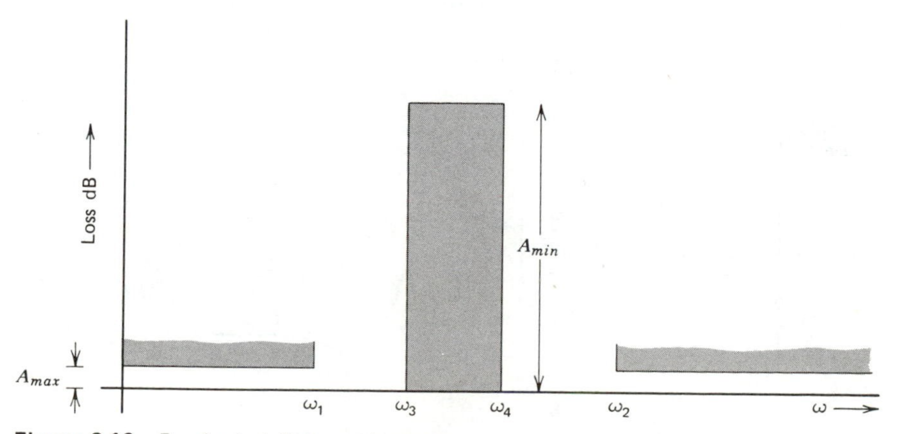

Figure 3.12 Band-reject filter requirements.

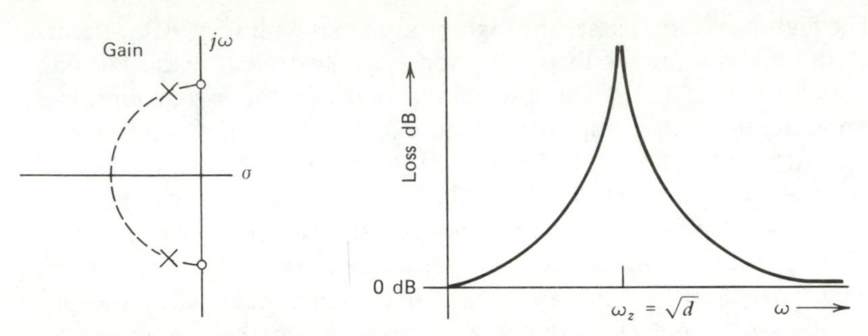

Figure 3.13 A second-order *BR* filter: (*a*) Pole-zero plot; (*b*) Loss.

frequency (Figure 3.13*a*). The loss at low frequencies and that at high frequencies approach unity, while the loss at the zero frequency $s = j\omega_z$ is infinite, as illustrated in Figure 3.13*b*.

Note that if $\omega_z \gg \omega_p$, Equation 3.5 represents a low-pass function with a null in the stopband, as shown in Figure 3.14*a*. In this case the loss at high frequencies is seen to be greater than that at low frequencies. Such a filter characteristic,

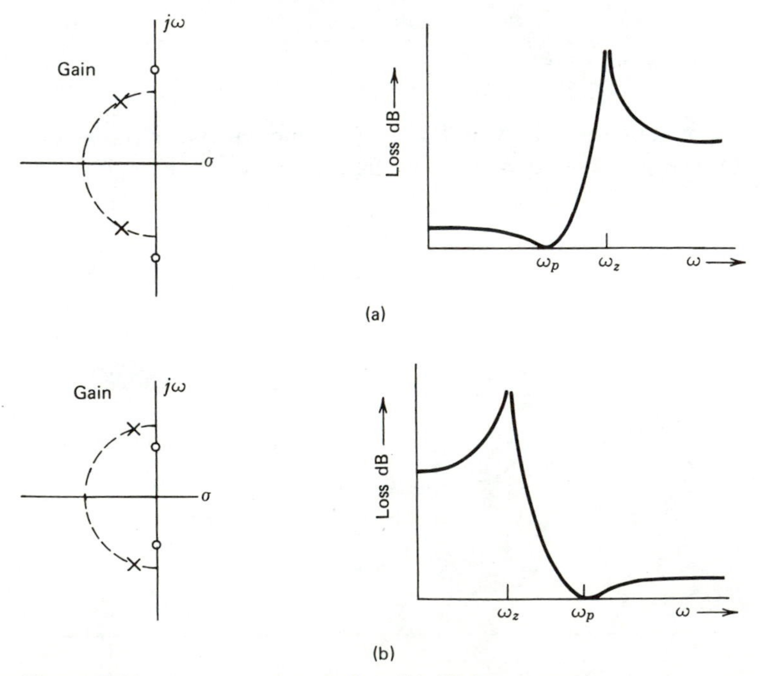

Figure 3.14 (*a*) Low-pass notch; (*b*) High-pass notch.

referred to as a **low-pass-notch,** is often used in the design of filters. Again if $\omega_z \ll \omega_p$, the function obtained has the **high-pass-notch** filter characteristic shown in Figure 3.14*b*.

The most common application of band-reject filters is in the removal of undesired tones from the frequency spectrum of a signal. One such application is in the system used for the billing of long-distance telephone calls. In a normal long-distance call, a single frequency signal tone is transmitted from the caller to the telephone office until the *end* of the dialing of the number. As soon as the called party answers, the signal tone ceases and the billing begins. The billing continues as long as the signal tone is absent. An exception to this system needs

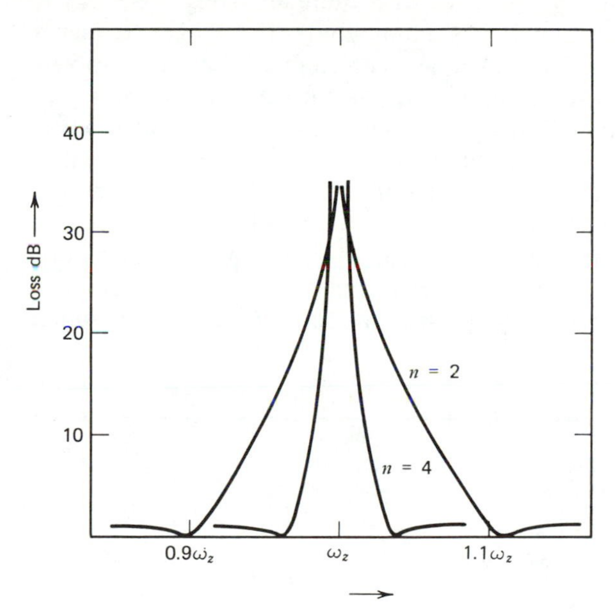

Figure 3.15 Typical band-reject filters used to reject a single tone in the voiceband.

to be made for long distance calls that are toll-free, such as calls to the operator for information. To prevent these calls from being billed, the signal tone is transmitted to the telephone office through the entire period of the call. However, since the signal tone is usually within the voice frequency band, it must be removed from the voice signal before being transmitted from the telephone office to the listener. A second-order *BR* filter that might be used to remove the signal tone is shown in Figure 3.15. Higher-order *BR* filters could be used to reduce the attenuation on voice frequencies in the neighborhood of the signal tone, as is illustrated in the figure.

3.1.5 GAIN EQUALIZERS

Gain equalizers are used to shape the gain versus frequency spectrum of a given signal. The shaping can take the form of a bump or a dip, that is, an emphasis or de-emphasis of a band of frequencies. Gain equalizers differ from the filter types discussed thus far, in that the shapes they provide are not characterized by a passband and a stopband. In fact, any gain versus frequency shape that does not fall into the four standard categories (*LP*, *HP*, *BP*, *BR*) will be considered a gain equalizer.

A familiar application of gain equalizers occurs in the recording and reproduction of music on phonograph records. High frequency background-hiss noise associated with the recording of sound is quite annoying. One way of alleviating this problem is to increase the amplitude of the high frequency signal, as shown in Figure 3.16. This is known as preemphasis. Another problem associated with phonograph recording is that, for normal levels of sound, the low frequencies require impractically wide excursions in the record grooves. These excursions can be reduced by attenuating the low frequency band as shown in the recording equalizer curve in Figure 3.16. In the playback system, which consists of a turntable and an amplifier, the high frequencies must be de-emphasized and the low frequencies boosted, as shown in the reproduction equalizer characteristic of Figure 3.16. After this equalization the reproduced sound will have the same frequency spectrum as that of the original source

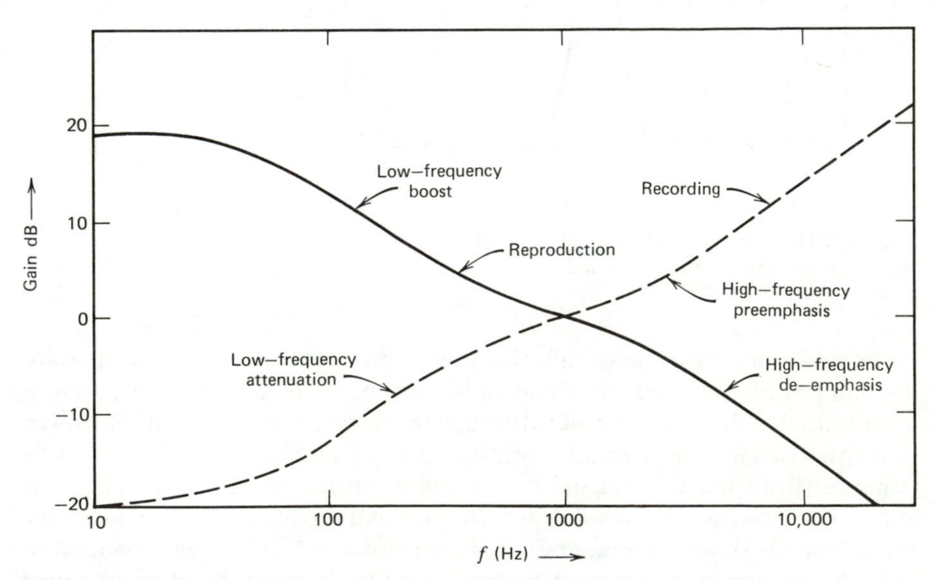

Figure 3.16 Equalization curves for phonograph recording and playback.

generated in the recording studio. To allow for different recording schemes, some high quality phonograph amplifiers are equipped with variable shape equalizers, which are most conveniently designed using active RC networks.

3.1.6 DELAY EQUALIZERS

Thus far we have discussed the gain (loss) characteristics of filters, but have not paid any attention to their phase characteristics. In many applications this omission is justifiable because the human ear is insensitive to phase changes. Therefore, in the transmission of *voice*, we need not be concerned with the phase characteristics of the filter function. However, in *digital* transmission systems, where the information is transmitted as square wave time domain pulses, the phase distortions introduced by the filter cause a variable delay and this cannot be ignored. Delay equalizers are used to compensate for the delay distortions introduced by filters and other parts of the transmission system.

An ideal delay characteristic is flat for all frequencies, as depicted in Figure 3.17. A digital pulse subjected to this flat delay characteristic will be translated

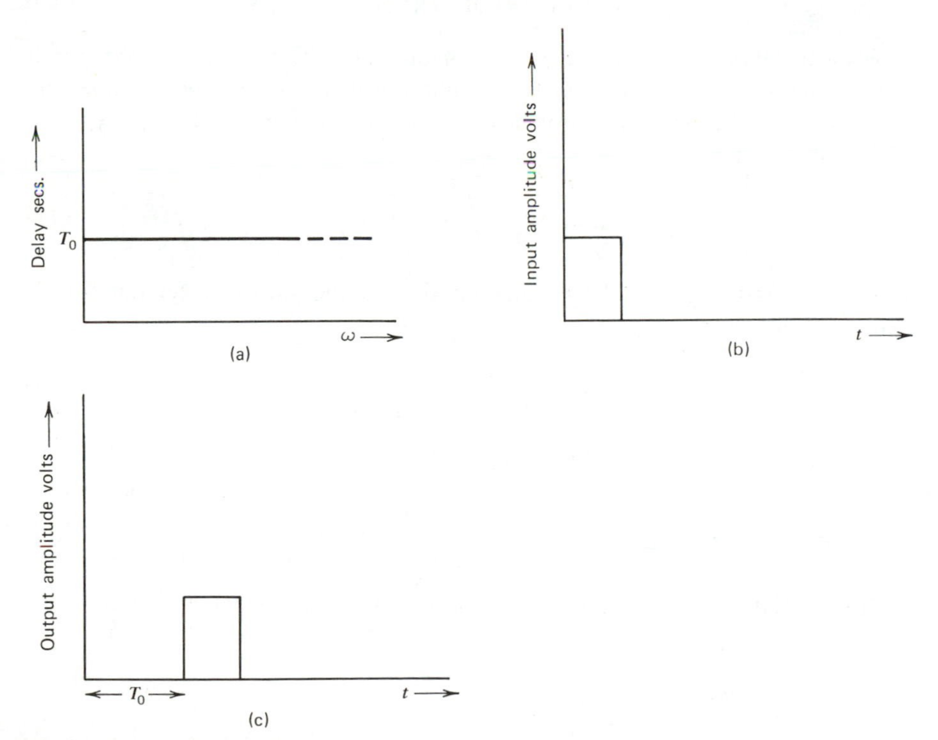

Figure 3.17 (*a*) The ideal delay characteristic; (*b*) Input pulse; (*c*) Output pulse.

on the time axis by T_0 seconds, but will otherwise be undistorted. Mathematically, the ideal delay characteristic is described by

$$V_O(t) = V_{IN}(t - T_0)$$

Taking Laplace transforms

$$V_O(s) = V_{IN}(s)e^{-sT_0}$$

The gain function is thus

$$H(s) = \frac{V_O(s)}{V_{IN}(s)} = e^{-sT_0} \tag{3.6}$$

Letting $s = j\omega$

$$H(j\omega) = e^{-j\omega T_0} \tag{3.7}$$

Thus, the amplitude and phase of this function are

$$\text{amplitude}\,(H(j\omega)) = |H(j\omega)| = 1 \tag{3.8}$$

$$\text{phase}\,(H(j\omega)) = \arg H(j\omega) = -\omega T_0 \tag{3.9}$$

This ideal delay characteristic has a constant amplitude, and the phase is a linear function of frequency. Observe that the delay T_0 can be obtained by differentiating the phase function with respect to ω. This, in fact, serves as a definition of delay:

$$\text{delay} = \frac{d}{d\omega}(-\phi(\omega)) \tag{3.10}$$

where $\phi(\omega)$ is the phase of the gain function. If the gain transfer function is expressed in factored form as

$$T(s) = \prod_{i=1}^{N} \frac{m_i s^2 + c_i s + d_i}{n_i s^2 + a_i s + b_i} \tag{3.11}$$

the phase, from Equation 2.22, is given by

$$\phi(\omega) = \sum_{i=1}^{N} \left[\tan^{-1} \frac{c_i \omega}{d_i - m_i \omega^2} - \tan^{-1} \frac{a_i \omega}{b_i - n_i \omega^2} \right] \tag{3.12}$$

By differentiation, the following expression is obtained for delay:

$$\begin{aligned} D &= \frac{d}{d\omega}(-\phi(\omega)) \\ &= \sum_{i=1}^{N} \left[-\frac{c_i(d_i + m_i\omega^2)}{(d_i - m_i\omega^2)^2 + c_i^2\omega^2} + \frac{a_i(b_i + n_i\omega^2)}{(b_i - n_i\omega^2)^2 + a_i^2\omega^2} \right] \end{aligned} \tag{3.13}$$

In general the delay of filters will not be flat, and will therefore need to be corrected. This correction is achieved by following the filter by a delay equalizer. The purpose of the delay equalizer is to introduce the necessary delay shape to make the total delay (of the filter and equalizer) as flat as possible. In addition, the delay equalizer must not perturb the loss characteristic of the filter; in other words, the loss characteristic must be flat over the frequency band of interest.

A second-order delay equalizer can be realized by the function

$$\frac{V_O}{V_{IN}} = \frac{s^2 - as + b}{s^2 + as + b} \tag{3.14}$$

The complex poles and zeros of this function are symmetrical about the $j\omega$ axis as shown in Figure 3.18. The gain of this function is

$$\begin{aligned} 20 \log_{10} \left| \frac{s^2 - as + b}{s^2 + as + b} \right|_{s = j\omega} &= 10 \log_{10}[(b - \omega^2)^2 + (-a\omega)^2] \\ &\quad - 10 \log_{10}[(b - \omega^2)^2 + (a\omega)^2] \\ &= 0 \text{ dB}. \end{aligned}$$

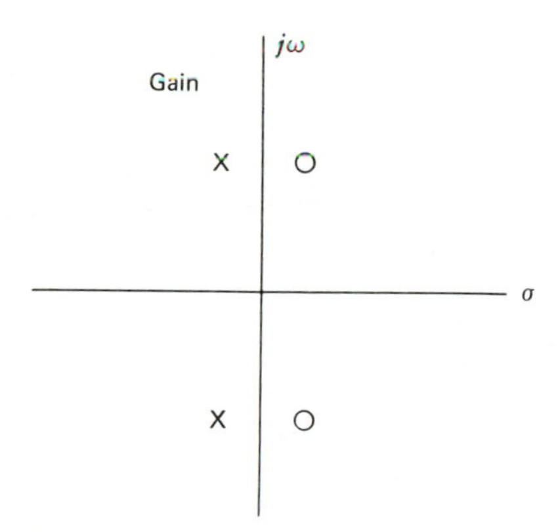

Figure 3.18 Pole-zero plot of a second-order delay equalizer.

The function is seen to have a flat gain of unity at all frequencies. For this reason it is often referred to as an **all-pass** function. The delay of the function will depend on the coefficients a and b and can be evaluated using Equation 3.13 or from the MAG computer program (Section 2.7).

An example of the application of delay equalizers is in the transmission of data on cables. The delay characteristic for a typical cable is shown in Figure 3.19. A second- (or higher-)order delay equalizer compensates for this distortion by introducing the complement of this delay shape as indicated in the figure.

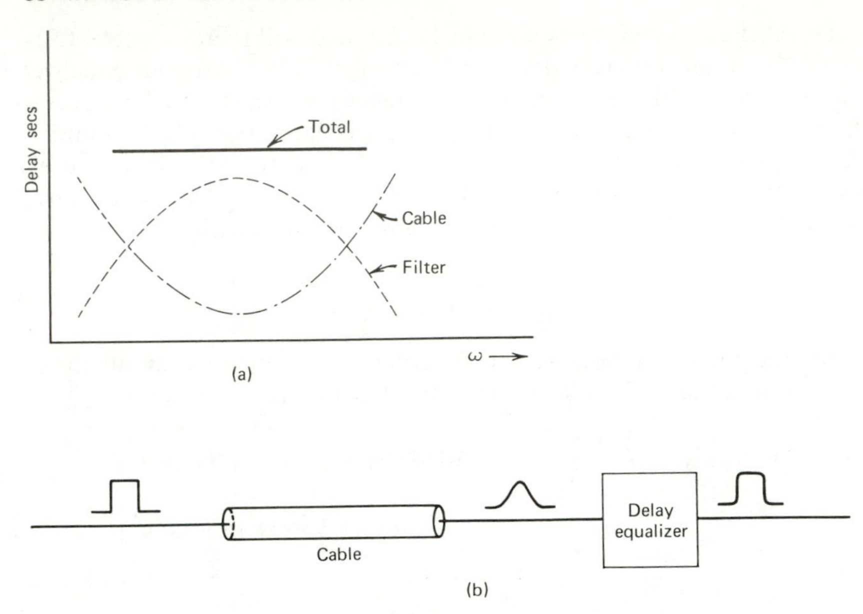

Figure 3.19 Equalization of cable delay.

The sum of the delays of the cable and equalizer will then be flat, if the parameters a and b are chosen properly. The choice of these parameters and the equalization of more complex delay shapes will be covered in the next chapter.

3.2 PASSIVE, ACTIVE, AND OTHER FILTERS

This section describes the different ways of building filters using passive RLC, active RC, electromechanical, digital, and microwave components. Of these, passive and active filters are the most used in voice and data communications, and their relative advantages will be discussed in some detail.

Passive filters use resistors, capacitors, and inductors. For audio frequency applications the use of the inductor presents certain problems. This is so because the impedance of a practical inductor deviates from its ideal value due to inherent resistance associated with its realization. Referring to the model of the inductor shown in Figure 3.20, the *quality factor* Q_L of the inductor is given by

$$Q_L = \frac{\omega L}{R} \tag{3.15}$$

R L

Figure 3.20 Model of a practical inductor.

The larger the resistance R, the lower the quality factor and the further from ideal the inductor is. To minimize distortion in the filter characteristic, it is desirable to use inductors with high quality factors. However, at frequencies below approximately 1 kHz, high quality inductors tend to become bulky and expensive. Attempts at miniaturizing the inductor have not met with much success.

Active filters overcome these drawbacks and, in addition, offer several other advantages. Active filters are realized using resistors, capacitors, and active devices, which are usually operational amplifiers. These devices can all be integrated, thereby allowing active RC filters to provide the following advantages:

- A reduction in size and weight.
- Increased circuit reliability, because all the processing steps can be automated.
- In large quantities the cost of an integrated circuit can be much lower than its equivalent passive counterpart.
- Improvement in performance because high quality components can be realized readily.
- A reduction in the parasitics, because of the smaller size.

Other advantages of active RC realizations that are independent of the physical implementation are:

- The design process is simpler than that for passive filters.
- Active filters can realize a wider class of functions.
- Active realizations can provide voltage gains; in contrast, passive filters often exhibit a significant voltage loss.

Among the drawbacks of active RC realizations is the finite bandwidth of the active devices, which places a limit on the highest attainable pole frequency. This maximum pole frequency limit decreases with the pole Q, which defines the sharpness of the filter characteristic. Considering both pole frequency and pole Q, a more accurate measure of the limitation on the op amp is the product of the pole Q and the pole frequency ($Q_p f_p$). With present-day technology and for most applications, reasonably good filter performance can be achieved for $Q_p f_p$ products up to approximately 500 kHz. Thus, for frequencies below 5 kHz, pole Q's up to 100 can be achieved. This is quite adequate for most voice and data applications. However, the high frequency limitation has barred the use of active filters much above 30 kHz. It should be emphasized that these bounds reflect the state of the present day technology and it is reasonable to expect that with advances in integrated circuit technology higher $Q_p f_p$ products will be realizable.

In contrast, passive filters do not have such an upper frequency limitation, and they can be used up to approximately 500 MHz. In this case the limitations at high frequencies are due to the parasitics associated with the passive elements. Another important criterion for comparing realizations is **sensitivity**, which is a measure of the deviation of the filter response due to variations in the elements, caused by environmental changes. In later chapters it will be shown that the sensitivity of passive realizations is much less than for active realizations. One other disadvantage of active filters is that they require power supplies while passive filters do not.

In conclusion it can be said that in voice and data communication systems, which represent a large percentage of all filter applications, the economic and performance advantages of active RC realizations far outweigh the above-mentioned disadvantages—and the modern engineering trend is to use active filters in most of these applications.

In the remainder of this section we will briefly describe the principles and applications of other ways of building filters using electromechanical, digital, and microwave components.

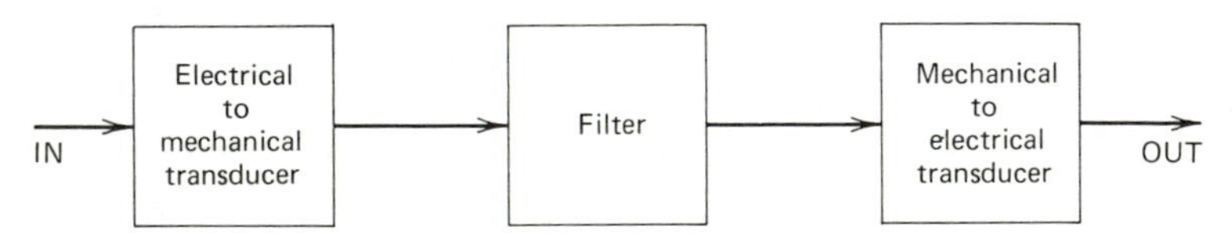

Figure 3.21 Block diagram of an electromechanical filter.

Electromechanical filters [2, 4] use mechanical resonances to accomplish the filtering of electric signals. A block diagram representation of such filters is shown in Figure 3.21. The electrical signal is converted to a mechanical vibration by a transducer and, after the filtering, the resultant mechanical vibration is converted back to an electrical signal. Electromechanical filters are classified according to whether or not separate devices are needed for the transducers and the filter. In one kind of electromechanical filter, known simply as a *mechanical* filter, the filtering is achieved by rod, disk, or plate resonators; and the transducer is usually a separate piezoelectric crystal. Because of the low damping readily achieved in mechanical vibrations, these filters can provide pole Q's up to 1500 and pole frequencies up to 500 kHz. In *monolithic-crystal* and *ceramic* type electromechanical filters the transducer action comes from the piezoelectric properties of the crystal or ceramic, which also serves as the resonant body. Therefore, in these two classes of electromechanical filters, the filtering and transducer action are achieved on the same device. These filters can be used up to extremely high pole frequencies and pole Q's. In particular, ceramic filters can achieve pole frequencies in the range of 0.1 MHz to 10 MHz and pole Q's from 30 to 1500, while monolithic-crystal filters go even higher, providing pole frequencies in the range of 5 MHz to 150 MHz and pole Q's from 1000 to 25,000. Electromechanical filters represent the only practical solution for applications requiring these high pole Q's. Typical applications are found in telephone carrier systems and in radio and TV transmission.

Let us next consider **digital** filters. A functional block diagram of a digital filter [1,7] is shown in Figure 3.22. The input analog signal is sampled at uniformly spaced intervals and the sampled values converted to binary words using an analog to digital converter. The binary number representation of the input signal is then filtered by the digital filter. The filtering operation involves numerical calculations and is accomplished using the types of circuit elements

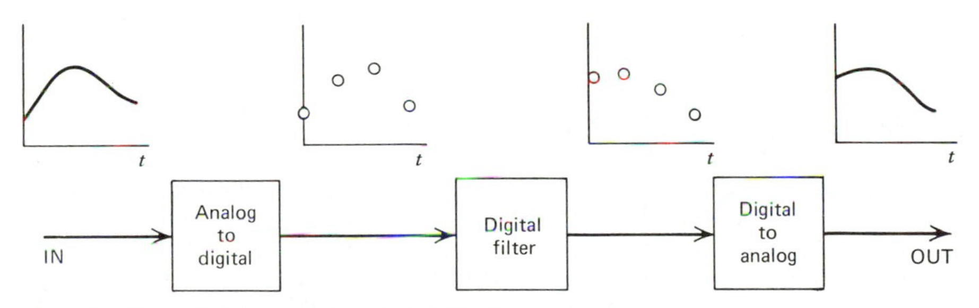

Figure 3.22 Block diagram of a digital filtering scheme.

used in a digital computer, namely, adders, multipliers, shift registers, and memory devices. The output word is finally converted back to an analog signal. Such a filtering scheme is particularly useful in applications where several channels of information need to be processed by the same filtering function. The sampled number representations of a number of channels can be interleaved to form *one* continuous string of numbers. This string can then be processed by one common filter. At the output the individual channels are then separated before being converted to their analog forms. The interleaving operation is called multiplexing while the separation at the output is called demultiplexing. By using this scheme the cost of the common digital hardware is shared by all the channels, so that the per-channel filtering cost could very well be economically competitive. Digital filters, therefore, are a reasonable alternative in applications where many voice or data signals need to be processed using the same filtering function. A unique characteristic of digital filters that makes them particularly suited to digital transmission applications is that these filters can be designed so that they introduce no delay distortion.

The last category we mention are **microwave** filters [2, 7], which are used for transmission at frequencies from approximately 200 MHz to 100 GHz. Physically, these filters could consist of distributed elements such as transmission lines or waveguides. Some areas of application for microwave filters are in radar, space satellite communication (e.g., the TELSTAR satellite), and telephone carrier systems.

3.3 CONCLUDING REMARKS

In this chapter we describe the gain versus frequency characteristics of various second-order filter functions. The subject of higher-order filter functions is discussed in the following chapters. In the next chapter we consider ways of finding rational functions to meet prescribed filter requirements—this is known as the approximation problem. The circuit realization, or synthesis, of the approximation functions using passive *RLC* networks is covered in Chapter 6. The active *RC* synthesis techniques are elaborated in Chapters 7 to 12.

FURTHER READING

1. L. P. Huelsman, *Active Filters: Lumped, Distributed, Integrated, Digital and Parametric*, McGraw-Hill, New York, 1970, Chapter 1.
2. Y. J. Lubkin, *Filter Systems and Design: Electrical, Microwave, and Digital*, Addison-Wesley, Reading, Mass., 1970, Chapters 10 and 11.
3. S. K. Mitra, *Analysis and Synthesis of Linear Active Networks*, Wiley, New York, 1969, Chapter 1.
4. G. S. Moschytz, "Inductorless filters: a survey; Part I. Electromechanical filters," *IEEE Spectrum*, August 1970, pp. 30–36.
5. G. S. Moschytz, "Inductorless filters: a survey; Part II. Linear active and digital filters," *IEEE Spectrum*, September 1970, pp. 63–76.
6. A. V. Oppenheim and R. W. Schafer, *Digital Signal Processing*, Prentice-Hall, Englewood Cliffs, N.J., 1975.
7. G. C. Temes and S. K. Mitra, eds., *Modern Filter Theory and Design*, Wiley, New York, 1973: Chapter 4, "Crystal and Ceramic Filters," G. S. Szentirmai; Chapter 5, "Mechanical Bandpass Filters," R. A. Johnson; Chapter 7, "Microwave Filters," E. G. Cristal; Chapter 12, "Digital Filters," R. M. Golden.
   Further references will be found in this book.

PROBLEMS

3.1 *Filter pole-zero patterns.* Identify the filter types corresponding to the gain function pole-zero patterns shown in Figure P3.1.

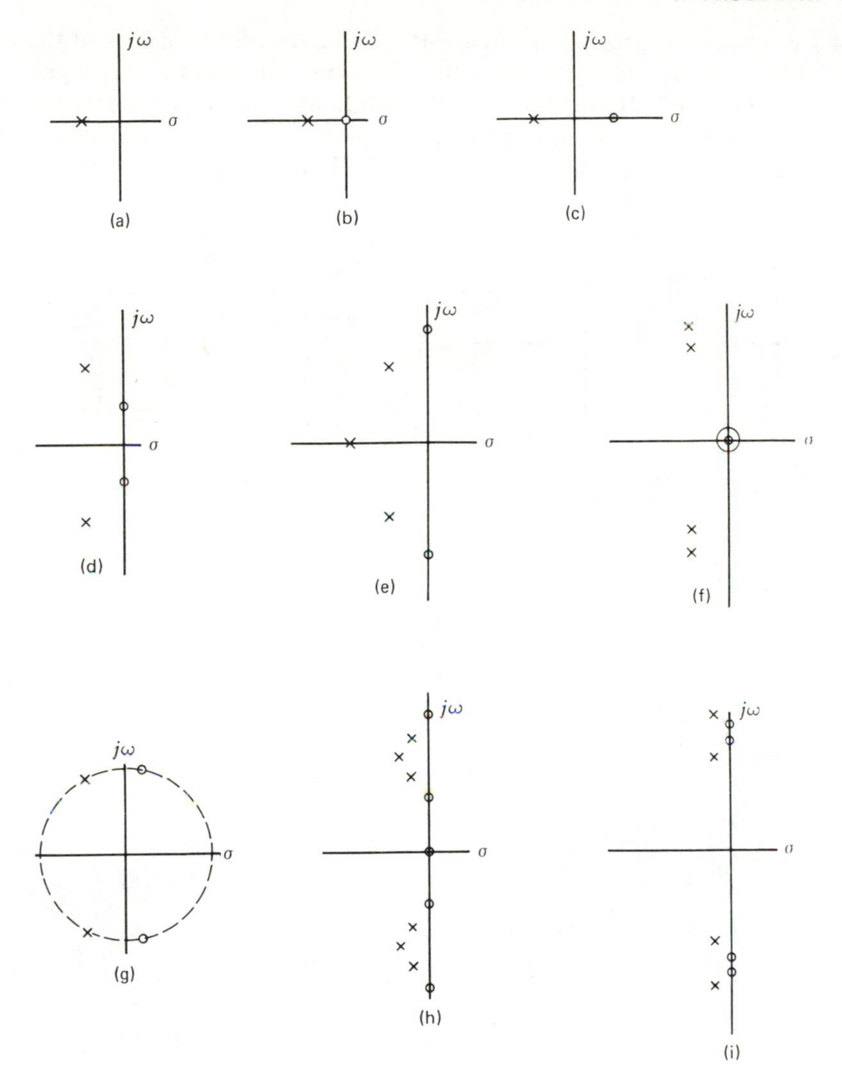

Figure P3.1

3.2 *Band-reject realization.* Show that a second-order band-reject filter can be realized as the sum of a second-order low-pass and a second-order high-pass function. Sketch a circuit implementation to realize a band-reject filter given circuits that can realize low-pass and high-pass functions.

3.3 Show that a second-order band-reject filter function can be expressed as the difference of unity and a second-order band-pass function. Use this idea to describe a circuit implementation of a second-order band-reject filter given a circuit that can realize band-pass functions.

3.4 *Band-pass realization.* In Figure P3.4*a*, the transfer functions of the two blocks are T_1 and T_2. (a) Show that the transfer function V_O/V_{IN} is given by the product $T_1 T_2$. (b) Show how the band-pass filter requirement of Figure P3.4*b* can be realized using this topology by choosing T_1 to realize a low-pass requirement and T_2 to realize a high-pass requirement.

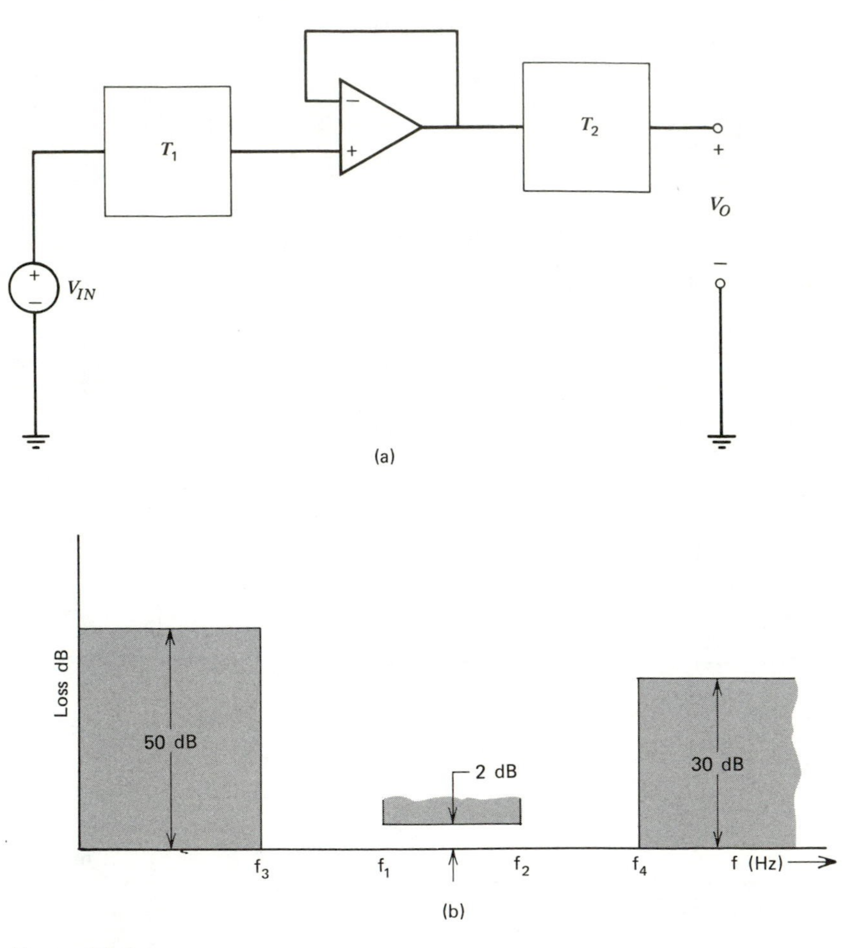

Figure P3.4

3.5 *Low-pass notch.* The *dc* gain of a second-order low-pass notch function is 0 dB, and the infinite frequency gain is 20 dB. Determine the ratio of the zero frequency to the pole frequency.

3.6 *Tone separation.* Three tones at frequencies f_1, f_2, and f_3 need to be isolated from each other. Show how this can be done by using two low-pass and two high-pass filters.

3.7 *Gain equalizer.* The second-order function,

$$\frac{s^2 + cs + b}{s^2 + as + b}$$

may be used as a gain equalizer to obtain a bump or a dip at the pole frequency. Such functions find application in cable transmission systems. Sketch the gain of this function for (a) $c = 2a$, (b) $c = a/2$.

3.8 *Delay of first-order function.* Find an expression for the delay of a real pole-zero pair represented by the gain function

$$T(s) = K\frac{s + z_1}{s + p_1}$$

3.9 *Delay evaluation.* Determine the delay at $\omega = 0$, $\omega = 1$, and $\omega = 5$ rad/sec for the following gain functions:

(a) $\dfrac{3}{s + 1}$

(b) $\dfrac{1}{s + 1} \cdot \dfrac{s + 2}{s + 4}$

(c) $\dfrac{s^2 - 3s + 9}{s^2 + 3s + 9}$

(d) $\dfrac{s^2 + s}{s^2 + s + 16}$

3.10 *MAG computer program.* Use the MAG computer program to compute the delay for the gain function

$$\frac{s}{s + 16} \cdot \frac{s^2 + 100}{s^2 + 10s + 2500}$$

at frequencies from 0.5 Hz to 10 Hz in steps of 0.5 Hz.

4. THE APPROXIMATION PROBLEM

As we mentioned in Chapter 3, the specifications for a filter are usually given in terms of loss requirements in the passband and the stopband. The **approximation problem** consists of finding a function whose loss characteristic lies within the permitted region. An additional constraint on the function chosen is that it be realizable using passive and/or active components. The required realizability conditions were covered in Chapter 2. It is also desirable to keep the order of the function as low as possible in order to minimize the number of components required in the design.

A method of approximation based on Bode plots will be described in Section 4.1. This method is suitable for low order, simple, filter designs. More complex filter characteristics are approximated by using some well-described rational functions whose roots have been tabulated. The most popular among these approximations are the Butterworth, Chebyshev, Bessel, and the elliptic (or Cauer) types. The delay characteristics of these functions and the design of delay equalizers are covered in Sections 4.5 and 4.6. These approximations are directly applicable to low-pass filters. However, they can also be used to design high-pass filters, and symmetrical band-pass and band-reject filters, by employing the frequency transformation functions described in Section 4.7. Finally, in Section 4.8, a computer program is presented for obtaining the Chebyshev approximation function for low-pass, high-pass, band-pass, and band-reject filters.

4.1 BODE PLOT APPROXIMATION TECHNIQUE

The Bode plots of the following functions were discussed in Chapter 2:

(a) Constant term.
(b) Root at the origin, corresponding to the factor s.
(c) A simple root, corresponding to the factor $s + \alpha$.
(d) A pair of complex roots, corresponding to the factor $s^2 + as + b$.

For some simple filters it is possible to use these sketches to fit the requirements. The procedure consists of estimating the pole and zero locations to fit the given requirements, as is illustrated by the following example.

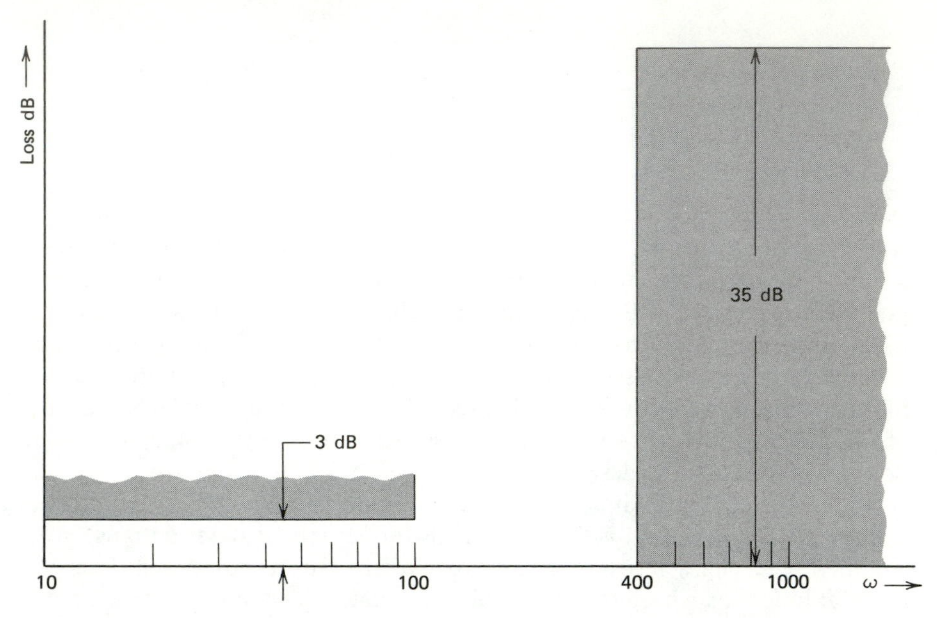

Figure 4.1 Requirements for Example 4.1.

Example 4.1
Approximate the low-pass requirement shown in Figure 4.1.

Solution
The first step is to estimate the order of the desired filter function. From the given requirements, the loss in the transition band from 100 to 400 rad/sec is seen to increase by 35 dB. This corresponds to a linear slope of 35 dB in two octaves, or 17.5 dB/octave. The high-frequency slope of the Bode asymptote of a second-order function is 12 dB/octave, while that of a third-order function is 18 dB/octave. It is therefore estimated that the given requirements will be satisfied by a third-order *LP* gain function of the form

$$T_{LP}(s) = \frac{K}{(s + a)\left(s^2 + \dfrac{\omega_p}{Q_p}s + \omega_p^2\right)} \tag{4.1}$$

where the constants K, a, ω_p, and Q_p are obtained from the filter requirements, as follows. A comparison of the requirements with the Bode plot of $1/(s + \alpha)$ (Figure 2.11, page 48) suggests that the real pole be located at the filter cutoff frequency, that is,

$$a = 100$$

Again, from the sketch of a second-order function shown in Figure 2.12, it is seen that the pole frequency should be made equal to the cutoff frequency, that is,

$$\omega_p = 100$$

The constant K is determined from the *dc* value of the function which, from the requirements, may be assumed to be 1 (i.e., 0 dB). From Equation 4.1, the dc value of $T_{LP}(s)$ is

$$T_{LP}(s)|_{s=0} = \frac{K}{a\omega_p^2} = 1$$

Thus

$$K = 10^6$$

The remaining unknown constant, Q_p, is obtained from the given loss at the cutoff frequency $\omega = 100$ rad/sec:

$$20 \log_{10} \left| \frac{K}{(s+a)\left(s^2 + \dfrac{\omega_p}{Q_p} s + \omega_p^2\right)} \right|_{s=j100} = -3$$

Substituting for K, a, and ω_p, and simplifying this expression

$$\frac{10^6}{|j100 + 100| \times \left| -(100)^2 + \dfrac{100}{Q_p}(j100) + (100)^2 \right|} = 0.7078$$

Solving for Q_p, we get

$$Q_p = 1$$

Thus, the desired *LP* function is

$$T_{LP}(s) = \frac{10^6}{(s + 100)(s^2 + 100s + 10^4)}$$

The MAG program was used to obtain the magnitude versus frequency plot of this function shown in Figure 4.2. It is seen that this function just meets the passband requirement, and that the stopband requirement is met with some margin to spare.

The Bode approximation method is useful for low-order, rough approximations. The more complex approximations associated with filter syntheses are usually obtained by using some well-known rational functions that are described in the next few sections. ■

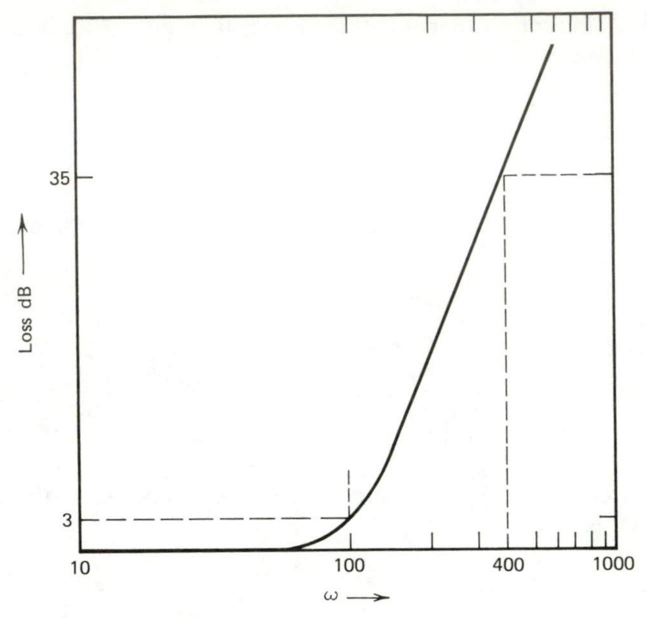

Figure 4.2 *LP* approximation function for Example 4.1.

4.2 BUTTERWORTH APPROXIMATION

In this section we consider the approximation of the low-pass requirements that were introduced in Chapter 3. Referring to Figure 4.3, the requirements are characterized by the passband from *dc* to ω_P, the stopband from ω_S to infinity, the maximum passband loss $A_{\max}$, and the minimum stopband loss $A_{\min}$.

The rational function *LP* approximations, which we describe in this and the next few sections, have the general form

$$|H(j\omega)|^2 = 1 + |K(j\omega)|^2 = 1 + \left|\frac{N(j\omega)}{D(j\omega)}\right|^2 \tag{4.2}$$

where $H(s)$ is the desired loss function and $K(s) = N(s)/D(s)$ is a rational function in s. The function $K(s)$ must be chosen so that its magnitude is small in the passband, to make the magnitude of $H(j\omega)$ close to unity. In the stopband the magnitude of $K(s)$ must be large in order to satisfy the stopband loss requirements. In particular, $K(s)$ may be chosen to be a polynomial of the form

$$K(s) = P_n(s) = a_0 + a_1 s + a_2 s^2 + \cdots a_n s^n \tag{4.3}$$

where the coefficients of the nth-order polynomial $P_n(s)$ are chosen so that the corresponding loss function $H(s)$ satisfies the given filter requirements. In this

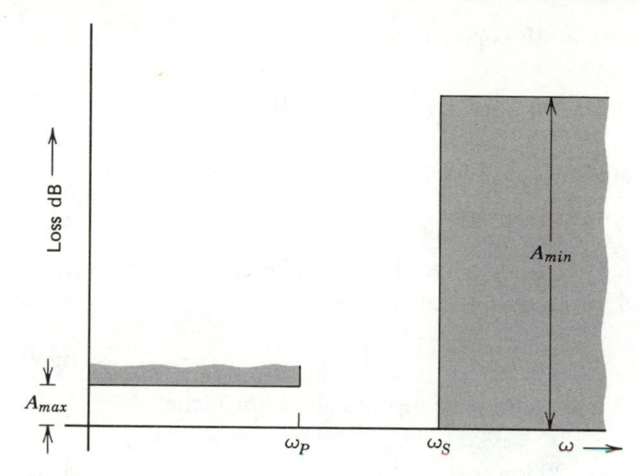

Figure 4.3 *LP* requirements.

section we will describe the well-known **Butterworth** approximation, which is characterized by

$$K(s) = P_n(s) = \varepsilon\left(\frac{s}{\omega_P}\right)^n \tag{4.4}$$

where ε is a constant, n is the order of the polynomial, and ω_P is the desired passband edge frequency. The corresponding loss function is

$$|H(j\omega)| = \left|\frac{V_{IN}(j\omega)}{V_O(j\omega)}\right| = \sqrt{1 + \varepsilon^2\left(\frac{\omega}{\omega_P}\right)^{2n}} \tag{4.5}$$

Let us study the characteristics of this approximation. At *dc*, from Equation 4.5, the loss is seen to be unity. The slope of the function at dc is obtained by expanding (4.5) as a binomial series. Near $\omega = 0$,

$$\varepsilon^2\left(\frac{\omega}{\omega_P}\right)^{2n} \ll 1$$

so

$$\left[1 + \varepsilon^2\left(\frac{\omega}{\omega_P}\right)^{2n}\right]^{1/2} = 1 + \frac{1}{2}\varepsilon^2\left(\frac{\omega}{\omega_P}\right)^{2n} - \frac{1}{8}\varepsilon^4\left(\frac{\omega}{\omega_P}\right)^{4n} + \frac{1}{16}\varepsilon^6\left(\frac{\omega}{\omega_P}\right)^{6n} + \cdots \tag{4.6}$$

This expansion shows that the first $2n - 1$ derivatives are zero at $\omega = 0$. Since $K(s)$ was chosen to be an nth order polynomial, this is the maximum number of derivatives that can be made zero. Thus the slope is *as flat as possible* at *dc*. For this reason the Butterworth approximation is also known as the **maximally flat** approximation.

From Equation 4.5 the loss in dB is given by

$$A(\omega) = 10 \log_{10}\left[1 + \varepsilon^2\left(\frac{\omega}{\omega_P}\right)^{2n}\right] \text{dB} \tag{4.7}$$

In particular, at the passband edge frequency $\omega = \omega_P$, the loss is

$$A(\omega_P) = 10 \log_{10}(1 + \varepsilon^2) \tag{4.8}$$

The filter requirements specify this loss to be $A_{\max}$ dB. Therefore, the parameter ε is related to the passband loss requirement $A_{\max}$ by

$$\varepsilon = \sqrt{10^{0.1A_{\max}} - 1} \tag{4.9}$$

At high frequencies ($\omega \gg \omega_P$) the loss asymptotically approaches

$$20 \log_{10} \varepsilon\left(\frac{\omega}{\omega_P}\right)^n \tag{4.10}$$

This loss is seen to increase with the order n. Equation 4.10 also shows that at high frequencies the loss slope is $6n$ dB/octave. Therefore, the stopband loss increases with the order n.

The fourth-order Butterworth loss function shown in Figure 4.4 illustrates the maximally flat characteristic of the passband and the monotonically increasing nature of the loss in the stopband. For design purposes it is convenient to plot a family of such characteristics for different n, against the normalized frequency

$$\Omega = \varepsilon^{1/n}\left(\frac{\omega}{\omega_P}\right) \tag{4.11}$$

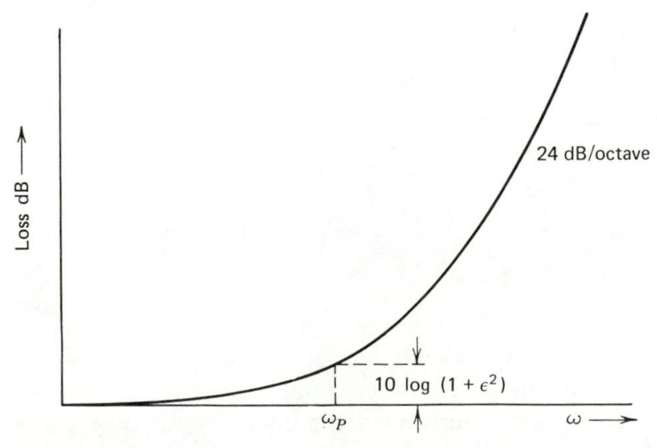

Figure 4.4 A fourth-order Butterworth *LP* approximation.

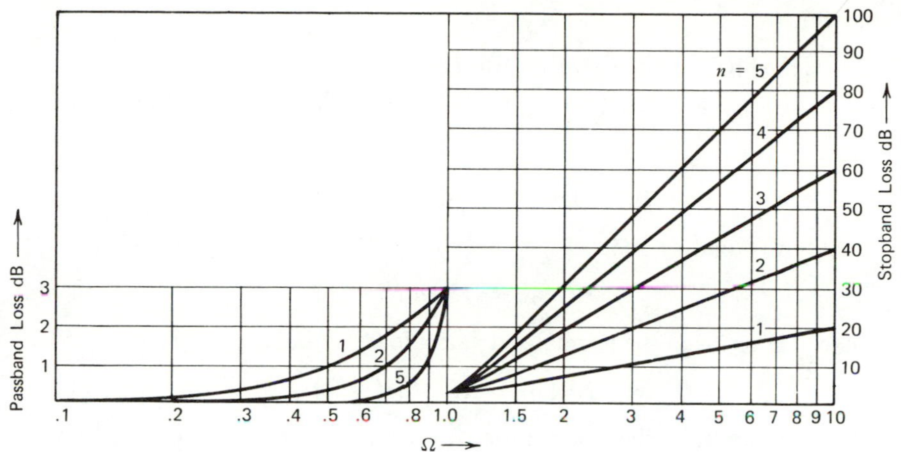

Figure 4.5 Loss of *LP* normalized Butterworth approximations.

In terms of this normalized frequency, the loss is given by

$$A(\Omega) = 10 \log_{10}(1 + \Omega^{2n}) \tag{4.12}$$

This function is plotted in Figure 4.5 for orders up to 5. The use of these plots is illustrated by the following example.

Example 4.2

Find the loss at $\omega_S = 40$ rad/sec for a fifth-order Butterworth filter that has a maximum loss of 1 dB at the passband edge frequency, $\omega_P = 10$ rad/sec.

Solution

From Equation 4.9

$$\varepsilon = \sqrt{10^{0.1} - 1} = 0.509$$

and from Equation 4.11

$$\Omega_S = (0.509)^{1/5}(4) = 3.49$$

The loss at $\Omega_S = 3.49$, from Figure 4.5, is approximately 55 dB. A more exact value for the loss can be obtained from Equation 4.12:

$$A(\omega) = 10 \log_{10}(1 + (3.49)^{10}) = 54.3 \text{ dB}$$ ■

Thus far we have been working with the magnitude of loss function, namely $|H(j\omega)|$. In the following we will show how the s domain loss function $H(s)$ is derived from the expression for $|H(j\omega)|$. First the function $H(j\omega)$ is expressed in terms of its real and imaginary parts, as

$$H(j\omega) = \text{Re } H(j\omega) + j \text{ Im } H(j\omega) \tag{4.13}$$

from which

$$|H(j\omega)|^2 = (\text{Re } H(j\omega))^2 + (\text{Im } H(j\omega))^2$$
$$= [\text{Re } H(j\omega) + j \text{ Im } H(j\omega)][\text{Re } H(j\omega) - j \text{ Im } H(j\omega)]$$

However,

$$H(-j\omega) = \text{Re } H(j\omega) - j \text{ Im } H(j\omega)$$

so the above equation reduces to

$$|H(j\omega)|^2 = H(j\omega)H(-j\omega) \tag{4.14a}$$

Similarly, in terms of the normalized variable Ω, we have

$$|H(j\Omega)|^2 = H(j\Omega)H(-j\Omega) \tag{4.14b}$$

This equation describes the transfer function for frequencies on the $j\Omega$ axis; it can be shown* that this leads to the more general relationship, true for all s:

$$|H(s)|^2 = H(s)H(-s) \tag{4.15}$$

where s is the normalized frequency variable $\Sigma + j\Omega$. Now the roots of $H(s)$ are the roots of $H(-s)$, reflected about the origin. Since the desired filter function must have all its poles in the left half s plane, we must associate the left half plane roots of $|H(s)|^2$ with $H(s)$, and the right half plane roots of $|H(s)|^2$ with $H(-s)$.

In particular, for the Butterworth approximation, from Equation 4.5

$$|H(j\Omega)|^2 = 1 + \Omega^{2n} = 1 + [-(j\Omega)^2]^n \tag{4.16}$$

Extending this to the s domain

$$|H(s)|^2 = 1 + (-s^2)^n \tag{4.17}$$

The roots of $|H(s)|^2$ are obtained by solving the equation

$$1 + (-s^2)^n = 0 \tag{4.18}$$

The solution of this equation is:

$$s_k = \exp\left[\frac{j\pi}{2}\left(\frac{2k + n - 1}{n}\right)\right] \quad \text{where } k = 1, 2, \ldots, 2n \tag{4.19}$$

These $2n$ roots are located on the unit circle and are equally spaced at π/n radian intervals. The s domain loss function is therefore given by

$$H(s) = \prod_j (s - s_j) \tag{4.20}$$

where s_j are the left half plane roots of (4.18).

* The basis of this assertion is known as *analytic continuation* in complex variable theory [1].

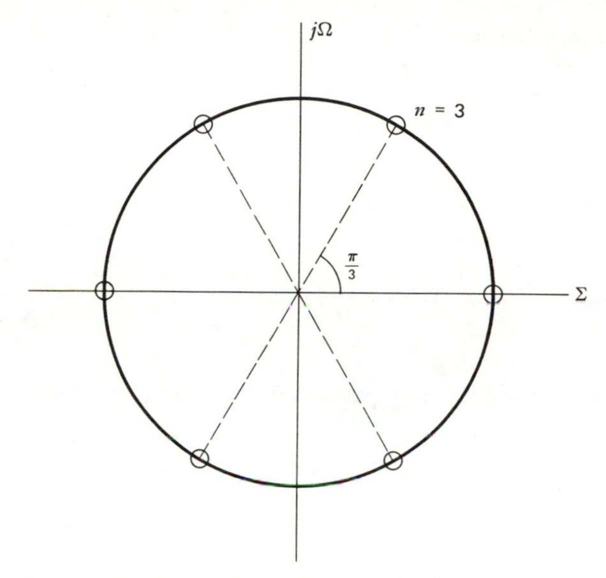

Figure 4.6 Roots of a third-order Butterworth.

Example 4.3
Find the approximation function for the third-order normalized Butterworth *LP* filter.

Solution
The six roots of $|H(s)|^2$ are located on the unit circle at $\pi/3$ radian intervals, as shown in Figure 4.6. The left half plane roots, to be associated with $H(s)$, are at

$$s = -1 \qquad \text{and} \qquad s = -0.5 \pm j0.866$$

The approximation function is therefore

$$\begin{aligned} H(s) &= (s + 1)(s + 0.5 + 0.866j)(s + 0.5 - 0.866j) \\ &= (s + 1)(s^2 + s + 1) \end{aligned}$$

■

The factored form of the normalized Butterworth polynomials for $n = 1$ to 5 are given in Table 4.1. To determine the Butterworth function for a filter whose cutoff frequency is ω_P and maximum passband loss is $A_{\max}$ dB, the polynomials given in Table 4.1 need to be denormalized by replacing

$$s \text{ by } s\left(\frac{\varepsilon^{1/n}}{\omega_P}\right) \tag{4.21}$$

Table 4.1 Butterworth Approximation Functions

n	$H(s)$
1	$s + 1$
2	$s^2 + 1.414s + 1$
3	$(s^2 + s + 1)(s + 1)$
4	$(s^2 + 0.76537s + 1)(s^2 + 1.84776s + 1)$
5	$(s^2 + 0.61803s + 1)(s^2 + 1.61803s + 1)(s + 1)$

Example 4.4

Find the Butterworth approximation for a low-pass filter whose requirements are characterized by

$$A_{max} = 0.5 \text{ dB} \qquad A_{min} = 12 \qquad \omega_P = 100 \qquad \omega_S = 400$$

Solution

From Equation 4.9

$$\varepsilon = \sqrt{(10)^{(0.1)(0.5)} - 1} = 0.35$$

The parameter n is determined by the loss at ω_S

$$A_{min} = 10 \log_{10}\left[1 + \varepsilon^2\left(\frac{\omega_S}{\omega_P}\right)^{2n}\right] \tag{4.22}$$

from which

$$10^{0.1A_{min}} = 1 + \varepsilon^2\left(\frac{\omega_S}{\omega_P}\right)^{2n}$$

which yields

$$n = \frac{\log_{10}\left(\dfrac{10^{0.1A_{min}} - 1}{\varepsilon^2}\right)}{\log_{10}\left(\dfrac{\omega_S}{\omega_P}\right)^2} \tag{4.23}$$

Substituting for the given values of ω_S, ω_P, A_{min}, and ε:

$$n = 1.73$$

Thus, a second-order function is required. From Table 4.1, the normalized function is

$$s^2 + 1.414s + 1$$

The desired function is obtained from this by replacing s by

$$s\left(\frac{\varepsilon^{1/n}}{\omega_P}\right) = 0.0059s$$

The resulting denormalized *LP* approximation function is

$$H(s) = \frac{s^2 + 239.6s + 28727.4}{28727.4}$$

The attenuation achieved by this function at $\omega_S = 400$, as calculated from Equation 4.22, is 15.1 dB. This exceeds the requirement by 3.1 dB. ■

4.3 CHEBYSHEV APPROXIMATION

The main feature of the Butterworth approximation is that the loss is maximally flat at the origin. Thus the approximation to a flat passband is very good at the origin but it gets progressively poorer as ω approaches ω_P. Moreover, the attenuation provided in the stopband is less than that attainable using some other polynomial types, such as the Chebyshev, which is described in this section. The increased stopband attenuation is achieved by changing the approximation conditions in the passband. The criterion used is to minimize the maximum deviation from the ideal flat characteristic. Pictorially, we attempt to get the equiripple characteristic shown in Figure 4.7. The **Chebyshev** polynomials, whose properties are developed in this section, are ideal for this purpose.

The nth-order Chebyshev function $C_n(\Omega)$ is defined as [3]

$$C_n(\Omega) = \cos(n \cos^{-1} \Omega) \qquad |\Omega| \leq 1 \tag{4.24a}$$

$$ = \cosh(n \cosh^{-1} \Omega) \qquad |\Omega| > 1 \tag{4.24b}$$

where, Ω is the normalized frequency,

$$\Omega = \frac{\omega}{\omega_P}$$

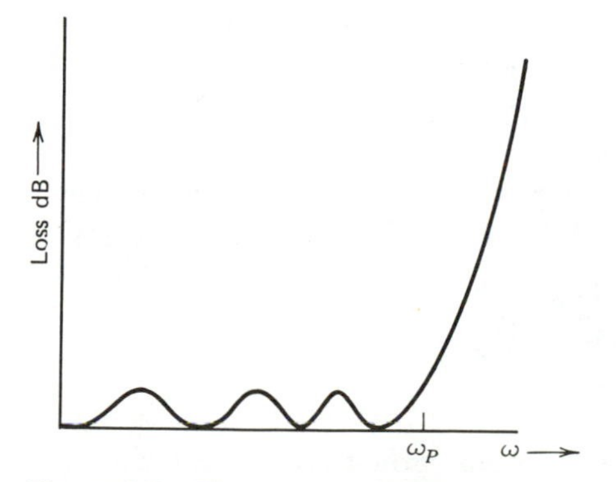

Figure 4.7 The equiripple passband characteristic.

The Chebyshev function can also be expressed as a polynomial in Ω, as shown in the following. From Equation 4.24a

$$C_{n+1}(\Omega) + C_{n-1}(\Omega) = \cos[(n+1)\cos^{-1}\Omega] + \cos[(n-1)\cos^{-1}\Omega]$$

Using the identity $\cos(A+B) + \cos(A-B) = 2\cos A \cos B$, the right-hand side reduces to:

$$2\cos(\cos^{-1}\Omega)\cos(n\cos^{-1}\Omega) = 2\Omega C_n(\Omega) \tag{4.25}$$

which yields the recursive relationship:

$$C_{n+1}(\Omega) = 2\Omega C_n(\Omega) - C_{n-1}(\Omega) \tag{4.26}$$

From Equation 4.24a we have

$$C_0(\Omega) = 1$$
$$C_1(\Omega) = \Omega$$

The higher-order polynomials are obtained from the recursive relationship of Equation 4.26:

$$\begin{aligned} C_2(\Omega) &= 2\Omega^2 - 1 \\ C_3(\Omega) &= 4\Omega^3 - 3\Omega \\ C_4(\Omega) &= 8\Omega^4 - 8\Omega^2 + 1 \\ C_5(\Omega) &= 16\Omega^5 - 20\Omega^3 + 5\Omega \text{ etc.} \end{aligned} \tag{4.27}$$

A plot of the Chebyshev functions using the above polynomial form shows that they do indeed have an equiripple characteristic in the band $-1 \leq \Omega \leq 1$ (Figure 4.8*a* and *b*).

The Chebyshev low-pass approximation function is obtained from the Chebyshev polynomials and is given by [3]:

$$|H(j\Omega)| = \frac{V_{IN}(j\omega)}{V_O(j\omega)} = \sqrt{1 + \varepsilon^2 C_n^2(\Omega)} \tag{4.28}$$

The loss functions for $n = 3$ and $n = 4$ are sketched in Figure 4.8*c* and *d*. Observe that the approximation functions ripple between a minimum of one and a maximum of $\sqrt{1+\varepsilon^2}$, for $|\Omega| \leq 1$; and that the number of minima of $|H(j\Omega)|$ in the band $-1 \leq \Omega \leq 1$ is equal to the order n. It can readily be shown that these properties apply to Chebyshev approximations of all orders [3].

Let us consider the loss of $H(j\Omega)$ at the passband edge frequency ω_P. Here, the normalized frequency Ω is unity, and $C_n(1) = 1$. Thus the passband loss is

$$A_{\max} = 10\log_{10}(1 + \varepsilon^2) \tag{4.29}$$

The passband ripple $A_{\max}$ therefore defines the parameter ε

$$\varepsilon = \sqrt{10^{0.1A_{\max}} - 1} \tag{4.30}$$

Given the ripple requirement $A_{\max}$ and the order n, the normalized Chebyshev loss can be plotted using Equation 4.24, 4.28, and 4.30. These functions are

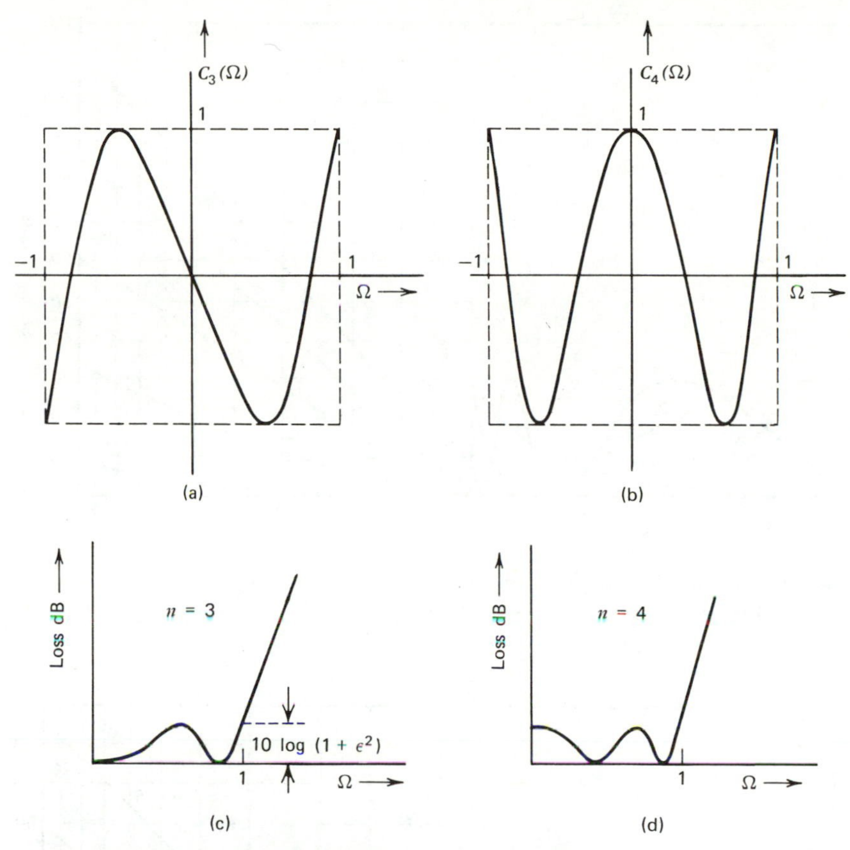

Figure 4.8 Plots of (*a*) Third-order Chebyshev function. (*b*) Fourth-order Chebyshev function. (*c*) Third-order Chebyshev *LP* approximation. (*d*) Fourth-order Chebyshev *LP* approximation.

plotted for $A_{max} = 0.25$ dB, 0.5 dB, and 1 dB in Figure 4.9, 4.10, and 4.11, respectively. The use of these plots is illustrated by the following example.

Example 4.5

Find the order needed for a Chebyshev low-pass filter whose requirements are characterized by

$$f_P = 2000 \text{ Hz} \qquad f_S = 5000 \text{ Hz} \qquad A_{max} = 1 \text{ dB} \qquad A_{min} = 35 \text{ dB}$$

Solution

The normalized frequency for the stopband edge frequency is

$$\Omega_S = f_S/f_P = 2.5$$

From Figure 4.11, we see that the order needed is 4. ■

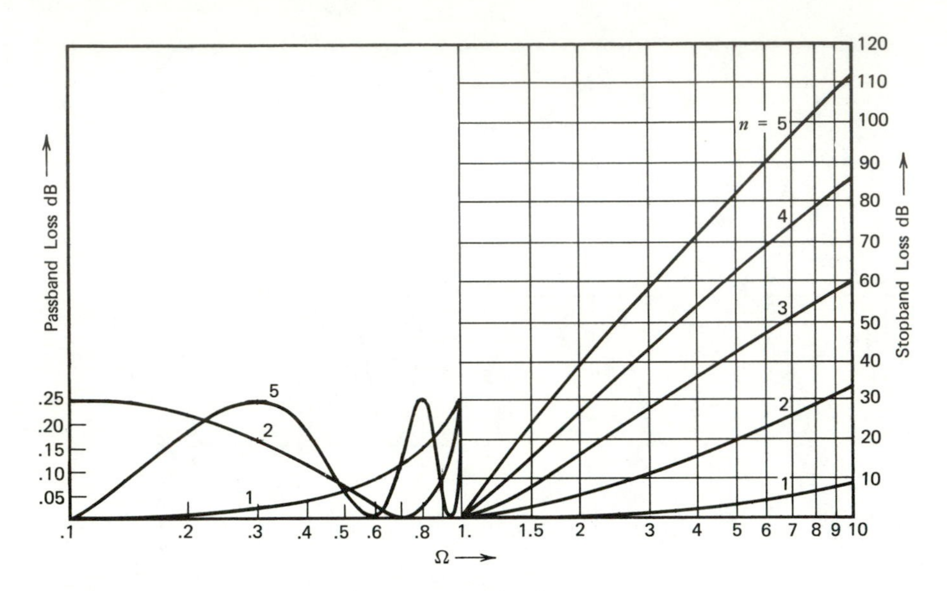

Figure 4.9 Loss of *LP* Chebyshev approximation for A_{max} = 0.25 dB.

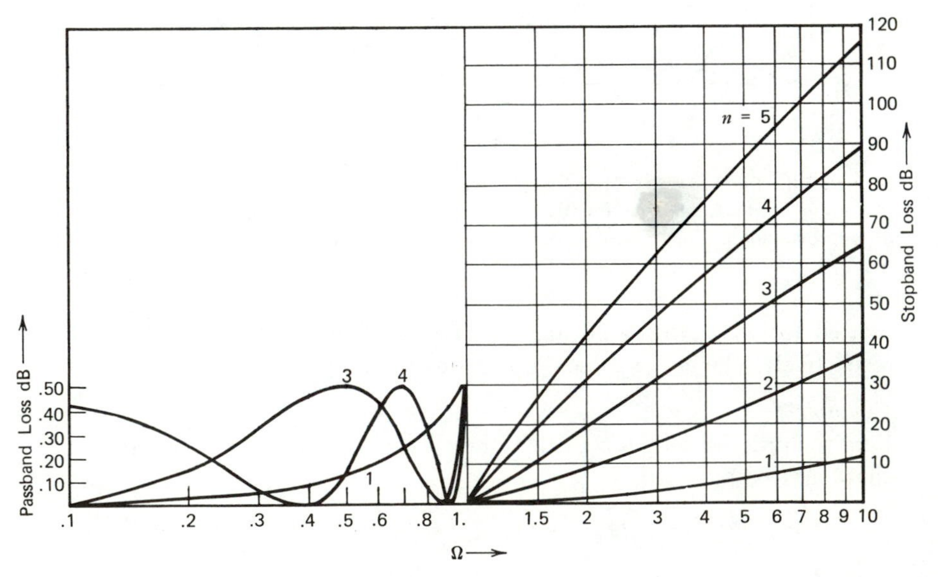

Figure 4.10 Loss of *LP* Chebyshev approximation for A_{max} = 0.50 dB.

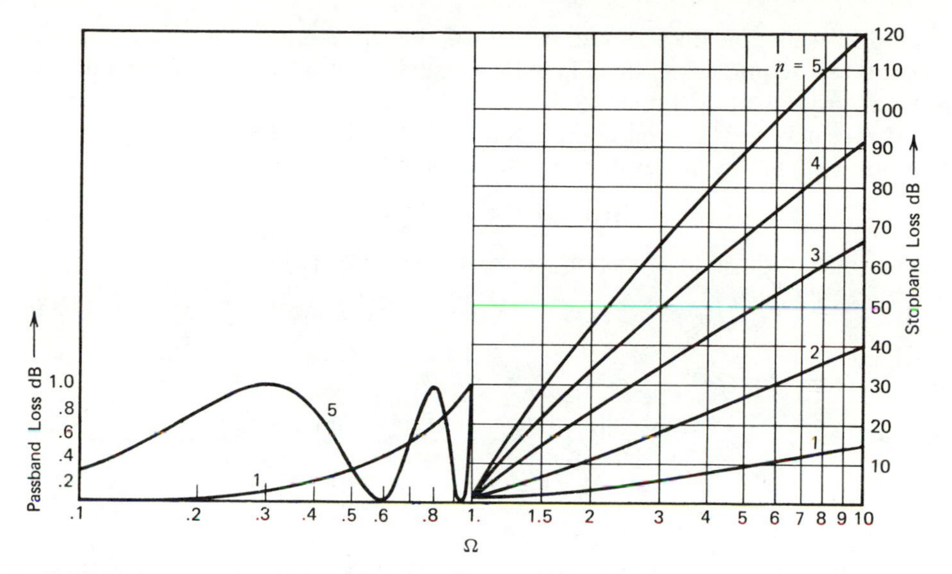

Figure 4.11 Loss of *LP* Chebyshev approximation for A_{max} = 1 dB.

One of the objectives in considering an equiripple passband was to improve on the stopband attenuation provided by the Butterworth approximation. Let us compare these two approximations for $\omega \gg \omega_P$. From Equation 4.7, the Butterworth attenuation for $\omega \gg \omega_P$ is approximately

$$20 \log_{10} \varepsilon\left(\frac{\omega}{\omega_P}\right)^n \tag{4.31}$$

The Chebyshev attenuation is obtained from Equation 4.28, where for $\omega \gg \omega_P$ (i.e., $\Omega_P \gg 1$) the term $\varepsilon C_n(\Omega) \gg 1$. Thus

$$A(\Omega)|_{\Omega \gg 1} \cong 20 \log_{10} \varepsilon C_n(\Omega) \tag{4.32}$$

From (4.27), for $\Omega \gg 1$

$$C_n(\Omega) \cong 2^{n-1}\Omega^n$$

Using this expression, (4.32) reduces to

$$A(\Omega)|_{\Omega \gg 1} = A\left(\frac{\omega}{\omega_P}\right)\bigg|_{\omega/\omega_P \gg 1} = 20 \log_{10}\left[\varepsilon\left(\frac{\omega}{\omega_P}\right)^n 2^{n-1}\right] \tag{4.33}$$

Comparing (4.31) and (4.33), it is seen that the Chebyshev approximation provides

$$20 \log(2)^{n-1} = 6(n-1) \text{ dB} \tag{4.34}$$

more attenuation than a Butterworth of the same order. Therefore, for the same loss requirements the Chebyshev approximation will usually require a lower order than the Butterworth.

Let us next find the roots of the function $H(s)$. As in the Butterworth case these roots are found by first evaluating the roots of $|H(s)|^2$, where

$$|H(s)|^2 = 1 + \varepsilon^2 C_n^2(\Omega)|_{\Omega = s/j} \tag{4.35}$$

The roots of the above function can be shown to be [3]

$$s_k = \sigma_k \pm j\omega_k \qquad k = 0, 1, 2, \ldots, 2n - 1 \tag{4.36}$$

where

$$\sigma_k = \pm\sin\frac{\pi}{2}\left(\frac{1 + 2k}{n}\right)\sinh\left(\frac{1}{n}\sinh^{-1}\frac{1}{\varepsilon}\right) \tag{4.37a}$$

$$\omega_k = \cos\frac{\pi}{2}\left(\frac{1 + 2k}{n}\right)\cosh\left(\frac{1}{n}\sinh^{-1}\frac{1}{\varepsilon}\right) \tag{4.37b}$$

As in the Butterworth approximation the n left half plane roots, corresponding to negative σ, are associated with $H(s)$. Furthermore, from 4.37a and b, it can easily be seen that

$$\left[\frac{\sigma_k}{\sinh\left(\frac{1}{n}\sinh^{-1}\frac{1}{\varepsilon}\right)}\right]^2 + \left[\frac{\omega_k}{\cosh\left(\frac{1}{n}\sinh^{-1}\frac{1}{\varepsilon}\right)}\right]^2 = 1 \tag{4.38}$$

which is the equation of an ellipse. Thus the roots of the Chebyshev approximation lie on an ellipse in the s plane, whose real and imaginary intercepts are indicated in Figure 4.12.

The Chebyshev approximation function can now be expressed in factored form, as

$$H(s) = \prod_j \frac{1}{K}(s - s_j) \tag{4.39}$$

where s_j are the left half plane roots of $|H(s)|^2$. The denominator constant K is adjusted to provide a loss of 0 dB at the passband minima. The factored form of the loss function for $A_{max} = 0.25$ dB, 0.5 dB, and 1 dB, for orders up to $n = 5$ are given in Table 4.2. The polynomials given in the tables apply to the normalized filter requirements, for which the passband edge frequency $\omega = \omega_P = 1$. For the general LP filter, with the passband edge at $\omega = \omega_P$, these polynomials need to be denormalized by replacing

$$s \text{ by } \frac{s}{\omega_P} \tag{4.40}$$

Table 4.2 Chebyshev Approximation Functions

(*a*) $A_{max} = 0.25$ dB

n	Numerator of $H(s)$	Denominator Constant K
1	$s + 4.10811$	4.10811
2	$s^2 + 1.79668s + 2.11403$	2.05405
3	$(s^2 + 0.76722s + 1.33863)(s + 0.76722)$	1.02702
4	$(s^2 + 0.42504s + 1.16195)(s^2 + 1.02613s + 0.45485)$	0.51352
5	$(s^2 + 0.27005s + 1.09543)(s^2 + 0.70700s + 0.53642)(s + 0.43695)$	0.25676

(*b*) $A_{max} = 0.5$ dB

n	Numerator of $H(s)$	Denominator Constant K
1	$s + 2.86278$	2.86278
2	$s^2 + 1.42562s + 1.51620$	1.43138
3	$(s^2 + 0.62646s + 1.14245)(s + 0.62646)$	0.71570
4	$(s^2 + 0.35071s + 1.06352)(s^2 + 0.84668s + 0.356412)$	0.35785
5	$(s^2 + 0.22393s + 1.03578)(s^2 + 0.58625s + 0.47677)(s + 0.362332)$	0.17892

(*c*) $A_{max} = 1$ dB

n	Numerator of $H(s)$	Denominator Constant K
1	$s + 1.96523$	1.96523
2	$s^2 + 1.09773s + 1.10251$	0.98261
3	$(s^2 + 0.49417s + 0.99420)(s + 0.49417)$	0.49130
4	$(s^2 + 0.27907s + 0.98650)(s^2 + 0.67374s + 0.27940)$	0.24565
5	$(s^2 + 0.17892s + 0.98831)(s^2 + 0.46841s + 0.42930)(s + 0.28949)$	0.12283

The use of these tables is illustrated by the following example.

Example 4.6

Find the low-pass approximation function for the filter requirements

$$\omega_P = 200 \qquad \omega_S = 600 \qquad A_{max} = 0.5 \text{ dB} \qquad A_{min} = 20 \text{ dB}$$

Solution

The normalized stopband edge frequency is $\Omega_S = 600/200 = 3$. From Figure 4.10, the required order is 3. The normalized *LP* function, $H_N(s)$, for $n = 3$ and

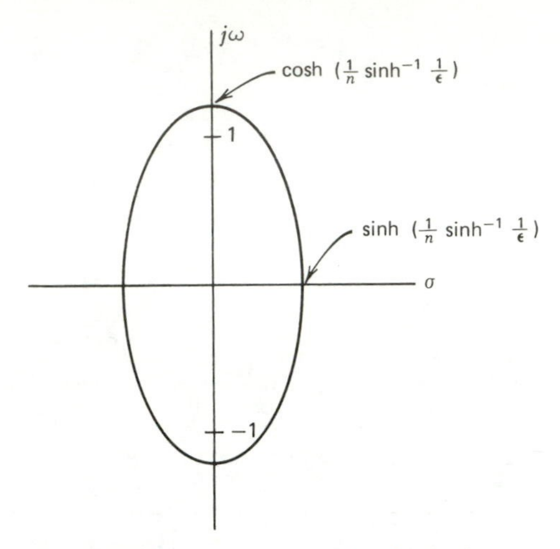

Figure 4.12 Locus of roots for Chebyshev approximation.

$A_{max} = 0.5$ dB is obtained from Table 4.2*b*:

$$H_N(s) = \frac{(s^2 + 0.62646s + 1.14245)(s + 0.62646)}{0.71570}$$

The desired third-order LP filter function $H(s)$, is obtained by denormalizing this function, by replacing s by $s/200$:

$$H(s) = \frac{(s^2 + 125.3s + 45698)(s + 125.3)}{5725600}$$ ■

4.4 ELLIPTIC APPROXIMATION

We have seen that the Chebyshev approximation, which has an equiripple passband, yields a greater stopband loss than the maximally flat Butterworth approximation. In both approximations the stopband loss keeps increasing at the maximum possible rate of $6n$ dB/octave for an nth order function. Therefore these approximations *provide increasingly more loss than the flat* A_{min} *needed* above the edge of the stopband. This source of inefficiency is remedied by the **elliptic** approximation (also known as the **Cauer** approximation).

The elliptic approximation is the most commonly used function in the design of filters. A typical elliptic approximation function is sketched in Figure 4.13. The distinguishing feature of this approximation function is that it has poles of attenuation in the stopband. Thus the elliptic approximation is a rational

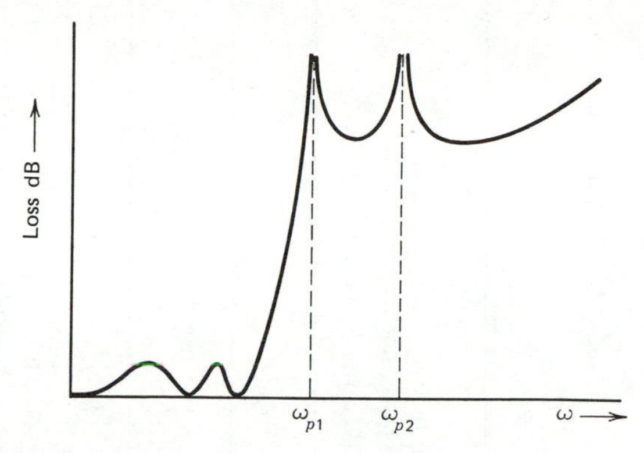

Figure 4.13 Typical loss characteristic of a *LP* elliptic approximation.

function with finite poles and zeros, while the Butterworth and Chebyshev are polynomials and as such have all their loss poles at infinity. In particular, in the elliptic approximation the location of the poles must be chosen to provide the equiripple stopband characteristic shown. The pole closest to the stopband edge (ω_{p_1}) significantly increases the slope in the transition band. The further poles (ω_{p_2} and infinity) are needed to maintain the required level of stopband attenuation. By using finite poles, the elliptic approximation is able to provide a considerably higher *flat* level of stopband loss than the Butterworth and Chebyshev approximations. Thus for a given requirement the elliptic approximation will, in general, require a lower order than the Butterworth or the Chebyshev. Since a lower order corresponds to less components in the filter circuit, the elliptic approximation will lead to the least expensive filter realization.

The mathematical development of the elliptic approximation is based on the rather complex theory of elliptic functions, which is beyond the scope of this book. The interested reader is referred to R. W. Daniel's text on approximation methods [3] for details. The poles and zeros of the *LP* elliptic approximation function have been tabulated for a large number of cases by Christian and Eisenmann [2]. A sample of some normalized elliptic *LP* filter functions, in factored form, is given in Table 4.3. In these tables the frequencies Ω are normalized to the passband edge frequency ω_P (i.e., $\Omega = \omega/\omega_P$). The denominator constant K, shown in the second column, is determined by the *dc* loss. Observe that a different table is needed for each value of Ω_S, where

$$\Omega_S = \frac{\omega_S}{\omega_P} = \frac{\text{stopband edge frequency}}{\text{passband edge frequency}}$$

Also different tables are needed for each $A_{\max}$.

Table 4.3 Elliptic Approximation Functions for $A_{max} = 0.5$ dB

(*a*) $\Omega_s = 1.5$

n	Denominator Constant K	Denominator of $H(s)$	Numerator of $H(s)$	A_{min}
2	0.38540	$s^2 + 3.92705$	$s^2 + 1.03153s + 1.60319$	8.3
3	0.31410	$s^2 + 2.80601$	$(s^2 + 0.45286s + 1.14917)(s + 0.766952)$	21.9
4	0.015397	$(s^2 + 2.53555)(s^2 + 12.09931)$	$(s^2 + 0.25496s + 1.06044)(s^2 + 0.92001s + 0.47183)$	36.3
5	0.019197	$(s^2 + 2.42551)(s^2 + 5.43764)$	$(s^2 + 0.16346s + 1.03189)(s^2 + 0.57023s + 0.57601)(s + 0.42597)$	50.6

(*b*) $\Omega_s = 2.0$

n	Denominator Constant K	Denominator of $H(s)$	Numerator of $H(s)$	A_{min}
2	0.20133	$s^2 + 7.4641$	$s^2 + 1.24504s + 1.59179$	13.9
3	0.15424	$s^2 + 5.15321$	$(s^2 + 0.53787s + 1.14849)(s + 0.69212)$	31.2
4	0.0036987	$(s^2 + 4.59326)(s^2 + 24.22720)$	$(s^2 + 0.30116s + 1.06258)(s^2 + 0.88456s + 0.41032)$	48.6
5	0.0046205	$(s^2 + 4.36495)(s^2 + 10.56773)$	$(s^2 + 0.19255s + 1.03402)(s^2 + 0.58054s + 0.52500)(s + 0.392612)$	66.1

(*c*) $\Omega_s = 3.0$

n	Denominator Constant K	Denominator of $H(s)$	Numerator of $H(s)$	A_{min}
2	0.083974	$s^2 + 17.48528$	$s^2 + 1.35715s + 1.55532$	21.5
3	0.063211	$s^2 + 11.82781$	$(s^2 + 0.58942s + 1.14559)(s + 0.65263)$	42.8
4	0.00062046	$(s^2 + 10.4554)(s^2 + 58.471)$	$(s^2 + 0.32979s + 1.063281)(s^2 + 0.86258s + 0.37787)$	64.1
5	0.00077547	$(s^2 + 9.8955)(s^2 + 25.0769)$	$(s^2 + 0.21066s + 1.0351)(s^2 + 0.58441s + 0.496388)(s + 0.37452)$	85.5

The following example illustrates the use of these tables.

Example 4.7

Find the elliptic *LP* filter function for the requirements given in Example 4.6:

$$\omega_P = 200 \qquad \omega_S = 600 \qquad A_{max} = 0.5 \text{ dB} \qquad A_{min} = 20 \text{ dB}$$

Solution

The stopband edge to passband edge ratio, Ω_S, is

$$\Omega_S = \frac{600}{200} = 3$$

From Table 4.3*c*, a second-order elliptic function will provide 21.5 dB of attenuation above $\Omega_S = 3$.

The normalized approximation loss function is

$$H_N(s) = \frac{s^2 + 1.35715s + 1.55532}{0.083974(s^2 + 17.48528)}$$

Denormalizing, by replacing s by $s/200$, we get the desired *LP* elliptic approximation function

$$H(s) = \frac{s^2 + 271.4s + 62212.8}{0.083974(s^2 + 699411)}$$

Observation

Comparing this result with the Chebyshev realization for the same requirements (Example 4.6), it is seen that the above elliptic approximation requires one lower order than the Chebyshev. ■

4.5 BESSEL APPROXIMATION

Thus far we have discussed the gain (loss) characteristics of filter functions, but have not paid any attention to their phase and delay characteristics. As mentioned in Chapter 3, in digital transmission systems, where the information is transmitted as pulses, the phase distortion introduced by the filter cannot be ignored. In this section we will present an approximation function, known as the **Bessel** approximation, which concentrates on the phase and delay characteristics.

Before describing the Bessel approximation let us first consider the delay characteristics of the Butterworth and Chebyshev approximations. The magnitude and delay of a fourth-order Chebyshev filter function ($A_{max} = 0.5$ dB), obtained by using the MAG program, are sketched in Figures 4.14*a* and *b*. The delay characteristic in the passband is far from flat, the high frequencies being delayed much more than the low frequencies. Considering the response to the rectangular step input shown in Figure 4.14*c*, the high frequencies are

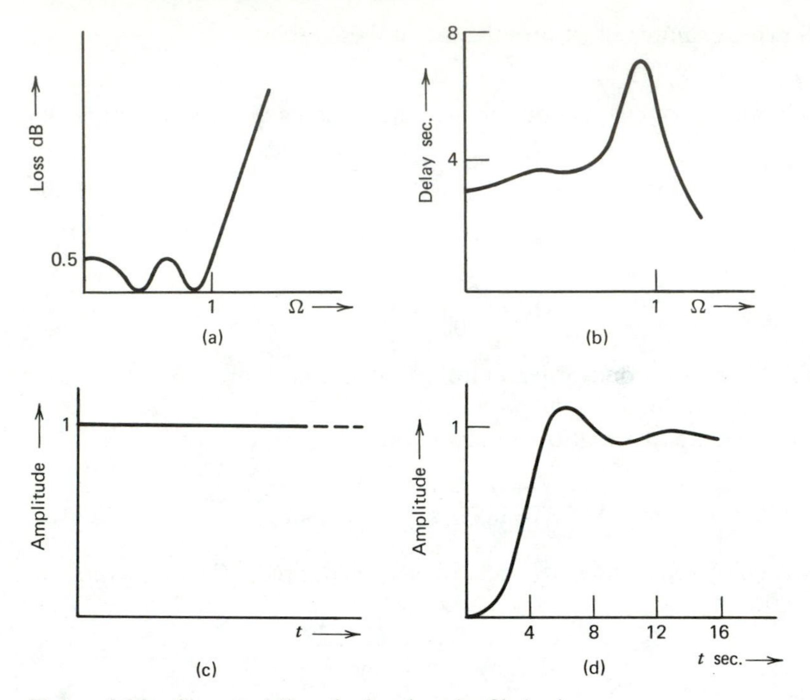

Figure 4.14 Characteristics of a fourth-order Chebyshev (A_{max} = 0.5 dB); (*a*) Loss. (*b*) Delay. (*c*) Step input. (*d*) Step response.

expected to appear at the output of the filter later than the low frequencies. Since the high frequencies control the sharp rising edge of the step, the rise time of the pulse* will be increased, as indicated in Figure 4.14*d*. When the high frequencies do arrive at the output they show up as a high frequency ringing in the step response.† Thus, it can be seen that this Chebyshev filter function would greatly deteriorate the time response of digital signals.

Let us next consider the Butterworth characteristic. In this case the magnitude characteristic is monotonic in the passband and the delay is relatively flat. Figure 4.15*a* and *b* show the magnitude and delay characteristics of a fourth-order Butterworth ($A_{max} = 3$ dB). The step response, shown in Figure 4.15*c*, has less ringing and the rise time is smaller than in the Chebyshev case. We observe that the smoother the magnitude characteristic the flatter is the delay characteristic. However, the smoother magnitude characteristic of the

* The rise time of a step response is commonly defined as the time required for the step response to rise from 10 to 90 percent of its final value.

† The time response was computed using the program CORNAP (Reference 12, Chapter 1).

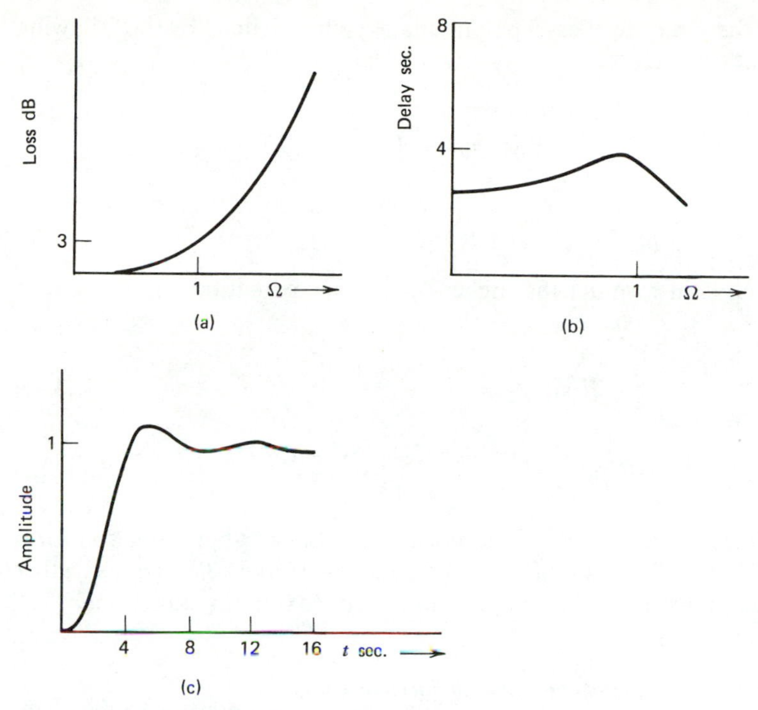

Figure 4.15 Characteristics of a fourth-order Butterworth (A_{max} = 3 dB); (*a*) Loss. (*b*) Delay. (*c*) Step response.

Butterworth approximation provides much less stopband attenuation than the equiripple Chebyshev approximation.

With these qualitative remarks we are ready to embark on the Bessel approximation, where the goal is to obtain as flat a delay characteristic as possible in the passband. The *loss* function for the ideal delay characteristic, from Equation 3.6, is

$$H(s) = e^{sT_0} \tag{4.41}$$

The Bessel approximation is a polynomial that approximates this ideal characteristic. In this approximation the delay at the origin is maximally flat, that is, as many derivatives as possible are zero at the origin. It is convenient to consider the approximation of the normalized function, with the *dc* delay $T_0 = 1$ second, that is,

$$H(s) = e^{s} \tag{4.42}$$

It can be shown that [3] the Bessel approximation to this normalized function is

$$H(s) = \frac{B_n(s)}{B_n(0)} \tag{4.43}$$

where $B_n(s)$ is the nth order Bessel polynomial which is defined by the following recursive equation

$$B_0(s) = 1$$
$$B_1(s) = s + 1$$

and

$$B_n(s) = (2n - 1)B_{n-1}(s) + s^2 B_{n-2}(s) \tag{4.44}$$

Using this recursion formula the higher-order approximations of e^s are seen to be

$$H(s)|_{n=2} = \frac{(s^2 + 3s + 3)}{3} \tag{4.45}$$

$$H(s)|_{n=3} = \frac{(s^3 + 6s^2 + 15s + 15)}{15} \tag{4.46}$$

and so on. The factored forms of the normalized Bessel approximation for n up to 5 are given in Table 4.4. If the low frequency delay is T_0 seconds (rather than 1 second) s must be replaced by sT_0 in the approximation functions.

Table 4.4 Bessel Approximation Functions in Factored Form

n	Numerator of $H(s)$	Denominator constant K
1	$s + 1$	1
2	$s^2 + 3s + 3$	3
3	$(s^2 + 3.67782s + 6.45944)(s + 2.32219)$	15
4	$(s^2 + 5.79242s + 9.14013)(s^2 + 4.20758s + 11.4878)$	105
5	$(s^2 + 6.70391s + 14.2725)(s^2 + 4.64934s + 18.15631)(s + 3.64674)$	945

The loss and delay of the Bessel approximations ($n = 1$ to 5) are sketched in Figures 4.16 and 4.17, respectively. In these figures the normalized frequency Ω is related to ω by

$$\Omega = \omega T_0 \tag{4.47}$$

Figure 4.17 shows that the higher the order n, the wider is the band of frequencies over which the delay is flat. The delay characteristics of the Bessel approximation are far superior to those of the Butterworth and the Chebyshev. As a result, the step response (Figure 4.18c) is also superior, having no overshoot. However, the flat delay is achieved at the expense of the stopband attenuation which, for the Bessel approximation, is even lower than that for the Butterworth.

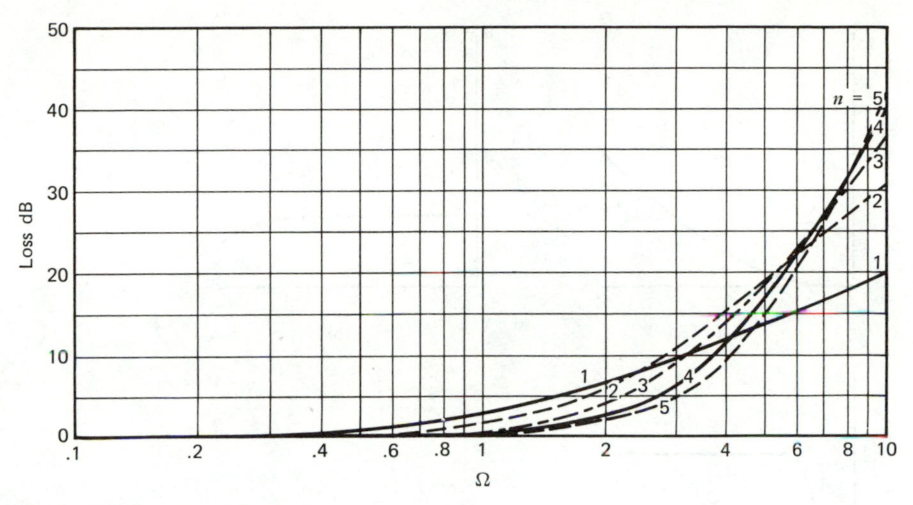

Figure 4.16 Loss of LP Bessel approximations.

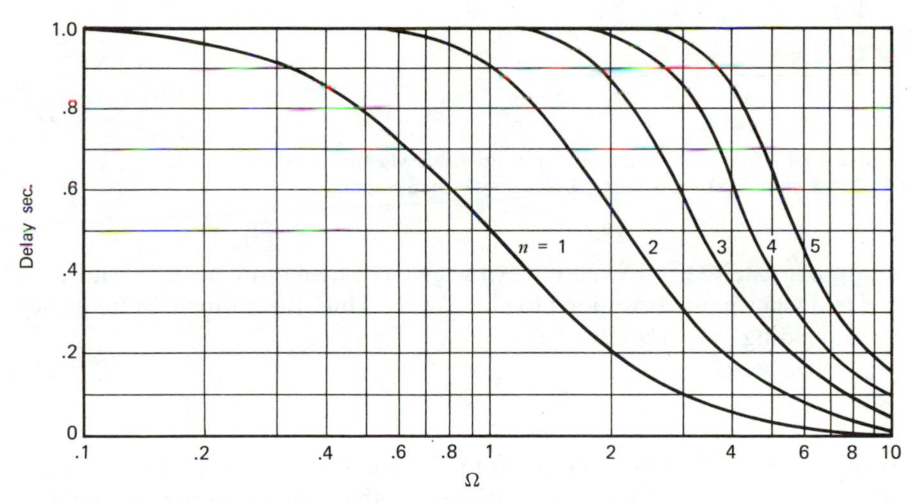

Figure 4.17 Delay of LP Bessel approximations.

Example 4.8

Find the LP Bessel approximation function for the following filter requirements:

(a) The delay must be flat within 1 percent of the dc value up to 2 kHz.
(b) The attenuation at 6 kHz must exceed 25 dB.

Solution

As a first attempt we try a fourth-order filter. From Figure 4.17, the fourth-order Bessel approximation is seen to have a delay that is flat to within 1 percent up

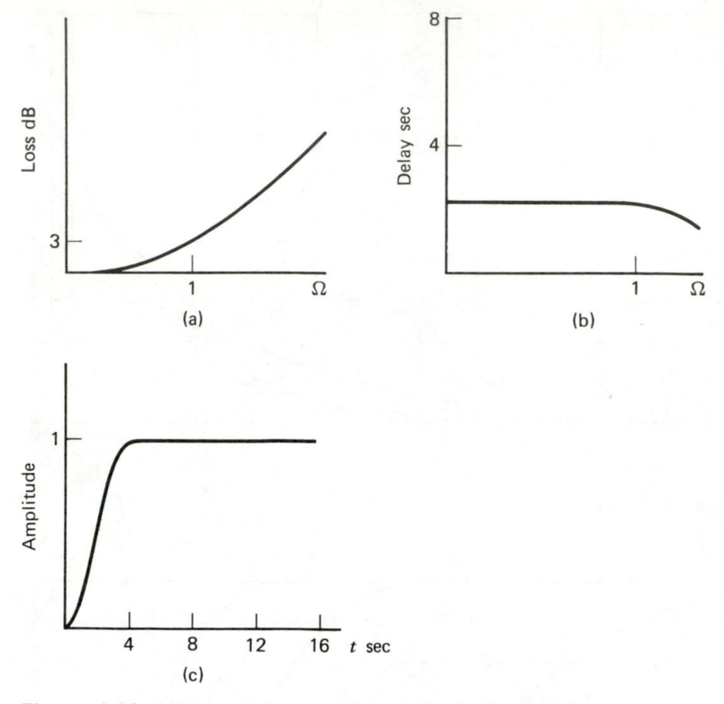

Figure 4.18 Characteristics of a fourth-order Bessel approximation (A_{max} = 3 dB); (*a*) Loss. (*b*) Delay. (*c*) Step response.

to approximately $\Omega = 1.9$. To satisfy the given delay requirements this normalized frequency must correspond to $\omega = 2$ kHz. Thus, the normalized frequency corresponding to 6 kHz must be

$$\Omega_S = \frac{6}{2}(1.9) = 5.7$$

At this frequency the attenuation, from Figure 4.16, is only 22 dB. Since the fourth-order approximation does not meet the loss requirements, we will next try a fifth-order function.

From Figure 4.17 the delay stays within 1 percent up to approximately $\Omega = 2.5$, for $n = 5$. If this frequency corresponds to $\omega = 2$ kHz, the normalized frequency corresponding to $\omega = 6$ kHz is

$$\Omega_S = \frac{6}{2}(2.5) = 7.5$$

From Figure 4.16, the fifth-order function is seen to provide 29.5 dB of attenuation at this frequency. The fifth-order Bessel approximation therefore satisfies both the delay and the loss requirements.

The normalized fifth-order function is listed in Table 4.4. This function is denormalized by replacing s by sT_0 where, from Equation 4.47,

$$T_0 = \frac{\Omega}{\omega} = \frac{2.5}{2\pi(2000)} = 1.989(10)^{-4} \text{ sec}$$

The resulting denormalized gain function is

$$\frac{V_O}{V_{IN}} = \frac{3.608(10^8)}{s^2 + 3.370(10^4)s + 3.608(10^8)} \cdot \frac{4.5894(10^8)}{s^2 + 2.338(10^4)s + 4.5894(10^8)}$$

$$\frac{1.8335(10^4)}{s + 1.8335(10^4)}$$

Observation

In comparison, a fifth-order Butterworth with a cutoff frequency of 2 kHz would provide approximately 55 dB of attenuation at 6 kHz. Thus, we see that while the Bessel approximation does provide a flat delay characteristic, its filtering action in the stopband is much worse than the Butterworth! The poor stopband characteristics of the Bessel approximation makes it an impractical approximation for most filtering applications. An alternate solution to the problem of attaining a flat delay characteristic is by the use of delay equalizers, which is the subject of the next section. ■

4.6 DELAY EQUALIZERS

In the preceding section we discussed the time response distortion resulting from the nonflat delay characteristics of filters. The Bessel approximation did yield a flat delay in the passband; however, its stopband attenuation proves to be inadequate for most filter applications. In this section we present an alternate way of obtaining flat delay characteristics without sacrificing attenuation in the stopband. The approach used is to first approximate the required loss using the Butterworth, Chebyshev, or elliptic functions. The delay of this approximation function will, of course, have ripples and bumps and will certainly not be flat (Figure 4.14 and 4.15). Therefore, some means is needed to compensate for the delay distortion introduced. As mentioned in Chapter 3 (page 87), this compensation can be achieved by following the filter circuit by delay equalizers. The purpose of the delay equalizer is to introduce the necessary delay shape to make the total delay (of the filter and equalizer) as flat as possible in the passband. Furthermore, the equalizer must not perturb the loss characteristic and therefore the loss of the equalizer must be flat for all frequencies.

In Chapter 3 it was shown that a second-order delay equalizer could be realized by the all-pass function:

$$\frac{V_O}{V_{IN}} = \frac{s^2 - as + b}{s^2 + as + b} \tag{4.48}$$

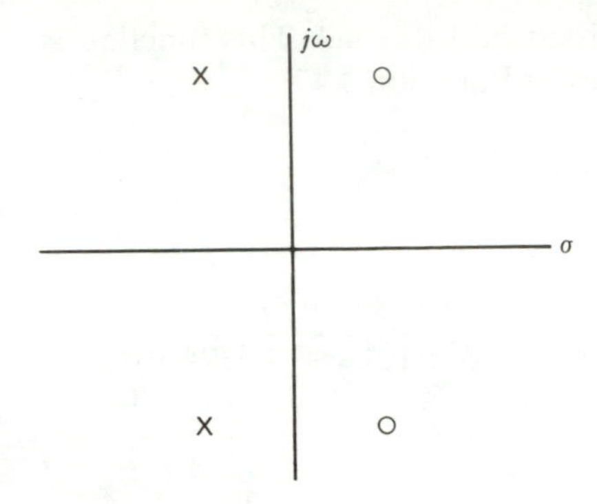

Figure 4.19 Poles and zeros of a second-order delay equalizer.

for which the pole-zero diagram is as shown in Figure 4.19. The delay of this function can be evaluated using Equation 3.13 or from the MAG computer program (Appendix D). The resulting delay characteristics for different values of pole $Q(=\sqrt{b}/a)$ are sketched in Figure 4.20. The normalizing frequency for these curves is the pole frequency ($\omega = \sqrt{b}$), that is

$$\Omega = \frac{\omega}{\sqrt{b}} \tag{4.49}$$

From these curves it can be seen that the delay bumps occur at or close to $\Omega = 1$ (i.e., $\omega = \sqrt{b}$) for $Q_p > 1$. The parameter b in (4.49) therefore determines the location of the delay bump. The sharpness of the bump is a function of the parameter a. The curves shown in Figure 4.20 are a sample of the large variety of delay shapes that can be generated by the second-order equalizer function. The general delay approximation problem consists of finding the minimum number of second-order delay functions of the form of Equation 4.48, whose delays add together to yield the desired delay characteristic. The form of the desired delay equalizer function is therefore

$$T(s) = \prod_{i=1}^{N} \frac{s^2 - a_i s + b_i}{s^2 + a_i s + b_i} \tag{4.50}$$

The number of delay sections N and their defining parameters (a_i, b_i) for approximating a given delay shape are usually obtained by computer optimization.

As an example, consider the equalization of the fourth-order Chebyshev delay function sketched in Figure 4.21. By using a computer optimization program, the parameters have been found for the two second-order delay sections sketched in the figure which will equalize the delay. The parameters

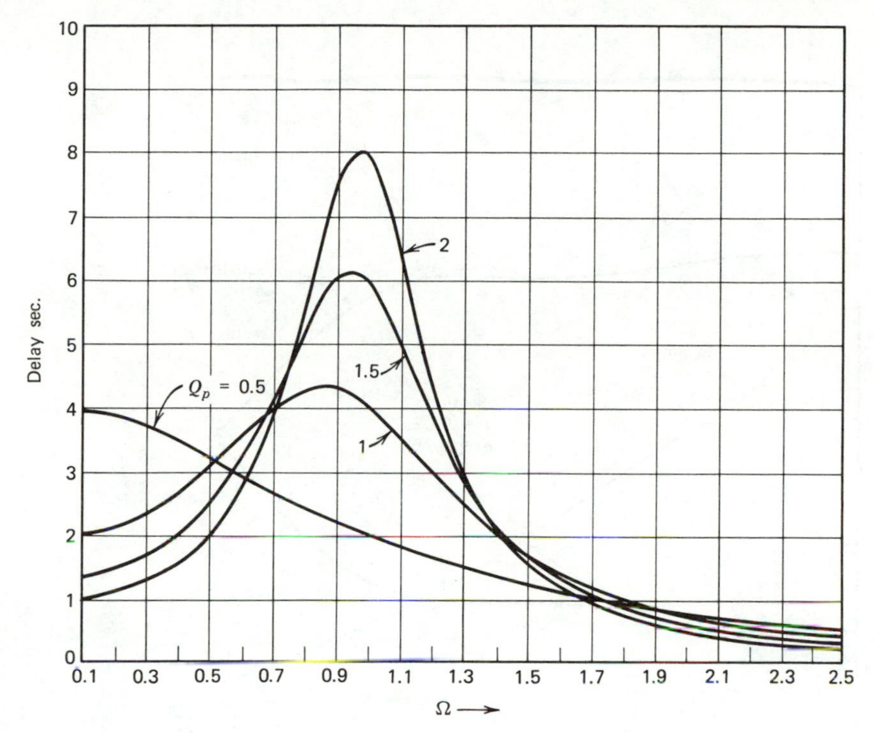

Figure 4.20 Delay of second-order delay equalizers for different pole Q's.

defining these two sections are

$$\begin{array}{lll} \text{section 1} & a_1 = 1.07 & b_1 = 0.334 \\ \text{section 2} & a_2 = 0.825 & b_2 = 0.756 \end{array}$$

The delay of the equalizer is the sum of the delay contributions of the two individual sections. As can be seen from the figure, the sum of the Chebyshev filter delay and the equalizer delay is essentially flat.* Also, since the delay equalizer has a flat gain of unity, it will not change the gain characteristic of the original fourth-order Chebyshev approximation.

Summarizing this section, the steps in the design of a filter with a flat passband delay are:

1. Obtain the Butterworth, Chebyshev, or elliptic approximation to satisfy the magnitude characteristics.
2. Equalize the passband delay of the filter obtained in step 1 by using second-order equalizer sections.

* Reasonable, but not this good, results could be obtained by trial and error.

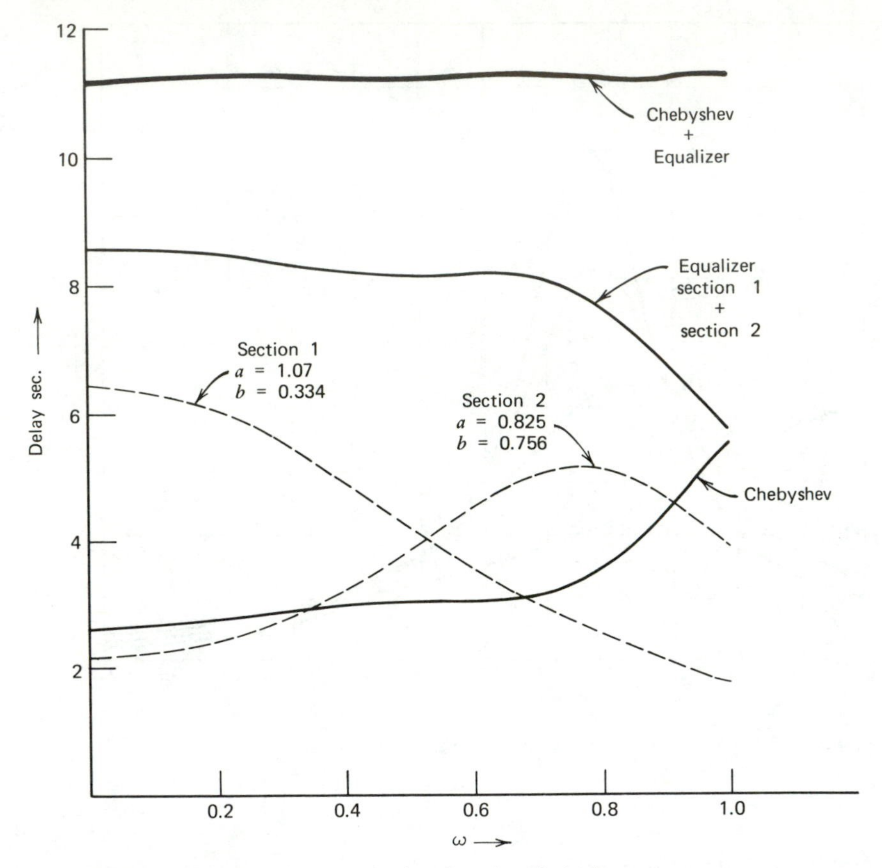

Figure 4.21 Delay equalization of a fourth-order Chebyshev (A_{max} = 0.25 dB, passband edge = 1 rad/sec).

4.7 FREQUENCY TRANSFORMATIONS

The approximations described in the last few sections were directly applicable to low-pass filters. In this section we show how these approximations can be adapted to high-pass filters, symmetrical band-pass filters, and symmetrical band-reject filters. A block diagram of the steps in the approximation of these filters is shown in Figure 4.22. The first step is to translate the given *HP*, *BP*, or *BR* requirement to a related low-pass requirement by using a frequency transformation function. The resulting low-pass requirement is then approximated using the methods described in the previous sections. Finally, the low-pass approximation function is transformed to the desired *HP*, *BP*, or *BR* approximation function. The details of this procedure are described in the following three subsections.

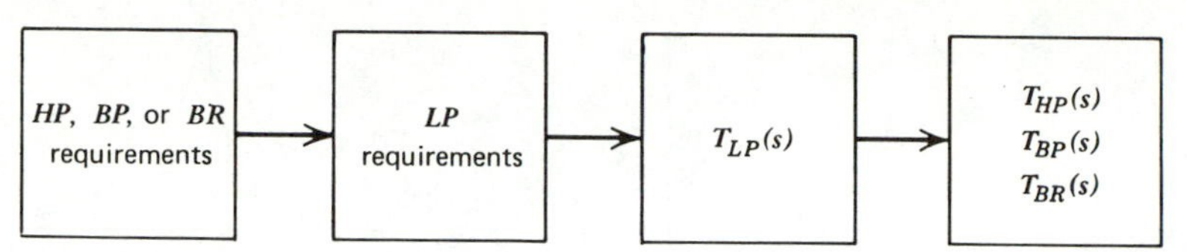

Figure 4.22 Block diagram of the frequency transformation procedure.

4.7.1 HIGH-PASS FILTERS

A high-pass filter function $T_{HP}(s)$, with a passband edge frequency ω_P, can be transformed to a low-pass function by the frequency transformation

$$S = \frac{\omega_P}{s} \tag{4.51}$$

This can be verified by applying the transformation to the HP filter function shown in Figure 4.23. For frequencies on the imaginary axis $S = j\Omega$ and $s = j\omega$, so

$$\Omega = -\frac{\omega_P}{\omega} \tag{4.52}$$

By using this equation the HP passband edge frequency ω_P is seen to be transformed to the LP frequency -1. Since frequency characteristics are symmetrical about the origin, the HP frequency $-\omega_P$ will transform to the LP frequency $+1$

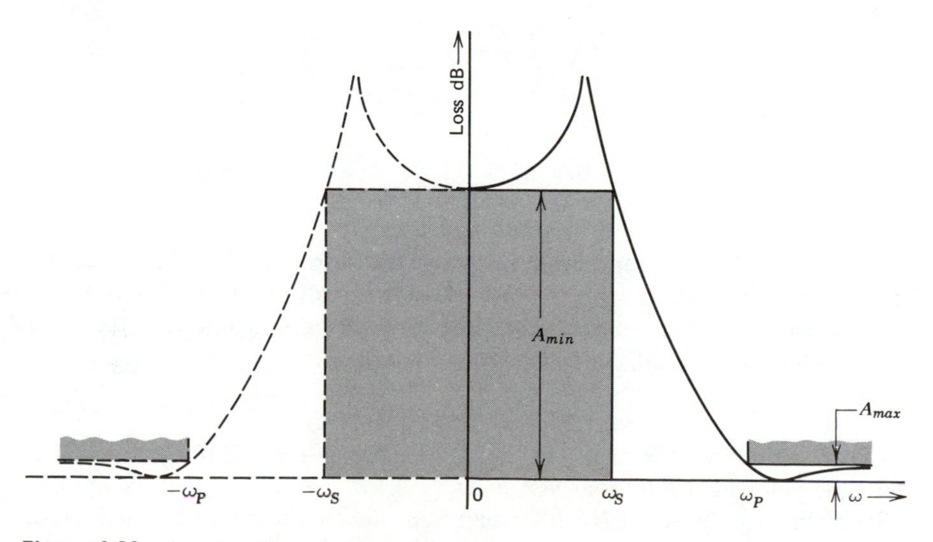

Figure 4.23 A typical high-pass function.

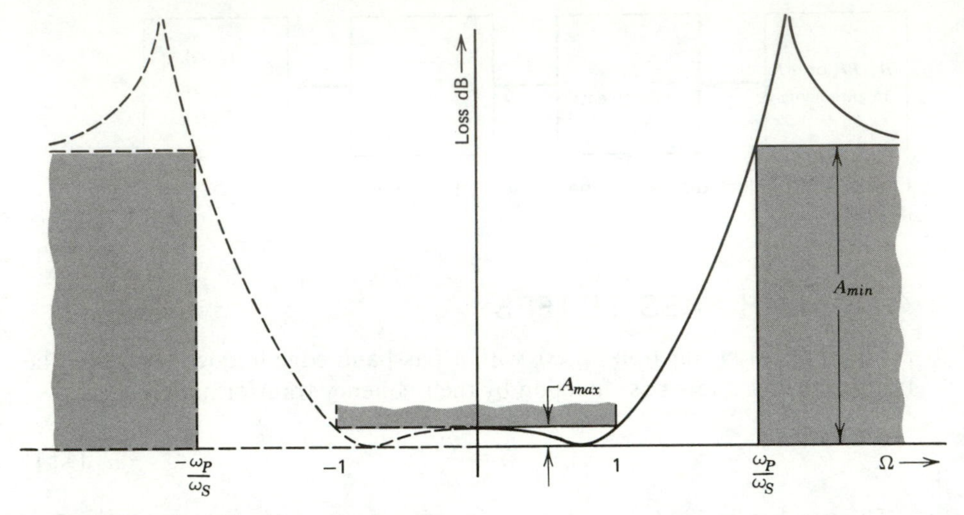

Figure 4.24 Normalized low-pass function.

(Figure 4.24). Applying the transformation to the *HP* stopband edge ω_S, to *dc*, and to infinity, the following relationship between the *HP* and *LP* frequencies are observed:

HP (ω)		*LP* (Ω)
$\pm\omega_P$	→	∓ 1
$\pm\omega_S$	→	$\mp \frac{\omega_P}{\omega_S}$
dc	→	∞
∞	→	dc

In general, the *HP* passband, from ω_P to ∞, transforms to the *LP* passband 0 to 1; and the *HP* stopband, 0 to ω_S, transforms to the *LP* stopband ω_P/ω_S to ∞. Therefore Equation 4.51 transforms a high-pass function, $T_{HP}(s)$, to a low-pass function, $T_{LP}(S)$, defined in the *S* plane:

$$T_{HP}(s) = T_{LP}(S)|_{S=\omega_P/s} \tag{4.53}$$

From this equation it is seen that the attenuation of the *HP* filter at $s = s_1$ is the same as for the *LP* filter at $S = \omega_P/s_1$.

To realize a given set of *HP* filter requirements characterized by A_{max}, A_{min}, ω_P, ω_S (Figure 4.23), we first transform these, using Equation 4.53, to the equivalent normalized LP requirements characterized by A_{max}, A_{min}, 1, ω_P/ω_S (Figure 4.24). These LP requirements are then approximated using the Butterworth, Chebyshev, elliptic, Bessel, or other functions, the choice depending on the filter application. Finally, the normalized low-pass filter function obtained, $T_{LP}(S)$, is transformed to the desired high-pass filter function by using Equation 4.53.

Example 4.9

Find a Butterworth approximation for the high-pass filter requirements characterized by

$$A_{min} = 15 \text{ dB} \qquad A_{max} = 3 \text{ dB} \qquad \omega_P = 1000 \qquad \omega_S = 500$$

Solution

The equivalent normalized LP filter requirements are

$$A_{min} = 15 \text{ dB} \qquad A_{max} = 3 \text{ dB} \qquad \Omega_P = 1 \qquad \Omega_S = 2$$

From Figure 4.5, it is seen that a third-order filter will meet these requirements. The normalized third-order LP Butterworth filter function, from Table 4.1, is

$$T_{LP}(S) = \frac{V_O(S)}{V_{IN}(S)} = \frac{1}{(S^2 + S + 1)(S + 1)}$$

The corresponding HP filter function $T_{HP}(s)$ is obtained by replacing S by $1000/s$:

$$T_{HP}(s) = \frac{s^3}{(s^2 + 1000s + 10^6)(s + 1000)}$$

This function will have a Butterworth characteristic (that is, will be maximally flat at infinity) and will meet the prescribed high-pass requirements. ■

4.7.2 BAND-PASS FILTERS

In this section we consider the approximation of band-pass filters having requirements as shown in Figure 4.25. Again, using frequency transformation techniques, the bandpass function $T_{BP}(s)$ can be transformed to a normalized low-pass function $T_{LP}(S)$. The frequency transformation that accomplishes this is

$$S = \frac{s^2 + \omega_0^2}{Bs} \tag{4.54}$$

where

$B = \omega_2 - \omega_1$ is the passband width of BP filter

$\omega_0 = \sqrt{\omega_1 \omega_2}$ is the center (geometric mean) of the passband

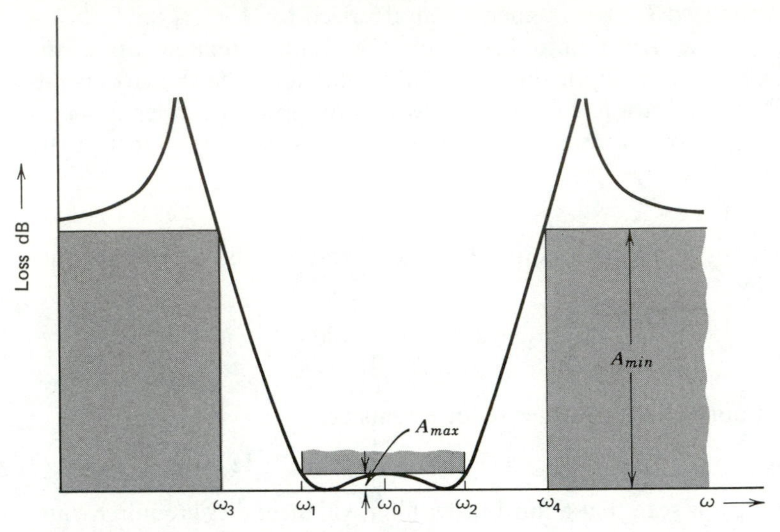

Figure 4.25 A typical band-pass function.

To show this, let us apply the transformation to the *BP* function shown in Figure 4.25. For frequencies on the imaginary axis, $s = j\omega$ and $S = j\Omega$, so

$$\Omega = -\frac{-\omega^2 + \omega_0^2}{(\omega_2 - \omega_1)\omega} \tag{4.55}$$

Using this equation, the center of the *BP* passband ω_0 (Figure 4.26) translates to

$$\Omega_0 = -\frac{-\omega_0^2 + \omega_0^2}{(\omega_2 - \omega_1)\omega_0} = 0 \tag{4.56}$$

The passband edge ω_1 is translated to the low-pass frequency

$$\Omega_1 = -\frac{-\omega_1^2 + \omega_0^2}{(\omega_2 - \omega_1)\omega_1} = -1 \tag{4.57}$$

and the passband edge ω_2 is translated to $\Omega_2 = +1$. The band-pass passband, ω_1 to ω_2, can be seen to be transformed to the frequency band -1 to $+1$.

Next, considering the stopband edge frequencies, we see that ω_3 is transformed to

$$\Omega_3 = -\frac{-\omega_3^2 + \omega_0^2}{(\omega_2 - \omega_1)\omega_3} \tag{4.58}$$

and ω_4 is transformed to

$$\Omega_4 = -\frac{-\omega_4^2 + \omega_0^2}{(\omega_2 - \omega_1)\omega_4} \tag{4.59}$$

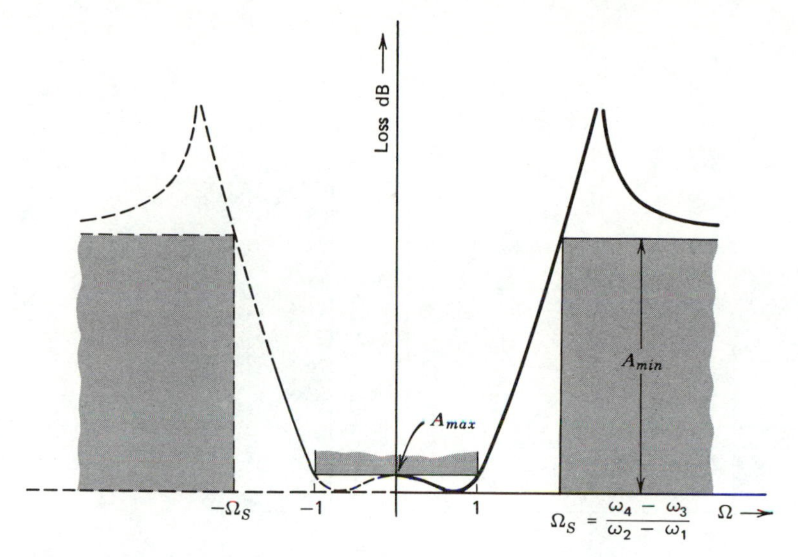

Figure 4.26 Normalized low-pass function.

If the stopband attenuation exhibits geometrical symmetry about the center frequency, that is, if

$$\omega_3 \omega_4 = \omega_0^2 \tag{4.60}$$

then, from (4.58), ω_3 and ω_4 will translate to the two edges of the low-pass passband:

$$\Omega_3 = -\frac{\omega_4 - \omega_3}{\omega_2 - \omega_1} \qquad \Omega_4 = \frac{\omega_4 - \omega_3}{\omega_2 - \omega_1} = -\Omega_3 \tag{4.61}$$

Thus the stopbands of the band-pass function are transformed to the low-pass stopbands from Ω_S to ∞ and $-\Omega_S$ to $-\infty$, where $\Omega_S = (\omega_4 - \omega_3)/(\omega_2 - \omega_1)$.

From the above discussion we see that the band-pass approximation function is related to the low-pass function by

$$T_{BP}(s) = T_{LP}(S)|_{S=(s^2+\omega_0^2)/Bs} \tag{4.62}$$

To realize the symmetrical *BP* requirements shown in Figure 4.25, first the normalized *LP* requirement characterized by (Figure 4.26)

$$A_{\max} \qquad A_{\min} \qquad \Omega_P = 1 \qquad \Omega_S = \frac{\omega_4 - \omega_3}{\omega_2 - \omega_1}$$

is approximated. The required band-pass function $T_{BP}(s)$ is then obtained from $T_{LP}(S)$ by using Equation 4.62.

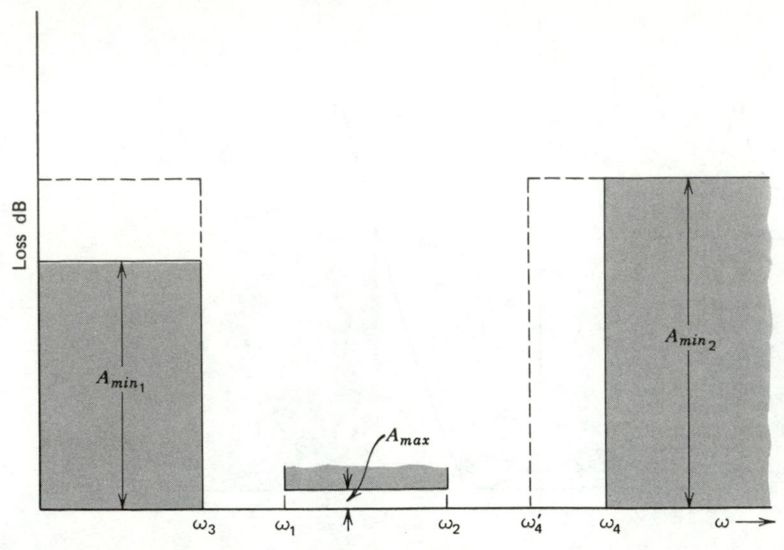

Figure 4.27 Nonsymmetrical *BP* requirements.

The above method can be adapted to nonsymmetrical *BP* requirements, as follows. Consider the *BP* requirements shown in Figure 4.27, where $A_{\min_1} \neq A_{\min_2}$ and $\omega_3\omega_4 \neq \omega_1\omega_2$. A new *BP* requirement (dotted lines in Figure 4.27) which does have geometrical symmetry can be generated by increasing the lower stopband attenuation to $A_{\min_2}$ and by decreasing ω_4 to ω_4' so that

$$\omega_3\omega_4' = \omega_1\omega_2 \tag{4.63}$$

The approximation function for this new, and more stringent, requirement will certainly meet the original nonsymmetrical requirements. However, the resulting approximation function will be more complex, and the corresponding circuit realization more expensive than is really necessary to meet the original *BP* requirements. More economical, direct methods of approximating nonsymmetrical requirements do exist [3], but their discussion is beyond the scope of this book.

Example 4.10

Find the elliptic approximation for the following band-pass requirements:

$$A_{\max} = 0.5 \text{ dB} \qquad A_{\min} = 20$$

$$\text{passband} = 500 \text{ Hz to } 1000 \text{ Hz}$$
$$\text{stopbands} = \text{dc to } 275 \text{ Hz and } 2000 \text{ Hz to } \infty$$

Solution

From the given information

$$\omega_1 = 2\pi 500 \quad \omega_2 = 2\pi 1000 \quad \omega_3 = 2\pi 275 \quad \omega_4 = 2\pi 2000$$

Observe that the stopband requirements are not symmetrical ($\omega_1\omega_2 \neq \omega_3\omega_4$). To obtain geometrical symmetry the frequency ω_4 can be decreased to the new frequency (Equation 4.63)

$$\omega_4' = \frac{\omega_1\omega_2}{\omega_3} = 2\pi 1818$$

The equivalent normalized *LP* requirements are then characterized by

$$A_{\max} = 0.5 \text{ dB} \quad A_{\min} = 20 \quad \Omega_P = 1 \quad \Omega_S = \frac{1818 - 275}{1000 - 500} = 3.08$$

From Table 4.3*c*, for $A_{\max} = 0.5$ dB and $\Omega_S = 3.0$, this *LP* requirement can be approximated by the second-order elliptic function

$$T_{LP}(S) = \frac{V_O(S)}{V_{IN}(S)} = \frac{0.083947(S^2 + 17.48528)}{(S^2 + 1.35715S + 1.55532)}$$

Using Equation 4.62, the *BP* function is obtained by replacing *S* by

$$\frac{s^2 + \omega_1\omega_2}{(\omega_2 - \omega_1)s} = \frac{s^2 + 4\pi^2(500000)}{2\pi 500s}$$

Making this substitution, the desired *BP* approximation function is

$$T_{BP}(s) = \frac{0.084(s^4 + 2.12(10)^8 s^2 + 3.89(10)^{14})}{s^4 + 4.26(10)^3 s^3 + 5.48(10)^7 s^2 + 8.42(10)^{10} s + 3.89(10)^{14}}$$

Observation

From Equation 4.62 we see that each LP pole (zero) transforms to two BP poles (zeros), thus the order of the BP filter function is twice that of the LP filter function. ■

4.7.3 BAND-REJECT FILTERS

In this section we consider the approximation of the symmetrical band-reject requirements shown in Figure 4.28. The stopband extends from ω_3 to ω_4, and the passbands extend from *dc* to ω_1 and ω_2 to infinity. The geometrical symmetry of the requirements imply that

$$\omega_1\omega_2 = \omega_3\omega_4 \tag{4.64}$$

The frequency transformation used to obtain the band-reject filter function $T_{BR}(s)$ from the equivalent normalized low-pass function $T_{LP}(S)$ is

$$S = \frac{Bs}{s^2 + \omega_0^2} \tag{4.65}$$

where

$B = \omega_2 - \omega_1$ is the passband width

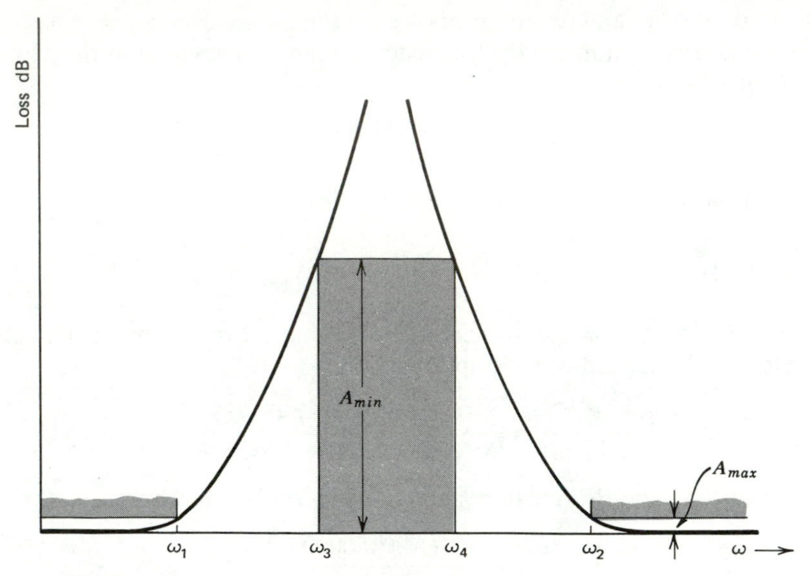

Figure 4.28 A typical band-reject function.

and

$$\omega_0 = \sqrt{\omega_1 \omega_2} \text{ is the center of the stopband}$$

Consider the application of the transformation of Equation 4.65 to the *BR* function shown in Figure 4.28. For frequencies on the imaginary axis, $s = j\omega$ and $S = j\Omega$, so

$$\Omega = \frac{(\omega_2 - \omega_1)\omega}{-\omega^2 + \omega_0^2} \tag{4.66}$$

Proceeding as in the band-pass case, it can be shown that the frequencies $\omega_0, \omega_1, \omega_2, \omega_3$, and ω_4 transform as follows:

BR (ω)		*LP* (Ω)
ω_0	$\rightarrow$	∞
ω_1	$\rightarrow$	$+1$
ω_2	$\rightarrow$	-1
ω_3	$\rightarrow$	$+\dfrac{\omega_2 - \omega_1}{\omega_4 - \omega_3}$
ω_4	$\rightarrow$	$-\dfrac{\omega_2 - \omega_1}{\omega_4 - \omega_3}$

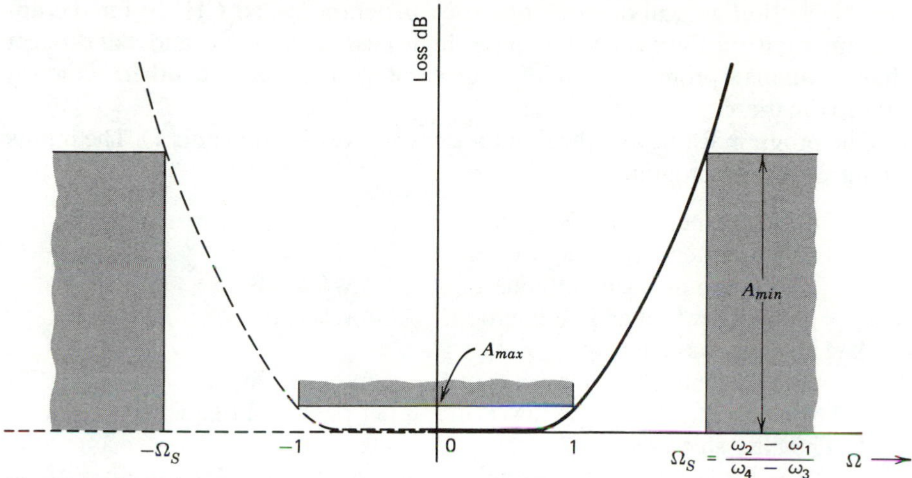

Figure 4.29 Normalized low-pass function.

Thus the band-reject stopband ω_3 to ω_4 is seen to transform to the low-pass stopbands Ω_S to ∞ and $-\Omega_S$ to $-\infty$, where $\Omega_S = (\omega_2 - \omega_1)/(\omega_4 - \omega_3)$; while the band-reject passbands transform to the low-pass band -1 to $+1$. The band-reject filter function $T_{BR}(s)$ is obtained from the low-pass filter function $T_{LP}(S)$, by using the transformation

$$T_{BR}(s) = T_{LP}(S)|_{S = Bs/(s^2 + \omega_0^2)} \tag{4.67}$$

To realize the *BR* requirements shown in Figure 4.28, we first approximate the *LP* requirements characterized by (Figure 4.29):

$$A_{\max} \quad A_{\min} \quad \Omega_P = 1 \quad \Omega_S = \frac{\omega_2 - \omega_1}{\omega_4 - \omega_3}$$

This *LP* requirement is approximated using the Butterworth, Chebyshev, elliptic, or Bessel approximation. Finally, the low-pass approximation function $T_{LP}(S)$ is transformed to the desired band-reject function using Equation 4.67.

4.8 CHEBYSHEV APPROXIMATION COMPUTER PROGRAM

The use of standard tables geatly facilitates the computations in obtaining approximation functions. However, even the most extensive tables can cover only a limited number of cases. If, for example, the passband requirement was $A_{\max} = 0.6$ dB, the tabular approach would constrain the designer to use the closest listed $A_{\max}$ (usually 0.5 dB). A computer program which simulates the equations describing the approximation steps does not have such a limitation.

In this section we will describe one such program (called CHEB) for the approximation of Chebyshev low-pass, high-pass, band-pass, and band-reject filters. Similar programs can of course be written for the other standard approximations.

The program listing and the input format is given in Appendix D. The inputs required for the program are:

1. The filter type *LP*, *HP*, *BP*, *BR*
2. The filter attenuation requirements
   - maximum passband attenuation AMAX dB
   - minimum stopband attenuation AMIN dB
3. The filter passband
   - For *LP* and *HP*: the passband edge frequency in Hz
   - For *BP* and *BR*: the lower and upper passband frequencies in Hz
4. The filter stopband
   - For *LP* and *HP*: the stopband edge frequency in Hz
   - For *BP* and *BR*: the lower and upper stopband frequencies in Hz
5. Frequencies for computation of gain
   - *FS* start frequency Hz
   - *FI* frequency increment Hz
   - *FF* final frequency Hz

The output of the program is the approximation function in the following factored form

$$T(s) = \frac{V_O}{V_{IN}} = \prod_{J=1}^{N} \frac{M(J)s^2 + C(J)s + D(J)}{N(J)s^2 + A(J)s + B(J)}$$

The program also computes and prints the gain of the filter at the specified frequencies.

Example 4.11

Find the Chebyshev approximation for a band-reject filter whose requirements are

$A_{max} = 0.3$ dB, $A_{min} = 50$ dB
lower passband edge = 200 Hz
upper passband edge = 1000 Hz
lower stopband edge = 400 Hz
upper stopband edge = 500 Hz

Solution

Using the CHEB program the filter function is found to be

$$T(s) = \frac{s^2 + 7895683}{s + 2297.72s + 32279472} \cdot \frac{s^2 + 7895683}{s^2 + 6892.5s + 7895682} \cdot \frac{s^2 + 7895683}{s^2 + 562s + 1931312}$$

From the program the calculated loss at the pass and stopband edges are

Freq (Hz)	Loss (dB)
200	0.3
1000	0.3
400	54.7
500	54.7

The approximation function is seen to meet the prescribed requirements with 4.7 dB of attenuation to spare at the stopband edges. ■

4.9 CONCLUDING REMARKS

In this chapter we have discussed the approximation of filter requirements using the Butterworth, Chebyshev, elliptic, and Bessel approximations. Of these, the elliptic is the most commonly used because it requires the lowest order for a given filter requirement. The synthesis of these functions using passive *RLC* circuits is discussed in Chapter 6; and the synthesis using active filters is covered in Chapters 8 to 11.

A few sample tables of approximations for the normalized low-pass requirements were presented in this chapter. The reader is referred to Christian and Eisenmann [2] for a more extensive set of tables for the Butterworth, Chebyshev, and elliptic approximations. The roots for higher-order Bessel approximations have been tabulated by Orchard [10]. The synthesis of delay equalizers is not so straightforward and will usually require a computer optimization program.

The transformation techniques described were applicable to the design of *HP* and symmetrical *BP* and *BR* filters. The approximation of arbitrary stopbands (Figure 3.2) and nonsymmetrical requirements is much more difficult and the reader is referred to Daniels [3] for methods to approximate such requirements.

FURTHER READING

1. N. Balbanian, T. A. Bickart, and S. Seshu, *Electrical Network Theory*, Wiley, New York, 1969, Appendix A2.8.
2. E. Christian and E. Eisenmann, *Filter Design Tables and Graphs*, Wiley, New York, 1966.
3. R. W. Daniels, *Approximation Methods for Electronic Filter Design*, McGraw-Hill, New York, 1974.

4. A. J. Grossman, "Synthesis of Tchebycheff parameter symmetrical filters," *Proc. IRE*, *45*, No. 4, April 1957, pp. 454–473.
5. E. A. Guillemin, *Synthesis of Passive Networks*, Wiley, London, 1957, Chapter 14.
6. J. L. Herrero and G. Willoner, *Synthesis of Filters*, Prentice-Hall, Englewood Cliffs, N.J., 1966, Chapter 6.
7. S. Karni, *Network Theory: Analysis and Synthesis*, Allyn and Bacon, Boston, Mass., 1966, Chapter 13.
8. Y. J. Lubkin, *Filters, Systems and Design: Electrical, Microwave, and Digital*, Addison-Wesley, Reading, Mass., 1970.
9. S. K. Mitra and G. C. Temes, Eds., *Modern Filter Theory and Design*, Wiley, New York, 1973, Chapter 2.
10. H. J. Orchard, "The roots of maximally flat delay polynomials," *IEEE Trans. Circuit Theory*, *CT-12*, No. 3, September 1965, pp. 452–454.
11. L. Weinberg, *Network Analysis and Synthesis*, McGraw-Hill, New York, 1962, Chapter 11.
12. M. E. Van Valkenburg, *Introduction to Modern Network Synthesis*, Wiley, New York, 1960, Chapter 13.
13. A. I. Zverev, *Handbook of Filter Synthesis*, Wiley, New York, 1967.

PROBLEMS

4.1 *Bode plot approximation.* A high-pass filter requirement is specified by the parameters $A_{max} = 1$ dB, $A_{min} = 28$ dB, $f_P = 3500$ Hz, $f_S = 1000$ Hz. Approximate this requirement using the Bode plots of first- and second-order high-pass functions. Compute the loss achieved by the approximation function at f_P and f_S.

4.2 *Butterworth loss.* A fourth-order LP Butterworth approximation function has a loss of 2 dB at 100 rad/sec. Compute the loss at 350 rad/sec. Verify your answer using Figure 4.5.

4.3 The passband loss of a fourth-order LP Butterworth filter function is 1 dB at 500 Hz. Beyond what frequency is the loss greater than 40 dB? Verify your answer using Figure 4.5.

4.4 Sketch the loss characteristic of a seventh-order *LP* Butterworth approximation for $\varepsilon = 0.1$, $\omega_P = 1$.

4.5 *Butterworth approximation.* A low-pass filter requirement is specified by $A_{max} = 1$ dB, $A_{min} = 35$ dB, $f_P = 1000$ Hz, $f_S = 3500$ Hz.

(a) Find the Butterworth approximation function needed.

(b) Determine the loss at 9000 Hz.

(c) Determine the pole Q's of the gain function.

4.6 *Butterworth, pole Q.* Show that the normalized *LP* Butterworth loss function can be expressed as

$$H(s) = \prod_{k=1}^{n/2} \left\{ s^2 - \left[2\cos\left(\frac{2k+n-1}{n}\right)\frac{\pi}{2} \right] s + 1 \right\}$$

for n even. Find the expression for n odd. Derive an expression for the maximum pole Q (of the gain function) for a given order. Verify your answer for $n = 4$ using Table 4.1.

4.7 *Butterworth, slope.* Prove that the slope of the *LP* Butterworth loss function $|H(j\omega)|$ at the passband edge frequency is

$$\frac{n\varepsilon^2}{\omega_p(1 + \varepsilon^2)^{1/2}}$$

4.8 *Butterworth filter.* The network shown can be used to realize a third-order function. Find values for L_1, C_1, and C_2 to realize the Butterworth filter function

$$\frac{V_O}{V_{IN}} = \frac{K}{s^3 + 2s^2 + 2s + 1}$$

Evaluate K for the solution obtained.

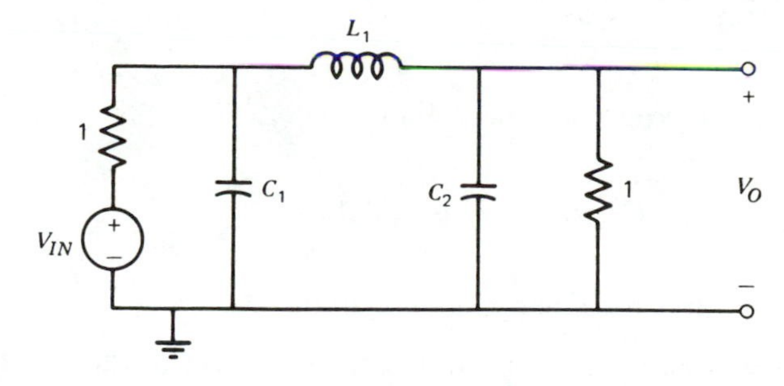

Figure P4.8

4.9 *Chebyshev approximation.* Find the Chebyshev approximation function needed to satisfy the following *LP* requirements:

$$A_{\max} = 0.25 \text{ dB} \qquad A_{\min} = 40 \text{ dB}$$
$$\omega_P = 1200 \text{ rad/sec} \qquad \omega_S = 4000 \text{ rad/sec}$$

Compute the loss obtained at the stopband edge frequency. By how much may the width of the passband be increased, the other requirements and the order remaining unchanged.

4.10 *Chebyshev vs Butterworth.* A fifth-order *LP* Chebyshev filter function has a loss of 72 dB at 4000 Hz. Find the approximate frequency at which a fifth-order Butterworth approximation exhibits the same loss, given that both approximations satisfy the same passband requirement.

4.11 *Chebyshev, order.* Prove that the order n for the normalized *LP* Chebyshev approximation is given by

$$n = \frac{\cosh^{-1}[(10^{0.1A_{\min}} - 1)/(10^{0.1A_{\max}} - 1)]^{1/2}}{\cosh^{-1}(\omega_S/\omega_P)}$$

4.12 *Chebyshev approximation.* Estimate the order of the Chebyshev approximation function needed to meet the filter requirement sketched in Figure P4.12. Obtain an expression for the transfer function. How would you change the coefficients to get 10 dB of gain in the passband, without changing the shape of the filter characteristic?

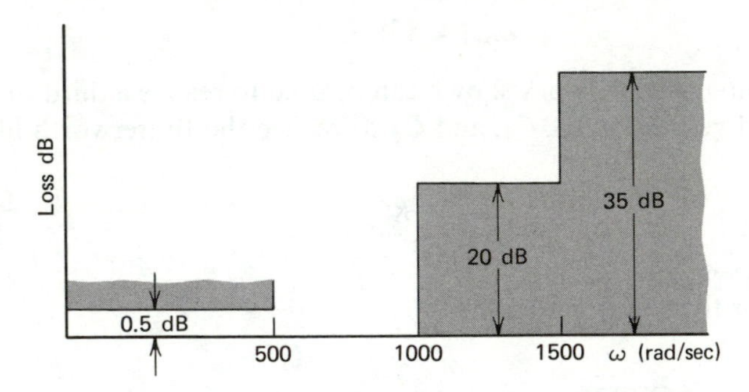

Figure P4.12

4.13 *Chebyshev polynomial properties.* Show that

$$C_n^2(\Omega) = \frac{1}{2}[1 + C_{2n}(\Omega)]$$

where $C_n(\Omega)$ is the nth-order Chebyshev polynomial.

4.14 *Chebyshev, passband max and min.* Find expressions for (a) the number; (b) the magnitudes; and (c) the locations of the maxima and minima in the passband of a normalized *LP* Chebyshev approximation loss function, in terms of n and ε. Use the results to sketch the passband and stopband loss characteristic of a fifth-order normalized *LP* Chebyshev approximation, given $\varepsilon = 0.3$.

4.15 *Chebyshev, slope.* Show that the slope of an nth-order normalized *LP* Chebyshev approximation function at the passband edge frequency is n times that of the Butterworth approximation of the same order, assuming both approximations satisfy the same passband requirement.

4.16 *Chebyshev, pole Q.* Find an expression for the pole Q of a normalized *LP* Chebyshev approximation (gain) function, in terms of σ_k and ω_k (Equation 4.37). Hence, determine the maximum pole Q for the case $n = 4$ and $\varepsilon = 1$.

4.17 *CHEB program.* Find the Chebyshev approximation function needed to meet the following requirements, by using the CHEB computer program: $A_{max} = 0.2$ dB, $A_{min} = 30$ dB, $f_P = 1$ kHz, $f_S = 2.5$ kHz.

4.18 *Elliptic approximation.* Use Table 4.3 to find an elliptic function approximation that will meet the low-pass requirements described by $A_{max} = 0.5$ dB, $A_{min} = 40$ dB, $\omega_P = 1000$ rad/sec, $\omega_S = 3200$ rad/sec. Compute the loss attained at the stopband edge frequency.

4.19 *Elliptic vs Chebyshev.* (a) Find the order of the Chebyshev approximation satisfying the requirements stated in Problem 4.18. (b) Determine the loss at $\omega_S = 3200$ rad/sec if the order used is that computed for the elliptic approximation.

4.20 *Elliptic vs Butterworth.* Repeat Problem 4.19 for the Butterworth approximation function.

4.21 *Maximally flat function.* Prove that for the function

$$\frac{1 + a_1\omega + a_2\omega^2}{1 + b_1\omega + b_2\omega^2 + b_3\omega^3}$$

to be maximally flat at the origin, the coefficients must satisfy the equalities $a_1 = b_1$, $a_2 = b_2$. (Hint: Divide denominator into numerator to obtain a Maclaurin series expansion.)

4.22 *Maximally flat delay.* Show that if the delay of the function

$$1 + a_1 s + a_2 s^2 + a_3 s^3$$

is maximally flat at the origin, and the *dc* delay is one second, then $a_1 = 1$, $a_2 = 2/5$, $a_3 = 1/15$. (*Hint*: use the result stated in Problem 4.21.)

4.23 *Bessel loss.* A fifth-order Bessel approximation function has a delay characteristic that is flat to within 5 percent of the *dc* delay up to 100 rad/sec. Determine the loss at (a) 250 rad/sec and (b) 600 rad/sec, using Figures 4.16 and 4.17.

4.24 *Bessel approximation.* A low-pass filter is required to have 3 dB of loss at the passband edge frequency of 1000 rad/sec, and at least 30 dB of loss at 4000 rad/sec. (a) Find the Bessel approximation function satisfying the given requirements. (b) Up to what frequency will the delay be flat to within 10 percent of the *dc* delay?

4.25 The delay of a *LP* Bessel approximation is required to be flat to within 5 percent of the *dc* delay of 1 msec upto the passband edge frequency of 1800 rad/sec, and the loss must exceed 40 dB at 14,000 rad/sec. Determine:
(a) The Bessel approximation function.
(b) The loss and delay at 1800 rad/sec.
(c) The loss and delay at 14,000 rad/sec.

4.26 *Bessel vs Butterworth.* Find the Butterworth approximation function needed to meet the loss requirements specified in Problem 4.25, for $A_{max} = 1$ dB. Determine the frequency up to which the delay stays within 5 percent of the *dc* delay, by using either the MAG computer program or Equation 3.13.

4.27 *Bessel vs Chebyshev.* Repeat Problem 4.26 using a Chebyshev approximation. Once again assume $A_{max} = 1$ dB.

4.28 *Delay equalizer.* Prove that the total area (from $\omega = 0$ to $\omega = \infty$) under any second-order delay curve is 2π radians. Use this result to estimate the minimum number of second-order delay sections needed to render flat the delay characteristic shown.

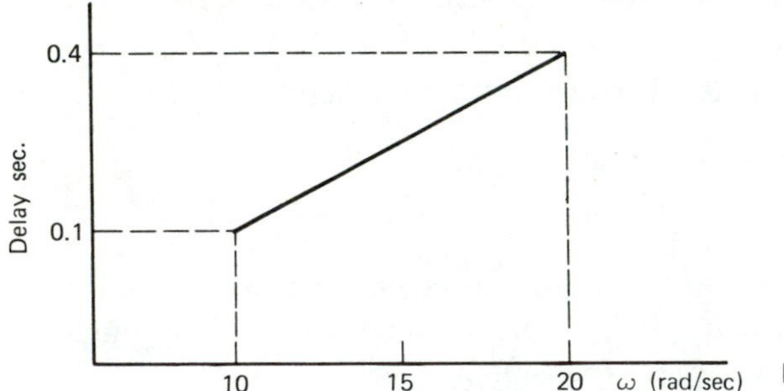

Figure P4.28

4.29 Use the MAG computer program (or Figure 4.20) to determine the a and b values of the delay equalizer section(s) needed to render the delay characteristic of Figure P4.28 flat to within ± 3 percent.

4.30 *Chebyshev high-pass, order.* Estimate the order of the Chebyshev approximation function needed to realize the high-pass requirement of Figure P4.30.

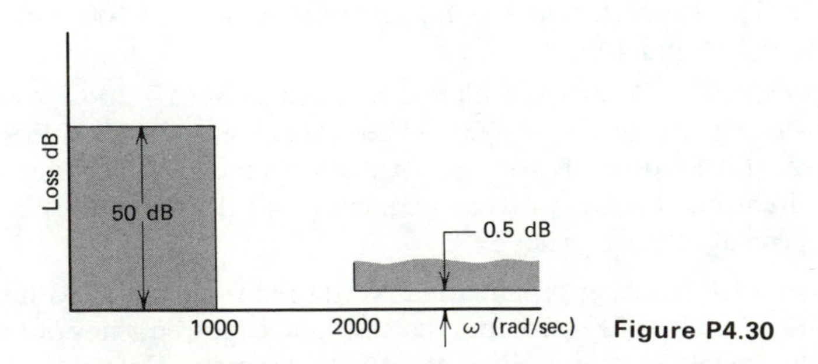

Figure P4.30

4.31 *Butterworth band-pass, order.* Estimate the order of the Butterworth approximation function needed to realize the symmetrical band-pass requirement of Figure P4.31.

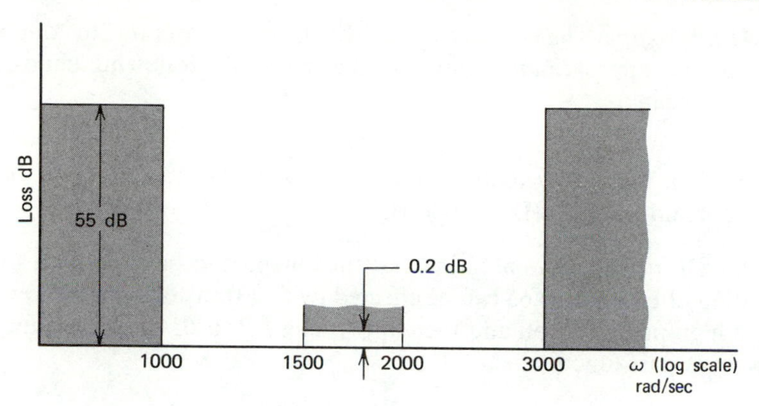

Figure P4.31

4.32 *Elliptic band-pass, order.* Repeat Problem 4.31 using the elliptic approximation.

4.33 *High-Pass approximation.* Find a Chebyshev approximation to satisfy the following high-pass requirements: $A_{max} = 0.5$ dB, $A_{min} = 20$ dB, $\omega_P =$ 3000 rad/sec, $\omega_S = 1000$ rad/sec.

4.34 *High-Pass poles.* Show that the *LP* to *HP* transformation leads to a set of high-pass poles that may be obtained by reflecting the normalized low-pass poles about a circle of radius ω_P.

4.35 *High-Pass approximation.* A high-pass Butterworth filter must have at least 45 dB of attenuation below 300 Hz, and the attenuation must be no more than 0.5 dB above 3000 Hz. Find the approximation function.

4.36 *Band-Pass approximation.* Find a Chebyshev approximation function for the following band-pass requirements:

$A_{max} = 0.5$ dB $\quad A_{min} = 15$ dB
passbands: 200 Hz to 400 Hz
stopbands: below 100 Hz and above 1000 Hz.

4.37 *CHEB program, band-pass.* Express the approximation function for Problem 4.36 as a product of biquadratics, by using the CHEB program.

4.38 *Band-Reject approximation.* Transform the following normalized low-pass function to a symmetrical band-reject function that has its center frequency at 1000 Hz and its low frequency passband edge at 100 Hz:

$$T_{LP}(S) = \frac{1}{S + 1}$$

4.39 *CHEB program, band-reject.* Use the CHEB program to obtain a Chebyshev approximation function that meets the following band-reject requirements:

$A_{max} = 0.2$ dB $A_{min} = 40$ dB
passbands: below 1000 Hz and above 6000 Hz
stopband: 2000 Hz to 3000 Hz.

4.40 *LP to BR transformation.* Show that the low-pass to band-reject transformation of Equation 4.65 can be effected by first transforming the low-pass to a high-pass function and then applying a *LP* to *BP* transformation on the high-pass function.

4.41 *Narrow-band LP to BP transformation.*

(a) Show that the *LP* to *BP* transformation of Equation 4.54 transforms a low-pass pole on the real axis to a pair of complex conjugate poles. Find the approximate location of the complex poles for the so-called narrow-band case when $\omega_0 \gg B$.

(b) For the narrow-band case, show that a low-pass pole at $-\Sigma - j\Omega$ transforms to an upper *s* plane pole at

$$-\frac{B}{2}\Sigma + j\left(\omega_0 - \frac{B}{2}\Omega\right)$$

4.42 Using the results of Problem 4.41, determine the band-pass function obtained by transforming a third-order normalized *LP* Butterworth approximation, given $\omega_0 = 1000$ rad/sec, $B = 100$ rad/sec, and $A_{max} = 3$ dB for the BP function. Plot the low-pass and band-pass pole-zero patterns and determine the pole *Q*'s for the low-pass and band-pass functions.

4.43 *Narrow-band LP to BR transformation.* Apply the *LP* to *BR* transformation of Equation 4.65 to plot the band-reject pole-zero locations obtained by transforming:

(a) A low-pass pole on the real axis.

(b) A pair of complex low-pass poles at $-\Sigma \pm j\Omega$ for the narrow-band case $\omega_0 \gg B$.

4.44 Use the results of Problem 4.43 to find a Chebyshev approximation for the symmetrical band-reject requirements:

$A_{max} = 0.25$ dB $A_{min} = 35$ dB
passband width = 100 rad/sec
stopband width = 25 rad/sec
center frequency $\omega_0 = 1000$ rad/sec.

Check the accuracy of the answer by comparing it with the exact band-reject function, obtained by using the CHEB program.

4.45 *Transitional Butterworth–Chebyshev.* The Transitional Butterworth–Chebyshev (TBC) approximation represented by

$$|H(j\Omega)|^2 = 1 + \varepsilon^2(\Omega)^{2k}C_{n-k}^2(\Omega) \qquad 0 \le k \le n$$

realizes a characteristic that is in between that of the Butterworth and the Chebyshev. The loss function reduces to the Butterworth for $k = n$, and to the Chebyshev for $k = 0$.

(a) Show that the TBC function provides $6(n - k - 1)$ dB more attenuation than the Butterworth for $\Omega \gg 1$.

(b) Compare the slopes of the Butterworth, Chebyshev, and TBC (for $k = 2$) at the passband edge frequency ($\Omega = 1$).

(c) Determine the number of derivatives of $|H(j\Omega)|^2$ that are zero at $\Omega = 0$ for a fourth-order Butterworth, Chebyshev, and the TBC ($k = 2$) approximations.

4.46 *Inverse Chebyshev.* The loss function for the Inverse Chebyshev approximation is given by

$$|H(j\omega)|^2 = 1 + \frac{1}{\varepsilon^2 C_n^2\left(\dfrac{1}{\omega}\right)}$$

where $C_n(\omega)$ is the nth-order Chebyshev polynomial. Show that:

(a) The loss function is maximally flat at the origin

(b) The loss function has an equiripple characteristic in the stopband. Find the frequencies of maxima and minima in the stopband.

(c) The minimum stopband loss is given by

$$20 \log_{10}\left(\frac{1 + \varepsilon^2}{\varepsilon^2}\right) \text{dB}$$

5. SENSITIVITY

In the preceding chapter we show how to obtain transfer functions that satisfy given filter requirements. The next step in the design process is the choice of a circuit and the determination of its element values—the synthesis step. As will be seen in the next few chapters on synthesis, one has a choice of many circuit realizations for a prescribed filter function. Given perfect components there would be little difference among the various realizations. In practice, real components will deviate from their nominal values due to the initial tolerances associated with their manufacture; the environmental effects of temperature and humidity; and chemical changes due to the aging of the components. As a consequence, the performance of the built filters will differ from the nominal design. One way to minimize this difference is to choose components with small manufacturing tolerances, and with low temperature, aging, and humidity coefficients. However, this approach will usually result in a circuit that is more expensive than is necessary. A more practical solution is to select a circuit that has a low **sensitivity** to these changes. The lower the sensitivity of the circuit, the less will its performance deviate because of element changes. Stated differently, the lower the sensitivity the less stringent will the requirements on the components be and, accordingly, the circuit becomes cheaper to manufacture. For this reason, sensitivity is one of the more important criteria used for comparing various circuit realizations. Since a good understanding of sensitivity is essential to the design of practical circuits, this subject is treated in this early chapter. More specifically, we will study the definitions of different kinds of sensitivities, and describe ways of evaluating the sensitivity of a circuit.

5.1 ω AND Q SENSITIVITY

In a qualitative sense, the sensitivity of a network is a measure of the degree of variation of its performance from nominal, due to changes in the elements constituting the network. As mentioned in Chapter 2, a biquadratic filter function can be expressed in terms of the parameters ω_p, ω_z, Q_p, Q_z, and K, as

$$T(s) = K \frac{s^2 + \dfrac{\omega_z}{Q_z} s + \omega_z^2}{s^2 + \dfrac{\omega_p}{Q_p} s + \omega_p^2} \tag{5.1}$$

In this section, we study the sensitivity of these biquadratic parameters to the elements and illustrate the evaluation of sensitivity by examples.

Let us first consider the sensitivity of the pole frequency ω_p to a change in a resistor R. Pole sensitivity is defined as the *per-unit* change in the pole frequency, $\Delta\omega_p/\omega_p$, caused by a per-unit change in the resistor, $\Delta R/R$. Mathematically

$$S_R^{\omega_p} = \lim_{\Delta R \to 0} \frac{\dfrac{\Delta\omega_p}{\omega_p}}{\dfrac{\Delta R}{R}} \tag{5.2}$$

$$= \frac{R}{\omega_p}\frac{\partial\omega_p}{\partial R} \tag{5.3}$$

This is equivalent to

$$S_R^{\omega_p} = \frac{\partial(\ln \omega_p)}{\partial(\ln R)} \tag{5.4}$$

Note that the cost of manufacturing a component is a function of the *percentage* change ($100 \times \Delta R/R$) rather than the absolute change (ΔR) of the component. For this reason it is desirable to measure sensitivity in terms of the *relative* changes in components, as is done in Equation 5.2.

The sensitivities of the parameters ω_z, Q_p, Q_z, and K to any element of the network are defined in a similar way:

$$S_C^{\omega_p} = \frac{C}{\omega_p}\frac{\partial\omega_p}{\partial C} \qquad S_R^{Q_p} = \frac{R}{Q_p}\frac{\partial Q_p}{\partial R} \qquad S_R^K = \frac{R}{K}\frac{\partial K}{\partial R} \quad \text{etc.} \tag{5.5}$$

Equation 5.4 can be used to develop some useful rules that simplify sensitivity calculations. The sensitivity of a parameter p to an element x is

$$S_x^p = \frac{x}{p}\frac{\partial p}{\partial x} = \frac{\partial(\ln p)}{\partial(\ln x)} \tag{5.6}$$

If p is not a function of x (e.g., $p = a$ constant), then

$$S_x^p = 0 \tag{5.7}$$

If $p = cx$, where c is a constant

$$S_x^{cx} = \frac{\partial(\ln cx)}{\partial(\ln x)} = \frac{\partial(\ln c)}{\partial(\ln x)} + \frac{\partial(\ln x)}{\partial(\ln x)} = 1 \tag{5.8}$$

Another useful relationship is

$$S_x^p = -S_x^{1/p} \tag{5.9}$$

This follows from Equation 5.4, since

$$-S_x^{1/p} = -\frac{\partial(\ln 1/p)}{\partial(\ln x)} = -\frac{\partial(-(\ln p))}{\partial(\ln x)} = S_x^p$$

In a similar way, we can show that

$$S_x^p = -S_{1/x}^p \tag{5.10}$$

Other useful relationships that can easily be proved are:

$$S_x^{p_1 p_2} = S_x^{p_1} + S_x^{p_2} \tag{5.11a}$$

$$S_x^{p_1/p_2} = S_x^{p_1} - S_x^{p_2} \tag{5.11b}$$

$$S_{x^n}^p = \frac{1}{n} S_x^p \tag{5.11c}$$

$$S_x^{p^n} = nS_x^p \tag{5.11d}$$

$$S_x^{p_1+p_2} = \frac{p_1 S_x^{p_1} + p_2 S_x^{p_2}}{p_1 + p_2} \tag{5.11e}$$

$$S_x^{cf(x)} = S_x^{f(x)} \tag{5.11f}$$

where c is independent of x, and $f(x)$ is a function of x.

The evaluation of sensitivity using these relationships is illustrated in the following examples.

Example 5.1

The transfer function V_O/I_{IN} for the passive circuit of Figure 5.1 can be shown to be

$$\frac{V_O}{I_{IN}} = \frac{1}{C} \frac{s}{s^2 + \dfrac{s}{RC} + \dfrac{1}{LC}} \tag{5.12}$$

Compute the sensitivites of K, ω_p, and Q_p to the passive elements.

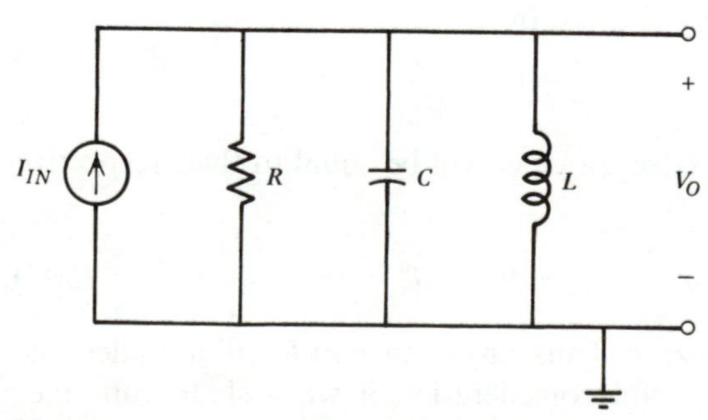

Figure 5.1 Passive circuit for Example 5.1.

Solution

From Equation 5.1, the biquadratic parameters are:

$$K = \frac{1}{C} \qquad \omega_p = \frac{1}{\sqrt{LC}} \qquad Q_p = R\sqrt{\frac{C}{L}}$$

Using the sensitivity relationships developed in the above section:

$$S_C^K = S_C^{1/C} = -S_C^C = -1$$
$$S_L^{\omega_p} = S_L^{1/\sqrt{LC}} = -S_L^{\sqrt{LC}} = -\tfrac{1}{2}S_L^{LC} = -\tfrac{1}{2}$$
$$S_C^{\omega_p} = -S_C^{\sqrt{LC}} = -\tfrac{1}{2}$$
$$S_R^{Q_p} = S_R^{R\sqrt{C/L}} = 1$$
$$S_C^{Q_p} = \tfrac{1}{2}S_C^{R^2C/L} = \tfrac{1}{2}$$
$$S_L^{Q_p} = \tfrac{1}{2}S_L^{R^2C/L} = -\tfrac{1}{2}S_L^{L/R^2C} = -\tfrac{1}{2}$$

All the other sensitivities are zero.

Observations

1. The magnitudes of the sensitivities are all less than one. A sensitivity of one implies that a 1 percent change in the element will cause a 1 percent change in the parameter (Equation 5.2). This is considered a low sensitivity. In Chapter 6 we see that, in general, passive ladder structures can always be designed to have low sensitivities.
2. Note that the sensitivities evaluated are equal to the exponent of the element. For instance, in the expression for pole Q

   $$Q_p = R^1C^{1/2}L^{-1/2}$$

   the sensitivities are

   $$S_R^{Q_p} = 1 \qquad S_C^{Q_p} = \tfrac{1}{2} \qquad S_L^{Q_p} = -\tfrac{1}{2}$$

   In general, if the parameter p is given by

   $$p = x_1^a x_2^b x_3^c$$

   then the sensitivity of p to x_1, x_2, and x_3 will be equal to their respective exponents, that is

   $$S_{x_1}^p = a \qquad S_{x_2}^p = b \qquad S_{x_3}^p = c \tag{5.13}$$

3. The sensitivity of ω_p to R is zero. Thus, any change in R will not affect the pole frequency. This is a useful consideration if we wish to tune (i.e., adjust) Q_p without affecting ω_p. ■

5.2 MULTI-ELEMENT DEVIATIONS

In the last section, we obtained an expression for the change in a biquadratic parameter due to a change in a particular circuit element. For instance, the change in a resistance causes the pole frequency to change by (Equation 5.2)

$$\Delta\omega_p = \lim_{\Delta R \to 0} S_R^{\omega_p} \frac{\Delta R}{R} \omega_p \tag{5.14}$$

For small deviations in R

$$\Delta\omega_p \cong S_R^{\omega_p} \frac{\Delta R}{R} \omega_p \tag{5.15}$$

This is the change due to one element. In general we will be interested in the change due to the simultaneous variation of all the elements in the circuit, as discussed in this section.

Consider, for example, the change in ω_p due to deviations of all the circuit elements x_j (where the elements can be resistors, capacitors, inductors, or the parameters describing the active device). The change $\Delta\omega_p$ may be obtained by expanding it in a Taylor series, as

$$\Delta\omega_p = \frac{\partial\omega_p}{\partial x_1}\Delta x_1 + \frac{\partial\omega_p}{\partial x_2}\Delta x_2 + \cdots \frac{\partial\omega_p}{\partial x_m}\Delta x_m$$
$$+ \text{ second- and higher-order terms}$$

where m is the total number of elements in the circuit. Since the changes in the components Δx_j are assumed to be small, the second- and higher-order terms can be ignored. Thus

$$\Delta\omega_p \cong \sum_{j=1}^{m} \frac{\partial\omega_p}{\partial x_j}\Delta x_j \tag{5.16}$$

To bring the sensitivity term into evidence, (5.16) may be written as

$$\Delta\omega_p \cong \sum_{j=1}^{m}\left(\frac{\partial\omega_p}{\partial x_j}\frac{x_j}{\omega_p}\right)\left(\frac{\Delta x_j}{x_j}\right)\omega_p$$
$$= \sum_{j=1}^{m} S_{x_j}^{\omega_p} V_{x_j}\omega_p \tag{5.17}$$

where $V_{x_j} = \Delta x_j/x_j$ is the per-unit change in the element x_j, and is known as the **variability** of x. From (5.17) the per-unit change in ω_p is

$$\frac{\Delta\omega_p}{\omega_p} = \sum_{j=1}^{m} S_{x_j}^{\omega_p} V_{x_j} \tag{5.18}$$

Similarly the per-unit changes in pole Q, ω_z, Q_z, and K, due to the simultaneous deviations of all the components, are given by

$$\frac{\Delta Q_p}{Q_p} = \sum_{j=1}^{m} S_{x_j}^{Q_p} V_{x_j} \qquad \frac{\Delta K}{K} = \sum_{j=1}^{m} S_{x_j}^{K} V_{x_j}$$
$$\frac{\Delta \omega_z}{\omega_z} = \sum_{j=1}^{m} S_{x_j}^{\omega_z} V_{x_j} \qquad \frac{\Delta Q_z}{Q_z} = \sum_{j=1}^{m} S_{x_j}^{Q_z} V_{x_j} \tag{5.19}$$

Example 5.2

The active RC circuit shown in Figure 5.2 realizes a second-order high-pass function. Find its transfer function V_O/V_{IN} and derive expressions for the sensitivity of ω_p and Q_p to elements R_1, R_2, C_1, C_2, and the amplifier gain A.

Solution

The nodal equations for the circuit are:

node 1:

$$V^-\left(\frac{k-1}{R_A} + \frac{1}{R_A}\right) - V_O\left(\frac{1}{R_A}\right) = 0$$

node 2:

$$V^+\left(sC_1 + \frac{1}{R_1} + \frac{1}{R_2 + \dfrac{1}{sC_2}}\right) - V_O\left(\frac{1}{R_2 + \dfrac{1}{sC_2}}\right) = V_{IN}sC_1$$

The positive and negative terminal voltages of the op amp are related by

$$V_O = (V^+ - V^-)A$$

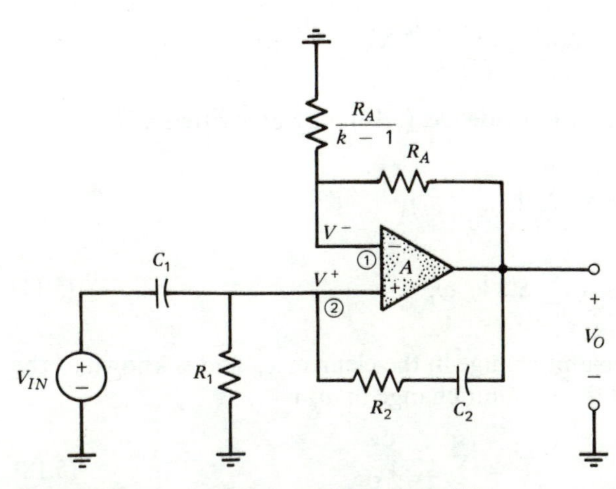

Figure 5.2 Active RC circuit for Example 5.3.

Solving these equations, the transfer function is found to be:

$$\frac{V_O}{V_{IN}} = \frac{s(s + 1/R_2C_2)k(1 + k/A)^{-1}}{s^2 + s\left(\dfrac{1}{R_1C_1} + \dfrac{1}{R_2C_2} + \dfrac{1}{R_2C_1}\left(1 - \dfrac{k}{1 + k/A}\right)\right) + \dfrac{1}{R_1R_2C_1C_2}} \tag{5.20}$$

Comparing this equation with Equation 5.1, the biquadratic parameters are seen to be:

$$\omega_p = R_1^{-1/2}R_2^{-1/2}C_1^{-1/2}C_2^{-1/2} \tag{5.21}$$

$$(\text{bw})_p = \frac{1}{R_1C_1} + \frac{1}{R_2C_2} + \frac{1}{R_2C_1}\left(1 - \frac{k}{1 + k/A}\right) \tag{5.22a}$$

$$Q_p = \frac{\omega_p}{(\text{bw})_p} \tag{5.22b}$$

The sensitivity of ω_p to the components R_1, R_2, C_1, and C_2 are equal to their exponents, that is,

$$S_{R_1}^{\omega_p} = S_{R_2}^{\omega_p} = S_{C_1}^{\omega_p} = S_{C_2}^{\omega_p} = -\tfrac{1}{2}$$

The sensitivity of Q_p to the components are evaluated as follows:

$$S_{R_1}^{Q_p} = S_{R_1}^{\omega_p/(\text{bw})_p} = S_{R_1}^{\omega_p} - S_{R_1}^{(\text{bw})_p}$$

From (5.21) and (5.22), using the relationship (5.11e):

$$S_x^{p_1 + p_2} = \frac{p_1 S_x^{p_1} + p_2 S_x^{p_2}}{p_1 + p_2}$$

we get:

$$S_{R_1}^{Q_p} = -\frac{1}{2} - \left(\frac{\dfrac{1}{R_1C_1}(-1) + \left(\dfrac{1}{R_2C_2} + \dfrac{1}{R_2C_1}\left(1 - \dfrac{k}{1 + k/A}\right)\right)(0)}{\dfrac{1}{R_1C_1} + \dfrac{1}{R_2C_2} + \dfrac{1}{R_2C_1}\left(1 - \dfrac{k}{1 + k/A}\right)}\right)$$

$$= -\frac{1}{2} + \frac{1}{R_1C_1} \cdot \frac{1}{(\text{bw})_p}$$

Similarly:

$$S_{R_2}^{Q_p} = -\frac{1}{2} + \frac{1}{(\text{bw})_p}\left[\frac{1}{R_2C_2} + \frac{1}{R_2C_1}\left(1 - \frac{k}{1 + k/A}\right)\right]$$

$$S_{C_1}^{Q_p} = -\frac{1}{2} + \frac{1}{(\text{bw})_p}\left[\frac{1}{R_1C_1} + \frac{1}{R_2C_1}\left(1 - \frac{k}{1 + k/A}\right)\right]$$

$$S_{C_2}^{Q_p} = -\frac{1}{2} + \frac{1}{R_2C_2} \cdot \frac{1}{(\text{bw})_p}$$

Next, considering the sensitivities to the amplifier gain:

$$S_A^{\omega_p} = 0$$

$$S_A^{Q_p} = -S_A^{(\text{bw})_p} = \frac{-1}{(\text{bw})_p}\left\{\frac{1}{R_2C_1}\left[\frac{-k}{1+k/A}\right]S_A^{-k/R_2C_1(1+k/A)}\right\}$$

$$= \frac{1}{(\text{bw})_p}\cdot\frac{1}{R_2C_1}\cdot\frac{k}{1+k/A}\cdot(-1)S_A^{(1+k/A)R_2C_1/k}$$

$$= \frac{1}{(\text{bw})_p}\cdot\frac{1}{R_2C_1}\cdot\frac{k}{1+k/A}\cdot(-1)\cdot\frac{\frac{k}{A}\cdot\frac{R_2C_1}{k}(-1)}{(1+k/A)R_2C_1/k}$$

$$= \frac{1}{A}\cdot\frac{1}{R_2C_1}\cdot\frac{k^2}{\left(1+\frac{k}{A}\right)^2}\cdot\frac{1}{(\text{bw})_p} \tag{5.23}$$

Observations

1. The expression for $S_A^{Q_p}$ suggests that this term can be reduced by increasing the amplifier gain. In particular, for an ideal op amp this sensitivity term becomes zero, that is, the pole Q becomes insensitive to the op amp gain.
2. Note that k depends on the resistors R_A and R_B ($k = 1 + R_A/R_B$), and is *not* a constant. The sensitivities of ω_p and Q_p to R_A/R_B can be computed by first evaluating the sensitivities to k (see Problem 5.6).
3. The sensitivities of ω_p and Q_p are related to their respective dimensions, as shown in the following. From the example, we see that

$$S_{R_1}^{Q_p} + S_{R_2}^{Q_p} = 0 \qquad S_{C_1}^{Q_p} + S_{C_2}^{Q_p} = 0 \tag{5.24a}$$

and

$$S_{R_1}^{\omega_p} + S_{R_2}^{\omega_p} = -1 \qquad S_{C_1}^{\omega_p} + S_{C_2}^{\omega_p} = -1 \tag{5.24b}$$

Now the pole frequency ω_p, which is of the form $1/RC$, has the dimensions

$$\dim(\omega_p) = [R]^{-1}[C]^{-1}$$

while the pole Q, given by $\omega_p/(\text{bw})_p$, is a dimensionless quantity. In other words

$$\dim(Q_p) = [R]^0[C]^0$$

Therefore, in this example it is seen that the summed sensitivity of a parameter to all the resistors (or capacitors) is equal to the dimension of resistance (or capacitance) for the parameter. This property can be shown to hold for all active RC networks. The general form of this so-called

dimensional homogeneity property* is

$$\sum S_{R_i}^{Q_p} = \sum S_{C_i}^{Q_p} = 0 \tag{5.25a}$$

$$\sum S_{R_i}^{\omega_p} = \sum S_{C_i}^{\omega_p} = -1 \tag{5.25b}$$

where the summation is over all the resistors (capacitors). ■

Example 5.3

Using the results of Example 5.2:

(a) Find the expressions for the per-unit change in ω_p and Q_p, given that the variability of every passive component is 0.01, and that of the amplifier gain is 0.5.

(b) Evaluate the per-unit change in ω_p and Q_p for the special case:

$$R_1 = R_2 = R \qquad C_1 = C_2 = C$$

$$Q_p = 20 \qquad \omega_p = 2\pi 10^4$$

$$A = 1000 \text{ (at 10 kHz)}$$

Solution

(a) The variabilities of the circuit element† are given to be

$$V_R = 0.01 \qquad V_C = 0.01 \qquad V_A = 0.5$$

Substituting in (5.18)

$$\frac{\Delta\omega_p}{\omega_p} = 0.01(S_{R_1}^{\omega_p} + S_{R_2}^{\omega_p} + S_{C_1}^{\omega_p} + S_{C_2}^{\omega_p}) + 0.5(S_A^{\omega_p})$$

Using Equation 5.25*b*, and recalling that $S_A^{\omega_p} = 0$, this reduces to

$$\frac{\Delta\omega_p}{\omega_p} = 0.01(-2) + 0.5(0) = -0.02$$

Similarly, from Equation 5.19, 5.23, and 5.25*a*:

$$\frac{\Delta Q_p}{Q_p} = 0.5S_A^{Q_p} = \frac{0.5}{AR_2C_1}\frac{k^2}{\left(1+\frac{k}{A}\right)^2}\frac{1}{(\text{bw})_p} \tag{5.26}$$

* This property follows from Euler's formula for homogeneous functions, [3] (page 222).

† In practice the variabilities are random numbers described by a mean and a standard deviation, as will be explained in Section 5.4.3.

(b) For the special case, from Equation 5.20

$$(\mathrm{bw})_p = \frac{1}{RC}\left(3 - \frac{k}{1 + \dfrac{k}{1000}}\right) \tag{5.27}$$

$$\omega_p = \frac{1}{RC} = 2\pi 10^4 \tag{5.28}$$

Dividing (5.28) by (5.27)

$$Q_p = \frac{1}{3 - \dfrac{k}{1 + \dfrac{k}{1000}}} = 20$$

which yields $k = 2.96$.
From (5.26)

$$\frac{\Delta Q_p}{Q_p} = \frac{0.5}{A} \frac{k^2}{\left(1 + \dfrac{k}{A}\right)^2} \frac{\omega_p}{(\mathrm{bw})_p} \tag{5.29}$$

Substituting for k, A, and Q_p in (5.29)

$$\frac{\Delta Q_p}{Q_p} = \frac{0.5}{1000} \frac{(2.96)^2}{(1 + 0.00296)^2} 20 = 0.087$$

which corresponds to an 8.7 percent increase in the pole Q.

Observation

The ω_p sensitivities are as low ($= 1/2$) as those observed for the passive network in Example 5.1; also, the contribution of the passive components to the pole Q is zero. However, the given variation in the amplifier gain causes the pole Q to change by 8.7 percent, which is far from negligible. We will see later that one of the major problems associated with active RC filters is the sensitivity to amplifier gain, and it will therefore be necessary to study methods for reducing this effect. ■

5.3 GAIN SENSITIVITY

Thus far, the effect of element deviations on the biquadratic parameters ω_p, Q_p, ω_z, Q_z, and K have been considered. However, filter requirements are usually stated in terms of the maximum allowable deviation in *gain* over specified bands of frequencies. In this section we show how this gain deviation is related to the biquadratic parameter sensitivities;* furthermore, we also suggest ways of adapting the design process to minimize the gain deviation.

Let us assume that the filter function has been factored into biquadratics, as

$$T(s) = \prod_{i=1}^{N} K_i \frac{s^2 + \dfrac{\omega_{z_i}}{Q_{z_i}} s + \omega_{z_i}^2}{s^2 + \dfrac{\omega_{p_i}}{Q_{p_i}} s + \omega_{p_i}^2} \tag{5.30}$$

The gain in dB is given by

$$\begin{aligned} G(\omega) &= 20 \log_{10} |T(j\omega)| \\ &= \sum_{i=1}^{N} 20 \log_{10} \left| s^2 + \frac{\omega_{z_i}}{Q_{z_i}} s + \omega_{z_i}^2 \right|_{s=j\omega} \\ &\quad - \sum_{i=1}^{N} 20 \log_{10} \left| s^2 + \frac{\omega_{p_i}}{Q_{p_i}} s + \omega_{p_i}^2 \right|_{s=j\omega} + 20 \log_{10} |K_i| \end{aligned} \tag{5.31}$$

Gain sensitivity is defined as the change in gain in dB† due to a per-unit change in an element (or parameter) x:

$$\begin{aligned} \mathcal{S}_x^{G(\omega)} &= \frac{\partial G(\omega)}{\dfrac{\partial x}{x}} \\ &= x \frac{\partial G(\omega)}{\partial x} \text{ dB} \end{aligned} \tag{5.32}$$

From this equation

$$\Delta G(\omega) = \lim_{\Delta x \to 0} \mathcal{S}_x^{G(\omega)} \frac{\Delta x}{x}$$

and for small changes in x

$$\Delta G(\omega) \cong \mathcal{S}_x^{G(\omega)} \frac{\Delta x}{x} \tag{5.33}$$

We are interested in the change in gain $\Delta G(\omega)$ (hereafter abbreviated as ΔG), due to the element variabilities V_{x_i}. Since the gain function is the sum of similar

* The approach presented is based on Hiberman [1].

† Note that

$$\mathcal{S}_x^{G(\omega)} = \frac{\partial G(\omega)}{\partial (\ln x)} = \frac{\partial (20 \log_{10} |T(j\omega)|)}{\partial (\ln x)} = 8.686 \frac{\partial (\ln |T(j\omega)|)}{\partial (\ln x)} = 8.686 S_x^{|T(j\omega)|}.$$

Therefore, the sensitivity of the gain function $S_x^{G(\omega)}$, defined in Equation 5.32, is proportional to the per-unit sensitivity (Equation 5.2) of the magnitude function, $S_x^{|T(j\omega)|}$.

second-order functions, it is only necessary to analyze one typical second-order function, the results of which can easily be extended to the summed expression of Equation 5.31. Let us consider the contribution to the gain deviation of the second-order numerator term

$$T(s) = s^2 + \frac{\omega_z}{Q_z} s + \omega_z^2 \tag{5.34}$$

The corresponding gain is

$$G(\omega) = 20 \log_{10} \left| s^2 + \frac{\omega_z}{Q_z} s + \omega_z^2 \right|_{s=j\omega} \tag{5.35}$$

Since $G(\omega)$ is a function of the variables ω_z and Q_z, a Taylor series expansion of ΔG will have the form:

$$\Delta G = \frac{\partial G}{\partial Q_z} \Delta Q_z + \frac{\partial G}{\partial \omega_z} \Delta\omega_z + \frac{\partial^2 G}{\partial Q_z^2} (\Delta Q_z)^2 + \frac{\partial^2 G}{\partial \omega_z^2} (\Delta\omega_z)^2$$
$$+ \frac{2\partial^2 G}{\partial \omega_z \partial Q_z} (\Delta Q_z \Delta\omega_z) + \cdots$$

For small changes in the components, the corresponding changes in Q_z and ω_z will be small, so that the second- and higher-order terms can be ignored. Then

$$\Delta G \cong \frac{\partial G}{\partial Q_z} \Delta Q_z + \frac{\partial G}{\partial \omega_z} \Delta\omega_z \tag{5.36}$$

Substituting for $\Delta\omega_z$ and ΔQ_z from (5.18) and (5.19), respectively, we get

$$\Delta G \cong \sum_{j=1}^{m} \left[Q_z \frac{\partial G}{\partial Q_z} S_{x_j}^{Q_z} V_{x_j} + \omega_z \frac{\partial G}{\partial \omega_z} S_{x_j}^{\omega_z} V_{x_j} \right] \tag{5.37}$$

From the definition of gain sensitivity this expression reduces to

$$\boxed{\Delta G = \sum_{j=1}^{m} (\mathcal{S}_{Q_z}^{G} S_{x_j}^{Q_z} V_{x_j} + \mathcal{S}_{\omega_z}^{G} S_{x_j}^{\omega_z} V_{x_j})} \tag{5.38}$$

This equation gives the change in the gain in dB due to the simultaneous variation in all the elements realizing the second-order function:

$$s^2 + \frac{\omega_z}{Q_z} s + \omega_z^2$$

The gain change for the complete transfer function of Equation 5.31 is obtained by adding the gain contributions of each of the second-order functions.

Let us study the terms in the summation of Equation 5.38. The component sensitivity terms, $S_{x_j}^{\omega_z}$ and $S_{x_j}^{Q_z}$, are evaluated as in Section 5.1, by analyzing the circuit. The gain deviation due to these terms will depend on the *circuit* chosen for the realization of the approximation function. The variability terms V_{x_j} were defined as the per-unit change in the components. These terms will depend on the *types of components* chosen for the circuit realization. The two remaining terms in Equation 5.38, namely $\mathcal{S}_{\omega_z}^{G}$ and $\mathcal{S}_{Q_z}^{G}$, are new. As shown in Section 5.4.1, these terms depend on the ω and Q parameters of the corresponding biquadratic. These biquadratic parameters are determined by the *approximation function*, which is derived from the given filter requirements.

Summarizing the above observations, the anticipated gain variation is affected by:

- The approximation function.
- The choice of circuit topology.
- The types of components used in the realization.

Each of these factors is discussed in detail in the following sections.

5.4 FACTORS AFFECTING GAIN SENSITIVITY

In this section we discuss the three major factors affecting the gain deviations—the approximation function, the circuit, and the components.

5.4.1 CONTRIBUTION OF THE APPROXIMATION FUNCTION

As in the last section, suppose the approximation function has been expressed as a product of biquadratics, each term being of the form:

$$T(s) = K \frac{s^2 + \dfrac{\omega_z}{Q_z} s + \omega_z^2}{s^2 + \dfrac{\omega_p}{Q_p} s + \omega_p^2} \tag{5.39}$$

The corresponding gain in dB is

$$G(\omega) = 20 \log_{10} \left| s^2 + \frac{\omega_z}{Q_z} s + \omega_z^2 \right|_{s=j\omega} - 20 \log_{10} \left| s^2 + \frac{\omega_p}{Q_p} s + \omega_p^2 \right|_{s=j\omega} + 20 \log_{10} |K| \tag{5.40}$$

The biquadratic parameters describing the approximation function contribute to the gain deviation expression (Equation 5.38) via the **biquadratic parameter sensitivity** terms:

$$\mathcal{S}^G_{\omega_z} \quad \mathcal{S}^G_{Q_z} \quad \mathcal{S}^G_{\omega_p} \quad \mathcal{S}^G_{Q_p} \quad \mathcal{S}^G_K$$

These sensitivities can be evaluated from the definition of gain sensitivity (5.32). For instance, consider the $\mathcal{S}^G_{\omega_z}$ term:

$$\begin{aligned}
\mathcal{S}^G_{\omega_z} &= \omega_z \frac{\partial G}{\partial \omega_z} = \omega_z \frac{\partial}{\partial \omega_z}\left(20 \log_{10}\left|-\omega^2 + \frac{\omega_z}{Q_z} j\omega + \omega_z^2\right|\right) \\
&= \omega_z \frac{\partial}{\partial \omega_z}\left\{10 \log_{10}\left[(-\omega^2 + \omega_z^2)^2 + \left(\frac{\omega_z \omega}{Q_z}\right)^2\right]\right\} \\
&= \omega_z \frac{\partial}{\partial \omega_z}\left\{\frac{10}{\ln(10)} \ln\left[(-\omega^2 + \omega_z^2)^2 + \left(\frac{\omega_z \omega}{Q_z}\right)^2\right]\right\} \\
&= \omega_z \frac{10}{\ln(10)} \frac{2(\omega_z^2 - \omega^2)2\omega_z + \dfrac{2\omega_z \omega^2}{Q_z^2}}{(\omega_z^2 - \omega^2)^2 + \left(\dfrac{\omega_z \omega}{Q_z}\right)^2}
\end{aligned}$$

$$\mathcal{S}^G_{\omega_z} = 8.686 \frac{2(1 - \Omega_z^2) + \left(\dfrac{\Omega_z}{Q_z}\right)^2}{(1 - \Omega_z^2)^2 + \left(\dfrac{\Omega_z}{Q_z}\right)^2} \text{ dB} \tag{5.41}$$

where Ω_z is the normalized frequency

$$\Omega_z = \frac{\omega}{\omega_z}$$

In a similar manner it can be shown that

$$\mathcal{S}^G_{Q_z} = -8.686 \frac{\left(\dfrac{\Omega_z}{Q_z}\right)^2}{(1 - \Omega_z^2)^2 + \left(\dfrac{\Omega_z}{Q_z}\right)^2} \text{ dB} \tag{5.42}$$

From Equation 5.32, $\mathcal{S}^G_K$ is given by

$$\mathcal{S}^G_K = \frac{\partial(20 \log_{10} K)}{\partial(\ln K)} = 8.686 \text{ dB} \tag{5.43}$$

Since the denominator of Equation 5.39 has the same form as the numerator, it can be seen that the $\mathcal{S}^G_{\omega_p}$ and $\mathcal{S}^G_{Q_p}$ terms will be the negatives of the $\mathcal{S}^G_{\omega_z}$ and

$\mathscr{S}^G_{Q_z}$ terms, respectively:

$$\mathscr{S}^G_{\omega_p} = -8.686 \frac{2(1-\Omega_p^2) + \left(\dfrac{\Omega_p}{Q_p}\right)^2}{(1-\Omega_p^2)^2 + \left(\dfrac{\Omega_p}{Q_p}\right)^2} \text{ dB} \tag{5.44}$$

$$\mathscr{S}^G_{Q_p} = 8.686 \frac{\left(\dfrac{\Omega_p}{Q_p}\right)^2}{(1-\Omega_p^2)^2 + \left(\dfrac{\Omega_p}{Q_p}\right)^2} \text{ dB} \tag{5.45}$$

where

$$\Omega_p = \frac{\omega}{\omega_p}$$

From Equations 5.42 to 5.45 it is seen that at a given frequency, the biquadratic parameter sensitivities depend on the pole Q and the zero Q. These are parameters that depend only on the approximation function which, as we well know, is determined by the filter requirements.

To see the relative magnitudes of the biquadratic parameter sensitivities, the terms $\mathscr{S}^G_{\omega_p}$ and $\mathscr{S}^G_{Q_p}$ are plotted against the normalized frequency Ω_p in Figure 5.3. Some important and easy to remember points on these sensitivity curves are:

(1) At the pole frequency $\omega = \omega_p$, or $\Omega_p = 1$, so

$$\mathscr{S}^G_{\omega_p} = -8.686 \text{ dB} \qquad \mathscr{S}^G_{Q_p} = 8.686 \text{ dB}$$

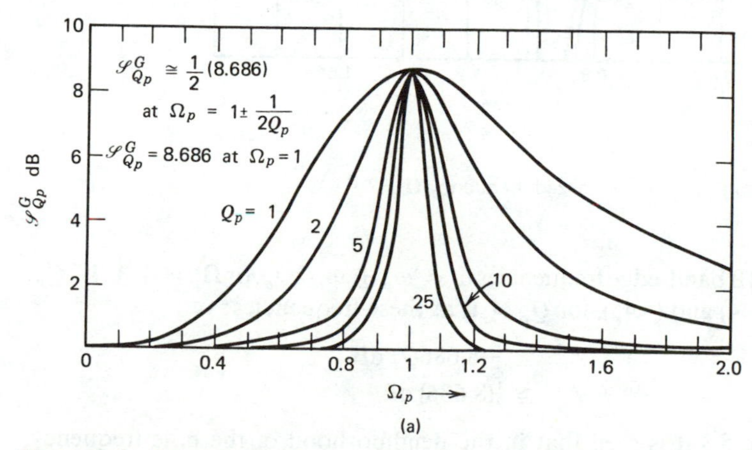

Figure 5.3 Plots of (*a*) $\mathscr{S}^G_{Q_p}$

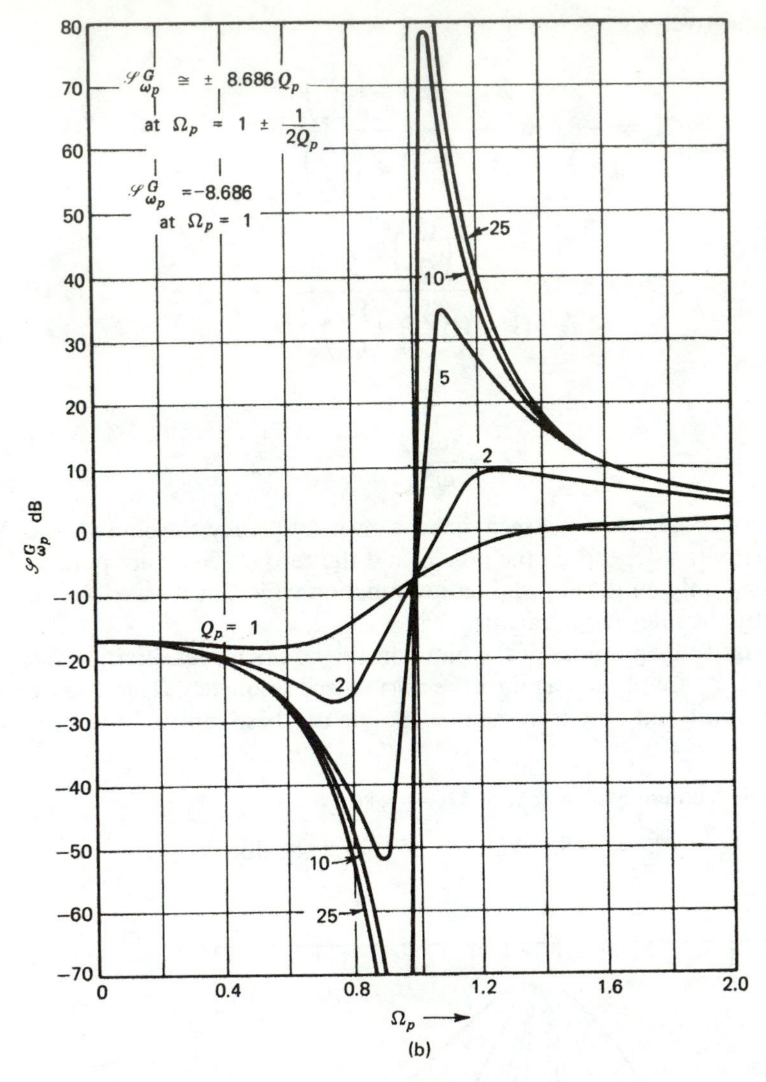

Figure 5.3 *(b)* $\mathcal{S}^G_{\omega_p}$ versus normalized frequency Ω.

(2) At the 3 dB band-edge frequencies $\omega = \omega_p \pm \omega_p/2Q_p$, or $\Omega_p = 1 \pm 1/2Q_p$. From (5.44) and (5.45), for $Q_p \gg 1$, at these frequencies:

$$\mathcal{S}^G_{\omega_p} \cong \pm 8.686 Q_p \text{ dB}$$
$$\mathcal{S}^G_{Q_p} \cong \tfrac{1}{2}(8.686) \text{ dB}$$

From Figure 5.3 it is seen that in the neighborhood of the pole frequency, the dominant sensitivity term is $\mathcal{S}^G_{\omega_p}$. This term increases with the pole Q, achieving its maximum value of approximately $\pm 8.686 Q_p$ at the 3 dB band-edge frequencies. From approximation theory, the pole Q increases with the slope of the filter characteristic in the transition band; therefore, the passband sensitivity of a network will increase as this slope increases.

Summarizing the above results:

- At a given frequency, the biquadratic parameter sensitivities depend only on the approximation function.
- The sensitivity in the passband increases with the pole Q.

Example 5.4

An active RC network is known to have the following transfer function:

$$T(s) = \frac{s^2 + \dfrac{1}{R_1 C_1 R_2 C_2}}{s^2 + \dfrac{1}{R_3 C_3} s + \dfrac{1}{R_1 C_1 R_4 C_4}}$$

The network is used to realize the band-reject filter function:

$$T'(s) = \frac{s^2 + 144}{s^2 + 0.8s + 16}$$

If every resistor and capacitor increases by one percent,* find

(a) The biquadratic parameter sensitivities at $\omega = 3.6$ rad/sec.
(b) The component sensitivities.
(c) The gain deviation at $\omega = 3.6$ rad/sec.

Solution

(a) From Equation 5.1, the biquadratic parameters are;

$$\omega_p = 4 \qquad Q_p = 5 \qquad \omega_z = 12 \qquad Q_z = \infty$$

At $\omega = 3.6$

$$\Omega_p = \frac{3.6}{4} = 0.9 \qquad \Omega_z = \frac{3.6}{12} = 0.3$$

From Equations 5.41 to 5.45, or from Figure 5.3:

$$\mathscr{S}^G_{\omega_p} = -52.3 \qquad \mathscr{S}^G_{Q_p} = 4.11 \qquad \mathscr{S}^G_{\omega_z} = 19.1 \qquad \mathscr{S}^G_{Q_z} \cong 0$$

* The components are assumed to have the same deviation only to keep the problem simple. In practice, component deviations are random, as will be discussed in Section 5.4.3.

(b) The sensitivities of the biquadratic parameters to the elements are*

$$S^{\omega_p}_{R_1, C_1, R_4, C_4} = -\tfrac{1}{2}$$
$$S^{Q_p}_{R_3, C_3} = 1$$
$$S^{Q_p}_{R_1, C_1, R_4, C_4} = -\tfrac{1}{2}$$
$$S^{\omega_z}_{R_1, C_1, R_2, C_2} = -\tfrac{1}{2}$$
$$S^{Q_z}_{R_i, C_i} = 0$$

(c) The change in gain, from Equation 5.38, is given by

$$\Delta G = \mathscr{S}^G_{\omega_p} V_x \sum_j S^{\omega_p}_{xj} + \mathscr{S}^G_{Q_p} V_x \sum_j S^{Q_p}_{xj} + \mathscr{S}^G_{\omega_z} V_x \sum_j S^{\omega_z}_{xj} + \mathscr{S}^G_{Q_z} V_x \sum_j S^{Q_z}_{xj}$$

The sensitivities of the gain to the biquadratic parameters were obtained in part a. From part b, or else from Equations 5.25, the sums of the sensitivities of the biquadratic parameters to the passive elements x_j are:

$$\sum_j S^{\omega_p}_{xj} = -2 \qquad \sum_j S^{Q_p}_{xj} = 0$$
$$\sum_j S^{\omega_z}_{xj} = -2 \qquad \sum_j S^{Q_z}_{xj} = 0$$

The variability V_x for the components is given to be 0.01. Substituting these values in the above expression for ΔG, we finally get

$$\begin{aligned}\Delta G &= -52.3(-2)(0.01) + 4.11(0.0)(0.01) + 19.1(-2)(0.01) \\ &\quad + 0.0(0.0)(0.01) \\ &= 0.664 \text{ dB}\end{aligned}$$

■

5.4.2 CHOICE OF THE CIRCUIT

Once the approximation function has been obtained, the next step is to realize it using active RC circuits. This is the synthesis step. As will be shown in the next few chapters, there are several possible circuits that can synthesize a given function. It will also be seen that the sensitivity of the biquadratic parameters to the components will be quite different for the different circuits. Needless to say, the circuits with the lower **component sensitivities** will be the more desirable ones.

Another important observation from Equation 5.38 is that the gain deviation depends on the number of elements, m, used in the circuit realization of the approximation function. The larger the number of elements, the more terms will be present in the summation and, in general, the larger will the gain deviation be.

* $S^P_{x_1, x_2} = c$ is an abbreviated notation used to represent the two equations $S^P_{x_1} = c$ and $S^P_{x_2} = c$.

Summarizing, the gain deviation increases with:

- The component sensitivities.
- The number of components used to synthesize the given function.

5.4.3 CHOICE OF COMPONENT TYPES

After the filter has been designed we will need to choose the types of resistors, capacitors, and op amps to be used in the manufacture of the circuit. Practical elements deviate from their nominal values due to manufacturing tolerances, temperature and humidity changes, and due to chemical changes that occur with the aging of the elements. In this section we show how the deviation in gain is related to these element deviations. To simplify the presentation, deviations in the resistors and capacitors only are considered, the op amp being assumed to be ideal.

Manufacture Tolerance

Due to the production process, resistor and capacitor values are spread about their nominal value. For a resistor, the production tolerance is represented by

$$R = R_0(1 + \gamma_R) \tag{5.46}$$

where R_0 is the nominal value, and γ_R is a random number whose range is the production tolerance in R. The random number γ_R will typically have a Gaussian or a uniform distribution characterized by a mean and a standard deviation (Appendix C). For instance, a resistor labeled as $1000 \pm 10\,\Omega$ refers to a resistor whose nominal value is $R_0 = 1000\,\Omega$; and the spread of the resistance value about the nominal value ranges from $-10\,\Omega$ to $+10\,\Omega$. The manufacturer may have also specified that the spread is Gaussian and the tolerance limits correspond to the $\pm 3\sigma$ points. This is equivalent to saying that the mean value of the random number γ_R is zero, and the standard deviation is defined by $\pm 3\sigma = \pm 10\,\Omega$ (i.e., $\sigma = 3.33\,\Omega$).

Environmental Effects

The environmental effects that cause the elements to deviate are temperature, humidity, and aging.

The deviation due to temperature changes is often approximately linear, that is, the value of a resistor R at ΔT°C above room temperature is given by

$$R = R_0(1 + \alpha_{TCR}\Delta T) \tag{5.47}$$

where R_0 is the value at room temperature. From the above expression, the temperature coefficient of resistance, α_{TCR}, is

$$\alpha_{TCR} = \frac{R - R_0}{R_0 \Delta T} \tag{5.48}$$

The units used for α_{TCR} are parts per million per degree Celsius (centigrade) or ppm/°C. For example, if a resistor has a temperature coefficient of 135 ppm/°C, then $\alpha_{TCR} = 135 \times 10^{-6}$. In practice, of course, α_{TCR} is a random number characterized by a mean and a standard deviation. The distribution function describing α_{TCR} is usually Gaussian.

Next consider the effect of aging. The change in the value of a resistor R with time can often* be represented by the relationship

$$R = R_0(1 + \alpha_{ACR}\sqrt{t}) \tag{5.49}$$

where t is the time in years after manufacture and α_{ACR} is the aging coefficient. This square root relationship implies that the resistor ages faster in the initial years, just after manufacture, than in later years. From (5.49), the aging coefficient α_{ACR}, is given by

$$\alpha_{ACR} = \frac{R - R_0}{R_0\sqrt{t}} \tag{5.50}$$

The units of α_{ACR} are ppm/yr. Just as for manufacturing tolerance and temperature coefficient, the aging coefficient is also a random number characterized by a mean and a standard distribution. The distribution function describing α_{ACR} is usually Gaussian.

Finally, the change due to humidity can usually be represented by the linear equation:

$$R = R_0(1 + \beta_R H) \tag{5.51}$$

where H is the relative humidity, and β_R is the humidity coefficient of the resistance.

The combined effects of initial tolerance, temperature, aging, and humidity are given by

$$R = R_0(1 + \gamma_R)(1 + \alpha_{TCR}\Delta T)(1 + \alpha_{ACR}\sqrt{t})(1 + \beta_R H) \tag{5.52}$$

For small deviations from the nominal value, the product terms can be ignored in the expansion of (5.52), so that

$$R \cong R_0(1 + \gamma_R + \alpha_{TCR}\Delta T + \alpha_{ACR}\sqrt{t} + \beta_R H) \tag{5.53}$$

Thus

$$\frac{R - R_0}{R_0} = \frac{\Delta R}{R} = \gamma_R + \alpha_{TCR}\Delta T + \alpha_{ACR}\sqrt{t} + \beta_R H \tag{5.54}$$

The above expression gives the per-unit change in a resistor at ΔT°C above room temperature, at a relative humidity of H, and t years after manufacture.

* This square root relationship holds for thin film tantalum nitride resistors.

The mean and standard deviation of the per-unit change in R are*

$$\mu\left(\frac{\Delta R}{R}\right) = \mu(V_R) = \mu(\gamma_R) + \mu(\alpha_{TCR})\Delta T + \mu(\alpha_{ACR})\sqrt{t} + \mu(\beta_R)H \quad (5.55)$$

$$\sigma\left(\frac{\Delta R}{R}\right) = \sigma(V_R) = \sqrt{\sigma^2(\gamma_R) + \sigma^2(\alpha_{TCR})\Delta T^2 + \sigma^2(\alpha_{ACR})t + \sigma^2(\beta_R)H^2} \quad (5.56)$$

Similar expressions can be derived for the per-unit change in capacitors.

Statistical Deviations in Gain

In this section the statistics of the gain deviation are related to the component deviations. From Equation 5.38, the gain deviation for the second-order numerator function of (5.34) is

$$\Delta G = \sum_j (\mathscr{S}^G_{\omega_z} S^{\omega_z}_{R_i} V_{R_i} + \mathscr{S}^G_{Q_z} S^{Q_z}_{R_i} V_{R_i}) + \sum_j (\mathscr{S}^G_{\omega_z} S^{\omega_z}_{C_j} V_{C_j} + \mathscr{S}^G_{Q_z} S^{Q_z}_{C_j} V_{C_j}) \quad (5.57)$$

where the first summation is over all the resistors and the second summation is over all the capacitors. If all the resistors are assumed to be of the same type then:

$$\mu\left(\frac{\Delta R_i}{R_i}\right) = \mu(V_{R_i}) = \mu(V_R)$$

$$\sigma\left(\frac{\Delta R_i}{R_i}\right) = \sigma(V_{R_i}) = \sigma(V_R)$$

and similar expressions hold for the variability of the capacitor. Then the μ and σ of ΔG are:

$$\mu(\Delta G) = \mu(V_R)\left[\mathscr{S}^G_{\omega_z} \sum_i S^{\omega_z}_{R_i} + \mathscr{S}^G_{Q_z} \sum_i S^{Q_z}_{R_i}\right] + \mu(V_C)\left[\mathscr{S}^G_{\omega_z} \sum_j S^{\omega_z}_{C_j} + \mathscr{S}^G_{Q_z} \sum_j S^{Q_z}_{C_j}\right] \quad (5.58)$$

and

$$\sigma^2(\Delta G) = \sum_i \{\sigma^2(V_R)(\mathscr{S}^G_{\omega_z})^2(S^{\omega_z}_{R_i})^2 + \sigma^2(V_R)(\mathscr{S}^G_{Q_z})^2(S^{Q_z}_{R_i})^2\}$$
$$+ \sum_j \{\sigma^2(V_C)(\mathscr{S}^G_{\omega_z})^2(S^{\omega_z}_{C_j})^2 + \sigma^2(V_C)(\mathscr{S}^G_{Q_z})^2(S^{Q_z}_{C_j})^2\} \quad (5.59)$$

* From Appendix C, if y is the algebraic sum of random variables x_i, that is, if $y = \sum_i a_i x_i$, then from Equations C.12 and C.13:

$$\mu(y) = \sum_i a_i \mu(x_i)$$

and

$$\sigma^2(y) = \sum_i a_i^2 \sigma^2(x_i)$$

The expression for $\mu(\Delta G)$ can be simplified using the dimensional homogeneity properties stated in Equation 5.25a and b. Substituting these relationships in (5.58)

$$\mu(\Delta G) = -\mu(V_R)\mathscr{S}^G_{\omega_z} - \mu(V_C)\mathscr{S}^G_{\omega_z} \tag{5.60}$$

This equation shows that *the mean change in gain is independent of the circuit chosen to realize the filter.*

If the mean change in resistance and capacitance is due to the mean change in temperature coefficient alone (all the other means being zero) then:

$$\mu(V_R) = \mu(\alpha_{TCR})\Delta T \qquad \text{and} \qquad \mu(V_C) = \mu(\alpha_{TCC})\Delta T$$

and (5.60) becomes

$$\mu(\Delta G) = -\mathscr{S}^G_{\omega_z}[\mu(\alpha_{TCR}) + \mu(\alpha_{TCC})]\Delta T \tag{5.61}$$

This equation suggests that to minimize the mean change in gain, the mean values of the resistor and capacitor temperature coefficients should have opposite signs, and their magnitudes should be as close to being equal as possible. In particular, if

$$\mu(\alpha_{TCR}) = -\mu(\alpha_{TCC})$$

then

$$\mu(\Delta G) = 0$$

that is, *the mean change in gain is reduced to zero.*

Next, considering the standard deviation, from Equation 5.59 it is seen that $\sigma^2(\Delta G)$ is a sum of squares; as such, it cannot be made zero. To reduce the standard deviation of ΔG, we must choose components with low spreads in the initial tolerances, and temperature, aging, and humidity coefficients. Of course the lower the spread required of the component values, the more expensive they will be.

Summarizing this section, the sensitivity to component types can be reduced by:

- Choosing resistors that have a mean temperature coefficient that is equal in magnitude but opposite in sign to that of the capacitors.
- Choosing components that have a low spread in their initial manufacturing tolerance, and in their temperature, aging, and humidity coefficients.

Example 5.5

A second-order band-pass function has the transfer function:

$$T(s) = \frac{\dfrac{1}{R_1C_1}s}{s^2 + \dfrac{1}{R_2C_1}s + \dfrac{1}{R_2R_3C_1C_2}} \tag{5.62}$$

The circuit is built using the following types of RC components:

Resistors:

Manufacturing tolerance	± 0.1 percent
Temperature Coefficient	10 ± 5 ppm/°C
Aging in 20 years	± 0.005 percent

Capacitors:

Manufacturing tolerance	± 0.1 percent
Temperature Coefficient	-10 ± 15 ppm/°C
Aging in 20 years	± 0.01 percent

Assume all the spreads to have a Gaussian distribution and that the limits correspond to the $\pm 3\sigma$ points. Compute the deviation in the gain at the pole frequency after 20 years aging in a surrounding where the temperature is 50°C above room temperature.

Solution

The given biquadratic function may be written in the form

$$\frac{Ks}{s^2 + \dfrac{\omega_p}{Q_p} s + \omega_p^2} \tag{5.63}$$

where

$$K = \frac{1}{R_1 C_1} \qquad \omega_p = \sqrt{\frac{1}{R_2 R_3 C_1 C_2}} \qquad Q_p = \sqrt{\frac{R_2 C_1}{R_3 C_2}} \tag{5.64}$$

The gain deviation for this function is given by an equation similar to Equation 5.38:

$$\Delta G = \sum_{j=1}^{m} [\mathcal{S}_{\omega_p}^{G} S_{x_i}^{\omega_p} V_{x_i} + \mathcal{S}_{Q_p}^{G} S_{x_i}^{Q_p} V_{x_i} + \mathcal{S}_{K}^{G} S_{x_i}^{K} V_{x_i}] \tag{5.65}$$

Since the components are of the same type, and the only random variable that has a nonzero mean is the temperature coefficient, $\mu(\Delta G)$ is given by (Equation 5.61)

$$\mu(\Delta G) = -\mathcal{S}_{\omega_p}^{G}[\mu(\alpha_{TCR}) + \mu(\alpha_{TCC})]\Delta T - \mathcal{S}_{K}^{G}[\mu(\alpha_{TCR}) + \mu(\alpha_{TCC})]\Delta T \tag{5.66}$$

Using Equation 5.59, $\sigma(\Delta G)$ is given by:

$$
\left.
\begin{aligned}
\sigma^2(\Delta G) &= \sigma^2(V_R)(\mathcal{S}^G_{\omega_p})^2 \sum_i (S^{\omega_p}_{R_i})^2 && \leftarrow \omega_p \\
&+ \sigma^2(V_R)(\mathcal{S}^G_{Q_p})^2 \sum_i (S^{Q_p}_{R_i})^2 \quad \text{Resistor terms} && \leftarrow Q_p \\
&+ \sigma^2(V_R)(\mathcal{S}^G_{K})^2 \sum_i (S^{K}_{R_i})^2 && \leftarrow K \\
&+ \sigma^2(V_C)(\mathcal{S}^G_{\omega_p})^2 \sum_j (S^{\omega_p}_{C_j})^2 && \leftarrow \omega_p \\
&+ \sigma^2(V_C)(\mathcal{S}^G_{Q_p})^2 \sum_j (S^{Q_p}_{C_j})^2 \quad \text{Capacitor terms} && \leftarrow Q_p \\
&+ \sigma^2(V_C)(\mathcal{S}^G_{K})^2 \sum_j (S^{K}_{C_j})^2 && \leftarrow K
\end{aligned}
\right. \tag{5.67}
$$

The component sensitivities are obtained from Equation 5.64:

$$S^{\omega_p}_{R_1} = 0 \qquad S^{\omega_p}_{R_2, R_3, C_1, C_2} = -\tfrac{1}{2}$$

$$S^{Q_p}_{R_1} = 0 \qquad S^{Q_p}_{R_2, C_1} = \tfrac{1}{2} \qquad S^{Q_p}_{R_3, C_2} = -\tfrac{1}{2}$$

$$S^{K}_{R_1, C_1} = -1 \qquad S^{K}_{R_2, R_3, C_2} = 0$$

The μ and σ of the variabilities of the components are obtained from the given information, as follows:

$$\mu(V_R) = \mu(\alpha_{TCR})\Delta T = 10(10)^{-6}50 = 0.0005$$

and $\sigma(V_R)$, from Equation 5.56, is given by

$$
\begin{aligned}
[3\sigma(V_R)]^2 &= (\text{Manufacturing } 3\sigma)^2 + (\text{Temperature } 3\sigma)^2 + (\text{Aging } 3\sigma)^2 \\
&= (0.001)^2 + [5 \times 50(10)^{-6}]^2 + (0.00005)^2 \\
&= 1.065(10)^{-6}
\end{aligned}
$$

Therefore $\sigma(V_R) = 3.44(10)^{-4}$. Similarly for the capacitor

$$\mu(V_C) = -10(50)10^{-6} = -0.0005$$

$$[3\sigma(V_C)]^2 = (0.001)^2 + (0.00075)^2 + (0.0001)^2$$

$$\sigma(V_C) = 4.18(10)^{-4}$$

The biquadratic parameter sensitivity terms $\mathcal{S}^G_K$, $\mathcal{S}^G_{\omega_p}$, and $\mathcal{S}^G_{Q_p}$ are evaluated using Equations 5.43, 5.44 and 5.45, respectively. At the pole frequency $\omega = \omega_p$, so $\Omega_p = 1$. Substituting for Ω_p in these equations:

$$\mathcal{S}^G_K = 8.686 \text{ dB} \qquad \mathcal{S}^G_{\omega_p} = -8.686 \text{ dB} \qquad \mathcal{S}^G_{Q_p} = 8.686 \text{ dB}$$

Substituting the computed values for the component sensitivities, the μ and σ for the component variabilities, and the biquadratic parameter sensitivity terms in Equations 5.66 and 5.67 for $\mu(\Delta G)$ and $\sigma^2(\Delta G)$, we finally get:

$$\mu(\Delta G) = 0 \text{ dB}$$

and

$$\begin{aligned}
\sigma^2(\Delta G) = {} & [3.44(10)^{-4}]^2(-8.686)^2(\tfrac{1}{4} + \tfrac{1}{4}) && \leftarrow \omega_p \\
& + [3.44(10)^{-4}]^2(8.686)^2(\tfrac{1}{4} + \tfrac{1}{4}) && \leftarrow Q_p \quad \text{Resistor terms} \\
& + [3.44(10)^{-4}]^2(8.686)^2(1) && \leftarrow K \\
& + [4.18(10)^{-4}]^2(-8.686)^2(\tfrac{1}{4} + \tfrac{1}{4}) && \leftarrow \omega_p \\
& + [4.18(10)^{-4}]^2(8.686)^2(\tfrac{1}{4} + \tfrac{1}{4}) && \leftarrow Q_p \quad \text{Capacitor terms} \\
& + [4.18(10)^{-4}]^2(8.686)^2(1) && \leftarrow K \\
= {} & 4.422(10)^{-5}
\end{aligned}$$

Hence

$$3\sigma(\Delta G) = 0.02 \text{ dB}$$

Therefore, the deviation in the gain at the center frequency has a Gaussian distribution given by

$$\Delta G = 0.0 \pm 0.02 \text{ dB}$$

where the limits refer to the $\pm 3\sigma$ points of the distribution.

Observations

1. Let us consider the gain deviations for the special case of $Q_p = 10$ (Figure 5.4*a*). The μ and σ for ΔG can be calculated just as in the example, by using Equations 5.66, 5.67, and the biquadratic sensitivity expressions (5.43), (5.44), and (5.45). The results of the computations are shown in Figure 5.4*b*. In this figure the $\mu \pm \sigma$, $\mu \pm 2\sigma$, and $\mu \pm 3\sigma$ points for ΔG have been plotted for a range of frequencies. Since the distribution is Gaussian, from Appendix C, the gain deviation of 99.74 percent of the circuits will be within the $\mu \pm 3\sigma$ boundaries. As expected from the discussions in Section 5.4.1, the gain deviation peaks at the 3 dB pass-band edge frequencies:

$$\Omega_p = 1 \pm \frac{1}{2Q_p} = 1 \pm 0.05$$

   and the deviation is seen to decrease as we move away from the passband.
2. From the above example it is seen that even for a simple second-order function, the computation of gain deviation becomes quite lengthy. In practice, for all but the simplest functions, it is desirable to use computer aids for evaluating the statistics of gain deviation. Some such computer algorithms are described in the following section. ■

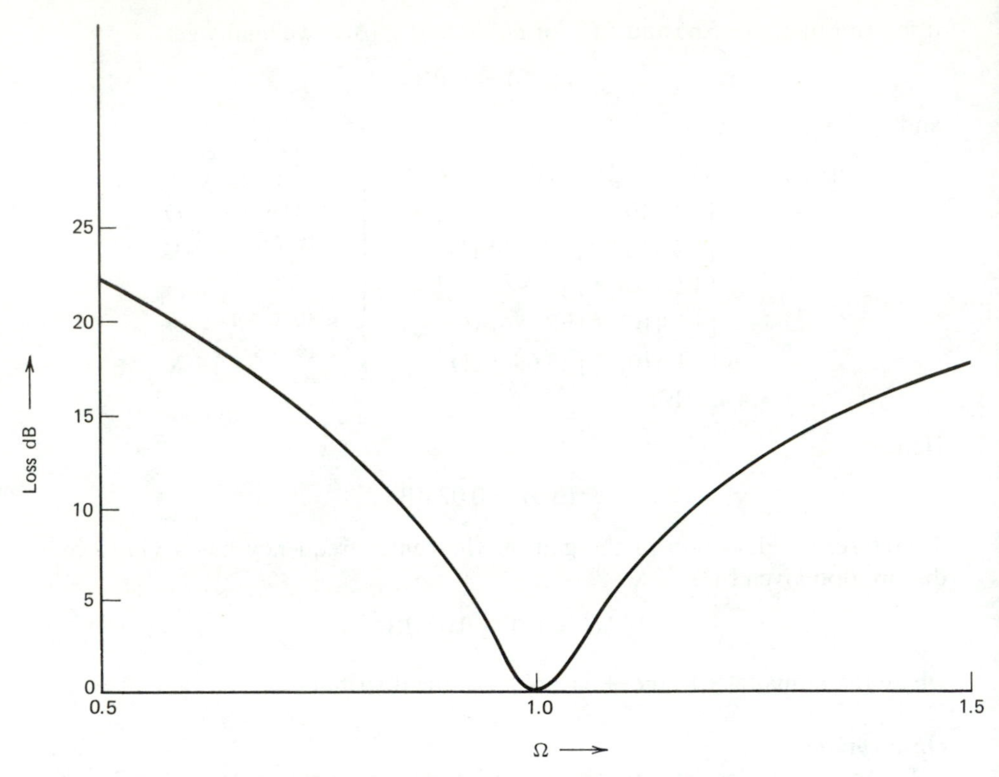

Figure 5.4 (*a*) Loss characteristic of band-pass function for Example 5.5 ($Q_p = 10$, $\omega_p = 1$).

5.5 COMPUTER AIDS

In this section we describe some computer aids for the evaluation of gain changes due to the statistical variations in the elements. In one approach, the input to the computer consists of a topological description of the elements and the percentage changes expected in the elements. The aglorithm for computing gain variation is based on Equation 5.33:

$$\Delta G = \sum_j \mathscr{S}_{x_j}^{G} \frac{\Delta x_j}{x_j} \tag{5.68}$$

The sensitivity term $\mathscr{S}_{x_j}^{G}$ is computed by perturbing the jth component (keeping all other components fixed) and comparing the deviated gain to the nominal gain, at all the frequencies of interest. This procedure is repeated for each element. Equation 5.68 then gives the desired gain change due to the simultaneous variation of all the components. This approach is based on the assumption that the changes in the components are small. In a modification of the above

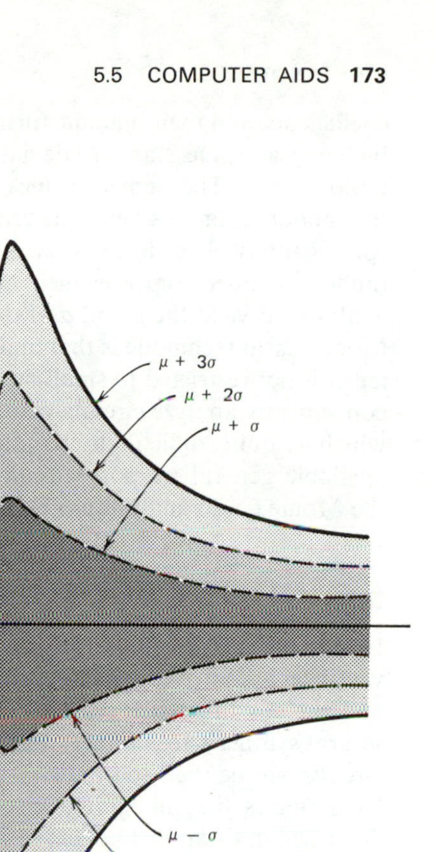

Figure 5.4 (*b*) Gain deviation versus normalized frequency.

approach, proposed by Director and Rohrer [6] (known as the adjoint-matrix approach), the sensitivities to all the components are evaluated by only two analyses of the circuit.

Another computer algorithm for computing gain changes is based on the so-called **Monte Carlo technique** [8]. In this approach the input is a topological or a functional description of the circuit, in which the elements of the circuit are described by their nominal value, their temperature, humidity, aging coefficients, and their manufacturing tolerances. These tolerances are described by the mean value, standard deviation, and the distribution function (Gaussian, uniform, etc.). The element values are chosen using equations similar to (5.52), the random numbers being generated by the computer according to the input specifications. The network is analyzed many times using different sets of random numbers for the elements. The responses for these several runs are then analyzed to yield the μ and σ of the gain deviation. A distinctive feature of the Monte Carlo technique is that, unlike the analytical approaches discussed thus far, it is not restricted to small deviations in the components. Moreover, it can conveniently analyze circuits with correlations among the random variables—which is quite difficult to handle by other methods. Among the presently available general purpose circuit analysis computer programs that provide the Monte Carlo analysis are SCEPTRE [7] and ASTAP [9].

5.6 CONCLUDING REMARKS

In this chapter the subject of sensitivity is discussed in some detail. The gain variation is shown to depend on the approximation function, the components, and the circuit used for the synthesis. In the next few chapters we describe several synthesis techniques; and, one of the major criteria for comparing the circuits will be their sensitivities. The components available using present-day technologies (i.e., integrated circuit, thin film, thick film, and discrete) are described in Chapter 13.

In most of the discussions, sensitivity to the parameters describing the op amp were not included, but only to avoid unwieldy analysis. In fact, the sensitivity of an active filter to the op amp is just as important as the passive elements, as will be demonstrated in the later chapters.

The methods of sensitivity analysis developed are directly applicable to the large class of active filters that are designed by factoring the filter function into biquadratics. This method, referred to as the cascade approach, is the most popular active synthesis technique, and will be covered in Chapters 7 to 10. Some other synthesis schemes, such as those using coupled structures (Chapter 11), do not use the biquadratic decomposition. The sensitivity analysis of these coupled structures is more difficult and is most conveniently done using the computer-aided Monte Carlo method described in Section 5.5.

FURTHER READING

Sensitivity Analysis

1. D. Hilberman, "An approach to the sensitivity and statistical variability of biquadratic filters," *IEEE Trans. Circuit Theory*, *CT-20*, No. 4, July 1973, pp. 382–390.
2. S. K. Mitra, *Analysis and Synthesis of Linear Active Networks*, Wiley, New York, 1969, Chapter 5.

3. G. S. Moschytz, *Linear Integrated Networks Fundamentals*, Van Nostrand, New York, 1974, Chapter 4.
4. G. S. Moschytz, *Linear Integrated Networks Design*, Van Nostrand, New York, 1975, Chapter 1.
5. R. Spence, *Linear Active Networks*, Wiley, London, 1970, Chapter 10.

Computer Aids

6. S. W. Director and R. A. Rohrer, "The generalized adjoint network and network sensitivities," *IEEE Trans. Circuit Theory*, *CT-16*, 1969, pp. 318–323.
7. H. W. Mathers, S. R. Sedore, and J. R. Sents, "Automated digital computer program for determining responses of electronic circuits to transient nuclear radiation, SCEPTRE," Owego, N.Y., *IBM Corp.*, *Technical Report* No. AXWL-TR-66-126, *1*, SCEPTRE Users' Manual, February 1967.
8. C. L. Semmelman, E. D. Walsh, and G. Daryanani, "Linear circuits and statistical design," *Bell System Tech. J.*, *50*, No. 4, April 1971, pp. 1149–1171.
9. W. T. Weeks et al., "Algorithms for ASTAP . . . A network analysis program," *IEEE Trans. Circuit Theory*, *CT-20*, No. 6, November 1973, pp. 628–634.

PROBLEMS

5.1 *Sensitivity relationships.* Prove the following relationships using the definition of sensitivity (Equation 5.6):

(a) $S_x^{p_1 p_2} = S_x^{p_1} + S_x^{p_2}$

(b) $S_x^{p_1 + p_2} = \dfrac{p_1 S_x^{p_1} + p_2 S_x^{p_2}}{p_1 + p_2}$

(c) $S_x^p = S_y^p \cdot S_x^y$

5.2 Use the sensitivity relationships of Problem 5.1 to show that:

(a) $S_x^{p^n} = n S_x^p$

(b) $S_x^{y+c} = \dfrac{y}{y+c} S_x^y$

(c) $S_{x^2}^p = \frac{1}{2} S_x^p$

(d) $S_{\sqrt{x}}^p = 2 S_x^p$

5.3 *ω and Q sensitivities.* The transfer function for an active *RC* circuit is

$$\frac{\dfrac{1}{R_2 R_3 C_1 C_2}}{s^2 + \dfrac{1}{R_1 C_1} s + \dfrac{1}{R_1 R_3 C_1 C_2}}$$

Identify the biquadratic parameters (ω_z, ω_p, Q_z, Q_p, K) and determine their sensitivities to the elements.

5.4 Consider the transfer function

$$\frac{\dfrac{1}{R_1C_1}s}{s^2 + \left(\dfrac{1}{R_1C_1} + \dfrac{1}{R_2C_2} - \dfrac{r_2}{r_1}\dfrac{1}{R_2C_2}\right)s + \dfrac{1}{R_1R_2C_1C_2}}$$

Compute the sensitivity of Q_p to the passive elements. (*Hint*: use Equation 5.11e.)

5.5 Compute the sensitivity of ω_p and Q_p to the amplifier gain A for the transfer function:

$$\frac{\dfrac{1}{R_2R_3C_1C_2}}{s^2 + \left(\dfrac{1}{R_1C_1} + \dfrac{1}{R_2C_2}\left(1 - \dfrac{1}{A}\right)\right)s + \dfrac{1}{R_2R_3C_1C_2}}$$

5.6 Compute the sensitivity of Q_p to R_A/R_B in Equation 5.20. (*Hint*: first use the approximation $(1 + k/A)^{-1} \cong 1 - k/A$.)

5.7 *Dimensional homogeneity*. In Problem 5.4 show that
(a) $S_{r_2}^{Q_p} = -S_{r_1}^{Q_p} = S_{r_2/r_1}^{Q_p}$
(b) Verify the dimensional homogeneity relationships (Equations 5.25) for Q_p and ω_p. Hence, show that $S_{R_2}^{Q_p}$ and $S_{C_2}^{Q_p}$ can be deduced from the expressions for $S_{R_1}^{Q_p}$ and $S_{C_1}^{Q_p}$.

5.8 *Multi-element deviations*. An active *RC* circuit has the band-pass transfer function

$$\frac{-\dfrac{1}{R_4C_1}s}{s^2 + \dfrac{1}{R_1C_1}s + \dfrac{1}{R_2R_3C_1C_2}}$$

Due to an increase in ambient (i.e., surrounding) temperature suppose the resistors increase by t_R percent and the capacitors increase by t_C percent.
(a) Show that the pole Q does not change.
(b) Show that the pole frequency ω_p changes by $(-t_R - t_C)$ percent.

5.9 Repeat Problem 5.8 for the transfer function of Problem 5.4.

5.10 *Worst-case change in ω_p and Q_p*. In Problem 5.8, suppose all the components can change by ± 1 percent. Compute the worst-case (i.e., maximum possible) per-unit change in ω_p and Q_p.

5.11 *Gain deviation at passband edges*. Show that the gain deviation for the function

$$\frac{1}{s^2 + \dfrac{\omega_p}{Q_p}s + \omega_p^2}$$

can be expressed in the form

$$\Delta G \cong \mathcal{S}_{Q_p}^{G} \frac{\Delta Q_p}{Q_p} + \mathcal{S}_{\omega_p}^{G} \frac{\Delta \omega_p}{\omega_p}$$

Hence, show that for $Q_p \gg 1$ the contribution of the ω_p term to the gain deviation is approximately $2Q_p$ times that of the Q_p term at the 3 dB passband edge frequencies $\omega = \omega_p(1 \pm 1/2Q_p)$.

5.12 *Gain deviation at pole frequency.* In Problem 5.11, show that the gain deviation at the pole frequency is given by

$$\Delta G = 8.686\left(\frac{\Delta Q}{Q} - \frac{\Delta \omega}{\omega}\right) \text{dB}$$

5.13 *Gain deviation.* A low-pass biquadratic function has a pole Q of 10, pole frequency of 100 rad/sec, and the dc gain is 0 dB. Compute the approximate change in gain at the pole frequency ω_p, at the passband edge frequencies $\omega = \omega_p(1 \pm 1/2Q_p)$, at *dc*, and at $2\omega_p$, for:

(a) A two percent increase in the pole Q.
(b) A two percent increase in the pole frequency.
(c) A two percent increase in the dc gain.

Sketch the gain deviation versus frequency for each case.

5.14 *Biquadratic parameter sensitivities.* Compute the biquadratic parameter sensitivities ($\mathcal{S}_{\omega_p}^{G}$ and $\mathcal{S}_{Q_p}^{G}$) at (a) $\omega = 6$ and (b) 9 rad/sec for the low-pass function

$$\frac{1000}{s^2 + 3s + 81}$$

by using Equation 5.44 and 5.45. Check your answer with Figure 5.3.

5.15 Consider the biquadratic function

$$\frac{s^2 + 2s + 100}{s^2 + 2s + 64}$$

(a) Identify the biquadratic parameters.
(b) Determine the biquadratic parameter sensitivities at $\omega = 4$, 6, and 10 rad/sec using Figure 5.3.
(c) Compute the biquadratic parameter sensitivities at $\omega = 6$ using Equations 5.41 to 5.45.

5.16 *Gain deviation.* The biquadratic transfer function for an active *RC* network is

$$\frac{s^2}{s^2 + \dfrac{1}{R_3 C_1} s + \dfrac{1}{R_1 R_2 C_1 C_2}}$$

The network is used to realize a high-pass filter with a pole frequency of 10 rad/sec and a pole Q of 5.

(a) Determine the component sensitivities.

(b) Find the biquadratic parameter sensitivities at 1, 9, 10, 11, and 20 rad/sec from Figure 5.3.

(c) Compute the deviation in gain at these frequencies for a one percent decrease in all the components caused by a decrease in the ambient temperature. Sketch the nominal and deviated gain characteristics.

5.17 An active RC network having the biquadratic transfer function given in Problem 5.4 is used to realize a band-pass filter with the following element values

$$C_1 = C_2 = 1 \qquad R_1 = R_2 = 0.01 \qquad r_2 = 1.9r_1$$

(a) Determine the biquadratic parameters K, ω_p, and Q_p.

(b) Find the biquadratic parameter sensitivities at $\omega = 10$, 90, 100, 110, and 200 rad/sec using Figure 5.3.

(c) Compute the deviation in gain at these frequencies due to a 2 percent increase in the resistor values and a 1 percent decrease in the capacitor values caused by a rise in the ambient temperature. Sketch the nominal and deviated gain characteristics.

5.18 *Resistance deviation, worst-case.* A 10 kΩ resistor has the following characteristics:

Manufacturing tolerance ± 0.5 percent
Temperature coefficient $+200 \pm 50$ ppm/°C
Aging coefficient ± 0.5 percent in 20 years

Determine the worst-case (i.e., maximum possible) deviation in ohms from the nominal value of the resistor:

(a) At the time of manufacture (room temperature).

(b) During a manufacturing test at 75°C.

(c) After 20 years, at 75°C.

5.19 *RC-product deviation, worst-case.* An active RC circuit having the band-pass function given in Problem 5.8 is built using components with the following temperature coefficients:

resistors 100 ± 20 ppm/°C; capacitors -150 ± 40 ppm/°C

For a 50°C rise in the ambient temperature:

(a) Determine the worst-case percentage deviation in the RC product.

(b) Determine the worst-case percentage deviation in the pole frequency and the pole Q.

(c) Repeat part (b) if the components *track*, that is, if all the resistors have the same T.C. and all the capacitors have the same T.C.

5.20 *Statistical deviation in resistance.* If the tolerance limits in Problem 5.18 refer to the $\pm 3\sigma$ points of a Gaussian distribution, compute the statistics (μ and σ) of the resistance spread $\Delta R/R$ during a manufacturing test at 75°C. Sketch the distribution function for $\Delta R/R$.

5.21 *Statistical deviation in Q_p and ω_p.* An active *RC* circuit for the realization of the low-pass transfer function given in Problem 5.3 uses resistors that have a T.C. of -120 ± 20 ppm/°C and capacitors that have a T.C. of $+100 \pm 50$ ppm/°C (the tolerance limits refer to the $\pm 3\sigma$ points of a Gaussian distribution). Compute the statistics of the per-unit change in pole frequency and pole Q due to a 50°C rise in the ambient temperature. Sketch the distributions of the per unit changes in pole Q and pole frequency.

5.22 *Statistical change in Q_p, ω_p, and gain.* Compute the statistics of the per-unit change in Q_p, ω_p, and K in Example 5.5. Use the results to determine the statistics of the gain deviation at the pole frequency. (*Hint*: see Problem 5.12.)

5.23 *Statistical deviation in gain.* The function of Example 5.5 is used to realize a band-pass filter with a pole Q of 5 and a pole frequency of 100 rad/sec. If the resistors and capacitors can change by ± 2 percent, sketch the statistics of the corresponding gain deviation in the passband (as in Figure 5.4*b*). The tolerance limits given refer to $\pm 3\sigma$ points of a Gaussian distribution.

5.24 The circuit of Problem 5.3 is used to realize a low-pass transfer function with a pole Q of 5 and a pole frequency of 1000 rad/sec. The components used for the realization have the characteristics described in Example 5.5. Compute the statistics of the gain deviation at 800, 900, 1000, 1100, and 1200 rad/sec. Sketch the gain deviation versus frequency, as in Figure 5.4*b*. (Use Figure 5.3 for finding the biquadratic parameter sensitivities.)

5.25 *Large deviation in components.* The expressions for gain deviation derived in this chapter assumed small changes in the components. To study the accuracy of Equation 5.38 for ΔG, consider a band-pass function with a pole Q of 5, pole frequency of 100 rad/sec, and center frequency gain of 0 dB, realized using the circuit of Example 5.5. Compute the change in gain at 110 rad/sec due to a one percent increase in R_3:

(a) Using Equation 5.38.

(b) By computing the gains of the nominal and deviated transfer functions.

Repeat the problem for a 10 percent increase in R_3.

5.26 *Biquadratic coefficient sensitivities.* The general biquadratic function is sometimes written in the form [1]

$$\frac{n_2 s^2 + n_1 s + n_0}{d_2 s^2 + d_1 s + d_0}$$

Show that the *biquadratic coefficient sensitivities* ($\mathcal{S}_{n_2}^G$, $\mathcal{S}_{n_1}^G$, $\mathcal{S}_{n_0}^G$) satisfy the identity

$$\mathcal{S}_{n_2}^G + \mathcal{S}_{n_1}^G + \mathcal{S}_{n_0}^G \equiv 8.686 \text{ dB}$$

A similar relationship holds for the denominator coefficients. (*Hint*: consider a one percent change in the coefficients n_2, n_1, and n_0.)

5.27 Show that the biquadratic coefficient sensitivities, defined in Problem 5.26, are related to the biquadratic parameter sensitivities by

$$\mathcal{S}_{\omega_z}^G = \mathcal{S}_{n_1}^G + 2\mathcal{S}_{n_0}^G$$
$$\mathcal{S}_{Q_z}^G = -\mathcal{S}_{n_1}^G$$

(*Hint*: consider a one percent change in ω_z for the first relationship, and a one percent change in Q_z for the second relationship.)

5.28 *Computer program.* Write a computer program to evaluate the biquadratic parameter sensitivities given by Equation 5.44 and 5.45. The inputs to the program should be ω_p, Q_p, and ω (the frequency of computation) and the outputs should be $S_{\omega_p}^G$ and $S_{Q_p}^G$.

6.

PASSIVE NETWORK SYNTHESIS

The subject of this chapter is the synthesis of transfer functions using *RLC* networks. A popular structure in the design of passive filters, consisting of an *LC* ladder terminated at both ends with resistors, will be studied in some detail. A salient feature of this so-called double-terminated ladder topology is its very low sensitivity to element variations. As is shown in Chapter 11, this topology is also used in the realization of low sensitivity active filters; indeed, some of these realizations are just an active-*RC* equivalent of the passive *LC* ladder filter. Since the synthesis of transfer functions requires a knowledge of driving point synthesis, we study this subject before embarking on transfer function synthesis.

6.1 SYNTHESIS BY INSPECTION

In this section we introduce the notion of synthesis by considering simple driving point functions that can be synthesized by inspection. Especially simple are functions that can be directly recognized as the sum of the impedances (or admittances) of resistors, capacitors, and inductors. These are illustrated in Figure 6.1. More complex network functions can be realized as a combination of these simple building blocks. The following examples illustrate the procedure.

Example 6.1
Synthesize

$$Z(s) = \frac{s^2 + 1}{2s}$$

Solution
The impedance may be written as

$$Z(s) = \frac{s}{2} + \frac{1}{2s}$$

which can be realized as a capacitor in series with an inductor, using block 1 (Figure 6.2). ■

Example 6.2
Synthesize

$$Z(s) = \frac{s + 1}{s + 4}$$

1. $Z = sL + \frac{1}{sC} = \frac{s^2LC + 1}{sC}$

2. $Z = sL + R$

3. $Z = R + \frac{1}{sC} = \frac{sCR + 1}{sC}$

4. $Z = sL + R + \frac{1}{sC} = \frac{s^2LC + sCR + 1}{sC}$

5. $Z = \frac{1}{sC + \frac{1}{sL}} = \frac{sL}{s^2LC + 1}$

6. $Z = \frac{1}{\frac{1}{sL} + \frac{1}{R}} = \frac{sRL}{R + sL}$

7. $Z = \frac{1}{sC + \frac{1}{R}} = \frac{R}{sCR + 1}$

8. $Z = \frac{1}{sC + \frac{1}{R} + \frac{1}{sL}} = \frac{s}{C\left(s^2 + s\frac{1}{RC} + \frac{1}{LC}\right)}$

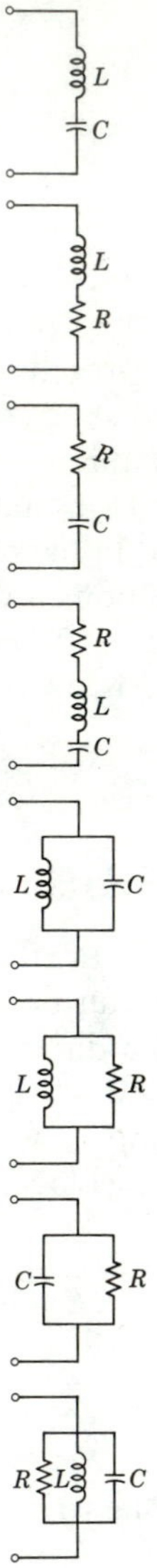

Figure 6.1 Some simple functions that can be synthesized by inspection.

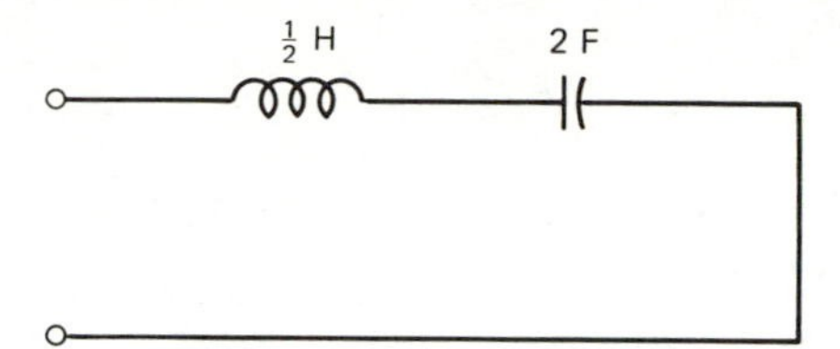

Figure 6.2 Circuit for Example 6.1.

Solution

One realization is obtained by writing $Z(s)$ as

$$Z(s) = \frac{s}{s+4} + \frac{1}{s+4}$$

which can be realized using blocks 6 and 7, as shown in Figure 6.3*a*. Alternately the corresponding admittance function may be expanded as

$$Y(s) = \frac{s+4}{s+1} = 1 + \frac{3}{s+1} = 1 + \frac{1}{\frac{s}{3} + \frac{1}{3}}$$

which is realized by the circuit of Figure 6.3*b*. Observe that this realization is superior to the circuit of Figure 6.3*a* in that it requires one less component. ■

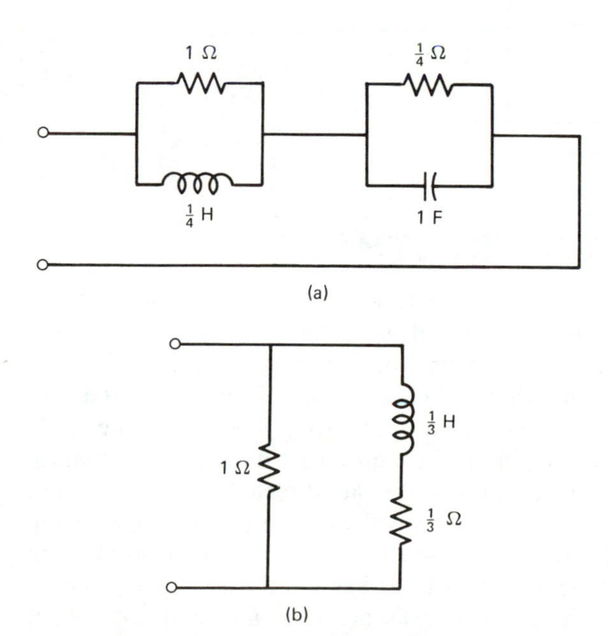

Figure 6.3 Circuit for Example 6.2: (*a*) using $Z(s)$. (*b*) using $Y(s)$.

Example 6.3
Synthesize

$$Z(s) = \frac{s + 1}{s(s + 2)}$$

Solution
This function is synthesized by first expanding it in partial fractions, as

$$Z(s) = \frac{\frac{1}{2}}{s} + \frac{\frac{1}{2}}{s + 2}$$

$$= \frac{1}{2s} + \frac{1}{2s + 4}$$

which is realized using the series circuit of Figure 6.4. A generalization of the partial fraction expansion for realizing driving point functions is described in the next section. ■

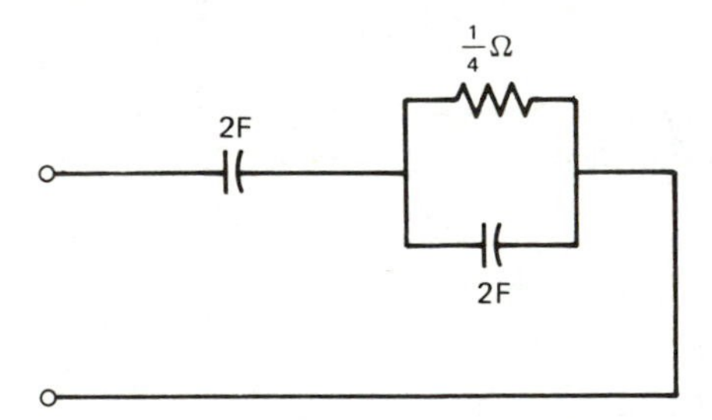

Figure 6.4 Circuit for Example 6.3.

6.2 DRIVING POINT SYNTHESIS

In Chapter 2, we had mentioned that if a rational driving point function is positive real (p.r.), it can always be realized using passive elements. This so-called sufficiency of the p.r. condition was first proved by Brune [3] who developed a procedure by which these functions could always be synthesized using resistors, capacitors, inductors, and transformers. A few years later Bott and Duffin [3] demonstrated a different procedure that did not entail the use of transformers. Both these classical approaches are directed to the realization of a general dp function. In this book, we will only be concerned with the synthesis of a subset of p.r. driving point functions that can be realized using *LC* or *RC* networks. The *LC* networks form the basis of the doubly terminated passive ladder filters, while the *RC* networks are used in the realization of active-*RC* filters. The synthesis procedure for these two-element type networks is quite straightforward and is elaborated in the following.

6.2.1 SYNTHESIS USING PARTIAL FRACTION EXPANSION

In this section we describe a method for synthesizing RC and LC driving point functions.* The method is based on the partial fraction expansion of the given function and was first recognized by Foster [10].

From Equation 2.17, an LC dp impedance function has the following general partial fraction expansion

$$Z_{LC}(s) = \frac{K_0}{s} + K_\infty s + \sum_i \frac{2K_i s}{s^2 + \omega_{p_i}^2} \tag{6.1}$$

The admittance $Y_{LC}(s)$ has a similar expansion. The expansion for an RC impedance function from Equation 2.14, is

$$Z_{RC}(s) = \frac{K_0}{s} + K_\infty + \sum_i \frac{K_i}{s + p_i} \tag{6.2}$$

and the expansion for $Y_{RC}(s)$ is

$$Y_{RC}(s) = K_0 + K_\infty s + \sum_i \frac{K_i s}{s + p_i} \tag{6.3}$$

Once the given dp function has been expressed in onc of the above forms, the individual terms in the expansion can readily be synthesized by inspection. For an impedance function, these blocks are connected in series to yield the complete network. For admittance functions, the fundamental circuit blocks realizing the individual terms are connected in parallel to complete the synthesis. The above synthesis procedure is illustrated by the following examples.

Example 6.4
Synthesize

$$Z(s) = \frac{s(s^2 + 2)}{(s^2 + 1)(s^2 + 3)}$$

Solution
The poles and zeros of this function lie on the imaginary axis, are simple, and alternate. These properties guarantee that the dp function can be expressed in the form of Equation 6.1 with positive K_i; and, therefore, ensure its realization as an LC network.

The dp function does not have a pole at the origin, nor does it have a pole at infinity. Thus the K_0 and K_∞ terms are both zero (from Section 2.3.3 page 43), and the partial fraction expansion has the form:

$$Z(s) = \frac{K_1 s}{s^2 + 1} + \frac{K_2 s}{s^2 + 3}$$

* The method is easily generalized to RL networks.

where K_1 and K_2 are given by

$$K_1 = \frac{s^2+1}{s} Z(s)\Bigg|_{s^2=-1}$$

$$= \frac{s^2+2}{s^2+3}\Bigg|_{s^2=-1} = \frac{1}{2}$$

$$K_2 = \frac{s^2+3}{s} Z(s)\Bigg|_{s^2=-3}$$

$$= \frac{s^2+2}{s^2+1}\Bigg|_{s^2=-3} = \frac{1}{2}$$

Thus

$$Z(s) = \frac{\frac{1}{2}s}{s^2+1} + \frac{\frac{1}{2}s}{s^2+3}$$

$$= \frac{1}{2s+\dfrac{2}{s}} + \frac{1}{2s+\dfrac{6}{s}}$$

which can easily be synthesized by inspection, as shown in Figure 6.5.

An alternate way to synthesize $Z(s)$ would be to obtain the partial fraction expansion of its reciprocal, $Y(s)$:

$$Y(s) = \frac{(s^2+1)(s^2+3)}{s(s^2+2)}$$

$$= \frac{K_1}{s} + \frac{K_2 s}{s^2+2} + K_3 s$$

where

$$K_1 = sY(s)\Big|_{s=0} = \frac{3}{2}$$

$$K_2 = \frac{s^2+2}{s} Y(s)\Bigg|_{s^2=-2} = \frac{1}{2}$$

$$K_3 = \frac{Y(s)}{s}\Bigg|_{s=\infty} = \frac{\left(1+\dfrac{1}{s^2}\right)\left(1+\dfrac{3}{s^2}\right)}{1\left(1+\dfrac{2}{s^2}\right)}\Bigg|_{s=\infty} = 1$$

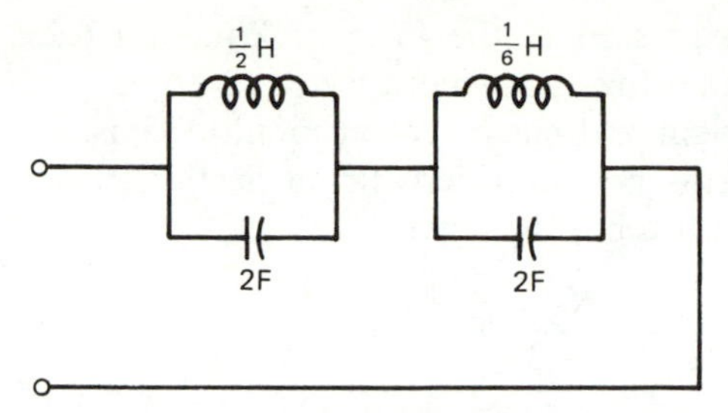

Figure 6.5 Circuit for Example 6.4 using $Z(s)$.

Thus

$$Y(s) = \frac{\frac{3}{2}}{s} + \frac{\frac{1}{2}s}{s^2 + 2} + s$$

$$= \frac{1}{\frac{2}{3}s} + \frac{1}{2s + \frac{4}{s}} + s$$

This expansion permits synthesis by inspection, to yield the circuit of Figure 6.6. ■

Example 6.5
Synthesize

$$Z(s) = \frac{s + 2}{(s + 1)(s + 3)}$$

Solution
The poles and zeros of this function lie on the negative real axis, are simple, and alternate. Also, the root closest to the origin ($s = -1$) is a pole, and the root furthest from the origin is a pole (at infinity). These properties guarantee that

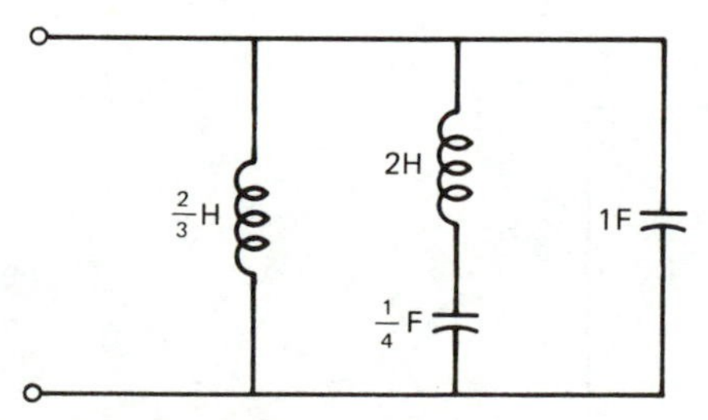

Figure 6.6 Alternate Circuit for Example 6.4 using $Y(s)$.

the dp impedance function can be expressed in the form of Equation 6.2, which ensures that it can be synthesized using an RC network.

The function is a constant at the origin and has a zero at infinity. Thus, in Equation 6.2 the K_0 and the K_∞ terms are both zero (from Section 2.3.2, page 40), and the partial fraction expansion has the form:

$$Z(s) = \frac{K_1}{s+1} + \frac{K_2}{s+3}$$

where

$$K_1 = Z(s)(s+1)|_{s=-1} = \tfrac{1}{2}$$
$$K_2 = Z(s)(s+3)|_{s=-3} = \tfrac{1}{2}$$

Thus

$$Z(s) = \frac{1}{2s+2} + \frac{1}{2s+6}$$

The circuit corresponding to this function is shown in Figure 6.7.

An alternate approach is to expand the admittance function:

$$Y(s) = \frac{(s+1)(s+3)}{s+2}$$

As in the case of the impedance function, the poles of this admittance function lie on the negative real axis, are simple, and alternate. Moreover, the root closest to the origin in this case is a zero ($s = -1$), and the root furthest from the origin is a pole (at infinity). These conditions are sufficient for an admittance function to be expressed in the form of Equation 6.3, which will always lead to an RC realization. Since the admittance function has both a constant at dc and a pole at infinity, its partial fraction expansion has the form:

$$Y(s) = K_0 + K_\infty s + \frac{K_1 s}{s+2}$$

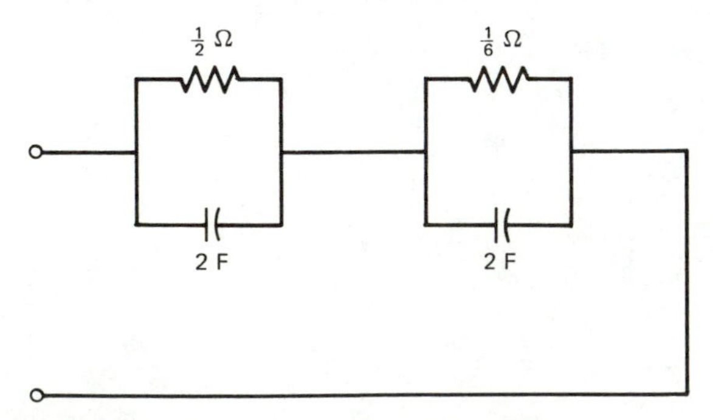

Figure 6.7 Circuit for Example 6.5 using $Z(s)$.

where

$$K_0 = Y(s)\Big|_{s=0} = \frac{3}{2}$$

$$K_\infty = \frac{Y(s)}{s}\Big|_{s=\infty} = 1$$

$$K_1 = Y(s)\,\frac{s+2}{s}\Big|_{s=-2} = \frac{1}{2}$$

Thus

$$Y(s) = \frac{3}{2} + s + \frac{\frac{1}{2}s}{s+2}$$

which leads to the RC network shown in Figure 6.8. ■

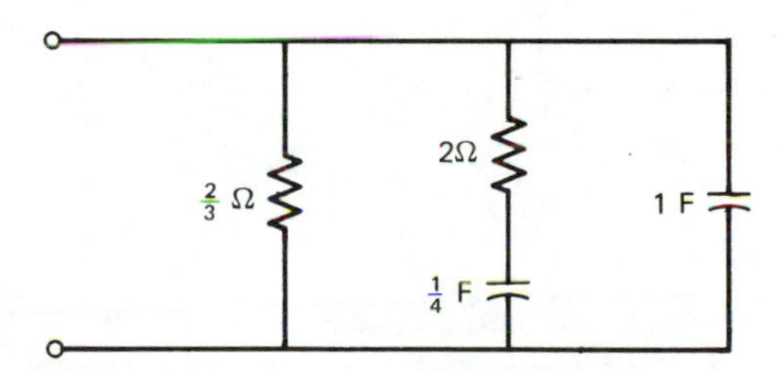

Figure 6.8 Alternate circuit for Example 6.5 using $Y(s)$.

6.2.2 SYNTHESIS USING CONTINUED FRACTION EXPANSION

An alternate dp synthesis technique, proposed by Cauer, is based on the continued fraction expansion of the given function [10]. The procedure consists of alternately removing series and shunt elements from $Z(s)$ until the driving point function is realized, as explained in the following.

Consider the synthesis of an LC dp function $Z(s)$, whose numerator is one degree higher than the denominator. The first step is to divide the denominator into the numerator in the form:

$$Z(s) = sL_1 + Z_1(s)$$

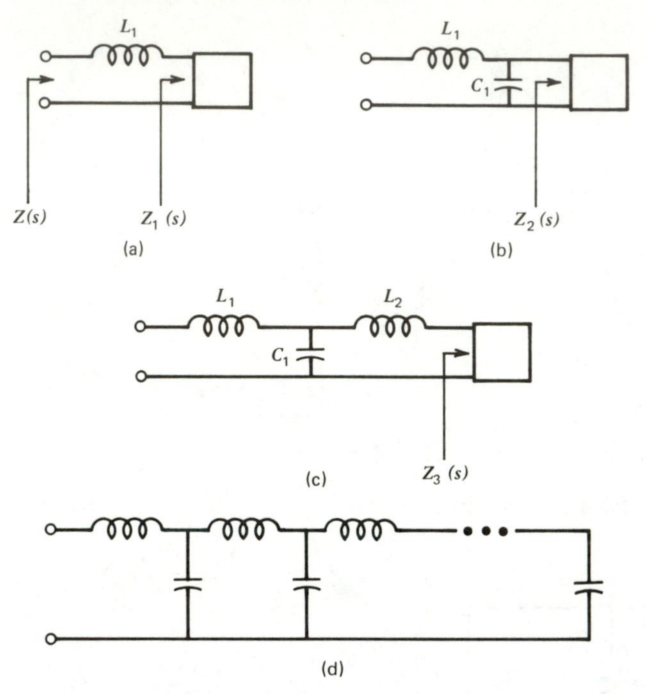

Figure 6.9 Steps in the Cauer synthesis.

where $Z_1(s)$ is the remainder term whose order (defined as the highest power of s in the function) is one lower than that of $Z(s)$.* Thus, $Z(s)$ can be realized as an inductor (L_1 henries) in series with a lower-order dp function $Z_1(s)$, as shown in Figure 6.9*a*. The subtraction of sL_1 from $Z(s)$ makes the degree of the denominator of $Z_1(s)$ one higher than its numerator. Therefore, we can repeat the above procedure on the admittance of the remainder function $Y_1(s) = 1/Z_1(s)$, expanding it as

$$Y_1(s) = sC_1 + Y_2(s)$$

The right-hand side of this expression represents a capacitor C_1 in parallel with an admittance function $Y_2(s)$, whose order is two lower than the original dp function $Z(s)$. Next, $Y_2(s)$ is inverted to get an impedance $Z_2(s)$ which is expanded to yield an inductor L_2 in series with a remainder term $Z_3(s)$. The order of $Z_3(s)$ will be three lower than $Z(s)$. This process of inversion and division is repeated until there is nothing more to be removed from the re-

* For instance if $Z(s) = (s^3 + 2s)/(s^2 + 1)$, the division yields

$$Z(s) = s + \frac{s}{s^2 + 1}$$

The order of the remainder function is one lower than that of $Z(s)$. Also, the degree of the denominator of the remainder is one higher than its numerator.

mainder term; at which point the synthesis is complete. For an nth-order dp function, n such divisions are needed. The form of the complete network is shown in Figure 6.9*d*.

This process of alternate removal of series and shunt elements is easily accomplished by performing a continued fraction expansion on $Z(s)$, as illustrated in the following example.

Example 6.6
Synthesize the following dp function:

$$Z(s) = \frac{s^3 + 2s}{s^2 + 1}$$

Solution
The impedance function may be written as

$$Z(s) = \frac{s^3 + 2s}{s^2 + 1} = s + \frac{s}{s^2 + 1} = s + Z_1(s)$$

The admittance $Y_1(s)$ is

$$Y_1(s) = \frac{s^2 + 1}{s} = s + \frac{1}{s}$$

This is realized by a 1 H inductor in parallel with a 1 F capacitor. The circuit is shown in Figure 6.10.

Observation

1. A convenient way of representing the above division steps is the continued fraction expansion of $Z(s)$:

$$Z(s) = s + \cfrac{1}{s + \cfrac{1}{s}}$$

For a general impedance function, the continued fraction expansion will have the form:

$$Z(s) = a_1 s + \cfrac{1}{a_2 s + \cfrac{1}{a_3 s + \cfrac{1}{a_4 s + \cdots}}} \tag{6.4}$$

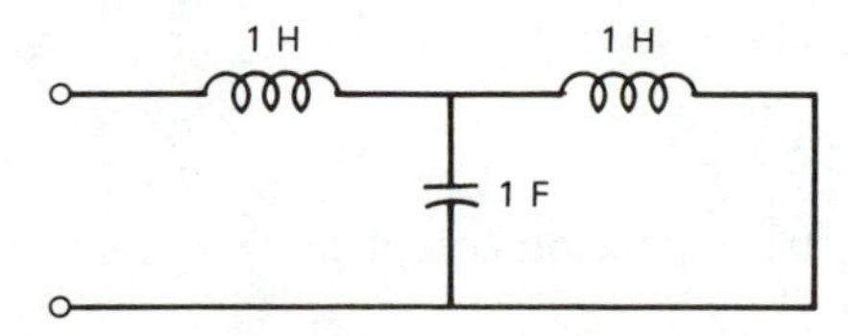

Figure 6.10 Circuit for Example 6.6.

A short, systematic way of obtaining the coefficients $(a_1, a_2, a_3 \ldots)$ in such an expansion, and hence the circuit elements in the ladder network, is shown below:

$$s^2 + 1\overline{)s^3 + 2s}(s = Z_1 \qquad L_1 = 1\text{ H}$$

$$\underline{s^3 + s}$$

$$s\overline{)s^2 + 1}(s = Y_1 \qquad C_1 = 1\text{ F}$$

$$\underline{s^2}$$

$$1\overline{)s}(s = Z_2 \qquad L_2 = 1\text{ H}$$

$$\underline{s}$$

$$\underline{0} \quad \text{expansion complete}$$

2. The procedure described is also applicable to RC and RL networks, where the appropriate shunt and series elements will need to be removed. The continued fraction technique can be applied to the numerator and denominator polynomials network functions in two ways, by arranging the polynomials in *descending* order or *ascending* order. These two ways yield two different circuits. The above comments are illustrated by the next two examples. ■

Example 6.7
Synthesize

$$Z(s) = \frac{s + 2}{s^2 + 4s + 3}$$

Solution

$$s + 2\overline{)s^2 + 4s + 3}(s = Y_1 \qquad C_1 = 1\text{ F}$$

$$\underline{s^2 + 2s}$$

$$2s + 3\overline{)s + 2}(\tfrac{1}{2} = Z_1 \qquad R_1 = \tfrac{1}{2}\,\Omega$$

$$\underline{s + \tfrac{3}{2}}$$

$$\tfrac{1}{2}\overline{)2s + 3}(4s = Y_2 \qquad C_2 = 4\text{ F}$$

$$\underline{2s}$$

$$3\overline{)\tfrac{1}{2}}(\tfrac{1}{6} = Z_2 \qquad R_2 = \tfrac{1}{6}\,\Omega$$

$$\underline{\tfrac{1}{2}}$$

$$\underline{0} \quad \text{expansion complete}$$

The circuit is shown in Figure 6.11. ■

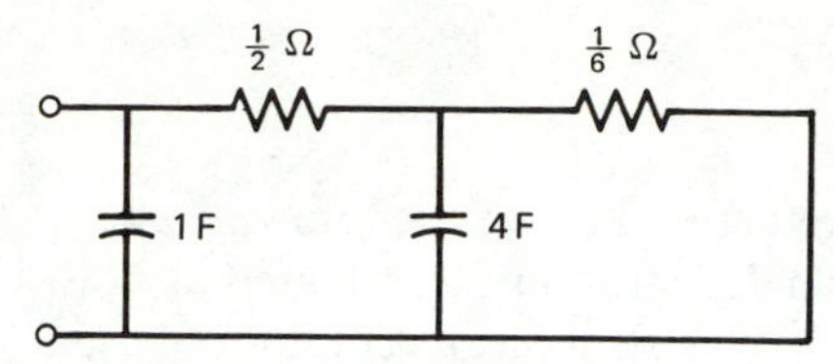

Figure 6.11 Circuit for Example 6.7.

Example 6.8

Synthesize the impedance function of Example 6.7 by the continued fraction expansion method, arranging the polynomials in ascending order.

Solution

The impedance function is written in the form:

$$Z(s) = \frac{2 + s}{3 + 4s + s^2}$$

The continued fraction expansion for this function is

$$2 + s\overline{)3 + 4s + s^2}\left(\tfrac{3}{2} = Y_1 \qquad R_1 = \tfrac{2}{3}\,\Omega\right.$$

$$\underline{2 + \tfrac{3}{2}s}$$

$$\tfrac{5}{2}s + s^2\overline{)2 + s}\left(\frac{4}{5s} = Z_1 \qquad C_1 = \tfrac{5}{4}\,\text{F}\right.$$

$$\underline{2 + \tfrac{4}{5}s}$$

$$\tfrac{1}{5}s\overline{)\tfrac{5}{2}s + s^2}\left(\tfrac{25}{2} = Y_2 \qquad R_2 = \tfrac{2}{25}\,\Omega\right.$$

$$\underline{\tfrac{5}{2}s}$$

$$s^2\overline{)\tfrac{1}{5}s}\left(\frac{1}{5s} = Z_2 \qquad C_2 = 5\,\text{F}\right.$$

$$\underline{\tfrac{1}{5}s}$$

$$\underline{0} \quad \text{expansion complete}$$

The complete circuit is shown in Figure 6.12. ■

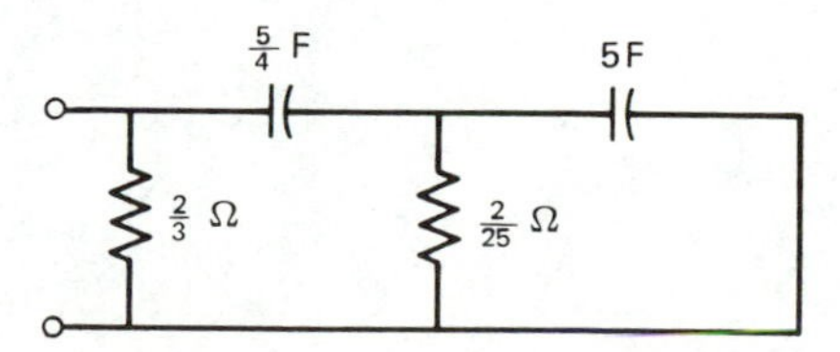

Figure 6.12 Circuit for Example 6.8.

6.3 LOW SENSITIVITY OF PASSIVE NETWORKS

In this section we consider a class of passive networks that are known to have low sensitivity. These networks consist of an LC structure terminated at both ends by resistors, as depicted in Figure 6.13. This structure is commonly used for realizing passive filters; it also forms the basis for the realization of some low-sensitivity active filters.

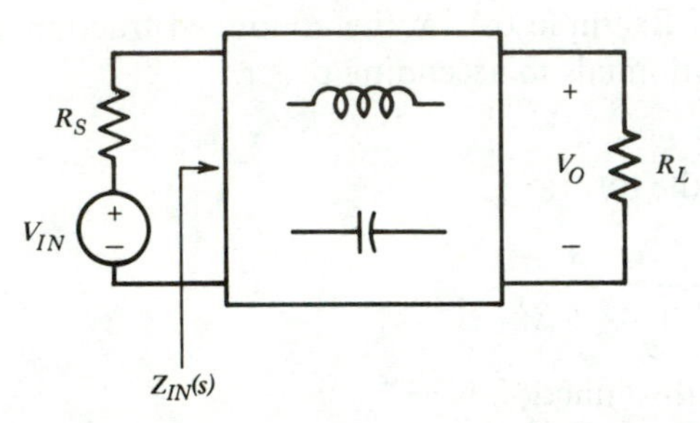

Figure 6.13 Resistively terminated LC network.

The following heuristic argument, due to Orchard [5], shows that these networks can indeed be designed to have a low sensitivity.

Consider the realization of a filter function with an equiripple passband (Figure 6.14). Such a filter function can be realized by a resistively terminated LC network. In addition, it is possible to realize the function so that the input voltage source delivers maximum power to the network at the passband minimum frequencies f_1, f_2, and f_3. At these frequencies, the input impedance Z_{IN} of the network must be equal to the complex conjugate of source impedance.

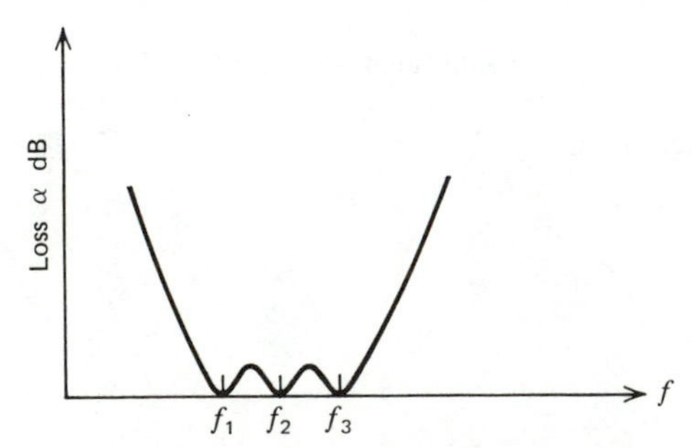

Figure 6.14 Equal minima passband filter characteristic.

If the source resistance is R_S, then

$$\text{Re}(Z_{IN}) = R_S \qquad \text{and} \qquad \text{Im}(Z_{IN}) = 0$$

at the loss minima. Since *LC* networks are lossless, all the power going into the network must be transmitted to the load R_L. Thus, at these loss minima the output power will be maximum; and since the output power is V_O^2/R_L, the output voltage is maximum. Consequently, the loss function V_{IN}/V_O will be minimum. This implies that any change in a component of the network N, whether it be an increase or a decrease, *can only result in an increase in the loss.* Hence at the loss minima:

$$\frac{\partial \alpha}{\partial x} = 0 \tag{6.5}$$

where

α is the loss in dB
x is any component in N.

This is illustrated by the loss versus component-change characteristic in Figure 6.15. From (6.5) and the definition of loss (gain) sensitivity (Equation 5.32), we have

$$\mathcal{S}_x^\alpha = x\frac{\partial \alpha}{\partial x} = 0$$

Therefore, the sensitivity is seen to be *zero* at each loss minimum in the passband. Since the loss is a smooth continuous function, it is reasonable to expect that the sensitivity at frequencies between the minima will remain small. Such a passive realization should therefore have a low sensitivity in the entire passband. The argument does not apply to the sensitivity in the stopband or the transition band. But this is not of great concern, because the requirements on the loss variations in the stopband and the transition band are not very stringent in most filter applications.

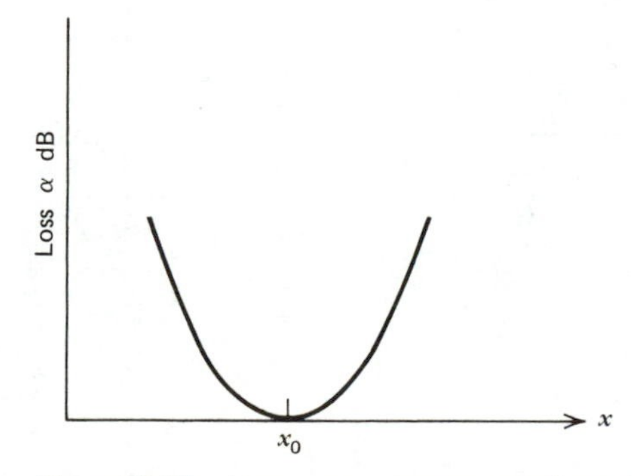

Figure 6.15 Loss versus component change.

6.4 TRANSFER FUNCTION SYNTHESIS

In this section we consider the synthesis of transfer functions using *LC* ladder networks terminated in resistors. The structure of the proposed realization is shown in Figure 6.16. First we consider the special case of a ladder network terminated by a resistor at one end, driven by an ideal voltage source at the other end. Although such singly terminated realizations are not often used in practice (because of their high sensitivity), the principles developed will help in the understanding of the synthesis of the popular doubly terminated ladder networks.

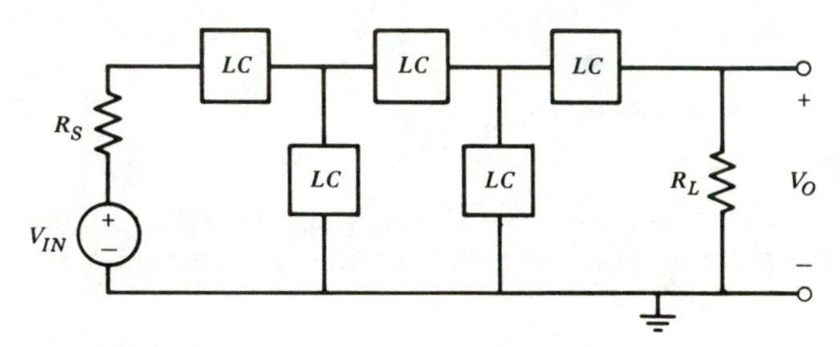

Figure 6.16 Ladder structure terminated by resistors at both ends.

6.4.1 SINGLY TERMINATED LADDER NETWORKS

The voltage transfer function of a lossless *LC* network terminated in a load Y_L (Figure 6.17) is derived in Appendix B. This transfer function, expressed in terms of the y parameters (Equation B.18), is

$$T(s) = \frac{-y_{21}}{Y_L + y_{22}}$$

$$= -\frac{\dfrac{y_{21}}{Y_L}}{1 + \dfrac{y_{22}}{Y_L}} \tag{6.6}$$

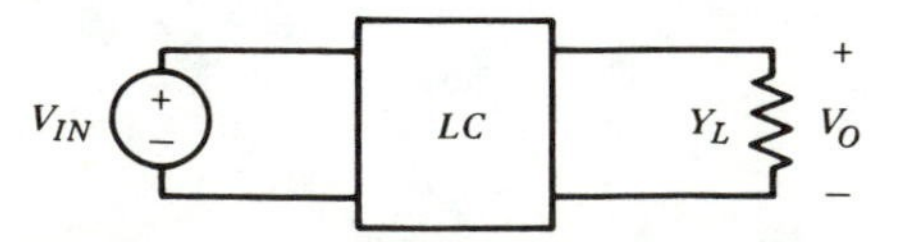

Figure 6.17 Singly terminated *LC* network.

Suppose the transfer function to be realized is given in the form:

$$\frac{V_O(s)}{V_{IN}(s)} = \frac{Q(s)}{P(s)} \tag{6.7}$$

It can be shown that the zeros of transmission of an LC network must be on the imaginary axis [10]; thereby requiring $Q(s)$ to be either an even or an odd polynomial in s. Equation 6.7 can therefore be written in the form:

$$\frac{V_O}{V_{IN}} = \frac{M_1}{M_2 + N_2} \qquad (N_1 = 0) \tag{6.8a}$$

or

$$= \frac{N_1}{M_2 + N_2} \qquad (M_1 = 0) \tag{6.8b}$$

where M_1 and N_1 are the even and odd parts of $Q(s)$, and M_2 and N_2 are the even and odd parts of $P(s)$. It can also be shown [10] that y_{21} must be an odd-rational function (i.e., the ratio of an odd over an even, or an even over odd polynomial). Thus, (6.8) can be written as

$$\frac{V_O}{V_{IN}} = \frac{\dfrac{M_1}{N_2}}{1+\dfrac{M_2}{N_2}} \qquad (N_1 = 0) \tag{6.9a}$$

or

$$\frac{V_O}{V_{IN}} = \frac{\dfrac{N_1}{M_2}}{1+\dfrac{N_2}{M_2}} \qquad (M_1 = 0) \tag{6.9b}$$

Comparing (6.9) with (6.6), the y parameters of the network can be identified as:*

$$\frac{y_{21}}{Y_L} = \frac{M_1}{N_2} \quad \text{and} \quad \frac{y_{22}}{Y_L} = \frac{M_2}{N_2} \qquad (N_1 = 0) \tag{6.10a}$$

or

$$\frac{y_{21}}{Y_L} = \frac{N_1}{M_2} \quad \text{and} \quad \frac{y_{22}}{Y_L} = \frac{N_2}{M_2} \qquad (M_1 = 0) \tag{6.10b}$$

* The negative sign associated with Equation 6.6 can be ignored, since the filter requirements are stated in terms of the magnitude of the approximation function.

From Equation 6.6, any circuit that realizes the parameters y_{21} and y_{22} constitutes a realization of the desired transfer-function $T(s)$. Since y_{22} is a dp admittance function, it can be realized using the Foster or Cauer methods developed in the last few sections. Also, any realization of y_{22} will automatically realize the *poles* of y_{21}, because the poles are determined by the network determinant, which is the same for all the y parameters. If we can arrange for the same network to also realize the *zeros* of y_{21}, we will have realized both the y parameters and, hence, the desired transfer function. How this is accomplished is shown in the following example.

Example 6.9
Realize the following voltage transfer function using an LC ladder terminated in a 1 Ω resistor:

$$T(s) = \frac{s}{s^4 + 3s^3 + 3s^2 + 3s + 1}$$

Solution
Since the numerator is odd ($M_1 = 0$), Equation 6.10b is used to identify y_{12} and y_{22}:

$$y_{21} = \frac{N_1}{M_2} = \frac{s}{s^4 + 3s^2 + 1} \qquad y_{22} = \frac{N_2}{M_2} = \frac{3s^3 + 3s}{s^4 + 3s^2 + 1}$$

y_{21} can be seen to have three zeros at infinity and one zero at the origin. Thus it is necessary to realize the admittance y_{22} so that these four zeros are attained. Now, a zero at the origin can be realized by a series capacitor or a shunt inductor (Figure 6.18, type I components), while a zero at infinity can be realized by a series inductor or a shunt capacitor (Figure 6.19, type II components). In this example, to realize the zero at the origin, one type I component is needed, and the three zeros at infinity require three type II components. One realization

Figure 6.18 Realization of zeros at the origin (Type I components).

Figure 6.19 Realization of zeros at infinity (Type II components).

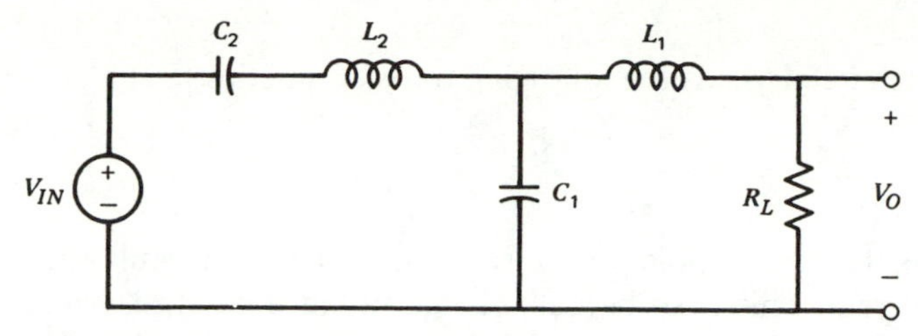

Figure 6.20 A realization for Example 6.9.

of the required zeros is indicated in Figure 6.20. If the given admittance parameter y_{22} is realized with the structure of Figure 6.20, the zeros of y_{21} will also have been realized.

Proceeding with the synthesis, the admittance function y_{22} is realized using the Cauer continued fraction expansion, as follows. The impedance function corresponding to y_{22} is

$$Z_1 = \frac{1}{y_{22}} = \frac{s^4 + 3s^2 + 1}{3s^3 + 3s}$$

The first inductor, L_1 in the desired structure of Figure 6.20, is obtained by dividing the numerator by the denominator:

$$Z_1 = \frac{s}{3} + \frac{2s^2 + 1}{3s^3 + 3s}$$

Thus $L_1 = \frac{1}{3}$ H, and the admittance of the remainder function is

$$Y_2 = \frac{3s^3 + 3s}{2s^2 + 1}$$

The shunt capacitor C_1 is obtained by dividing the numerator by the denominator:

$$Y_2 = \tfrac{3}{2}s + \frac{\frac{3}{2}s}{2s^2 + 1}$$

Thus $C_1 = \frac{3}{2}$ F, and the remainder impedance function is

$$Z_3 = \frac{2s^2 + 1}{\frac{3}{2}s}$$

$$= \frac{4s}{3} + \frac{2}{3s}$$

This is realized using a series branch consisting of an inductor L_2 and a capacitor C_2, where

$$L_2 = \tfrac{4}{3}\text{ H} \qquad C_2 = \tfrac{3}{2}\text{ F}$$

Observations

1. An analysis of the circuit yields the following transfer function:

$$T(s) = \frac{\frac{3}{2}s}{s^4 + 3s^3 + 3s^2 + 3s + 1}$$

   This function has the desired poles and zeros; but the constant multiplier is seen to be different. This is so because the admittance realized, being the ratio of the odd to even parts of the denominator ($y_{22} = N_2/M_2$), is totally independent of the prescribed numerator constant. Therefore, the above synthesis technique can only realize the desired transfer function within a constant multiplier.
2. The circuit topology for the realization is certainly not unique. The only constraint on the topology is that it exhibits the one zero at the origin and the three zeros at infinity.
3. The method described in this example is applicable to any transfer function with zeros at infinity and/or at the origin. If, however, y_{21} has finite zeros on the $j\omega$ axis, the method does not apply. Then we must use the zero shifting technique described in the next section. ■

6.4.2 ZERO SHIFTING TECHNIQUE

In this section we consider the synthesis of transfer functions with finite zeros. In the general ladder structure a zero of transmission is generated by any series branch that acts as an open circuit, or by any shunt branch that acts as a short circuit. Two zero producing sections are shown in Figure 6.21. The series circuit of Figure 6.21*a* is an open circuit at the antiresonant frequency ($\omega_0 = 1/\sqrt{LC}$), while the shunt circuit of Figure 6.21*b* is a short circuit at the resonant frequency ($\omega_0 = 1/\sqrt{LC}$).

From Equation 6.6 it is seen that y_{21} has its zeros at the zeros of transmission. Therefore, to realize the zeros of y_{21} we have to remove the appropriate resonant or antiresonant sections in the synthesis of the admittance parameter y_{22}. The problem that arises in attempting this is that y_{22} does not, in general, have its zeros at the zeros of transmission. What we need to do, therefore, is to somehow

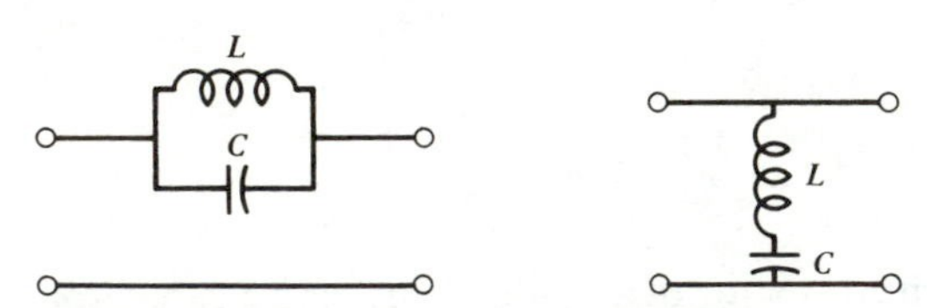

Figure 6.21 Zero producing sections:
(*a*) Series antiresonance. (*b*) Shunt resonance.

manipulate y_{22} so that it does exhibit a zero of transmission, which can then be removed as a series antiresonant or shunt resonant branch. This is accomplished by a procedure known as the *zero shifting technique.*

Consider the reactance function plot of a typical LC impedance function shown in Figure 6.22. (The following discussion is equally applicable to admittance functions.) The function shown has a pole at infinity that can be removed as a series inductor sL, where L is the residue of the pole at infinity

$$L = \left.\frac{Z(s)}{s}\right|_{s=\infty} \tag{6.11}$$

Consider what happens if a part of the residue at infinity is removed, in the form of an inductor αL, where $\alpha < 1$. The reactance of the inductor $\omega(\alpha L)$ is also plotted in Figure 6.22. The figure shows that the remainder function, after αL is removed, will have its zeros shifted toward infinity. In particular, the zero closest to infinity is shifted more than the ones closer to the origin. The amount by which a zero can be shifted depends on the value of α; that is, on what fraction of the residue at infinity is removed. In the limit, if the pole at infinity is removed completely ($\alpha = 1$), then the highest frequency zero is moved all the way to infinity. Thus, the zeros of an impedance function can be moved toward infinity by a partial removal of the pole at infinity. In an analogous manner it is possible to move the zeros of an impedance function toward the origin by a partial removal of a pole at the origin, in the form of a series capacitor. This is illustrated

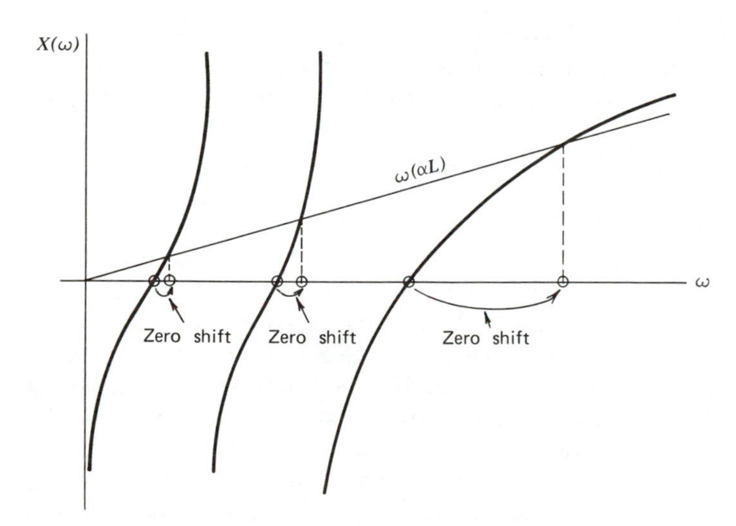

Figure 6.22 Zero shifting by partial removal of residue of a pole at infinity.

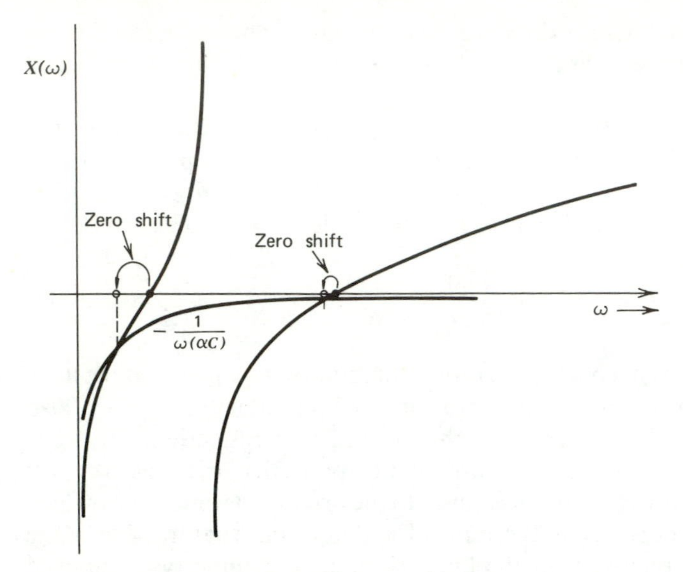

Figure 6.23 Zero shifting by partial removal of residue of a pole at origin.

in Figure 6.23. The residue at the origin is given by

$$\frac{1}{C} = Z(s)s|_{s=0} \tag{6.12}$$

A partial removal of the residue corresponds to removing a series capacitor of value αC, with $\alpha < 1$. Note that the zero closest to the origin is the one that moves the most. This zero can be moved all the way to the origin by removing the pole at the origin completely ($\alpha = 1$).

Let us return to the problem of synthesizing the admittance function y_{22} so that it exhibits the zeros of y_{21}. As mentioned previously, the given admittance function y_{22} will not have its zeros at the desired zeros of transmission. However, it will have a pole at the origin or at infinity, part of which can be removed so that one of the zeros of the remainder admittance is moved to a zero of transmission. This zero can then be realized by using one of the zero producing sections of Figure 6.21. The new remainder function can now be maneuvered in a similar fashion, by removing a part of the residue at infinity (or at the origin) so that a zero is located at the second zero of transmission—which can be realized by another zero producing section. The process is repeated on the remainder functions until all the finite zeros of transmission are realized. Finally, we are left with a function where all of the zeros are at infinity and/or at the origin. This last remainder function is easily realized by a Cauer continued function expansion, as in the last section. It is important to note that we may not remove all of the residues of the poles at the origin and infinity until all the finite zeros of transmission are realized.

The above procedure is illustrated by the following example.

Example 6.10

Realize the following transfer function using an LC network terminated in a 0.5 Ω resistance

$$T(s) = \frac{s^2 + 4}{s^3 + 2s^2 + 2s + 2}$$

Solution

Since the numerator is even ($N_1 = 0$), Equation 6.10a is used to identify the y parameters:

$$\frac{y_{21}}{2} = \frac{M_1}{N_2} = \frac{s^2 + 4}{s^3 + 2s} \qquad \frac{y_{22}}{2} = \frac{M_2}{N_2} = \frac{2s^2 + 2}{s^3 + 2s}$$

The admittance y_{22} must be realized so that zeros of y_{21} ($s = \pm j2$, $s = \infty$) are realized. A pole-zero plot of y_{21} and the impedance function $Z_1 = 1/y_{22}$ is shown in Figure 6.24. Using the zero shifting technique, a part of the residue of the pole of Z_1 at infinity is removed, so that its zero at $s = \pm j\sqrt{2}$ is shifted to $s = \pm j2$. If the residue removed is k_1, then the remainder impedance must satisfy the relationship:

$$(Z_1 - k_1 s)|_{s=\pm j2} = 0$$

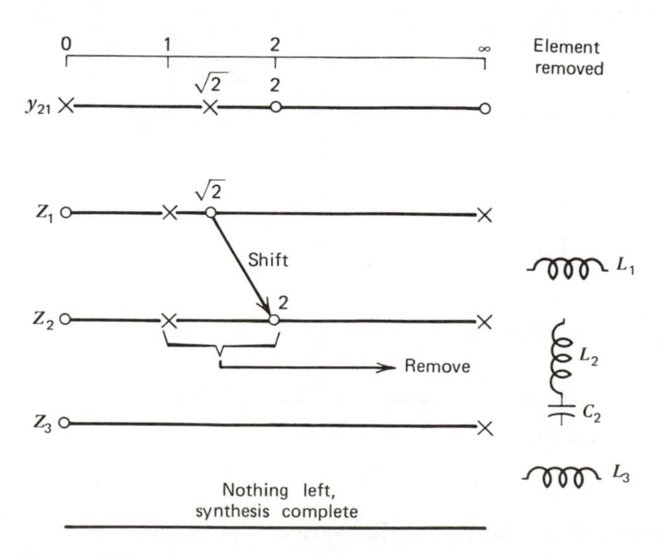

Figure 6.24 Pole-zero plot for Example 6.10.

The inductor L_1 that needs to be removed is therefore given by:

$$L_1 = k_1 = \left.\frac{Z_1}{s}\right|_{s^2=-4}$$

$$= \left.\frac{s^2+2}{4(s^2+1)}\right|_{s^2=-4} = \tfrac{1}{6}\text{ H}$$

The remainder function Z_2 is

$$Z_2 = Z_1 - sL_1$$

$$= \frac{s^3+2s}{4(s^2+1)} - \frac{s}{6} = \frac{s(s^2+4)}{12(s^2+1)}$$

As expected the remainder function has a zero at $s = \pm j2$ (Figure 6.24). The corresponding admittance $Y_2 = 1/Z_2$ has a pole at $s = \pm j2$, which can be realized by using the partial fraction expansion:

$$Y_2 = \frac{12(s^2+1)}{s(s^2+4)} = \frac{9s}{s^2+4} + \frac{3}{s}$$

The first term in this expansion is realized by the shunt elements L_2 and C_2 as shown in Figure 6.25, where

$$L_2 = \tfrac{1}{9}\text{ H} \qquad C_2 = \tfrac{9}{4}\text{ F}$$

The remainder $Y_3 = 3/s$ is realized by an inductor $L_3 = \frac{1}{3}$ H. This inductor realizes the zero of y_{21} at infinity. The complete circuit is shown in Figure 6.25. ■

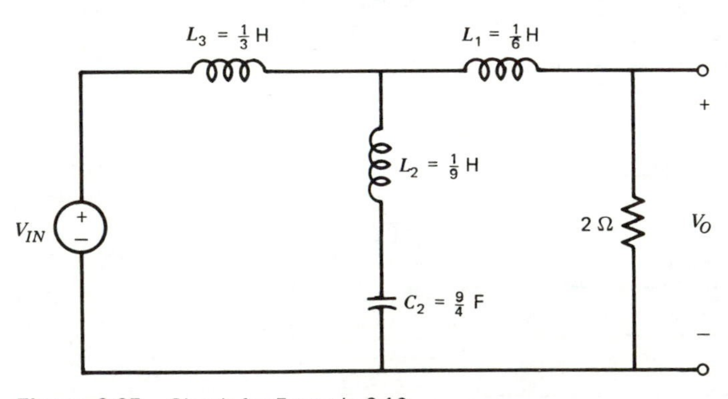

Figure 6.25 Circuit for Example 6.10.

Example 6.11

Synthesize an LC network terminated in 1 Ω that satisfies the normalized low-pass requirements shown in Figure 6.26.

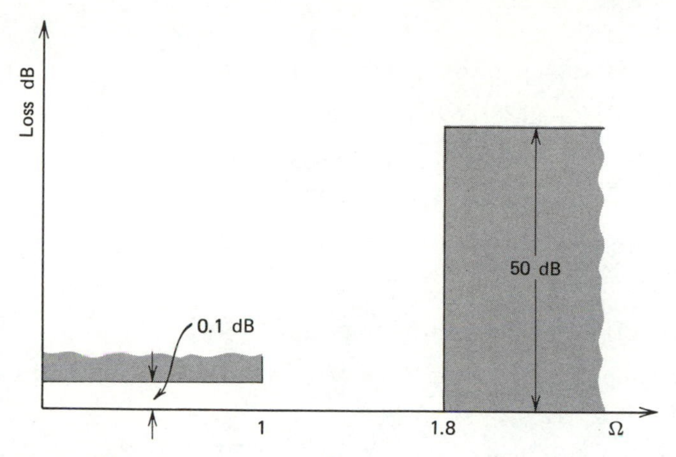

Figure 6.26 Requirements for Example 6.11.

Solution

In this problem, as in many practical situations, the choice of the approximation function has been left to the designer. From Chapter 4, we know that the elliptic approximation will lead to the lowest-order filter; hence, this is the approximation type we will use. The required function may be found using standard tables. For example, from Christian and Eisenmann* [1] (page 100, $\Omega_s =$ 1.78829), the elliptic approximation is found to be:

$$T(s) = \frac{(s^2 + 3.476896154)(s^2 + 8.227391422)}{55.3858(s + 0.60913)(s^2 + 0.263147s + 1.166357185)(s^2 + 0.85422659s + 0.7269594794)}$$

This function has a passband ripple of 0.0988 dB, minimum stopband loss of 53.10 dB and a transition band ratio of 1.788, thereby meeting the requirements with a margin to spare. The transfer function has two finite zeros at $\pm j1.864643$ and $\pm j2.868343$ and a zero at infinity.

Since the numerator is even, the y parameters that need to be realized are

$$y_{21} = \frac{M_1}{N_2} \qquad y_{22} = \frac{M_2}{N_2}$$

Expanding $T(s)$ and separating into even and odd parts, after much computation, we get

$$M_2 = 1.7265s^4 + 2.47783s^2 + 0.5164779$$
$$N_2 = s^5 + 2.79873s^3 + 1.571316s$$

* Alternately from Zverev [11] (page 216, $\Omega_K = 1.788$).

which can be factored to yield

$$y_{22} = \frac{1.7265[s^2 + (0.50305)^2][s^2 + (1.08729)^2]}{s[s^2 + (0.881669)^2][s^2 + (1.42179)^2]}$$

$$y_{21} = \frac{[s^2 + (1.864643)^2][s^2 + (2.868343)^2]}{s[s^2 + (0.881669)^2][s^2 + (1.42179)^2]}$$

The admittance y_{22} must be realized so as to also generate the zeros of y_{21}, which are at

$$s = 0, \pm j1.864643 \qquad \text{and} \qquad s = \pm j2.868343$$

A pole-zero plot of the impedance function $Z_1 = 1/y_{22}$ is shown in Figure 6.27*a*. Note that Z_1 has zeros at

$$s = 0, \pm j0.881669 \qquad \text{and} \qquad \pm j1.42179$$

Z_1 also has a pole at infinity. We can remove part of the residue of this pole at infinity so that the remainder impedance Z_2 has a zero at $s = \pm j2.868343$. If the residue removed is k_1, then the remainder function must satisfy the relationship:

$$(Z_1 - k_1 s)|_{s=\pm j2\cdot 868343} = 0$$

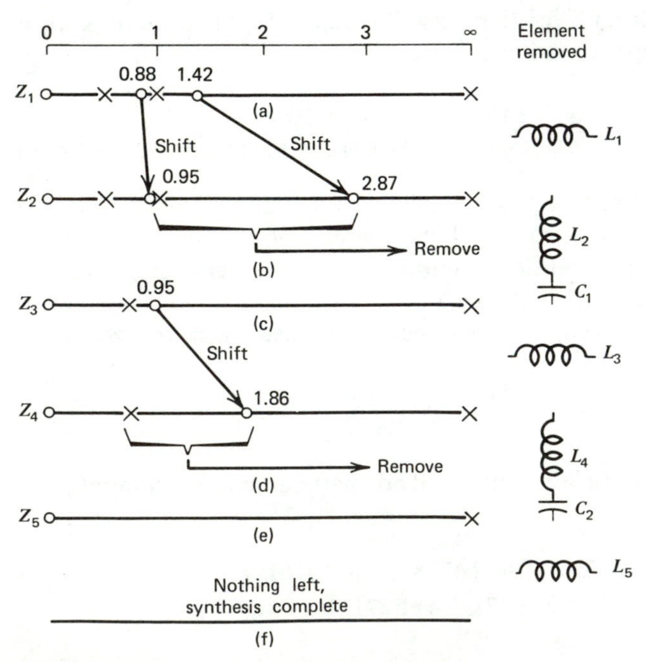

Figure 6.27 Pole-zero plot for Example 6.11.

The inductor L_1 that needs to be extracted is therefore given by

$$L_1 = k_1 = \left.\frac{Z_1}{s}\right|_{s^2=-(2.868343)^2} = 0.4769 \text{ H}$$

The remainder function Z_2 will have a zero at $s = \pm j2.868343$ (Figure 6.27*b*):

$$\begin{aligned} Z_2 &= Z_1 - 0.4769s \\ &= \frac{s[s^2 + (0.881669)^2][s^2 + (1.42179)^2]}{1.7265[s^2 + (0.50305)^2][s^2 + (1.08729)^2]} - 0.4769s \\ &= \frac{0.17662s[s^2 + (2.868343)^2][s^2 + (0.952045)^2]}{1.7265[s^2 + (0.50305)^2][s^2 + (1.08729)^2]} \end{aligned}$$

The admittance $Y_2 = 1/Z_2$ has a pole at $s = \pm j2.868343$, which can be separated from Y_2 by using the partial fraction expansion:

$$Y_2 = \frac{k_2 s}{s^2 + (2.868343)^2} + Y_3$$

where

$$k_2 = \left.\frac{[s^2 + (2.868343)^2]}{s} Y_2\right|_{s^2=-(2.868343)^2} = 9.1175$$

In the above expansion for Y_2, the term

$$\frac{9.1175s}{s^2 + (2.868343)^2} = \cfrac{1}{0.10967s + \cfrac{1}{1.10819s}}$$

can be realized as a shunt LC resonant branch as shown in Figure 6.28*a*. At this stage we have realized one of the zeros of transmission (at $s = \pm j2.868343$). The remainder function Y_3 must be synthesized in such a way as to realize the remaining zeros of transmission which are at $s = \infty$ and $s = \pm j1.864643$. The remainder is given by

$$Y_3 = Y_2 - \frac{9.1175s}{s^2 + (2.8686343)^2} = \frac{0.6577[s^2 + (0.73522)^2]}{s[s^2 + (0.95204)^2]}$$

The poles and zeros of $Z_3 = 1/Y_3$ are sketched in Figure 6.27*c*. Proceeding as before, a part of the residue of the pole at infinity is removed in order to shift the finite zero of Z_3 to $s = \pm j1.864643$. The required residue is given by

$$k_3 = \left.\frac{Z_3}{s}\right|_{s^2=-(1.864643)^2} = 1.3310$$

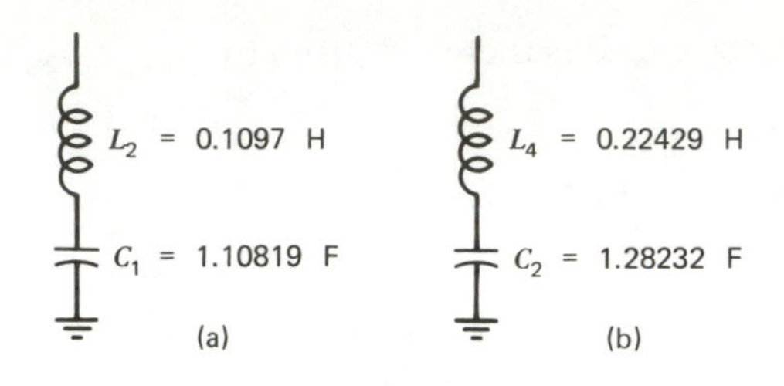

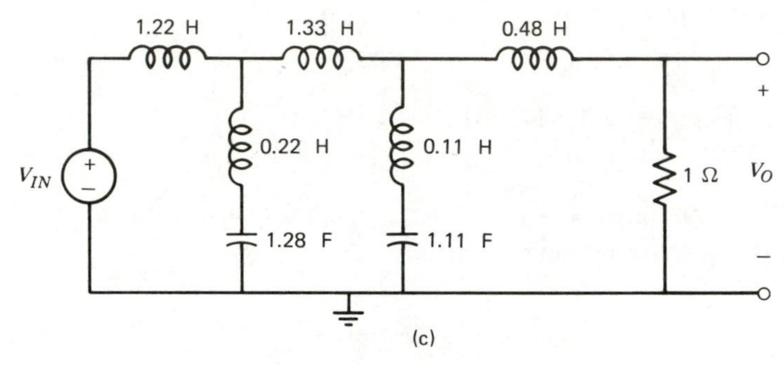

Figure 6.28 Realization for Example 6.11: (*a*) Realization of zero at $s = \pm j2.868343$. (*b*) Realization of zero at $s = \pm j1.864643$. (*c*) Complete circuit.

corresponding to a series inductor $L_3 = 1.3310$ H. The remainder $Z_4 = Z_3 - 1.3310s$ reduces to

$$Z_4 = \frac{0.12461s[s^2 + (1.864643)^2]}{0.6577[s^2 + (0.73522)^2]}$$

As expected, Z_4 has a zero at $s = \pm j1.864643$. The admittance $Y_4 = 1/Z_4$ has a pole at $s = \pm j1.8645$, which is separated from Y_4 by using the partial fraction expansion:

$$Y_4 = \frac{k_4 s}{[s^2 + (1.864643)^2]} + Y_5$$

where

$$k_4 = \frac{[s^2 + (1.864643)^2]}{s} Y_4 \Bigg|_{s^2 = -(1.864643)^2} = 4.4585$$

In the expansion for Y_4, the term

$$\frac{4.4585s}{s^2 + (1.864643)^2} = \frac{1}{0.22429s + \dfrac{1}{1.28232s}}$$

is synthesized using the shunt resonant branch shown in Figure 6.28*b*. This branch provides the zero of transmission at $s = \pm j1.864643$. The remainder admittance Y_5 is

$$Y_5 = \frac{0.6577[s^2 + (0.73522)^2]}{0.12461s[s^2 + (1.864643)^2]} - \frac{4.4585s}{[s^2 + (1.864643)^2]} = \frac{1}{1.2182s}$$

The corresponding impedance $Z_5 = 1/Y_5$ has a pole at infinity (Figure 6.27*e*), which is easily realized by the series inductor $L_5 = 1.2182$ H. This completes the realization of y_{22}, with the required zeros of y_{21}. The complete circuit is sketched in Figure 6.28*c*.

Observations

1. Even though the inductors and capacitors can only be manufactured to within approximately $\frac{1}{2}$ percent, the calculations need to be carried to several decimal places. This is because errors in the computation grow with the number of steps in the synthesis. In fact, for high-order filters it becomes necessary to use double precision (approximately 20 digits) computer algorithms.
2. In this example a different circuit could have been obtained by reversing the order of the finite zero extraction (first $\pm j1.864643$ then $\pm j2.868343$). In general, varying the order of zero extraction leads to different circuit realizations. However, in higher-order filter designs, some of the sequences of zero extractions may result in negative element values and must therefore be rejected. It may even happen that every sequence yields some negative elements. In these cases the zero shifting technique, as described above, does not work and we must resort to more complex structures [4]. Usually, though, more than one realization will exist. The criteria used in choosing between alternate realizations are (a) the sensitivity of the circuit to component changes, (b) the number of inductors used in the realization, and (c) the spreads in the element values.
3. The element values for this passive filter can be found in standard tables. For example, see Zverev [11] page 217 for $\Omega_K = 1.788$ and $K^2 = \infty$.

■

6.4.3 DOUBLY TERMINATED LADDER NETWORKS

With the background developed thus far on singly terminated networks, we are now ready to consider the synthesis of transfer functions using *LC* ladder networks with resistive terminations at both ends (Figure 6.29).

As in singly terminated networks, the synthesis of the transfer function will be reduced to the synthesis of a derived driving point immittance function. First, the *z* (or *y*) parameters are obtained in terms of the given transfer function. These parameters are then synthesized using the zero shifting technique.

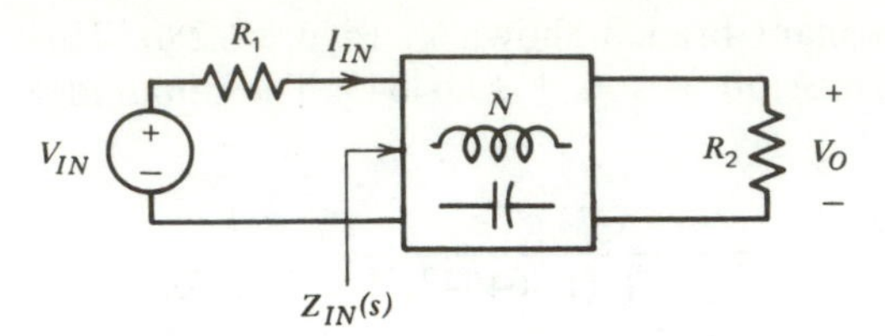

Figure 6.29 Doubly terminated *LC* network.

Referring to Figure 6.29, the given transfer function is $T(s) = V_O(s)/V_{IN}(s)$ and the terminating resistors are R_1 and R_2. The first step is to find an expression for $Z_{IN}(s)$, the input impedance of the network N terminated in R_2. The fundamental idea used in deriving $Z_{IN}(s)$ is that an LC network is lossless. Consequently, the power going into the network N must be equal to the power leaving N. If the input impedance is written as

$$Z_{IN}(j\omega) = R_{IN}(j\omega) + jX_{IN}(j\omega) \tag{6.13}$$

the power going into N is

$$P_{IN}(j\omega) = |I_{IN}(j\omega)|^2 R_{IN}(j\omega) = \left|\frac{V_{IN}(j\omega)}{R_1 + Z_{IN}(j\omega)}\right|^2 R_{IN}(j\omega) \tag{6.14}$$

The power leaving N is delivered to the load resistor R_2, and is given by

$$P_O(j\omega) = \left|\frac{V_O(j\omega)}{R_2}\right|^2 R_2 \tag{6.15}$$

Equating the expressions for P_{IN} and P_O, we get

$$\left|\frac{V_{IN}(j\omega)}{R_1 + Z_{IN}(j\omega)}\right|^2 R_{IN}(j\omega) = \left|\frac{V_O(j\omega)}{R_2}\right|^2 R_2 \tag{6.16}$$

from which

$$\left|\frac{V_O(j\omega)}{V_{IN}(j\omega)}\right|^2 = \frac{R_2 R_{IN}(j\omega)}{|R_1 + Z_{IN}(j\omega)|^2} \tag{6.17}$$

Now, the following relationship can be verified by expanding the right-hand side of the equation:

$$4R_1 R_{IN}(j\omega) = |R_1 + Z_{IN}(j\omega)|^2 - |R_1 - Z_{IN}(j\omega)|^2 \tag{6.18}$$

Substituting for $R_{IN}(j\omega)$ in (6.17), we get

$$\left|\frac{V_O(j\omega)}{V_{IN}(j\omega)}\right|^2 = \frac{R_2}{4R_1}\left[1 - \left|\frac{R_1 - Z_{IN}(j\omega)}{R_1 + Z_{IN}(j\omega)}\right|^2\right] \tag{6.19}$$

This equation can be written in terms of the **transducer function** $H(j\omega)$ and the **characteristic function** $K(j\omega)$ which are defined as

$$H(j\omega) = \sqrt{\frac{R_2}{4R_1}\frac{V_{IN}(j\omega)}{V_O(j\omega)}} \tag{6.20}$$

and

$$|K(j\omega)|^2 = |H(j\omega)|^2 - 1 \tag{6.21}$$

The transducer function is proportional to the loss of the network $V_{IN}(j\omega)/V_O(j\omega)$; and the characteristic function is a measure of how close the squared magnitude of the loss function is to unity. The names chosen for these functions are related to their properties, which are discussed in detail in [9].

From (6.20) and (6.21)

$$\left|\frac{V_O(j\omega)}{V_{IN}(j\omega)}\right|^2 = \frac{R_2}{4R_1}\frac{1}{|H(j\omega)|^2} = \frac{R_2}{4R_1}\left[1 - \left|\frac{K(j\omega)}{H(j\omega)}\right|^2\right] \tag{6.22}$$

Comparing this equation with (6.19), we get

$$\left|\frac{K(j\omega)}{H(j\omega)}\right|^2 = \left|\frac{R_1 - Z_{IN}(j\omega)}{R_1 + Z_{IN}(j\omega)}\right|^2 \tag{6.23}$$

Extending this function to the s domain by analytic continuation:*

$$\frac{K(s)K(-s)}{H(s)H(-s)} = \frac{R_1 - Z_{IN}(s)}{R_1 + Z_{IN}(s)} \cdot \frac{R_1 - Z_{IN}(-s)}{R_1 + Z_{IN}(-s)} \tag{6.24}$$

One way of separating $K(s)/H(s)$ from the above function is†

$$\frac{K(s)}{H(s)} = \frac{R_1 - Z_{IN}(s)}{R_1 + Z_{IN}(s)} \tag{6.25}$$

This equation may be rearranged to yield

$$Z_{IN}(s) = R_1 \frac{1 - \dfrac{K(s)}{H(s)}}{1 + \dfrac{K(s)}{H(s)}} \tag{6.26}$$

Expressing the numerators of K and H in terms of their respective even and odd parts, as

$$H = \frac{(H_e + H_o)}{P} \qquad K = \frac{(K_e + K_o)}{P}$$

* Analytic continuation was also used in Chapter 4 to obtain $H(s)$ from $H(j\omega)$.

† In [7] it is shown that the poles of $K(s)/H(s)$ must be the left half plane poles of the function given by (6.24). The zeros of $K(s)/H(s)$ however, may be chosen from the left and/or right half plane zeros of (6.24).

and substituting in (6.26), we get the following expression for the input impedance of the network:

$$Z_{IN}(s) = R_1 \frac{H_e + H_o - K_e - K_o}{H_e + H_o + K_e + K_o} \tag{6.27}$$

Thus far we have derived an expression for $Z_{IN}(s)$ in terms of the two functions $H(s)$ and $K(s)$, both of which can be evaluated from the given approximation function $T(s)$. Proceeding as in the singly terminated case, we next relate this expression for $Z_{IN}(s)$ to the z and y parameters of the network.

From Appendix B (Equation B.14), the input impedance of a network in terms of the z parameters is

$$Z_{IN}(s) = \frac{\dfrac{\Delta z}{R_2} + z_{11}}{\dfrac{z_{22}}{R_2} + 1} \tag{6.28}$$

where

$$\Delta z = z_{11}z_{22} - z_{12}z_{21} \tag{6.29}$$

Now z_{11} and z_{22}, since they are dp functions of an LC network, must be odd-rational functions (Section 2.3.3, page 43). With this in mind, Equation 6.27 is rearranged in the form of Equation 6.28, as either

$$Z_{IN}(s) = \frac{R_1 \dfrac{H_o - K_o}{H_o + K_o} + R_1 \dfrac{H_e - K_e}{H_o + K_o}}{\dfrac{H_e + K_e}{H_o + K_o} + 1} \tag{6.30a}$$

or

$$Z_{IN}(s) = \frac{R_1 \dfrac{H_e - K_e}{H_e + K_e} + R_1 \dfrac{H_o - K_o}{H_e + K_e}}{\dfrac{H_o + K_o}{H_e + K_e} + 1} \tag{6.30b}$$

Comparing (6.28) with (6.30), the z parameters are identified as

$$z_{11} = R_1 \frac{H_e - K_e}{H_o + K_o} \qquad z_{22} = R_2 \frac{H_e + K_e}{H_o + K_o} \tag{6.31a}$$

Alternately, comparing (6.28) and (6.30b)

$$z_{11} = R_1 \frac{H_o - K_o}{H_e + K_e} \qquad z_{22} = R_2 \frac{H_o + K_o}{H_e + K_e} \tag{6.31b}$$

These equations identify the impedance parameters z_{11} and z_{22} in terms of the even and odd parts of H and K. As in the singly terminated case, the synthesis of the approximation function is now reduced to that of realizing z_{11} (or z_{22}) in such a way that the zeros of transmission (which are also the zeros of z_{12} and of z_{21}) are also realized. This step entails the use of the zero shifting technique.

Alternately, we could consider the input admittance, which is given by (Equation B.17)

$$Y_{IN}(s) = \frac{y_{11} + \Delta y R_2}{1 + y_{22} R_2} \tag{6.32}$$

Comparing this equation with the inverse of Equation 6.27, the y parameters are identified as

$$y_{11} = \frac{1}{R_1} \frac{H_e + K_e}{H_o - K_o} \qquad y_{22} = \frac{1}{R_2} \frac{H_e - K_e}{H_o - K_o} \tag{6.33a}$$

or

$$y_{11} = \frac{1}{R_1} \frac{H_0 + K_o}{H_e - K_e} \qquad y_{22} = \frac{1}{R_2} \frac{H_o - K_o}{H_e - K_e} \tag{6.33b}$$

In this case the synthesis of the given transfer function reduces to the realization of either y_{11} or y_{22}, so as to exhibit the zeros of transmission.

Finally, we will show that it is desirable to scale the transducer function $H(j\omega)$ (i.e., multiply it by an appropriate constant) to ensure a realization that has a low sensitivity. Recall from Section 6.3 that the low sensitivity attributed to doubly terminated *LC* networks is achieved when the power transferred through the network is a maximum at the frequencies of loss minima. Therefore, at these minima, we want the maximum power available from the source to be delivered to the load. Now the source will deliver its maximum power when the real part of the input impedance of the *LC* network is equal to the source resistance R_1, and the imaginary part of the input impedance is zero. Then, the power from the source is

$$P_{IN}(j\omega)_{\max} = \left| \frac{V_{IN}(j\omega)}{R_1 + R_1} \right|^2 R_1 = \frac{|V_{IN}(j\omega)|^2}{4R_1} \tag{6.34}$$

The power delivered to the load, from Equation 6.15, is

$$P_O(j\omega) = \left| \frac{V_O(j\omega)}{R_2} \right|^2 R_2$$

Equating these two powers, we obtain the following relationship that must hold true at the frequencies of loss minima:

$$\frac{|V_{IN}(j\omega)|^2}{4R_1} = \frac{|V_O(j\omega)|^2}{R_2}$$

or

$$\left|\frac{V_{IN}(j\omega)}{V_O(j\omega)}\right|^2 \frac{R_2}{4R_1} = 1 \tag{6.35}$$

When compared with Equation 6.20, this equation implies that the transducer function must satisfy the relationship

$$|H(j\omega)|^2 = 1 \tag{6.36}$$

at the loss minima. Thus, to achieve the low sensitivity feature of doubly terminated *LC* networks, Equation 6.36 must be satisfied at the frequencies of loss minima. It may be mentioned that for the standard approximation functions described in Chapter 4 (Butterworth, Chebyshev, Bessel, and elliptic), the characteristic function $K(s)$ is constrained to have zeros at the passband minima. From relationship (6.21) between $H(s)$ and $K(s)$, we see that in these cases Equation 6.36 will be satisfied, and therefore the resulting circuit will always exhibit maximum power transfer.

A summary of the steps for the synthesis of doubly terminated *LC* networks is given below:

1. From the given filter requirements, determine the approximation function. The methods and criteria for choosing the approximation function were discussed in Chapters 4 and 5. The function will be described by poles, zeros, and a constant multiplier.
2. Determine the transducer function $H(s)$ and the characteristic function $K(s)$. For the standard approximations (Butterworth, Chebyshev, Bessel, and elliptic), these functions are uniquely defined by the filter requirements, and may be obtained from standard tables.

   For other than the standard functions, $H(s)$ and $K(s)$ may be obtained in one of the following two ways:

   (a) The transducer function is obtained from the transfer function using Equation 6.20. Next, the magnitude of $H(j\omega)$ is evaluated in the passband using, say, the MAG program, and $H(s)$ is scaled so that

   $$20 \log_{10}|H(j\omega)| = 0 \text{ dB} \tag{6.37}$$

   at the frequencies of loss minima. Notice that the constant multiplier so obtained depends only on Equation 6.37 and is not at all related to the constant associated with the approximation step or to the resistor values R_1 or R_2. The poles and zeros of $H(j\omega)$ are the same as those for the approximation loss function. Finally, the characteristic function $K(s)$ is evaluated from the relations

   $$|K(j\omega)|^2 = |H(j\omega)|^2 - 1 \tag{6.38}$$

   $$K(s)K(-s) = |K(j\omega)|^2\big|_{j\omega=s} \tag{6.39}$$

(b) Alternatively, the characteristic function $K(s)$ could be identified from the given filter requirements and the transducer function $H(s)$ derived from $K(s)$ using (6.38). Indeed, this approach is more convenient and is almost universally used. The details and advantages of this approach are explained by Temes in [9].

3. Using Equation 6.31 and 6.33, find the immittance parameters y_{11}, y_{22}, z_{11}, and z_{22}.

4. Synthesize any one of the 4 immittance parameters using the dp synthesis techniques, so as to exhibit the zeros of transmission. The zero shifting technique will be needed for this synthesis step.

Suppose, for example, that the open circuit input impedance z_{11} is synthesized and the realization is as shown in Figure 6.30*a*. This network gives all the element values except for the load termination.

5. The terminating resistance is determined as follows. Realize z_{22} using the reverse topology, as shown in Figure 6.30*b*. Depending on the resistance R_2, this network will be found to be an impedance scaled version of the original network of Figure 6.30*a*. In order that the forward and reverse realization be identical it will be necessary to impedance scale the circuit of Figure 6.30*b* by some factor α. The scaling changes the load resistance R_2 to R_2/α. The complete network is shown in Figure 6.30*c*, where the source resistance is R_1 and load resistance is R_2/α. If the filter requirements specify that the load resistance *must* be R_2, we will need to use an ideal transformer at the load end, as shown in Figure 6.30*d*. It should be mentioned that since transformers are expensive components the designer will usually accept the scaled load resistance, rather than use a transformer.

6. Determine the voltage transfer function. This is obtained using the relationship:

$$T(s) = \frac{V_O(s)}{V_{IN}(s)} = \sqrt{\frac{(R_2/\alpha)}{4R_1}}\frac{1}{H(s)} \tag{6.40}$$

In general, the constant multiplier associated with the synthesized transfer function will differ from that obtained in the approximation step. However, the difference in the constant multiplier only affects the flat loss of the filter, without affecting the frequency characteristics. This flat loss can be changed, if so desired, by following the filter with a flat gain amplifier.

The above procedure is illustrated by the following examples.

Example 6.12

Realize the following second-order bandpass transfer function using an *LC* ladder network terminated in a 1 Ω source resistance and a 2-Ω load resistance.

$$T(s) = \frac{V_O(s)}{V_{IN}(s)} = \frac{2s}{s^2 + s + 1}$$

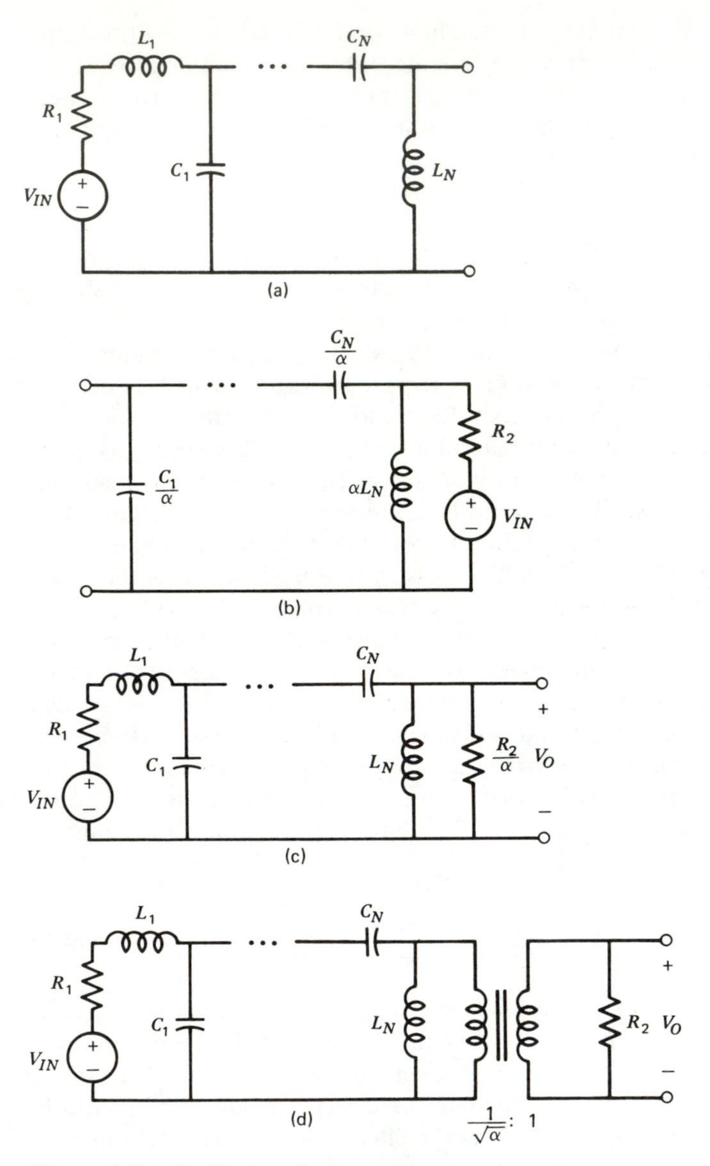

Figure 6.30 Realization of z parameters: (*a*) Forward realization of z_{11}. (*b*) Reverse realization of z_{22}. (*c*) Terminated network; (*d*) Terminated network with ideal transformer.

Solution

The transducer function $H(s)$ is given by

$$H(s) = \frac{V_{IN}(s)}{V_O(s)} = C\left(\frac{s^2 + s + 1}{s}\right)$$

The frequency at which this function has a minimum loss is $s = \pm j$. The constant C is chosen so that the loss at this frequency is 1 (Equation 6.36). Thus

$$C^2 \left| \frac{s^2 + s + 1}{s} \right|^2_{s = \pm j} = C^2 \left| \frac{-1 + j + 1}{j} \right|^2 = 1$$

from which $C = 1$. The next step is to find the characteristic function $K(j\omega)$. Using Equation 6.38

$$\begin{aligned} K(j\omega)K(-j\omega) &= |H(j\omega)|^2 - 1 \\ &= \left| \frac{-\omega^2 + j\omega + 1}{j\omega} \right|^2 - 1 \\ &= \frac{\omega^4 - 2\omega^2 + 1}{\omega^2} = \frac{(\omega^2 - 1)^2}{\omega^2} \end{aligned}$$

From Equation 6.39

$$K(s)K(-s) = \frac{(\omega^2 - 1)^2}{\omega^2}\bigg|_{\omega = s/j} = \frac{(s^2 + 1)^2}{-s^2} = \left(\frac{s^2 + 1}{s}\right)\left(\frac{s^2 + 1}{-s}\right)$$

from which the characteristic function may be identified as

$$K(s) = \frac{s^2 + 1}{s}$$

The even and odd parts of the numerators of $H(s)$ and $K(s)$ are

$$\begin{aligned} H_e &= s^2 + 1 \qquad & H_0 &= s \\ K_e &= s^2 + 1 \qquad & K_0 &= 0 \end{aligned}$$

It is convenient to realize the impedance parameter z_{11} given by Equation 6.31b

$$z_{11} = R_1 \frac{H_o - K_o}{H_e + K_e} = \frac{s}{2(s^2 + 1)}$$

This impedance function must be realized to exhibit the zeros of the transfer function, which are at

$$s = 0 \qquad \text{and} \qquad s = \infty$$

The circuit realization of the impedance function z_{11} is readily seen to be an inductance L_1 in parallel with a capacitance C_1, where

$$L_1 = \tfrac{1}{2}\ \text{H} \qquad \text{and} \qquad C_1 = 2\ \text{F}$$

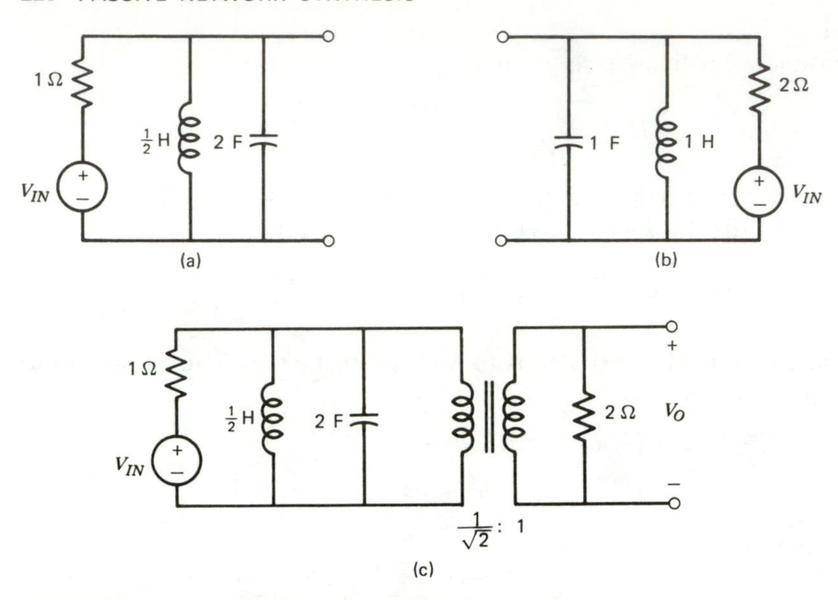

Figure 6.31 Realization for Example 6.12: (*a*) Forward realization of z_{11}. (*b*) Reverse realization of z_{22}. (*c*) Complete circuit.

The inductor provides the zero at the origin and the capacitor provides the zero at infinity. At this stage we have the forward realization of z_{11}, as shown in Figure 6.31*a*. To determine the terminating resistance, we must synthesize z_{22}, which is given by Equation 6.31b

$$z_{22} = R_2 \frac{H_o + K_o}{H_e + K_e} = 2 \frac{s}{2(s^2 + 1)}$$

The circuit for the reverse realization is shown in Figure 6.31*b*. It is seen that the elements in the reverse realization need to be scaled by $\frac{1}{2}$ to match the forward realization. But this makes the terminating resistance 1 Ω. To achieve the 2 Ω load resistance an ideal transformer with a turns ratio of $1/\sqrt{2}$:1 is needed, as shown in the complete circuit of Figure 6.31*c*.

The transfer function realized by this circuit, from Equation 6.40, is

$$T(s) = \frac{0.5s}{s^2 + s + 1}$$

This result can easily be verified by analyzing the circuit. ■

Example 6.13

Synthesize the low-pass characteristic of Example 6.11 (Figure 6.26), using an *LC* network terminated at both ends with a 1 Ω resistor.

Solution

As in Example 6.11, the elliptic approximation function will be used. For these approximations both the transducer function $H(s)$ and the characteristic function $K(s)$ are available in standard tables. From Christian and Eisenmann [1] (page 100, $\Omega_S = 1.78829$) these functions are

$$H(s) = \frac{55.3858(s + 0.60913)(s^2 + 0.263147s + 1.166357185)(s^2 + 0.85422659s + 0.7269594794)}{(s^2 + 3.476896154)(s^2 + 8.227391422)}$$

and

$$K(s) = \frac{55.3858s(s^2 + 0.38869988)(s^2 + 0.91978266)}{(s^2 + 3.476896154)(s^2 + 8.227391422)}$$

Expanding the numerators of $H(s)$ and $K(s)$, and factoring the even and odd parts, we get:

$$H_e = 1.7265(s^2 + (0.50305)^2)(s^2 + (1.08729)^2)$$
$$H_o = s(s^2 + (0.881669)^2)(s^2 + (1.42179)^2)$$
$$K_e = 0$$
$$K_o = 55.3858s(s^2 + (0.623458)^2)(s^2 + (0.959053)^2)$$

The immittance we choose to realize is

$$z_{11} = \frac{H_e - K_e}{H_o + K_o}$$

Substituting the above expressions for H_e, H_o, K_e, and K_o and factoring, we get

$$z_{11} = \frac{[s^2 + (0.50305)^2][s^2 + (1.08729)^2]}{1.1584s[s^2 + (1.15180)^2][s^2 + (0.852618)^2]}$$

This impedance function must be realized to exhibit the zeros of transmission, which are at $s = \infty$, $\pm j1.864643$, and $\pm j2.868343$. The admittance function $Y_1 = 1/z_{11}$ has a pole at infinity. Proceeding just as in Example 6.11, a part of the residue of this pole at infinity is removed to create a zero at $s = \pm j2.868343$. The required residue is given by

$$k_1 = \left.\frac{Y_1}{s}\right|_{s^2 = -(2.868343)^2} = 1.06722$$

The admittance corresponding to this partially removed residue is that of a shunt capacitor $C_1 = 1.06722$ F. The remainder admittance Y_2 has a zero at $s = \pm j2.868343$:

$$Y_2 = Y_1 - sC_1 = \frac{0.09118s[s^2 + (2.868343)^2][s^2 + (1.031318)^2]}{[s^2 + (0.50305)^2][s^2 + (1.08729)^2]}$$

The impedance $Z_2 = 1/Y_2$ has a pole at $s = \pm j2.868343$, which is removed from Z_2 by the partial fraction expansion

$$Z_2 = \frac{k_2 s}{s^2 + (2.868343)^2} + Z_3$$

where

$$k_2 = \frac{s^2 + (2.868343)^2}{s} Z_2 \Bigg|_{s^2 = -(2.868343)^2} = 10.4540$$

In the above expansion for Z_2, the impedance

$$\frac{10.454s}{s^2 + (2.868343)^2} = \cfrac{1}{0.095657s + \cfrac{1}{1.2706s}}$$

is realized by the antiresonant branch shown in Figure 6.32*a*. The remainder impedance Z_3 is

$$Z_3 = Z_2 - \frac{10.454s}{s^2 + (2.868343)^2} = \frac{0.51327[s^2 + (0.881464)^2]}{s[s^2 + (1.031318)^2]}$$

$Y_3 = 1/Z_3$ has a pole at infinity. A part of the residue of this pole at infinity is

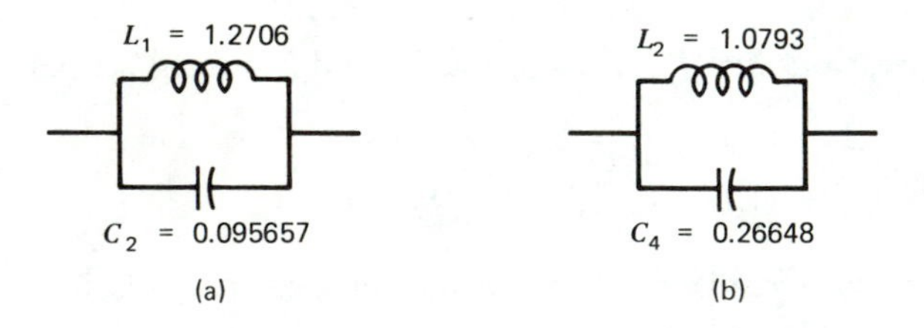

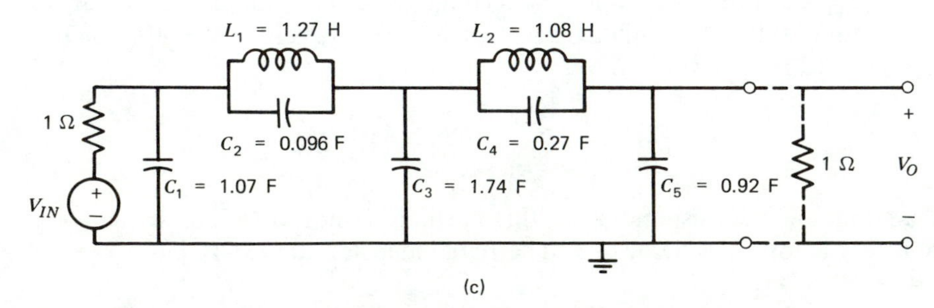

Figure 6.32 Realization for Example 6.13: (*a*) Realization of zero at $s = \pm j2.868343$. (*b*) Realization of zero at $s = \pm j1.864643$. (*c*) Complete circuit.

removed to create a zero at $s = \pm j1.864643$. This residue is given by

$$k_3 = \left.\frac{Y_3}{s}\right|_{s^2=-(1.864643)^2} = 1.74145$$

The admittance corresponding to this term is that of a shunt capacitor $C_3 = 1.74145$ F. The remainder admittance Y_4 has a zero at $s = \pm j1.864643$:

$$Y_4 = Y_3 - 1.74145s = \frac{0.20683s[s^2 + (1.86463)^2]}{[s^2 + (0.881464)^2]}$$

The impedance $Z_4 = 1/Y_4$ has a pole at $s = \pm j1.864643$, which is removed as an LC antiresonant branch as follows

$$Z_4 = \frac{k_4 s}{s^2 + (1.864643)^2} + Z_5$$

where

$$k_4 = \left.\frac{[s^2 + (1.864643)^2]}{s} Z_4\right|_{s^2=-(1.864643)^2} = 3.7526$$

The term in the partial fraction expansion

$$\frac{3.7526s}{s^2 + (1.864643)^2} = \frac{1}{0.26648s + \dfrac{1}{1.0793s}}$$

yields the antiresonant branch shown in Figure 6.32*b*. The remainder Z_5 is

$$Z_5 = Z_4 - \frac{3.7526s}{s^2 + (1.864643)^2} = \frac{1}{0.92396s}$$

This term corresponds to the capacitor $C_5 = 0.92396$ F, which realizes the zero at infinity. The circuit realizing z_{11} is shown in Figure 6.32*c* (solid lines). Let us next realize z_{22} from the reverse direction to check if we need an ideal transformer. From Equation 6.31*a*, z_{22} is seen to be equal to z_{11}, since $K_e = 0$. The first element we remove is the shunt capacitor C_5 (which corresponds to a partial removal of the pole of $Y_1 = 1/z_{22}$ at infinity), to create a zero at $s = \pm j1.864643$:

$$C_5 = \left.\frac{Y_1}{s}\right|_{s^2=-(1.864643)^2} = 0.92396 \text{ F}$$

Since this value of C_5 is the same as for the forward realization, we may conclude that the reverse realization of z_{22} will be the same as the forward realization (i.e., $\alpha = 1$) and a transformer will not be needed. The complete circuit, with the 1 Ω terminations, is shown in Figure 6.32*c*. ■

6.5 CONCLUDING REMARKS

This chapter provides an introduction to the synthesis of passive networks. This subject has received much attention by circuit theorists and mathematicians, and the literature abounds with books and papers on the realizability, approximation, and synthesis of passive filters.

The driving point synthesis techniques we described are restricted to functions that can be realized using two-element type networks (*RC*, *LC* or *RL*). Methods for realizing general dp functions, using *RLC*'s and, possibly, transformers are described in [4]. The structure we used for the synthesis of transfer functions, namely the doubly terminated ladder, is very popular in the design of passive filters. This structure serves to realize transfer functions with poles in the left half s plane and zeros on the $j\omega$ axis, thereby allowing the synthesis of the standard approximation functions (Butterworth, Chebyshev, Bessel, and elliptic). An important feature of these ladder structures is that they can be designed to yield a very low sensitivity in the passband. Other topologies that can be used to synthesize transfer functions are the lattice and the parallel ladder structures [4]. Complex zeros off the $j\omega$ axis can also be realized using passive *RLC* networks, which sometimes require transformers [4].

The low sensitivity of passive ladder structures will be exploited in a later chapter for the synthesis of active *RC* filters (Chapter 11). There we describe some active filters, whose topologies are derived from the passive ladder realization, in an attempt to retain the low sensitivity property. However, it will be seen that the active realizations will always require more components than the parent passive network. The active equivalent of an inductor, for example, requires at least one op amp, one capacitor, and two resistors. Thus, even if the active circuit could exactly simulate the passive structure, this increased component count would make the sensitivity of the active realization higher than the passive equivalent. For this reason it is safe to say that passive doubly terminated ladder structures will exhibit a lower sensitivity than any equivalent active realization.* Another feature of passive filters is that they do not have the frequency limitation that is inherent in active filter designs. In fact, passive *RLC* filters can easily be used up to 500 MHz.

The one major objection to passive filter realizations is that they need inductors—and inductors are large, expensive, elements that cannot be implemented using integrated circuit technologies. The disadvantages of the inductor were severe enough to have generated the field of active filters which employ resistors, capacitors, and an active device. At the present time, nearly all the filters designed for voice and data communication systems (below 30 kHz) use the active *RC* approach. In the next few chapters we become familiar with

* Assuming that the quality of the components used in the passive realization are comparable to those used in the active realization.

the synthesis, the advantages, and the problems associated with the design of active filters.

FURTHER READING

1. E. Christian and E. Eisenmann, *Filter Design Tables and Graphs*, Wiley, New York, 1966.
2. P. R. Geffe, *Simplified Modern Filter Design*, Rider, New York, 1963.
3. E. A. Guillemin, *Synthesis of Passive Networks*, Wiley, London, 1957.
4. S. Karni, *Network Theory: Analysis and Synthesis*, Allyn and Bacon, Boston, Mass., 1966, Chapters 10 and 11.
5. H. J. Orchard, "Inductorless Filters," *Electronics Letters*, *2*, September, 1966, pp. 224–225.
6. R. Saal and E.Ulbrich, "On the design of filters by synthesis," *IEEE Trans. Circuit Theory, CT-5,* No. 4, December, 1958, pp. 284-327.
7. J. L. Stewart, *Circuit Theory and Design*, Wiley, New York, 1956, Chapter 6.
8. J. E. Storer, *Passive Network Synthesis*, McGraw-Hill, New York, 1957.
9. G. C. Temes and S. K. Mitra, eds., *Modern Filter Theory and Design*, New York, 1973, Chapter 3.
10. M. E. Van Valkenburg, *Introduction to Modern Network Synthesis*, Wiley, New York, 1960, Chapter 14.
11. A. I. Zverev, *Handbook of Filter Synthesis*, Wiley, New York, 1967.

PROBLEMS

6.1 *Analysis by inspection.* Write expressions for the *dp* impedances of the networks shown, by inspection (i.e., without solving equations). Do not simplify the expressions.

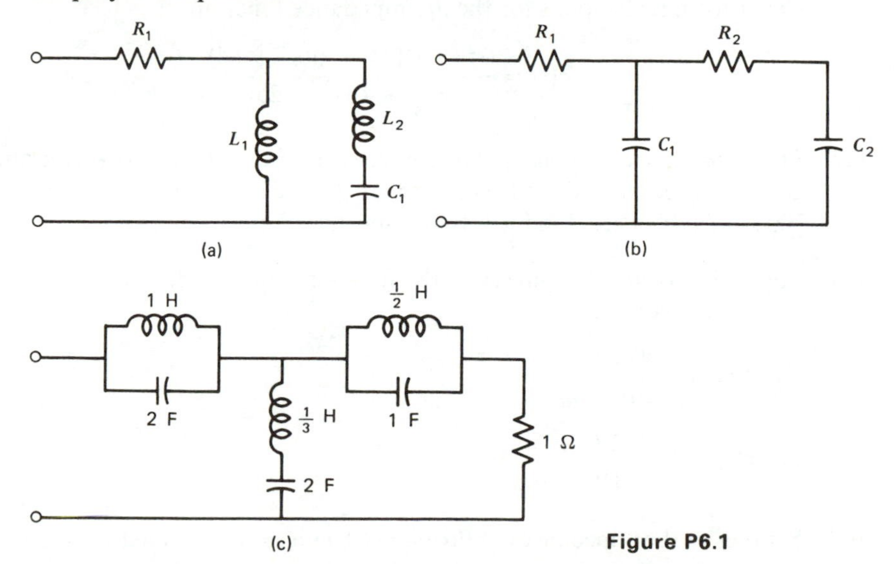

Figure P6.1

6.2 *Synthesis by inspection.* Synthesize the following *dp* immittance functions by inspection:

(a) $Z(s) = \dfrac{2s + 3}{2s + 4}$ (use three elements)

(b) $Z(s) = s + \dfrac{1}{2s + \dfrac{1}{3s}}$

(c) $Y(s) = \dfrac{2s}{3s^2 + 1}$

6.3 *Foster LC synthesis.* Synthesize the following *dp* impedance functions as a sum of series impedances, using the Foster expansion:

(a) $Z(s) = \dfrac{(s^2 + 1)(s^2 + 4)}{s(s^2 + 2)}$

(b) $Z(s) = \dfrac{s(s^2 + 2)(s^2 + 4)}{(s^2 + 1)(s^2 + 3)}$

6.4 Synthesize the impedances of Problem 6.3 as a sum of parallel admittances.

6.5 *Cauer LC synthesis.* Find a ladder structure realization for the impedances of Problem 6.3, by using the Cauer expansion.

6.6 *Foster and Cauer LC.* Sketch (but do not synthesize) two Foster and two Cauer form realizations for the *dp* impedance function

$$Z(s) = \frac{(s^2 + 1)(s^2 + 9)(s^2 + 64)}{s(s^2 + 4)(s^2 + 25)}$$

6.7 *LC dp impedance synthesis.* The impedance of an *LC* network is infinite at 20 rad/sec and 40 rad/sec; zero at 30 rad/sec and 50 rad/sec; and 200 Ω at 10 rad/sec. Find a network meeting these specifications.

6.8 *Foster RC synthesis.* Synthesize the following *dp* impedances as a sum of series impedances:

(a) $Z(s) = \dfrac{(s + 1)(s + 3)}{s(s + 2)}$

(b) $Z(s) = \dfrac{(s + 2)(s + 6)}{(s + 1)(s + 4)}$

6.9 Synthesize the impedances of Problem 6.8 as a sum of parallel admittances.

6.10 *Cauer RC synthesis.* Use the Cauer expansion to find ladder structure realizations for the impedance functions:

(a) $Z(s) = \dfrac{s^2 + 6s + 4}{2s^2 + 4s}$

(b) $Z(s) = \dfrac{s^2 + 4s + 3}{s^3 + 6s^2 + 8s}$

6.11 The pole-zero pattern for an RC impedance function is shown in Figure P6.11. It is also given that $Z(0) = 6$. Synthesize the function as a ladder network.

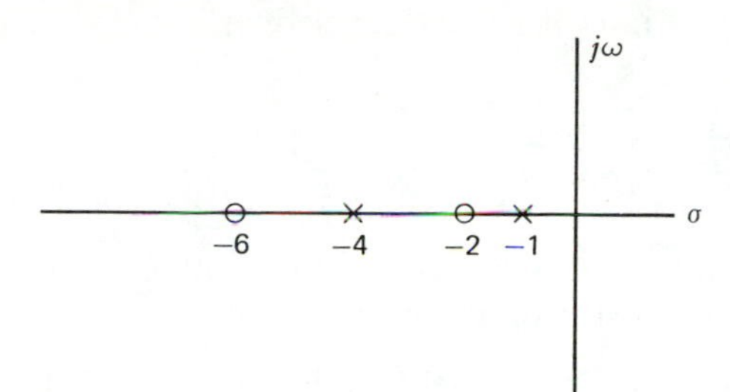

Figure P6.11

6.12 *Impedance synthesis, zero-shifting.* Consider the impedance function:

$$Z(s) = \frac{4s^2 + 1}{8s^3 + 3s}$$

(a) Synthesize the function using a ladder structure.

(b) Synthesize the function so that it exhibits a zero at $s = \pm j\sqrt{2}$, using the structure shown. Identify the zero-shifting and zero-producing elements.

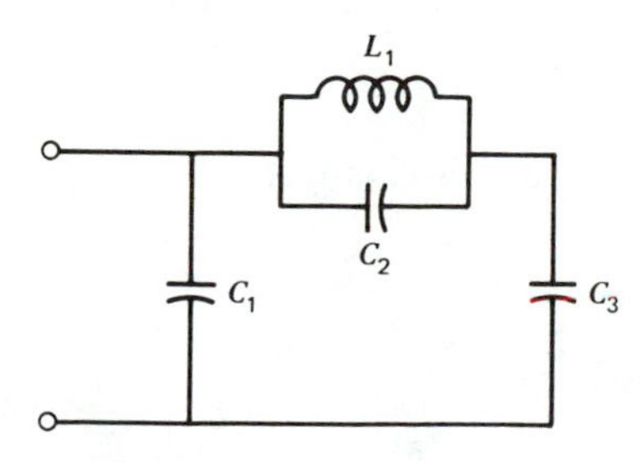

Figure P6.12

6.13 Synthesize the impedance function

$$Z(s) = \frac{s(s^2 + 4)}{(s^2 + 1)(s^2 + 5)}$$

so that it exhibits a zero at $s = \pm j\sqrt{2}$, using the circuit structure shown. Sketch the pole-zero diagram at each step of the synthesis.

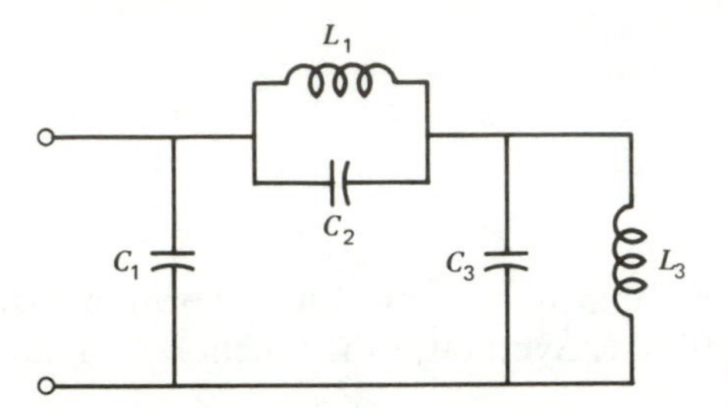

Figure P6.13

6.14 *LC transfer function, all-pole.* Find an *LC* ladder network characterized by the *z* parameters:

$$z_{11} = \frac{s^4 + 4s^2 + 3}{s^5 + 7s^3 + 10s} \qquad z_{12} = \frac{1}{s^5 + 7s^3 + 10s}$$

Indicate which elements produce the zeros of transmission at the origin and which produce zeros of transmission at infinity.

6.15 Repeat Problem 6.14 for an *LC* network that has the same z_{11}, but with z_{12} given by

$$z_{12} = \frac{s^2}{s^5 + 7s^3 + 10s}$$

6.16 Repeat Problem 6.15 for

$$z_{12} = \frac{s^4}{s^5 + 7s^3 + 10s}$$

6.17 *LC transfer function, finite zeros.* Use the *LC* ladder topology shown to synthesize the *z* parameters

$$y_{11} = \frac{s^2 + 1}{s(s^2 + 2)} \qquad y_{12} = \frac{s^2 + 3}{s(s^2 + 2)}$$

What is the transfer function of the network if it is terminated by a 1Ω load resistance?

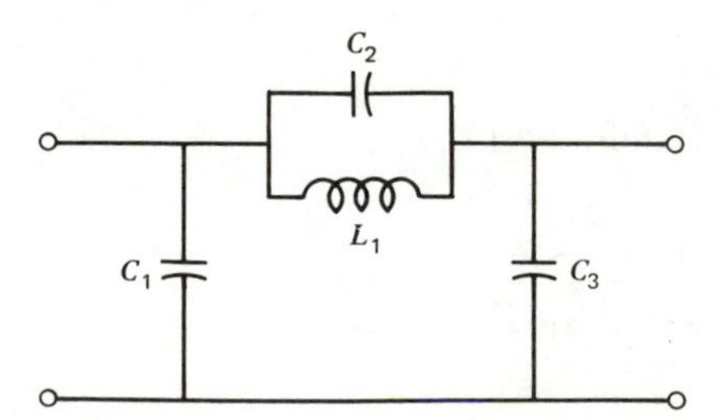

Figure P6.17

6.18 Synthesize an LC ladder network which has the z parameters

$$z_{11} = \frac{(s^2 + 1)(s^2 + 3)}{s(s^2 + 2)} \qquad z_{12} = \frac{(s^2 + 4)(s^2 + 5)}{s(s^2 + 2)}$$

using the topology shown. Sketch the pole-zero pattern at each step, and indicate the zero shifting and zero producing elements.

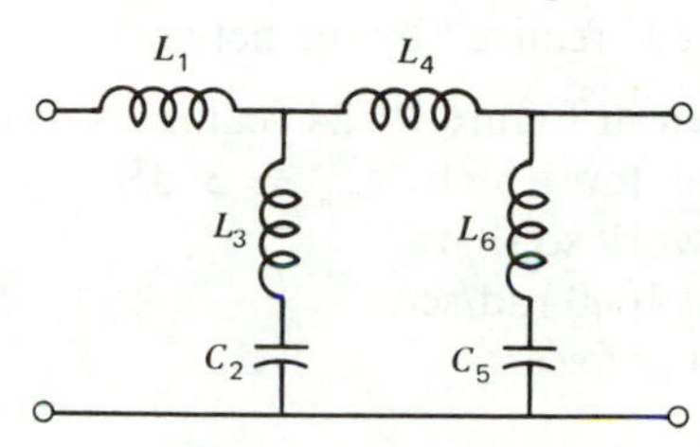

Figure P6.18

6.19 *LC ladder topology prediction.* Sketch (but do not synthesize) an LC ladder topology for the realization of the z parameters plotted in Figure P6.19.

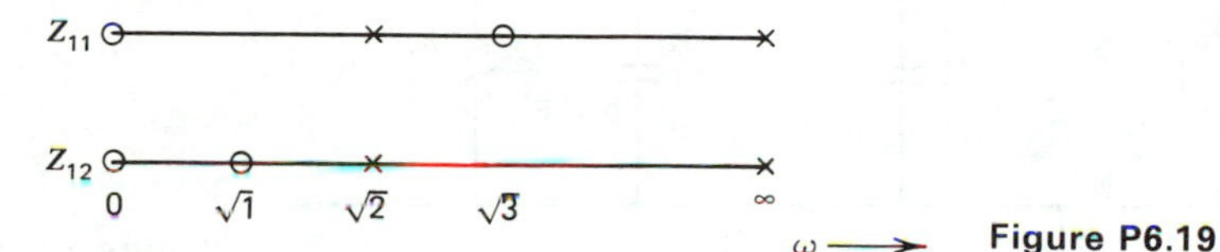

Figure P6.19

6.20 Repeat Problem 6.19 for the z parameters of Figure P6.20.

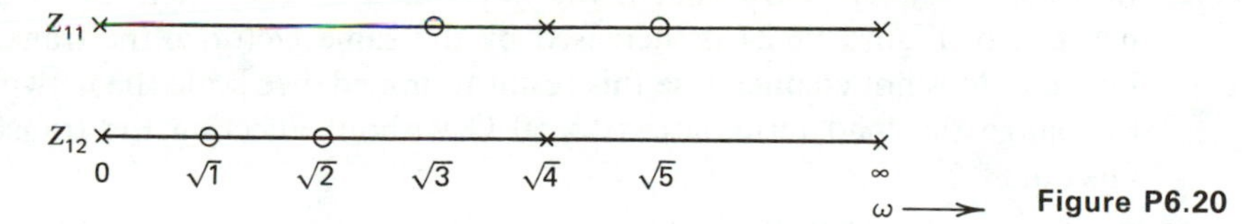

Figure P6.20

6.21 Repeat Problem 6.19 for the z parameters of Figure P6.21.

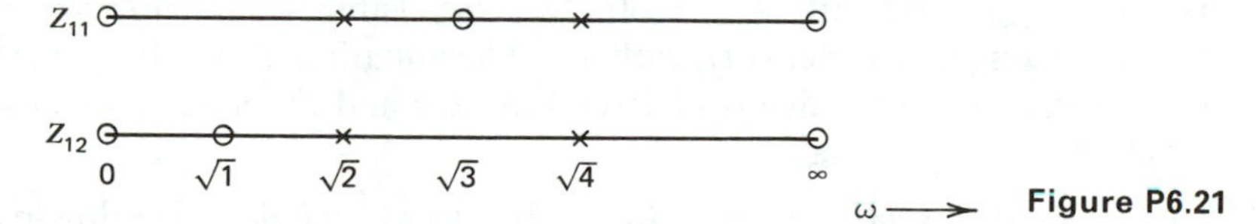

Figure P6.21

6.22 Repeat Problem 6.19 for the z parameters of Figure P6.22.

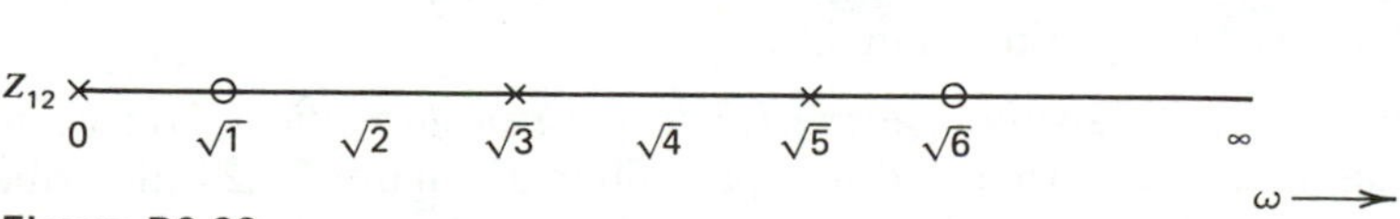

Figure P6.22

6.23 *Singly-terminated LP Butterworth filter.* Consider the normalized *LP* Butterworth approximation function

$$\frac{V_O}{V_{IN}} = \frac{K}{s^3 + 2s^2 + 2s + 1}$$

Synthesize this function using an *LC* network terminated at one end with a 1 Ω resistor. Determine the constant *K* realized by the network.

6.24 *Frequency scaling.* The network shown in Figure P6.24 realizes a fourth-order low-pass Butterworth function for which $A_{max} = 3$ dB, $\omega_P = 1$ rad/sec. Scale the elements of the network so that:
(a) The passband edge frequency is at 1000 rad/sec.
(b) The passband loss is 1 dB at 1000 rad/sec
(*Hint*: use Equation 4.21.)

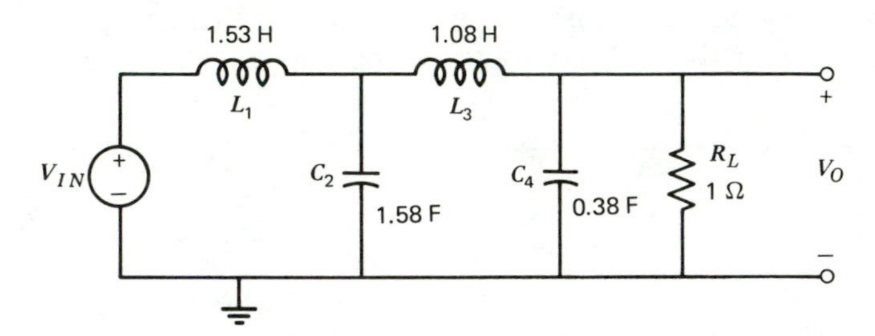

Figure P6.24

6.25 *Impedance scaling.* Show that if the impedance of each element in the network of Figure P6.24 is increased by the same factor α, the transfer function does not change. Use this result to impedance scale the network to change the load termination to 50 Ω without affecting the transfer function.

6.26 *Singly-terminated LP elliptic filter.* Find an elliptic function approximation for the low-pass requirements characterized by: $\omega_P = 1$ rad/sec, $\omega_S = 2$ rad/sec, $A_{max} = 0.5$ dB, $A_{min} = 30$ dB. Use Table 4.3. Synthesize the function using an *LC* network with a 1 Ω termination. Scale the elements so that the passband edge is at 10,000 rad/sec and the load termination is 600 Ω.

6.27 *High-pass filter synthesis using LP to HP transformation.* The low-pass filter of Figure P6.24 realizes a fourth-order Butterworth function for which $A_{max} = 3$ dB and $\omega_P = 1$ rad/sec. Transform the elements to realize a fourth-order high-pass filter for which $A_{max} = 3$ dB and $\omega_P = 100$ rad/sec. (*Hint*: use Equation 4.51.)

6.28 *Band-pass filter synthesis using LP to BP transformation.* Transform the elements in the normalized low-pass filter of Figure P6.24 (described in Problem 6.27) to realize an eighth-order band-pass filter for which the passband width is $B = 100$ rad/sec, center frequency is $\omega_0 = 1000$ rad/sec, and $A_{max} = 3$ dB. (*Hint*: use Equation 4.62.)

6.29 *Singly-terminated high-pass filter.* Find an LC network terminated by a load resistor of 600 Ω satisfying the high-pass Chebyshev filter requirements

$$\omega_P = 3000 \text{ rad/sec} \qquad \omega_S = 1000 \text{ rad/sec}$$
$$A_{max} = 0.5 \text{ dB} \qquad A_{min} = 25 \text{ dB}$$

(*Hints*: first synthesize the normalized LP function as given by Table 4.2, with a 1 Ω termination. Then impedance scale the network as in Problem 6.25. Finally use the LP to HP transformation of Equation 4.51.)

6.30 *Transducer and characteristic functions.* Identify the transducer function $H(s)$ and the characteristic function $K(s)$ for:

(a) A fourth-order low-pass Butterworth approximation function for which $A_{max} = 3$ dB, $\omega_P = 1$ rad/sec.
(b) A fourth-order low-pass Butterworth approximation function for which $A_{max} = 1$ dB, $\omega_P = 2$ rad/sec.
(c) A third-order low-pass Chebyshev approximation function for which $A_{max} = 0.25$ dB, $\omega_P = 1$ rad/sec.

6.31 *Double terminated LP filter.* Synthesize a third-order LP Butterworth approximation for which $A_{max} = 3$ dB and $\omega_P = 1$ rad/sec, using an LC ladder network terminated at both ends with 1 Ω resistors.

6.32 A low-pass Chebyshev filter is required to meet the following specifications:

$$f_P = 1200 \text{ Hz} \qquad f_S = 2400 \text{ Hz}$$
$$A_{max} = 0.5 \text{ dB} \qquad A_{min} = 25 \text{ dB}$$

Synthesize the function using an LC ladder network terminated at both ends with 600 Ω resistors. (*Hint*: first synthesize the normalized function with 1 Ω terminations.)

6.33 *Double terminated BP filter.* A band-pass filter is required to meet the following specifications:

$$A_{max} = 3 \text{ dB} \qquad A_{min} = 20 \text{ dB}$$

passband: 1000 rad/sec to 2000 rad/sec
stopbands: below 500 rad/sec and above 4000 rad/sec

Synthesize the Butterworth approximation function for these requirements using an LC ladder network terminated at both ends with 50 Ω resistors.

6.34 *Double terminated elliptic filter*. The elliptic approximation for the low-pass requirements shown is characterized by (see [1], page 49, $\Omega_S = 2.5593$):

$$H(s) = \frac{4.886(s + 1.05443)(s^2 + 0.84976s + 1.66129)}{(s^2 + 8.5589)}$$

$$K(s) = \frac{4.886s(s^2 + 0.76528)}{(s^2 + 8.5589)}$$

(a) Realize these LP requirements using the topology shown.

(b) Transform the elements, using Equation 4.65, to realize the band-reject requirements shown in Figure P6.34*c*. Use 600 Ω terminations.

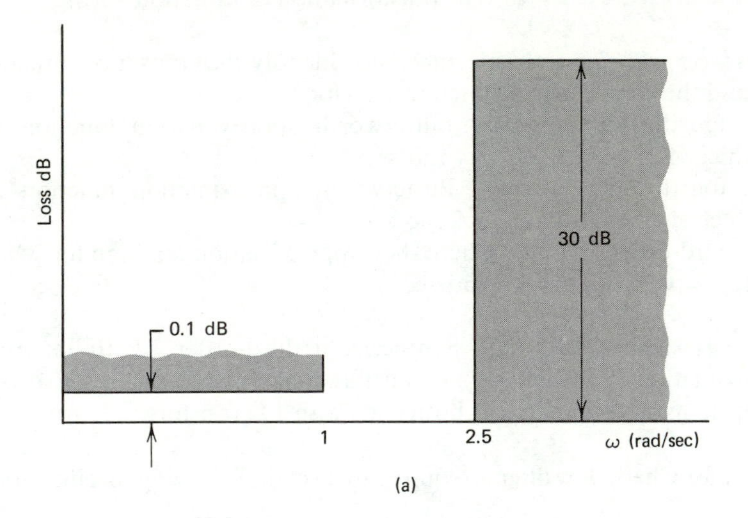

(a)

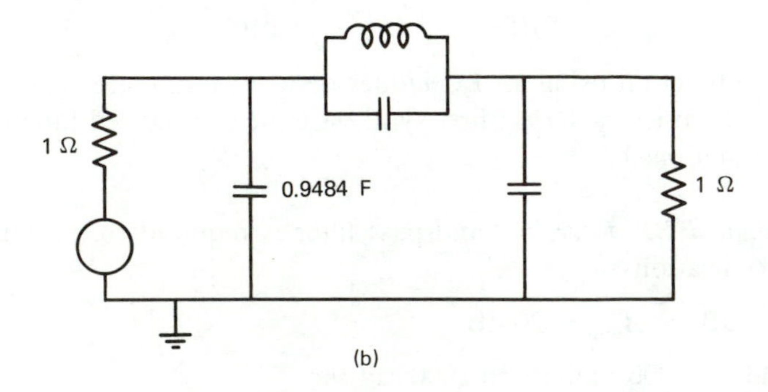

(b)

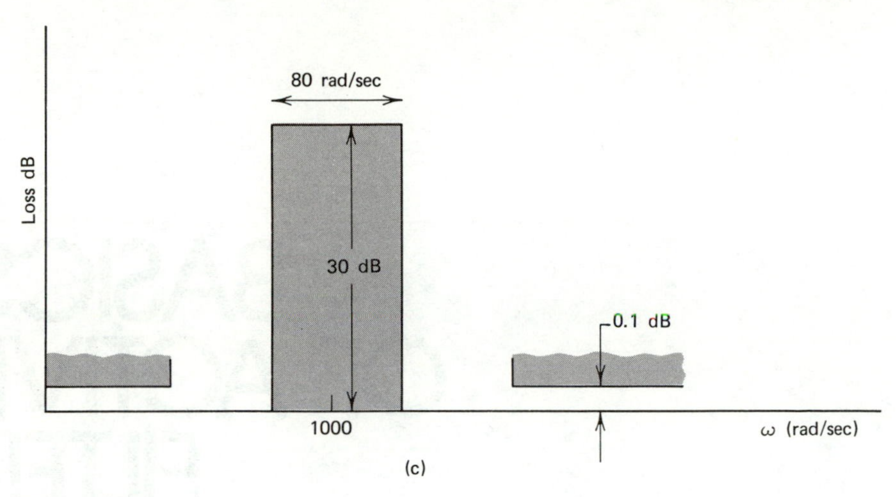

Figure P6.34

7.

BASICS OF ACTIVE FILTER SYNTHESIS

In the preceding chapter we discussed the synthesis of transfer functions using passive components. In this chapter we discuss some basic principles related to the synthesis of active networks. Broadly speaking, the topologies used in active network synthesis can be classified into two groups, namely, *cascaded* and *coupled*. The fundamental block used in these topologies is the biquadratic function, which was introduced in Section 2.6. Because there exist several active realizations of the biquadratic function, it is useful to further categorize these realizations themselves. In this chapter we discuss the classification, some fundamental properties, and the principles of synthesis of the commonly used single-amplifier realizations of the biquadratic. The detailed design considerations of cascaded and coupled active filters are the subject of the remainder of this book.

7.1 FACTORED FORMS OF THE APPROXIMATION FUNCTION

The approximation step yields a transfer function of the form:

$$T(s) = K\frac{(s - z_1)(s - z_2)\cdots(s - z_n)}{(s - p_1)(s - p_2)\cdots(s - p_m)} \tag{7.1}$$

As mentioned in Chapter 2, this function can be factored into biquadratics, as

$$T(s) = \prod_{i=1}^{N} K_i \frac{m_i s^2 + c_i s + d_i}{n_i s^2 + a_i s + b_i} \qquad \begin{pmatrix} m_i = 1 \text{ or } 0 \\ n_i = 1 \text{ or } 0 \end{pmatrix} \tag{7.2}$$

This form of the approximation function is quite general in that it can represent either real or complex roots. In the case of a real pole, for instance, we set $n = 0$ and $a = 1$,* so that the denominator reduces to

$$s + b$$

Similarly, for a real zero, $m = 0$ and $c = 1$. For complex poles we set $n = 1$, and for complex zeros $m = 1$. Thus, a complex pole-zero pair is represented by

$$T(s) = K\frac{s^2 + cs + d}{s^2 + as + b} \tag{7.3}$$

* The subscripts are dropped when considering only one biquadratic.

Table 7.1 Second-Order Filter Functions

Filter function	Transfer function
Low-pass	$K\dfrac{1}{s^2 + as + b}$
High-pass	$K\dfrac{s^2}{s^2 + as + b}$
Band-pass	$K\dfrac{s}{s^2 + as + b}$
Band-reject	$K\dfrac{s^2 + d}{s^2 + as + b}$
Delay equalizer	$\dfrac{s^2 - as + b}{s^2 + as + b}$

Recall from Chapter 2 that this biquadratic can also be expressed in terms of the parameters K, ω_z, ω_p, Q_z, and Q_p, as (Equation 2.39):

$$T(s) = K\frac{s^2 + \dfrac{\omega_z}{Q_z}s + \omega_z^2}{s^2 + \dfrac{\omega_p}{Q_p}s + \omega_p^2} \tag{7.4}$$

The second-order filter functions discussed in Chapter 3 are easily obtained as special cases of Equation 7.2, as shown in Table 7.1. For example, the low-pass function is obtained from Equation 7.2 with

$$m = 0 \qquad c = 0 \qquad d = 1 \qquad n = 1$$

As mentioned in Section 3.1.4, the function representing the band-reject filter also realizes a low-pass filter with a zero in the stopband, known as a low-pass-notch filter. For this case, the zero frequency must be higher than the pole frequency, which means that $d > b$. Similarly, a high-pass-notch filter is realized when $d < b$.

7.2 THE CASCADE APPROACH

In the last section it was shown that the approximation function could be expressed as a product of biquadratics. For brevity, $T(s)$ can be written as

$$T(s) = \prod_{i=1}^{N} T_i(s) \tag{7.5}$$

where $T_i(s)$ is of the form

$$T_i(s) = K_i \frac{m_i s^2 + c_i s + d_i}{n_i s^2 + a_i s + b_i} \tag{7.6}$$

The **cascade** approach, as the name implies, consists of realizing each of the biquadratics by an appropriate circuit and connecting these circuits in cascade. Referring to Figure 7.1*a*, the output voltage of the block T_1 is

$$V_{O_1} = T_1 V_{IN}$$

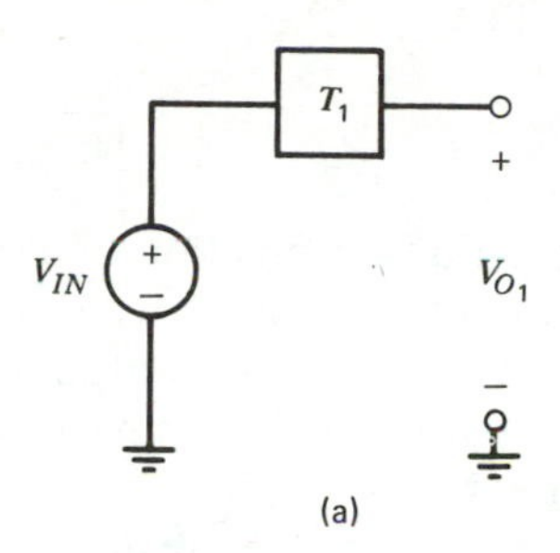

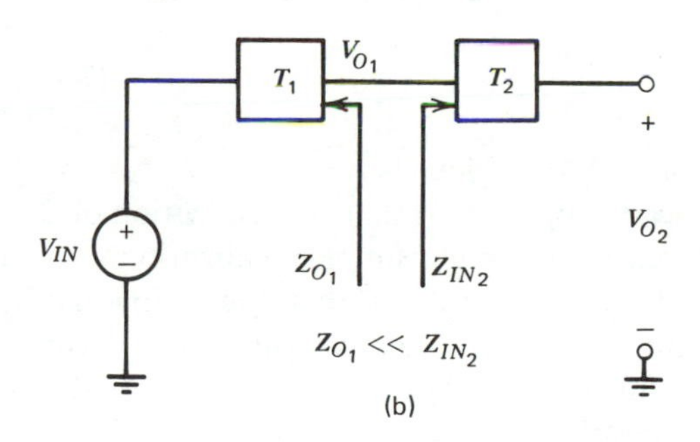

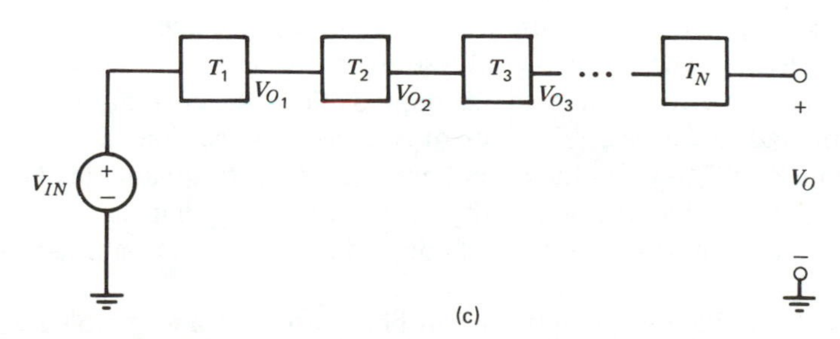

Figure 7.1 The cascaded topology: (*a*) One biquad. (*b*) Two biquads. (*c*) *N* biquads.

Now consider the effect of connecting the block T_2 to the output of T_1, as in Figure 7.1*b*. If the output impedance Z_{O_1} of T_1 is negligibly small compared with the input impedance Z_{IN_2} of T_2, then the output voltage of T_1 will not be loaded down (i.e., reduced) when T_2 is connected to it. Under this condition, the input voltage to T_2 is V_{O_1}, and the output of T_2 is

$$V_{O_2} = T_2 V_{O_1} = T_1 T_2 V_{IN}$$

Extending the argument to the cascade of N sections (Figure 7.1*c*), the output voltage V_O is

$$V_O = T_1 T_2 T_3 \cdots T_N V_{IN}$$

from which

$$\frac{V_O}{V_{IN}} = T_1 T_2 T_3 \cdots T_N = \prod_{i=1}^{N} T_i \tag{7.7}$$

which is the desired function of Equation 7.5. Thus, the transfer function of a cascade of networks is the product of the individual transfer functions, provided that the input impedance of each network is very large compared with the output impedance of the preceding network. This condition is readily satisfied in most op amp circuit realizations, as will be seen in later chapters.

Summarizing then, we see that a general transfer function can be realized by a cascade of circuit blocks, each of which realizes the biquadratic function

$$K \frac{ms^2 + cs + d}{ns^2 + as + b} \tag{7.8}$$

These circuit blocks are commonly referred to as **biquads**.

One important advantage of the cascade approach is that the realization of a higher-order transfer function is reduced to the much simpler realization of a general second-order function. An additional feature is that the individual biquads are totally isolated, so that *any change in one biquad does not affect any other biquad*. This property is useful in the adjusting (i.e., tuning) of the network's performance at the time of manufacture.

In a second family of structures the individual biquadratic blocks are coupled to each other via feedback paths. An example of these so-called **coupled** structures is shown in Figure 7.2. The synthesis of these structures is more complex than for the cascaded case, since a change in one biquad affects the currents and voltages in all the biquads. Moreover, this lack of isolation between the blocks makes their tuning more difficult. On the other hand, one distinct advantage of using coupled structures is that the sensitivity is usually lower than for the equivalent cascaded realization. The subject of coupled active filters is studied in Chapter 11.

The biquad is the basic building block used in both cascaded and coupled realizations. Therefore, biquad circuits are of fundamental importance in the design of active filters, and we devote Chapters 8 to 10 to this subject.

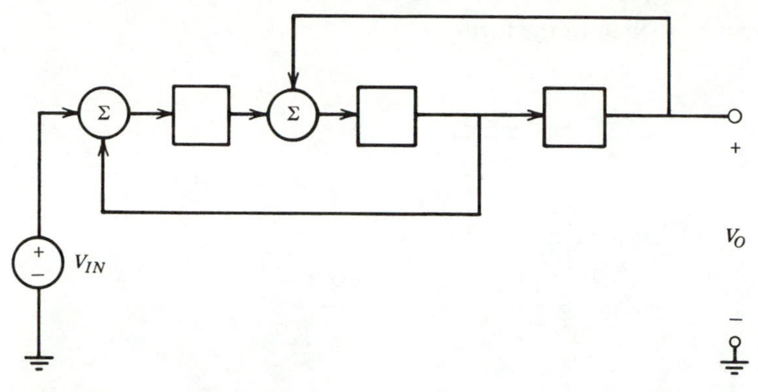

Figure 7.2 A coupled structure.

7.3 REAL POLES AND ZEROS

In this section we describe two simple circuits that can be used for the realization of real poles and zeros. The first is the inverting amplifier structure shown in Figure 7.3*a*. The transfer function for this structure, assuming the amplifier gain $A = \infty$, is

$$T(s) = -\frac{Z_2}{Z_1} \tag{7.9}$$

As an example, consider the realization of the function :

$$T(s) = -K\frac{s+d}{s+b}$$

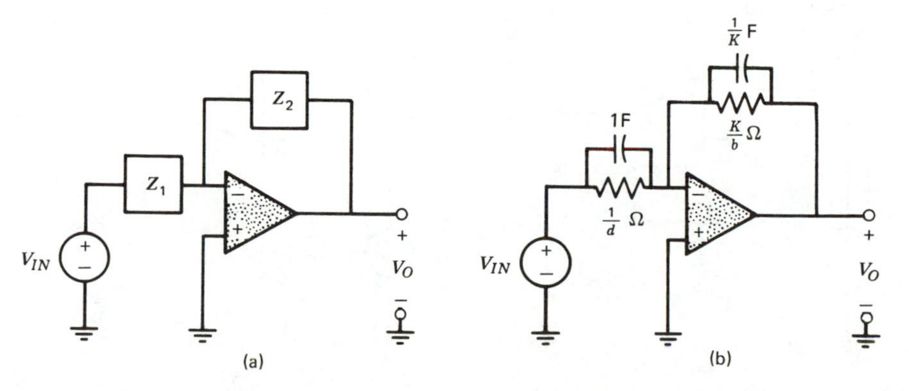

Figure 7.3 (*a*) The inverting amplifier structure. (*b*) Example.

This function can be written in the form:

$$-\frac{\dfrac{K}{s+b}}{\dfrac{1}{s+d}}$$

Then Z_1 and Z_2 are identified as

$$Z_1 = \frac{1}{s+d} \qquad Z_2 = \frac{1}{\dfrac{s}{K} + \dfrac{b}{K}}$$

The resulting circuit realization is easily seen to be that of Figure 7.3*b*.

Next, consider the noninverting structure of Figure 7.4*a*. Assuming an ideal op amp, the transfer function is

$$T(s) = 1 + \frac{Z_2}{Z_1} \tag{7.10}$$

Suppose the function to be realized is

$$T(s) = \frac{s+4}{s+1}$$

This function can be written as

$$1 + \frac{3}{s+1}$$

Comparing this with (7.10), Z_1 and Z_2 can be identified as

$$Z_1 = 1 \qquad Z_2 = \frac{1}{\dfrac{s}{3} + \dfrac{1}{3}}$$

The resulting active *RC* realization is shown in Figure 7.4*b*.

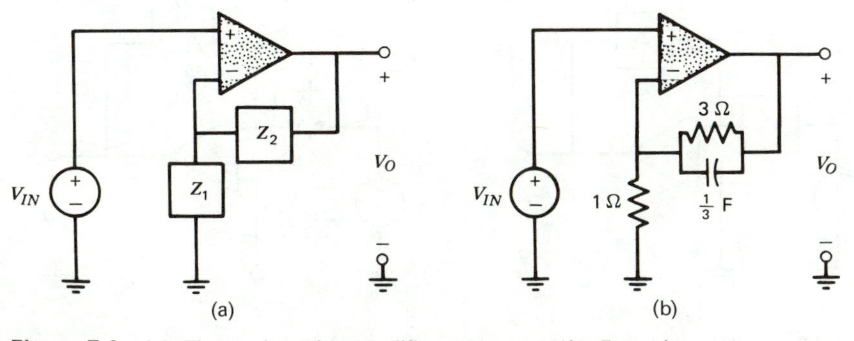

Figure 7.4 (*a*) The noninverting amplifier structure. (*b*) Example.

7.4 BIQUAD TOPOLOGIES

In this section we introduce the commonly used single-amplifier biquad topologies. These structures require an *RC* network in conjunction with one op amp. As mentioned in Chapter 2, the transfer function of an *RC* network will have poles on the negative real axis, while the zeros can be anywhere in the *s* plane.* Since the general biquad circuit must realize complex poles as well as complex zeros, the op amp must somehow be used to realize complex poles in spite of the fact that the *RC* network poles are real. There are many circuits that can accomplish this. The majority of these circuits can be classified into two basic categories, namely, the negative feedback topology and the positive feedback topology. This classification is based on which input terminal of the op amp the *RC* network is connected to.† The fundamental properties of these two topologies are discussed in the remainder of this section.

7.4.1 NEGATIVE FEEDBACK TOPOLOGY

The negative feedback topology is so called because the *RC* network associated with it provides a feedback path to the negative input terminal of the op amp. The topology is sketched in Figure 7.5. The transfer function of this general structure can be characterized in terms of the *feedforward* and *feedback* transfer functions of the passive *RC* network, defined as

$$\text{Feedforward Transfer Function} = T_{FF} = \left.\frac{V_1}{V_2}\right|_{V_3=0} \tag{7.11}$$

$$\text{Feedback Transfer Function} = T_{FB} = \left.\frac{V_1}{V_3}\right|_{V_2=0} \tag{7.12}$$

Analyzing the circuit, the output voltage is given by

$$V_O = (V^+ - V^-)A$$

where, by superposition

$$V^- = T_{FF}V_{IN} + T_{FB}V_O$$

Noting that $V^+ = 0$, we get

$$V_O = -V^-A = -(T_{FF}V_{IN} + T_{FB}V_O)A$$

Thus

$$T_V = \frac{V_O}{V_{IN}} = -\frac{T_{FF}}{T_{FB} + 1/A} \tag{7.13}$$

* The *RC* networks used in this book have one terminal each of the input and output ports connected to ground. For this class of *RC* networks, the zeros cannot lie on the positive real axis of the *s* plane (see footnote in Section 2.4).

† Different approaches to the categorization of biquads are described in Sedra [6] and Mitra [4].

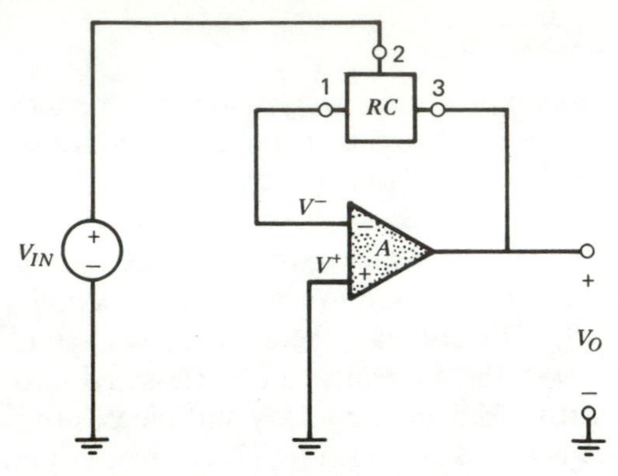

Figure 7.5 The negative feedback topology.

If the op amp is assumed to be ideal, so that $A = \infty$, the transfer function reduces to

$$\frac{V_O}{V_{IN}} = -\frac{T_{FF}}{T_{FB}} \tag{7.14}$$

In this equation, T_{FF} and T_{FB} may be written as:

$$T_{FF} = \frac{N_{FF}}{D_{FF}} \qquad \text{and} \qquad T_{FB} = \frac{N_{FB}}{D_{FB}}$$

where N_{FF} and N_{FB} represent the zeros of the RC network, observed from different ports. The denominators D_{FF} and D_{FB} are obtained from the nodal determinant of the RC network, which is independent of the ports used for input and output. Thus

$$D_{FF} = D_{FB} = D$$

and the transfer function of the negative feedback topology becomes

$$\boxed{T_V = -\frac{N_{FF}}{N_{FB}}} \tag{7.15}$$

Note that the real poles of the RC network do not contribute to the transfer function. Instead, the poles of the transfer function are determined by the zeros of the RC feedback network, which, as mentioned before, can be complex.

Summarizing, in the negative feedback topology:

- The zeros of the feedback network determine the poles of the transfer function.

- The zeros of the feedforward network determine the zeros of the transfer function.
- The poles and zeros can be complex; however, for a stable network the poles cannot lie in the right-half s plane.
- The poles of the RC network do not contribute to the transfer function (assuming the op amp to be ideal).

7.4.2 POSITIVE FEEDBACK TOPOLOGY

The positive feedback topology is shown in Figure 7.6. The topology is so called because the RC network provides a feedback path to the positive terminal of the op amp. (Note, however, that a part of the output voltage is also fed back to the negative terminal via the resistors r_1 and r_2, which constitute a potential divider. Thus, in a sense, this is really a mixed feedback topology.)

The feedforward and feedback transfer functions are defined just as in the negative feedback topology:

$$T_{FF} = \left.\frac{V_1}{V_2}\right|_{V_3=0} \qquad T_{FB} = \left.\frac{V_1}{V_3}\right|_{V_2=0}$$

Analyzing the circuit, the output voltage is given by

$$V_O = (V^+ - V^-)A$$

where

$$V^- = \frac{V_O r_1}{r_1 + (k-1)r_1} = \frac{V_O}{k}$$

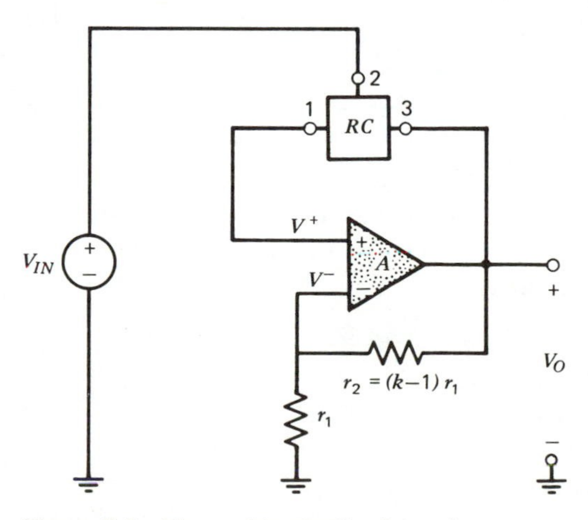

Figure 7.6 The positive feedback topology.

and

$$V^+ = V_{IN} T_{FF} + V_O T_{FB}$$

Thus

$$V_O = \left(V_{IN} T_{FF} + V_O T_{FB} - \frac{V_O}{k} \right) A$$

from which

$$T_V = \frac{V_O}{V_{IN}} = \frac{k T_{FF}}{1 - k T_{FB} + \dfrac{k}{A}} \tag{7.16}$$

If the op amp is assumed to be ideal, $A = \infty$, then

$$T_V = \frac{k T_{FF}}{1 - k T_{FB}} \tag{7.17}$$

As mentioned before, T_{FF} and T_{FB} have the same poles, so these functions may be written as

$$T_{FF} = \frac{N_{FF}}{D} \qquad \text{and} \qquad T_{FB} = \frac{N_{FB}}{D}$$

where

N_{FF} and N_{FB} represent the zeros of an *RC* network which can be complex, and
D represents the poles of an *RC* network which must be real.

From Equation 7.17, the transfer function T_V is given by

$$\boxed{T_V = \frac{k N_{FF}}{D - k N_{FB}}} \tag{7.18}$$

The zeros of T_V are determined by N_{FF}, while the poles of the T_V are the roots of

$$D - k N_{FB}$$

The presence of the term $-kN_{FB}$ in this expression permits the realization of complex poles. For example, if the *RC* feedback network is chosen to be a second-order band-pass function of the form

$$T_{FB} = \frac{s}{s^2 + as + b}$$

the poles of T_V will be given by

$$\begin{aligned} D - k N_{FB} &= (s^2 + as + b) - ks \\ &= s^2 + (a - k)s + b \end{aligned} \tag{7.19}$$

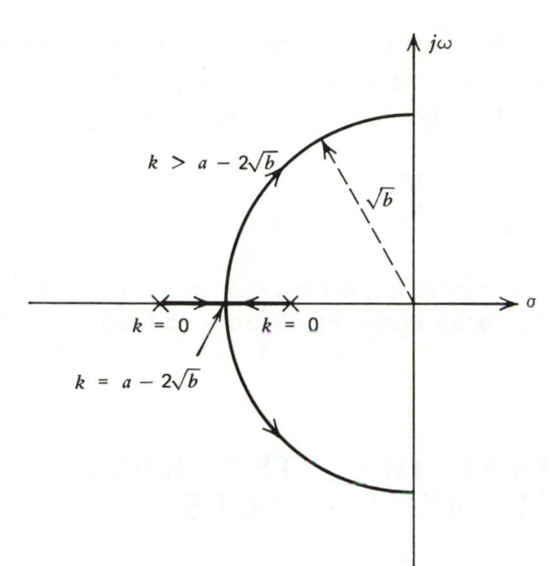

Figure 7.7 Root locus for Equation 7.19.

The roots of this equation are

$$s_1, s_2 = \frac{-(a-k) \pm \sqrt{(a-k)^2 - 4b}}{2} \tag{7.20}$$

For $k = 0$, the roots correspond to the poles of the RC network which lie on the negative real axis, as shown in Figure 7.7. As k is increased the two roots move toward each other, and finally coalesce to form a double root on the real axis for

$$k = a - 2\sqrt{b} \tag{7.21}$$

This expression for k makes the discriminant in (7.20) zero, and the location of the double roots is given by

$$s_1, s_2 = -\sqrt{b} \tag{7.22}$$

In Equation 7.20, if

$$k > a - 2\sqrt{b} \tag{7.23}$$

the roots become complex. Thus, expression 7.23 gives the condition for realizing complex poles. The locus of the complex poles for increasing k is shown in Figure 7.7.* From Equation 7.20, as k increases, the real parts of the poles

* Such a plot is known as a root locus.

decrease, while the magnitudes of the poles stay the same, that is, the poles move toward the imaginary axis on a circle whose radius is $\sqrt{b}$. Thus, by properly choosing the RC network and k, the poles can be located anywhere in the left half s plane.

Summarizing, for the positive feedback topology:

- The zeros of the transfer function are the zeros of the feedforward RC network, which can be complex.
- The poles of the transfer function can be located anywhere in the left half s plane, being determined by the poles of the RC network and the factor $k = 1 + r_2/r_1$.

7.5 COEFFICIENT MATCHING TECHNIQUE FOR OBTAINING ELEMENT VALUES

Let us assume that a suitable circuit topology has been obtained for the synthesis of the given biquadratic function. The next step in the synthesis procedure is to determine the element (R and C) values of this circuit. This is done as follows. The circuit is analyzed in terms of the R's and C's, the op amp being assumed to be ideal. By comparing the function so obtained with the given biquadratic, we can equate corresponding coefficients of s to generate a set of equations. The solution to these equations yields the required R and C values. The method, known as the **coefficient matching** technique, is illustrated by the following example.

Example 7.1

Synthesize the following low-pass function using the positive feedback circuit shown in Figure 7.8*a*:

$$T(s) = \frac{2b}{s^2 + as + b} \tag{7.24}$$

where a and b are positive constants.

Solution

The RC network used in this circuit is drawn separately in Figure 7.8*b*. The nodal equations of this RC network are

$$\begin{bmatrix} \dfrac{1}{R_1} + \dfrac{1}{R_2} + sC_1 & -\dfrac{1}{R_2} \\ -\dfrac{1}{R_2} & \dfrac{1}{R_2} + sC_2 \end{bmatrix} \begin{bmatrix} V_x \\ V_1 \end{bmatrix} = \begin{bmatrix} sC_1 & \dfrac{1}{R_1} \\ 0 & 0 \end{bmatrix} \begin{bmatrix} V_3 \\ V_2 \end{bmatrix} \tag{7.25}$$

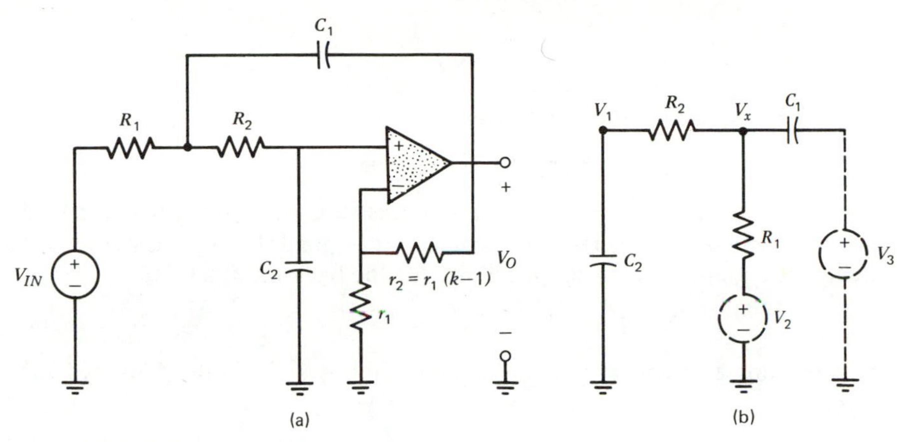

Figure 7.8 (*a*) Positive feedback circuit for Example 7.1. (*b*) *RC* circuit.

From this equation the feedback and feedforward transfer functions are found to be

$$T_{FB} = \left.\frac{V_1}{V_3}\right|_{V_2=0} = \frac{s/R_2C_2}{s^2 + s\left(\dfrac{1}{R_1C_1} + \dfrac{1}{R_2C_1} + \dfrac{1}{R_2C_2}\right) + \dfrac{1}{R_1R_2C_1C_2}} \tag{7.26}$$

$$T_{FF} = \left.\frac{V_1}{V_2}\right|_{V_3=0} = \frac{1/R_1R_2C_1C_2}{s^2 + s\left(\dfrac{1}{R_1C_1} + \dfrac{1}{R_2C_1} + \dfrac{1}{R_2C_2}\right) + \dfrac{1}{R_1R_2C_1C_2}} \tag{7.27}$$

Therefore, from Equation 7.18, the transfer function of the active circuit is

$$T_V = \frac{k/R_1R_2C_1C_2}{s^2 + s\left(\dfrac{1}{R_1C_1} + \dfrac{1}{R_2C_1} + \dfrac{1-k}{R_2C_2}\right) + \dfrac{1}{R_1R_2C_1C_2}} \tag{7.28}$$

where

$$k = 1 + \frac{r_2}{r_1} \tag{7.29}$$

The coefficients of like powers in s in (7.24) and (7.28) are equated to yield the following relationships:

$$\frac{1 + r_2/r_1}{R_1R_2C_1C_2} = 2b \tag{7.30}$$

$$\frac{1}{R_1C_1} + \frac{1}{R_2C_1} - \frac{r_2/r_1}{R_2C_2} = a \tag{7.31}$$

$$\frac{1}{R_1 R_2 C_1 C_2} = b \tag{7.32}$$

In the above equations the unknowns are

$$r_1, r_2, R_1, R_2, C_1 \text{ and } C_2$$

Now, the solution of three equations requires three independent parameters. Since we have six independent parameters, it is possible to *fix* three prior to solving the equations. One simple choice for the fixed parameters* is

$$C_1 = 1 \qquad C_2 = 1 \qquad r_1 = 1$$

The remaining elements are then obtained as follows. Dividing (7.30) by (7.32), we get

$$1 + \frac{r_2}{r_1} = 2 \qquad \text{so} \qquad r_2 = r_1 = 1$$

Substituting the values of C_1, C_2, r_1, and r_2 in Equation 7.31

$$R_1 = \frac{1}{a}$$

Finally, from Equation 7.32, the remaining element R_2 is given by

$$R_2 = \frac{a}{b}$$

Thus, one set of element values that will synthesize the given transfer function is:

$$C_1 = 1\text{F} \qquad C_2 = 1\text{F} \qquad r_1 = 1\,\Omega \qquad r_2 = 1\,\Omega \qquad R_1 = \frac{1}{a}\,\Omega \qquad R_2 = \frac{a}{b}\,\Omega$$

Observations

1. The element values are in ohms and farads and as such are definitely not practical to implement. However, these values can easily be scaled to yield practical values as explained in Section 7.7.
2. In the above synthesis, the choice of the fixed parameters was arbitrary. In practice, this choice is dictated by other design considerations, such as the sensitivity of the network and the spread in the values of the elements. These and other practical design matters will be discussed in the next few chapters.
3. The synthesis was based on the op amp being ideal, that is, $A = \infty$. In most simple filter designs the effect of op amp imperfections is quite

* This choice is not completely arbitrary in that some choices do not lead to a solution. For example, the choice $C_1 = 1$, $C_2 = 1$, $r_2 = 2r_1$ makes Equation 7.30 and 7.32 inconsistent and, therefore, does not allow a solution.

small and can be neglected. However, in applications where the requirements are stringent (high pole Q's, high frequencies, tight tolerances on the passband performance) it becomes necessary to consider the effects of the op amp.

4. The numerator coefficient, $2b$, was chosen so that the synthesis equations would yield a consistent solution. Let us now consider the synthesis of the more general low-pass function

$$T(s) = \frac{d}{s^2 + as + b} \tag{7.33}$$

where the term d is allowed to be any positive constant.

Proceeding as in the example, the coefficients of like power in s in Equation 7.28 and 7.33 are equated to obtain the following relationships:

$$\frac{1 + r_2/r_1}{R_1 R_2 C_1 C_2} = d \tag{7.34}$$

$$\frac{1}{R_1 C_1} + \frac{1}{R_2 C_1} - \frac{r_2/r_1}{R_2 C_2} = a \tag{7.35}$$

$$\frac{1}{R_1 R_2 C_1 C_2} = b \tag{7.36}$$

As before, we let $C_1 = C_2 = 1$ and $r_1 = 1$. Then the above synthesis equations can be solved to yield the following expressions for R_1, R_2, and r_2:

$$R_2 = \frac{2(d/b - 2)}{-a \pm \sqrt{a^2 + 4b(d/b - 2)}}$$

$$R_1 = \frac{1}{R_2 b}$$

$$r_2 = \frac{d}{b} - 1$$

It can be seen that r_2 is negative for $d/b < 1$; and, R_2 becomes complex whenever

$$\frac{d}{b} < 2 \qquad \text{and} \qquad \left|4b\left(\frac{d}{b} - 2\right)\right| > a^2$$

Therefore, we can conclude that given the pole location (as determined by a and b), the realizable range for d is restricted. However, d only affects the level of the output voltage, but not its frequency characteristic. In view of this fact, what is often done is to match only the coefficients that determine the pole location (a and b), as in the main example. As we saw, this yields a real solution for all values of a and b. If necessary, the gain constant (K in Equation 7.3) can later be increased or decreased to any desired value by using circuit techniques described in the next section. ■

7.6 ADJUSTING THE GAIN CONSTANT

In this section we discuss ways of altering the gain constant associated with active circuit realizations. Suppose a circuit realizes the transfer function

$$T(s) = K\frac{s^2 + cs + d}{s^2 + as + b} \tag{7.37}$$

and the desired transfer function is

$$T'(s) = \alpha K\frac{s^2 + cs + d}{s^2 + as + b} \tag{7.38}$$

where α is any positive constant. We wish to incorporate the factor α into the active circuit realization.

One simple way of introducing the factor α is to follow the original circuit by the inverting amplifier circuit shown in Figure 7.9. The original transfer function is then scaled by the transfer function of the inverting structure, which is

$$T_I(s) = -\frac{\alpha R_S}{R_S} = -\alpha \tag{7.39}$$

The minus sign is usually of no concern because neither the magnitude nor the delay of the transfer function is affected by it. Simple though this scheme is, the expense of an extra amplifier makes it a relatively impractical solution. Alternate solutions, which do not require an additional amplifier, are presented in the discussion that follows.

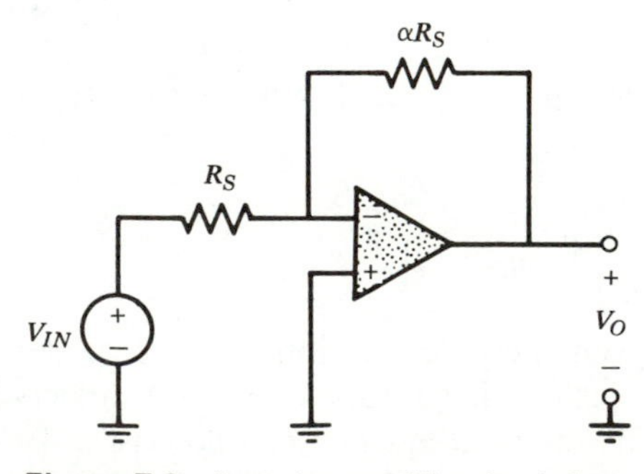

Figure 7.9 Inverting amplifier circuit for scaling.

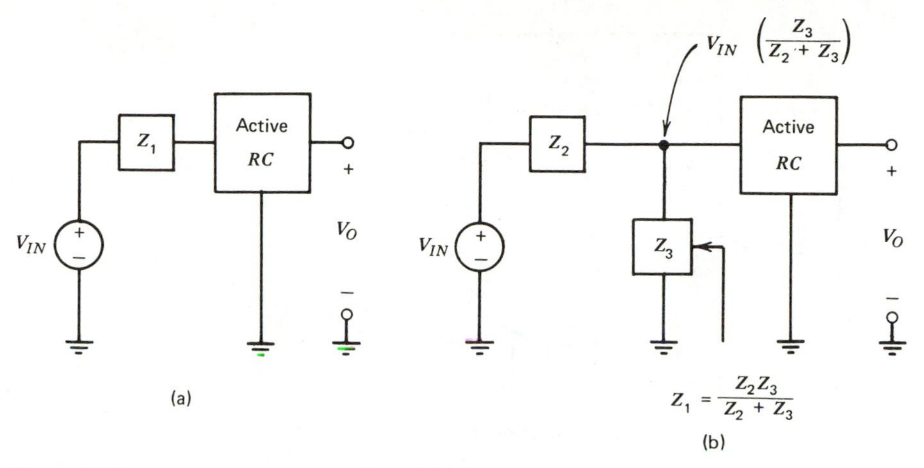

Figure 7.10 Input attenuation: (*a*) Original circuit. (*b*) Circuit with input attenuation.

First consider the case of $\alpha < 1$, when the output voltage of the given circuit is larger than desired. In this case the input voltage can be attenuated using the potential divider scheme shown in Figure 7.10. Suppose, for instance, the input voltage source drives the active RC network via an impedance Z_1 (Figure 7.10*a*). If the impedance Z_1 is replaced by the potential divider Z_2 and Z_3, the input voltage is attenuated by

$$\alpha = \frac{Z_3}{Z_2 + Z_3} \tag{7.40}$$

For the two circuits of Figure 7.10*a* and 7.10*b* to be equivalent, Z_2 and Z_3 should be chosen so that the parallel combination of their impedances equals the original input impedance, that is

$$Z_1 = \frac{Z_2 Z_3}{Z_2 + Z_3} \tag{7.41}$$

Equations 7.40 and 7.41 can easily be solved for Z_2 and Z_3. The above method of reducing the gain constant is referred to as **input attenuation**.

For example, consider the synthesis of the function

$$T(s) = \frac{b}{s^2 + as + b} \tag{7.42}$$

From Example 7.1, we know that the circuit of Figure 7.8*a* can be used to realize the function given by Equation 7.24. Comparing Equations 7.24 and 7.42, the factor α is seen to be 0.5. This factor can be implemented by replacing R_1

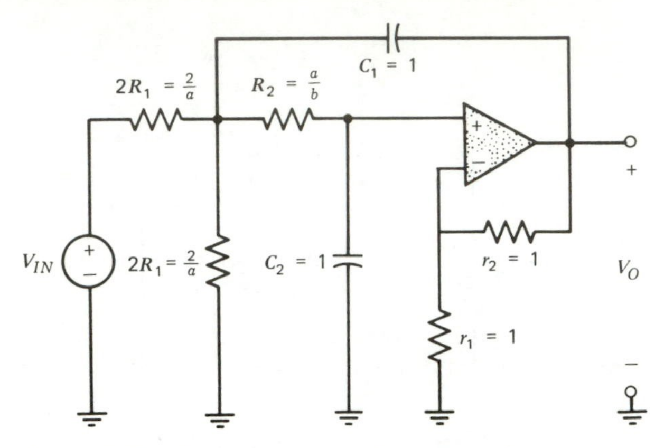

Figure 7.11 Circuit for realizing Equation 7.42.

by the potential divider shown in Figure 7.11, where

$$Z_2 = Z_3 = 2R_1 = \frac{2}{a}$$

Next, consider the case when $\alpha > 1$, that is, when the output voltage of the original circuit is lower than desired. In this case the scaling can be achieved by attenuating the output voltage as shown in Figure 7.12. Suppose the original circuit, represented by the general schematic of Figure 7.12*a*, has the transfer function $T(s) = V_O/V_{IN}$. The introduction of the attenuation α (Figure 7.12*b*) at the output of the op amp effectively decreases the amplifier gain to A/α. For moderate α's, this decrease in the amplifier gain will not affect the transfer function appreciably. Consequently, in Figure 7.12*b*

$$\frac{V_{O_1}}{V_{IN}} = T(s)$$

Thus, the output voltage V_{O_1} in this circuit is the same as V_O in Figure 7.12*a*. Therefore, the voltage obtained at the output of the op amp, αV_{O_1}, is the original output voltage V_O scaled up by α. The output voltage can be attenuated by α by using the potential divider network shown in Figure 7.12*c*, where

$$\alpha \cong \frac{R_A + R_B}{R_B}$$

This expression is approximate in that it assumes that the input impedance of the RC network, Z_{RC}, is large compared to R_B, and is therefore neglected in evaluating α. Obviously the approximation will improve as the potential divider resistors are made smaller. However, the sum of the resistors $(R_A + R_B)$ may not be made less than a certain minimum value, which is dictated by the maximum current the op amp can deliver (Chapter 12). The above method of scaling

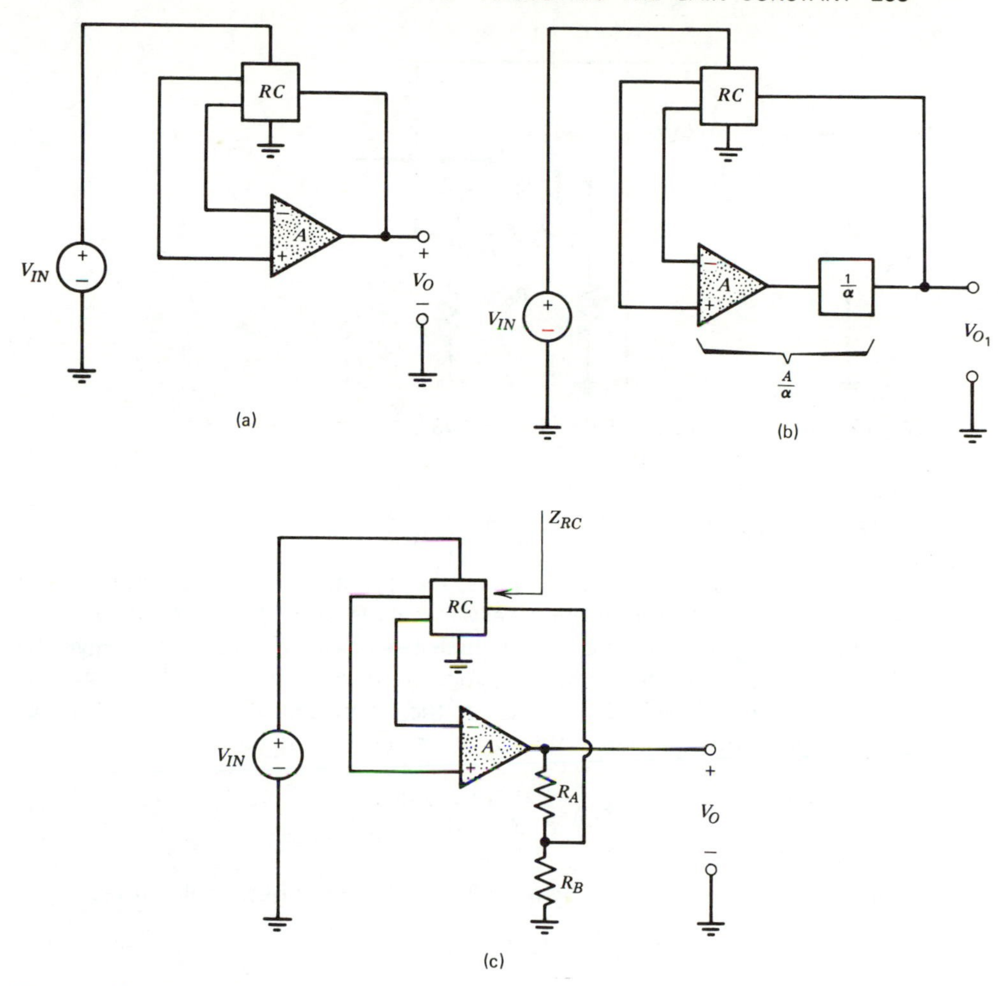

Figure 7.12 Gain enhancement: (*a*) Original circuit. (*b*) Circuit with output attenuation. (*c*) Resistive potential divider realization of output attenuation α.

is known as **gain enhancement**. An alternate, exact, gain enhancement technique is described in Problem 7.21.

As an example of the use of gain enhancement, consider the synthesis of

$$\frac{4b}{s^2 + as + b} \tag{7.43}$$

Comparing this function with Equation 7.24, the scale factor is seen to be $\alpha = 2$, and the circuit realization is as shown in Figure 7.13, with $R_A = R_B = 1$.

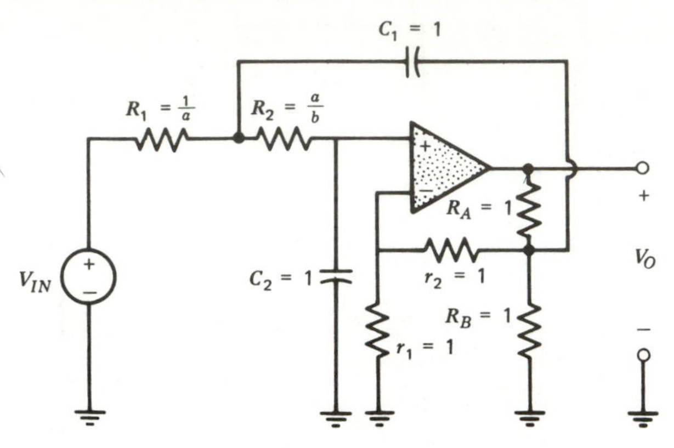

Figure 7.13 Circuit realization for Equation 7.43.

7.7 IMPEDANCE SCALING

The computations required to solve the synthesis equations are greatly simplified by choosing convenient, easy-to-handle values for the fixed elements. After obtaining the nominal design, **impedance scaling** is used to change the element values of the circuit, in order to make the circuit practically realizable. To explain the method, let us consider the circuit realization of the transfer function given by Equation 7.24 for $a = 100$ and $b = 10{,}000$:

$$T(s) = \frac{20{,}000}{s^2 + 100s + 10{,}000} \tag{7.44}$$

The element values obtained by using the realization of Example 7.1 are

$$C_1 = 1\text{F} \qquad C_2 = 1\text{F} \qquad R_1 = 0.01\ \Omega \qquad R_2 = 0.01\ \Omega \qquad r_1 = r_2 = 1\ \Omega$$

These elements are definitely not practical to implement. To see how these values can be scaled, consider the transfer function in terms of the R's and C's, given by Equations 7.28 and 7.29:

$$T(s) = \frac{\dfrac{1 + r_2/r_1}{R_1 R_2 C_1 C_2}}{s^2 + s\left(\dfrac{1}{R_1 C_1} + \dfrac{1}{R_2 C_1} - \dfrac{r_2}{r_1} \cdot \dfrac{1}{R_2 C_2}\right) + \dfrac{1}{R_1 R_2 C_1 C_2}} \tag{7.45}$$

Observe that the R's and C's always occur as an RC product, or as a ratio of resistors. Therefore, an increase in all the R's by a factor α, with a corresponding decrease in all the C's by this same factor, will leave the RC products and the resistor ratios unchanged, and hence the transfer function will not be affected.

For example, if the impedance scaling factor α is chosen to be 10^7, the following practically realizable element values result:

$$C_1 = 0.1\ \mu\text{F} \qquad C_2 = 0.1\ \mu\text{F} \qquad R_1 = 100\ \text{k}\Omega \qquad R_2 = 100\ \text{k}\Omega$$

The two elements r_1 and r_2, since they occur as a ratio, may be scaled independently. A practical choice for these two resistors is

$$r_1 = 10\ \text{k}\Omega \qquad r_2 = 10\ \text{k}\Omega$$

This example demonstrates how a network can be synthesized with easy-to-handle element values, which can then be impedance scaled to yield practical components. This technique will be frequently used in the next chapters on active network synthesis.

7.8 FREQUENCY SCALING

Frequency scaling is used to shift the frequency response of a filter to a different part of the frequency axis. This is useful in designing filters using normalized frequency requirements, such as those given in standard tables. One example of frequency scaling that we have already encountered is in the denormalization of an *LP* transfer function, which has a cutoff frequency of 1 rad/sec, to realize a *LP* function with cutoff frequency at ω_P rad/sec. From Chapter 4 (page 112), we recall that the desired transfer function was obtained by replacing s by s/ω_P in the normalized function. In this section we show how frequency scaling may be applied directly to the elements of an active *RC* circuit.

To illustrate the procedure consider, once again, the transfer function given by Equation 7.44. Suppose we wish to realize this same low-pass characteristic shifted up along the frequency axis by a factor of 5. The desired transfer function is

$$T\left(\frac{s}{5}\right) = \frac{20{,}000}{\left(\frac{s}{5}\right)^2 + 100\left(\frac{s}{5}\right) + 10{,}000} \tag{7.46}$$

Comparing this with (7.45)

$$T\left(\frac{s}{5}\right) = \frac{\dfrac{1 + r_2/r_1}{R_1R_2C_1C_2}}{\left(\frac{s}{5}\right)^2 + \left(\frac{s}{5}\right)\left[\dfrac{1}{R_1C_1} + \dfrac{1}{R_2C_1} - \dfrac{r_2}{r_1}\dfrac{1}{R_2C_2}\right] + \dfrac{1}{R_1R_2C_1C_2}} \tag{7.47}$$

where the resistors and capacitors are determined by the realization of (7.44). Equation 7.47 may be written in the form:

$$T\left(\frac{s}{5}\right) = \frac{\dfrac{\dfrac{1 + r_2/r_1}{R_1R_2C_1C_2}}{5 \times 5}}{s^2 + s\left(\dfrac{\dfrac{1}{R_1C_1}}{5} + \dfrac{\dfrac{1}{R_2C_1}}{5} - \dfrac{r_2}{r_1}\dfrac{\dfrac{1}{R_2C_2}}{5}\right) + \dfrac{\dfrac{1}{R_1R_2C_1C_2}}{5 \times 5}} \tag{7.48}$$

From this equation it is seen that the scaling can be achieved by decreasing all the resistors by 5, or else by decreasing all the capacitors by 5. Thus, Equation 7.46 may be synthesized by dividing the capacitor values in the realization of Equation 7.44 by 5. The element values for the scaled circuit are:

$$C_1 = 0.02\ \mu\text{F} \qquad C_2 = 0.02\ \mu\text{F} \qquad R_1 = 100\ \text{k}\Omega$$
$$R_2 = 100\ \text{k}\Omega \qquad r_1 = 10\ \text{k}\Omega \qquad r_2 = 10\ \text{k}\Omega$$

In general, the frequency response of a given active filter can be scaled up by a factor α by decreasing all the capacitors (or resistors) by the factor α.

7.9 CONCLUDING REMARKS

In this chapter we introduced the negative feedback and positive feedback topologies for the realization of the basic building block of active filters, the biquad. These two topologies will be studied in greater detail in Chapters 8 and 9. Another biquad circuit that is quite popular in active filters uses three op amps and is based on an analog computer type simulation of a second-order system. This three-amplifier biquad will be discussed in Chapter 10. These various biquad realizations are used for the synthesis of filter functions in both the cascaded and the coupled topologies.

FURTHER READING

1. A. Budak, *Passive and Active Network Analysis and Synthesis*, Houghton Mifflin, Boston, Mass., 1974, Chapter 9.
2. N. Fliege, "A new class of second-order RC-active filters with two operational amplifiers," *NTZ.*, *26*, No. 4, April 1973, pp. 279–282.
3. I. M. Horowitz and G. R. Banner, "A unified survey of active RC synthesis techniques," *Proc. Nat. Electron. Conf. (NEC)*, *23*, 1967, pp. 257–261.
4. S. K. Mitra, "Filter design using integrated operational amplifiers," *1969 WESCON Tech. Papers.* Also reprinted in S. K. Mitra, *Active Inductorless Filters*, IEEE Press, New York, 1971, pp. 31–41.

5. G. S. Moschytz, *Linear Integrated Networks Design*, Van Nostrand, New York, 1975, Chapter 2.
6. A. S. Sedra, "Generation and classification of single amplifier filters," *Int. J. of Circuit Theory and Appl.*, 2, 1974, pp. 51–67.
7. G. C. Temes and S. K. Mitra, eds., *Modern Filter Theory and Design*, Wiley, New York, 1973, Chapter 8.
8. G. E. Tobey, J. G. Graeme, and L. P. Huelsman, *Operational Amplifiers and Applications*, McGraw-Hill, New York, 1971, Chapter 8.

PROBLEMS

7.1 *Real-pole synthesis.* Synthesize the following real-pole transfer functions using active RC circuits, with no more than two op amps:

(a) $-\dfrac{3s}{s+4}$

(b) $-\dfrac{(s+8)}{(s+3)(s+4)}$

(c) $\dfrac{s}{s+4} + \dfrac{s+2}{s+3}$

(d) $\dfrac{8}{s^2+6s+8}$

7.2 *Negative feedback topology.* Identify the feedback and feedforward transfer functions for the inverter circuit of Figure 7.3*a*. Hence, determine the transfer function V_O/V_{IN}, assuming the op amp gain is A.

7.3 Consider the negative feedback biquad circuit shown in Figure P7.3.

(a) Identify the feedback and feedforward transfer functions and, hence, determine the transfer function V_O/V_{IN}, assuming an ideal op amp.

(b) Repeat part (a) assuming an op amp with a finite gain A.

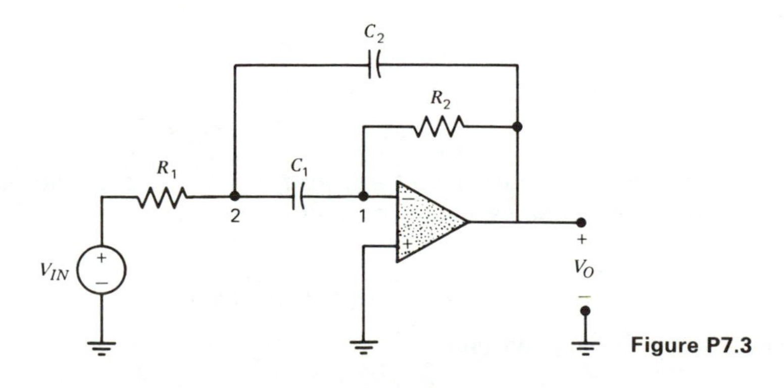

Figure P7.3

7.4 Determine the transfer function V_O/V_{IN} for the circuit of Figure P7.3, by writing the node equations for nodes 1 and 2. Assume an ideal op amp.

7.5 Repeat Problem 7.4, assuming the op amp gain to be A.

7.6 *Positive feedback topology.* Determine the transfer function for the positive feedback biquad circuit of Figure P7.6 in terms of the feedback and feed-forward transfer functions:

(a) Assuming an ideal op amp.

(b) Assuming the op amp gain is A.

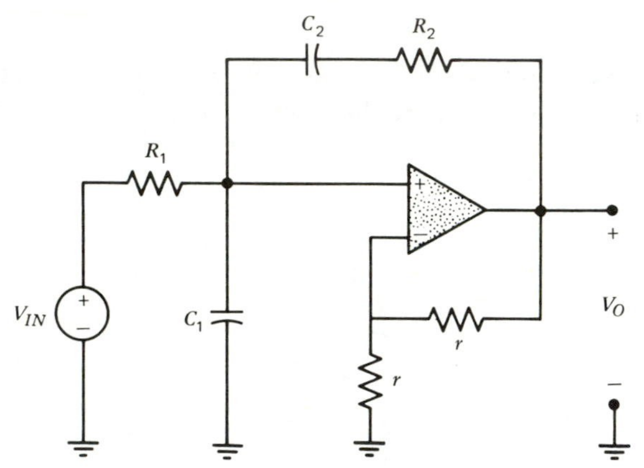

Figure P7.6

7.7 Determine the transfer function for the circuit of Figure P7.7 by writing node equations for nodes 1 and 2. Assume an ideal op amp.

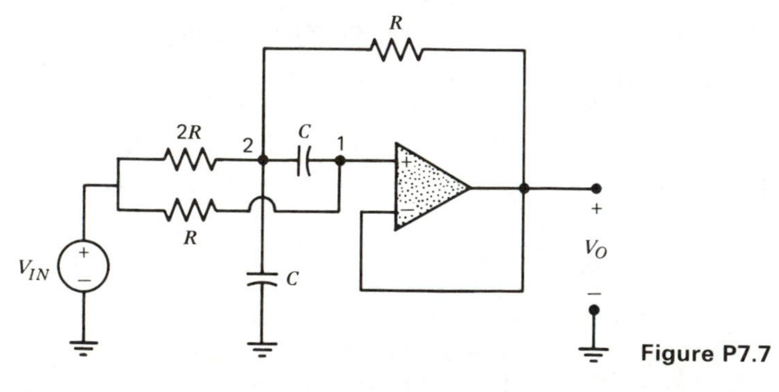

Figure P7.7

7.8 *Synthesis of biquads using coefficient matching.* Synthesize the following low-pass transfer function using the circuit of Figure 7.8:

$$T(s) = \frac{K}{(s^2 + 4s + 64)(s^2 + 3s + 81)}$$

where K is an arbitrary constant.

7.9 Synthesize the band-pass transfer function

$$\frac{Ks^2}{(s^2 + 3s + 81)(s^2 + 4s + 64)}$$

using the topology shown. Determine the gain constant for the solution. (*Hint*: choose $C_1 = C_2 = 1$, $R_1 = R_2 = R$.)

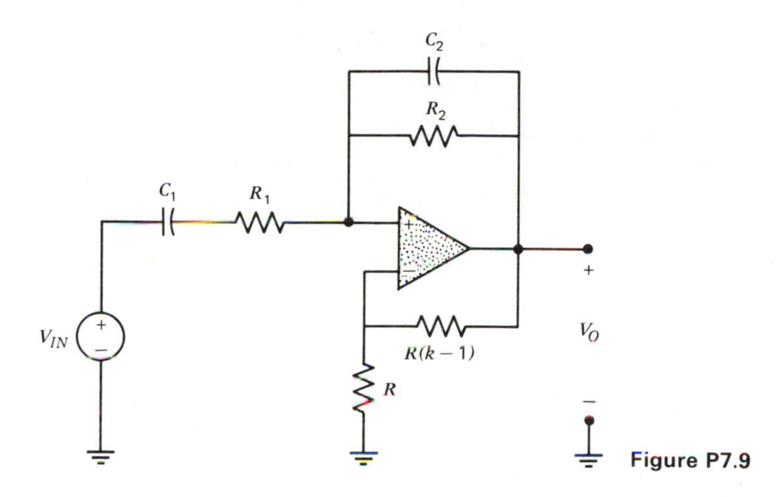

Figure P7.9

7.10 *Adjusting the gain constant*. Synthesize the following transfer functions:

(a) $\dfrac{20}{s^2 + 2s + 100}$

(b) $\dfrac{600}{s^2 + 2s + 100}$

7.11 *Impedance and frequency scaling*. A certain active-*RC* network realizes a low-pass transfer function with a passband edge frequency at 1000 Hz. How do the filter characteristics change if the resistors are increased by the factor α_R, and the capacitors are increased by the factor α_C, where:
(a) $\alpha_R = 1$, $\alpha_C = 6$
(b) $\alpha_R = 100$, $\alpha_C = 0.01$
(c) $\alpha_R = 5$, $\alpha_C = 0.1$

7.12 Consider the transfer function of Equation 7.44. Synthesize this function, using practical element values, for each of the following cases:
(a) $C_1 = C_2 = 0.047\ \mu F$.
(b) The gain constant is decreased by a factor of 4.
(c) The pole frequency is changed to 100 Hz.

7.13 *High-pass filter*. Synthesize the high-pass transfer function

$$\frac{s^2}{s^2 + 200s + 640{,}000}$$

using the topology shown, with practical element values. (*Hint*: choose $C_1 = C_2 = 1, k = 1$.)

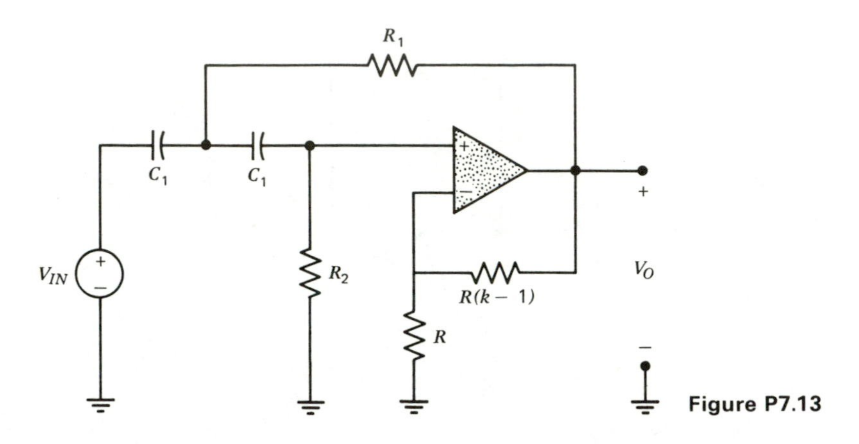

Figure P7.13

7.14 *Butterworth filter*. Synthesize a third-order low-pass Butterworth approximation function characterized by $A_{\max} = 3$ dB, $\omega_P = 10{,}000$ rad/sec, *dc* gain = 0 dB. Use the circuits of Figures 7.3*a* and 7.8 with practical element values.

7.15 *Chebyshev filter*. Use the circuit of Figure 7.8 to realize the Chebyshev low-pass requirements sketched in Figure P7.15.

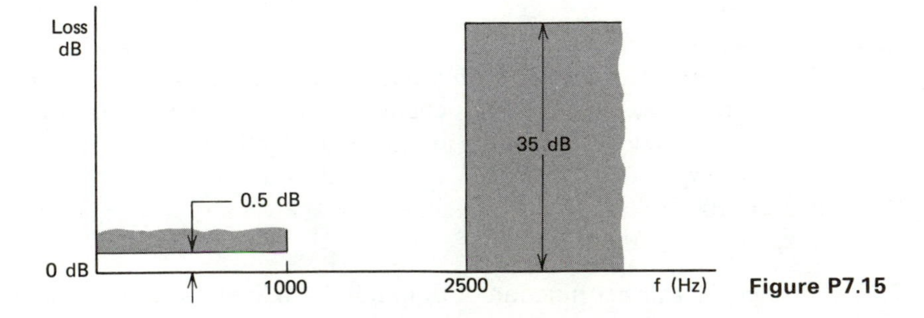

Figure P7.15

7.16 *Pole-zero cancellation biquad*. A generalization of the negative feedback topology is shown in the figure.

(a) Find an expression for the transfer function V_O/V_{IN} in terms of the feedback and feedforward transfer functions of the two RC networks, assuming an ideal op amp.
(b) How are the poles and zeros of the transfer function related to those of the RC networks?
(c) Explain how a general biquadratic transfer function (with complex poles and zeros) can be realized, given that the RC networks can realize second-order functions.

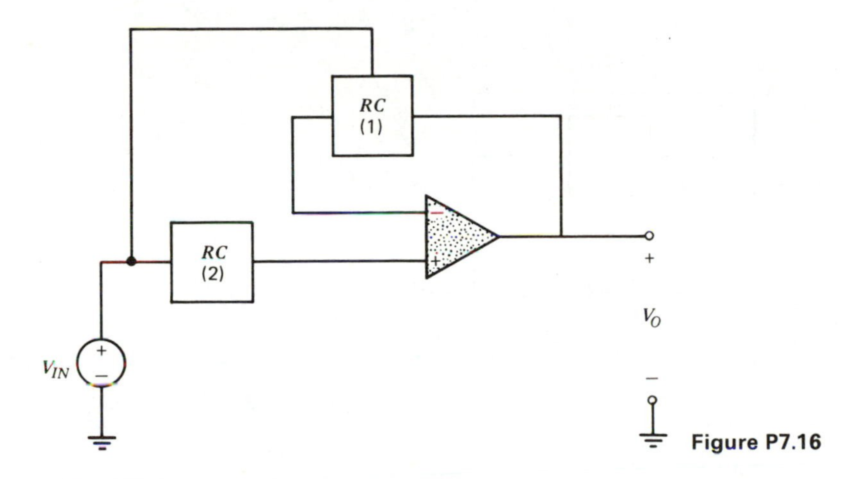

Figure P7.16

7.17 *Positive-cum-negative feedback topology.* Determine the transfer function for the positive-cum-negative feedback topology shown in terms of the feedback and feedforward transfer functions of the RC networks. Assume the gain of the op amp is A. Show that the positive and negative feedback topologies described in Section 7.4 are special cases of this general topology.

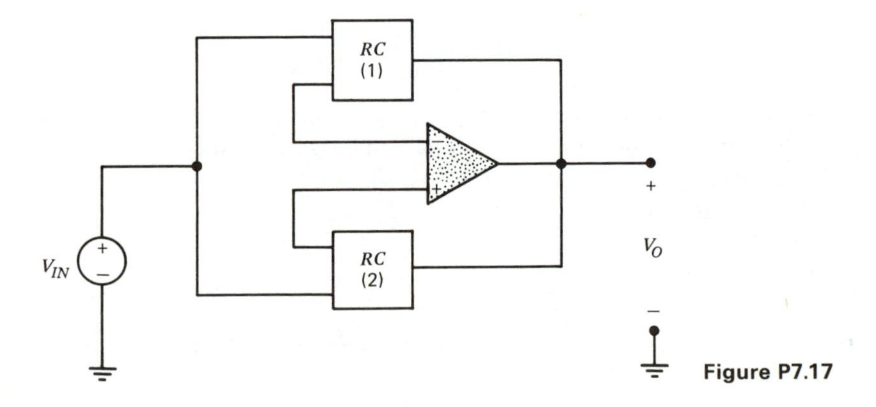

Figure P7.17

7.18 *$-K$ topology.* The schematic of a topology using a finite negative gain amplifier is shown in Figure P7.18.

(a) Find an expression for the transfer function V_O/V_{IN} in terms of the amplifier gain K and the feedforward and feedback transfer functions of the RC network.

(b) Sketch the root locus of the poles of V_O/V_{IN} for increasing K, if the feedforward transfer function is

$$T_{FF} = \frac{b}{s^2 + as + b}$$

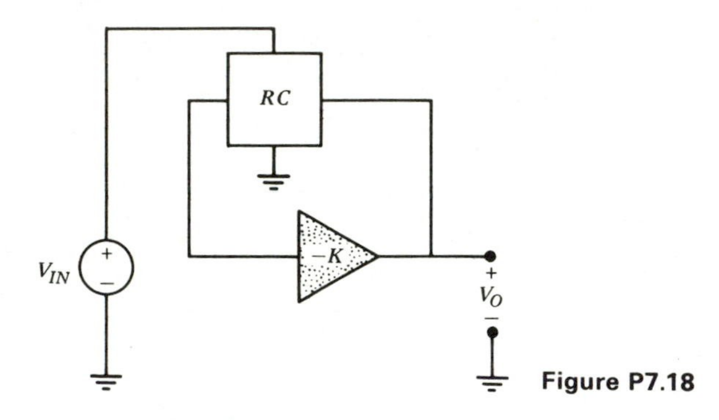

Figure P7.18

7.19 *$(1 - T)$ topology.* Suppose a single op amp RC network has the transfer function $T(s)$. Hilberman* has shown that if the input and ground leads of the network are interchanged, the resulting transfer function is $1 - T(s)$. Verify this for:

(a) The inverting amplifier circuit of Figure 7.3*a*.

(b) The noninverting amplifier circuit of Figure 7.4*a*.

7.20 Suppose a single op amp RC circuit realizes the function:

$$T(s) = \frac{c_1 s + c_2}{s^2 + as + b}$$

Choose the positive constants c_1 and c_2 to realize the following biquadratic filter functions, using the $[1 - T(s)]$ topology described in Problem 7.19:

(a) Band-reject.

(b) All-pass.

(c) High-pass notch.

(d) High-pass.

* D. Hilberman, "Input and ground as complements in active filters," *IEEE Trans. Circuit Theory*, *20* No. 5, September 1973, pp. 540–547.

7.21 *Gain enhancement using element splitting.* Show that if the impedance Z_1 in the general topology of Figure P7.21*a* is replaced by the potential divider αZ_1 and $\alpha Z_1/(\alpha - 1)$, as shown in Figure P7.21*b*, the gain constant of the transfer function is enhanced by the factor α. Unlike the approximate technique described in Section 7.6, this is an exact method for attaining gain enhancement. However, the method often requires extra capacitors, which add to the cost of the filter.

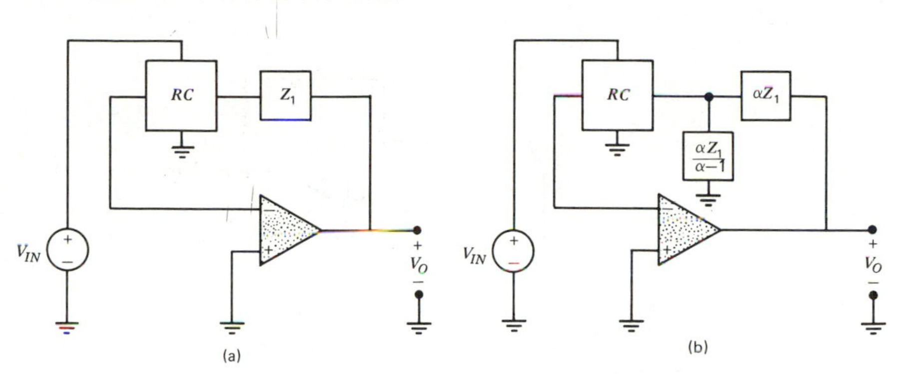

Figure P7.21

7.22 Use the element-splitting gain enhancement technique to realize the transfer function of Equation 7.43.

7.23 Show how the gain constant associated with the circuit shown can be increased by a factor of 5 using element-splitting gain enhancement.

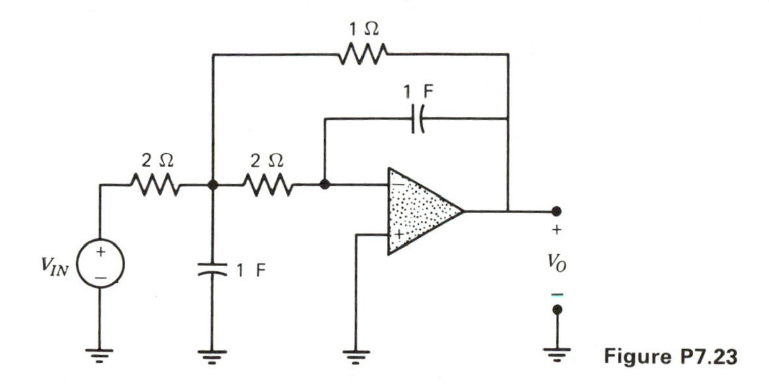

Figure P7.23

7.24 *Impedance synthesis.* Show that the topology of Figure P7.24 realizes the impedance function

$$Z(s) = \frac{R_F}{T(s)}$$

Hence, find an active-RC realization for the impedance function

$$Z(s) = 1000\left(\frac{s + 400}{s + 800}\right)$$

These active-impedance networks are used to balance the impedances of voice transmission cables in hybrid repeaters.

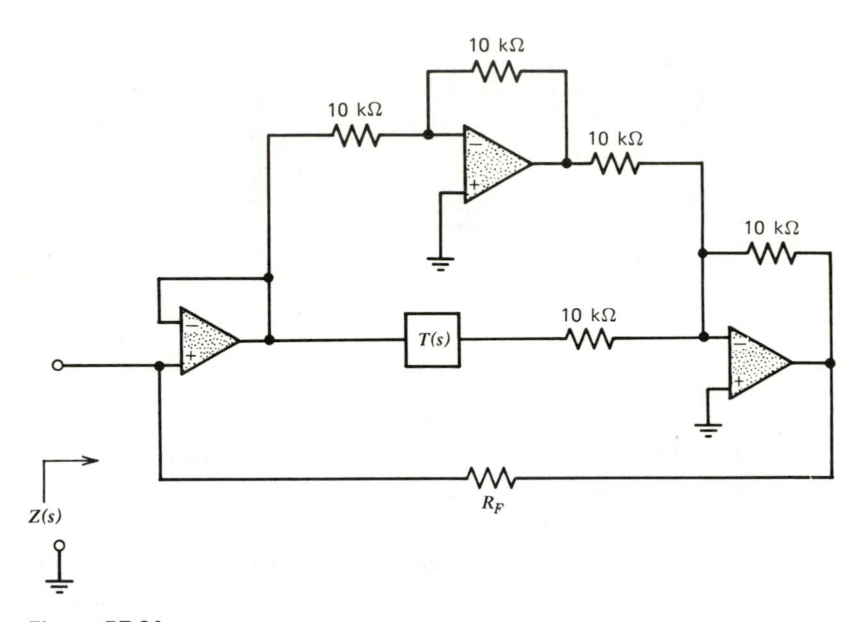

Figure P7.24

8.

POSITIVE FEEDBACK BIQUAD CIRCUITS

In this chapter we present the design of single amplifier biquad circuits based on the positive feedback topology. In particular, we will discuss the types of RC circuits that may be used in this topology, and the solution of the resulting synthesis equations. The choice between alternate realizations for a given filter function will, in general, be based on their respective sensitivities. In solving the synthesis equations, the op amp will be assumed to be ideal. This assumption does not affect the nominal circuit design to any significant degree. However, the sensitivity of the circuits to the finite gain of a real op amp does play a significant part in the design procedure, as is shown in Section 8.2. A detailed study of the effects of real amplifiers in filter design is covered in Chapter 12.

8.1 PASSIVE RC CIRCUITS USED IN THE POSITIVE FEEDBACK TOPOLOGY

As discussed in Chapter 7 the transfer function of the basic positive feedback topology of Figure 8.1, assuming an ideal op amp, is (Equation 7.17)

$$T_V = \frac{kT_{FF}}{1 - kT_{FB}} \tag{8.1}$$

where

$$T_{FF} \text{ is the feedforward transfer function } \left.\frac{V_1}{V_2}\right|_{V_3=0}$$

$$T_{FB} \text{ is the feedback transfer function } \left.\frac{V_1}{V_3}\right|_{V_2=0}$$

and

$$k = 1 + \frac{r_2}{r_1}$$

Recalling that the denominators of T_{FF} and T_{FB} are the same, so that $T_{FF} = N_{FF}/D$ and $T_{FB} = N_{FB}/D$, the transfer function can be written as

$$T_V = \frac{kN_{FF}}{D - kN_{FB}} \tag{8.2}$$

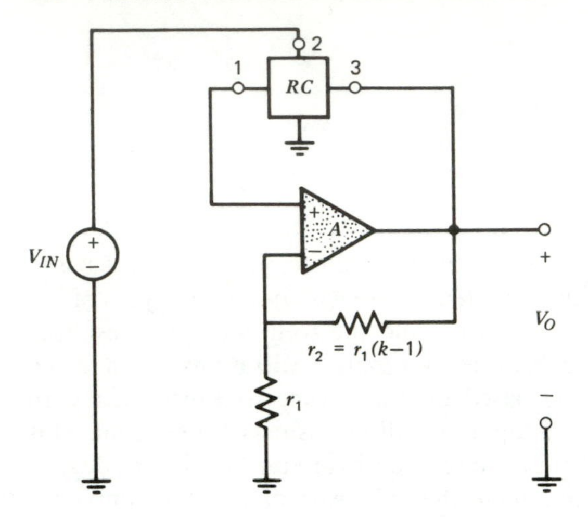

Figure 8.1 Positive feedback topology.

Since D represents the poles of an RC network, they must be on the negative real axis. It is shown in Chapter 7 that one form of T_{FB} that renders the poles of T_V complex is the band-pass function

$$T_{FB} = \frac{s}{s^2 + as + b} \tag{8.3}$$

Some RC networks that can realize this band-pass function are shown in Figure 8.2 [9]. The circuits shown are canonic, in the sense that they all use two capacitors, which is the minimum number of reactive elements needed to realize a second-order function. The band-pass nature of the function requires a series capacitor to provide attenuation at low frequencies, and a shunt capacitor to provide the high frequency attenuation.

The zeros of T_V are formed by introducing the input signal in the RC network. This must be done so that the poles of the RC network are not perturbed. Since the input is usually a low impedance grounded voltage source, the only permissible places where the input can be introduced without disturbing the poles are at element terminals that may be lifted from ground. These terminals are marked with a 2 (or 2′, 2″) in the circuits of Figure 8.2. If the input is introduced at more than one such terminal, the numerator of T_V can be obtained by superposition, considering each input separately.

In the next few sections we will consider the design of some positive feedback circuits using the RC circuits of Figure 8.2.

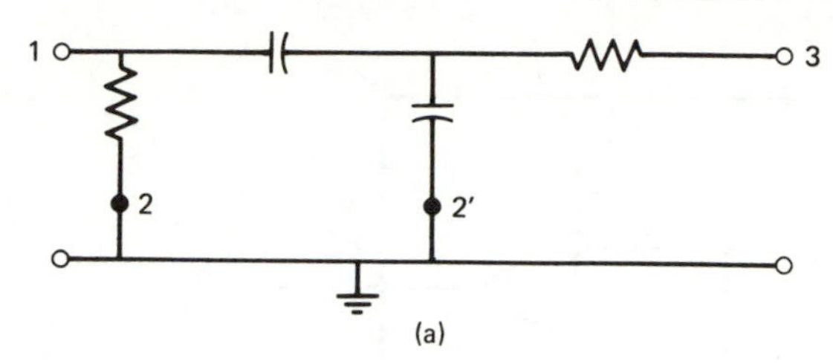

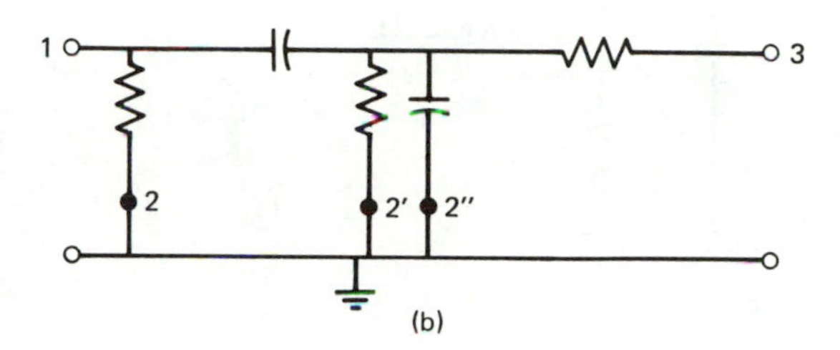

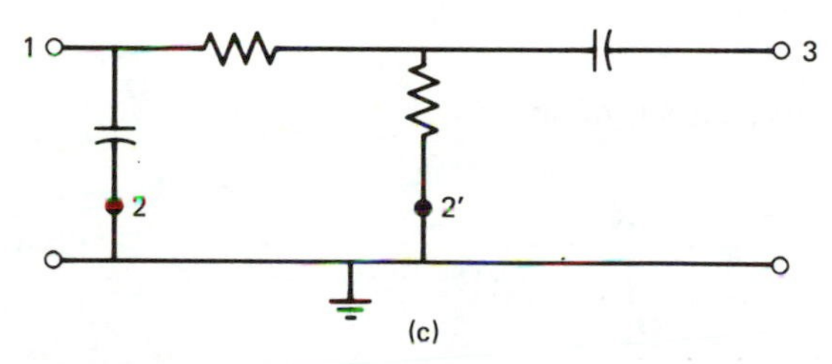

Figure 8.2 Passive *RC* circuits for positive feedback topology.

8.2 SALLEN AND KEY LOW-PASS CIRCUIT

Sallen and Key developed many circuits based on the positive feedback topology [7]. One of these is the low-pass circuit shown in Figure 8.3, which was analyzed in Example 7.1. This circuit uses the *RC* circuit of Figure 8.2*c*, with the input introduced at node 2′.

From Equation 7.28, the transfer function of the active circuit, assuming an ideal op amp, is

$$T_V = \frac{k/R_1R_2C_1C_2}{s^2 + s\left(\dfrac{1}{R_1C_1} + \dfrac{1}{R_2C_1} + \dfrac{1-k}{R_2C_2}\right) + \dfrac{1}{R_1R_2C_1C_2}} \tag{8.4}$$

where

$$k = 1 + \frac{r_2}{r_1}$$

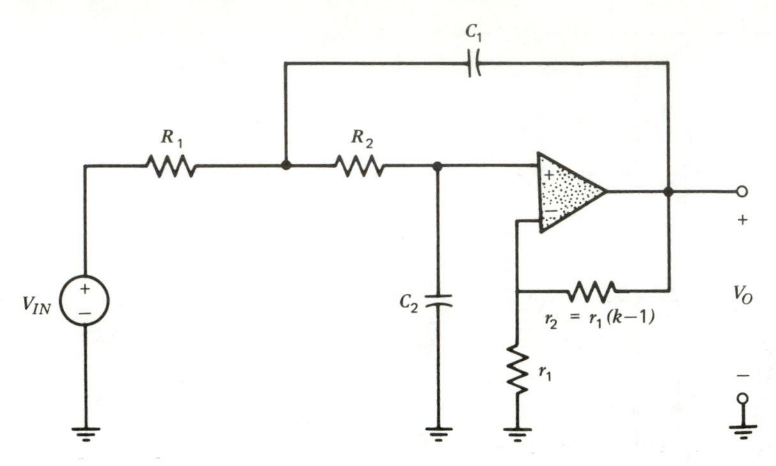

Figure 8.3 Low-pass Sallen and Key circuit.

Consider the synthesis of the low-pass function

$$T_{LP} = \frac{K}{s^2 + \dfrac{\omega_p}{Q_p}s + \omega_p^2} \tag{8.5}$$

As mentioned in Chapter 5, one of the major criteria in the evaluation of a circuit is its sensitivity. Therefore, let us first evaluate the sensitivities of the biquadratic parameters K, ω_p, and Q_p to the passive elements R_1, R_2, C_1, C_2, and k. The sensitivity to the op amp characteristics will be taken up in the latter part of this section.

From (8.4) and (8.5)

$$\omega_p = \sqrt{\frac{1}{R_1R_2C_1C_2}} \tag{8.6}$$

$$Q_p = \frac{\omega_p}{(\text{bw})_p} = \frac{\sqrt{\dfrac{1}{R_1R_2C_1C_2}}}{\dfrac{1}{R_1C_1} + \dfrac{1}{R_2C_1} + \dfrac{1-k}{R_2C_2}} \tag{8.7}$$

$$K = \frac{k}{R_1R_2C_1C_2} \tag{8.8}$$

Using the sensitivity relationships developed in Chapter 5 (Section 5.1), we get

$$S^{\omega_p}_{R_1, R_2, C_1, C_2} = -\tfrac{1}{2} \tag{8.9a}$$

$$S^{\omega_p}_{k} = 0 \tag{8.9b}$$

$$S^{Q_p}_{R_1} = S^{\omega_p}_{R_1} - S^{(\mathrm{bw})_p}_{R_1} = -\frac{1}{2} - \frac{\dfrac{1}{R_1 C_1}(-1)}{(\mathrm{bw})_p}$$

$$= -\frac{1}{2} + Q_p\sqrt{\frac{R_2 C_2}{R_1 C_1}} \tag{8.9c}$$

Similarly

$$S^{Q_p}_{R_2} = -\frac{1}{2} + Q_p\left(\sqrt{\frac{R_1 C_2}{R_2 C_1}} + (1-k)\sqrt{\frac{R_1 C_1}{R_2 C_2}}\right) \tag{8.9d}$$

$$S^{Q_p}_{C_1} = -\frac{1}{2} + Q_p\left(\sqrt{\frac{R_1 C_2}{R_2 C_1}} + \sqrt{\frac{R_2 C_2}{R_1 C_1}}\right) \tag{8.9e}$$

$$S^{Q_p}_{C_2} = -\frac{1}{2} + (1-k)Q_p\sqrt{\frac{R_1 C_1}{R_2 C_2}} \tag{8.9f}$$

$$S^{Q_p}_{r_2} = -S^{Q_p}_{r_1} = -(1-k)Q_p\sqrt{\frac{R_1 C_1}{R_2 C_2}} \tag{8.9g}$$

$$S^{K}_{R_1, R_2, C_1, C_2} = -1 \tag{8.9h}$$

$$S^{K}_{r_2} = -S^{K}_{r_1} = -\left(\frac{1-k}{k}\right) \tag{8.9i}$$

Our objective is to choose the element values in the synthesis procedure so as to make these sensitivities as low as possible. As in Example 7.1, the elements R_1, R_2, C_1, C_2, and k are chosen to satisfy Equations 8.6 and 8.7, the scale factor K being arbitrary. Since there are five elements to satisfy two equations, three of these can be fixed. As is shown in the following, one choice that makes most of the sensitivity terms low is

$$k = 1 \qquad R_1 = R_2 = 1 \tag{8.10a}$$

The remaining elements are then given by

$$C_1 = \frac{2Q_p}{\omega_p} \qquad C_2 = \frac{1}{2\omega_p Q_p} \tag{8.10b}$$

Hereafter this synthesis will be referred to as *Design 1*. With this choice of elements, the sensitivity expressions of Equation 8.9 reduce to the values given in Table 8.1. The sensitivities are all seen to be quite low. However, a practical

Table 8.1 Sensitivities of Sallen and Key LP Filter

	Design 1	Design 2	Design 3 (Saraga)
$S^{\omega_p}_{R_1,R_2,C_1,C_2}$	$-\frac{1}{2}$	$-\frac{1}{2}$	$-\frac{1}{2}$
$S^{\omega_p}_{k}$	0	0	0
$S^{Q_p}_{R_1}$	0	$-\frac{1}{2}+Q_p$	$-\frac{1}{2}+0.58Q_p$
$S^{Q_p}_{R_2}$	0	$\frac{1}{2}-Q_p$	$\frac{1}{2}-0.58Q_p$
$S^{Q_p}_{C_1}$	$\frac{1}{2}$	$-\frac{1}{2}+2Q_p$	$\frac{1}{2}+0.58Q_p$
$S^{Q_p}_{C_2}$	$-\frac{1}{2}$	$\frac{1}{2}-2Q_p$	$-\frac{1}{2}-0.58Q_p$
$S^{Q_p}_{r_2}=-S^{Q_p}_{r_1}$	0	$2Q_p-1$	$0.58Q_p$
$S^{K}_{R_1,R_2,C_1,C_2}$	-1	-1	-1
$S^{K}_{r_2}=-S^{K}_{r_1}$	0	$1-\dfrac{1}{3-1/Q_p}$	$\frac{1}{4}$

disadvantage of this design is that it requires capacitors of widely differing values ($C_1/C_2 = 4Q_p^2$). For all but very low pole Q circuits ($Q < 5$ approximately) this limitation makes the design unattractive from a manufacturing standpoint.

An alternate choice of parameters, which requires equal valued capacitors, is (*Design 2*):

$$C_1 = C_2 = 1 \qquad R_1 = R_2 = R \tag{8.11a}$$

Using Equations 8.6 and 8.7 the remaining elements are given by

$$R_1 = R_2 = \frac{1}{\omega_p} \qquad k = 3 - \frac{1}{Q_p} \tag{8.11b}$$

The resulting sensitivity expressions for this design are also listed in Table 8.1. Observe that the magnitude of the Q_p sensitivities are higher than in Design 1.

Finally, consider the following choice of parameters (Design 3), originally proposed by Saraga,* that makes all the Q_p sensitivity terms lower than in Design 2:

$$C_2 = 1 \qquad C_1 = \sqrt{3}Q_p \qquad \frac{R_2}{R_1} = \frac{Q_p}{\sqrt{3}} \tag{8.12a}$$

With this choice the remaining elements are

$$R_2 = \frac{1}{\sqrt{3}\omega_p} \qquad R_1 = \frac{1}{Q_p\omega_p} \qquad k = \frac{4}{3} \tag{8.12b}$$

and the sensitivities are as given in Table 8.1, under *Design 3*. The magnitudes of the Q_p sensitivities are seen to be appreciably lower than in Design 2.

* In [8], Saraga shows that the best compromise between active and passive sensitivities is a value of k between 1 and 4/3. Henceforth the design corresponding to $k = 4/3$ will be referred to, simply, as the Saraga design.

Example 8.1

Synthesize a second-order *LP* filter to have a pole frequency of 2 kHz and a pole Q of 10, using the Saraga design of the Sallen and Key circuit (Design 3). Also compute the component sensitivities for Designs 1, 2, and 3.

Solution

The pole frequency and pole Q are given to be $\omega_p = 2\pi(2000)$ rad/sec and $Q_p = 10$, respectively. From Equation 8.5, the desired *LP* function is therefore

$$\frac{K}{s^2 + 2\pi(200)s + (2\pi 2000)^2}$$

The element values for the Saraga design were shown to be (Equations 8.12):

$$C_2 = 1 \qquad C_1 = \sqrt{3}Q_p \qquad R_2 = \frac{1}{\sqrt{3}\omega_p} \qquad R_1 = \frac{1}{Q_p\omega_p} \qquad k = \frac{4}{3}$$

Substituting the values of ω_p and Q_p we get

$$C_2 = 1 \qquad C_1 = 10\sqrt{3} \qquad R_2 = \frac{1}{\sqrt{3}(2\pi)(2000)} \qquad R_1 = \frac{1}{2\pi(20{,}000)}$$

and the term $k = 1 + r_2/r_1 = 4/3$ can be realized using

$$r_2 = 1 \qquad r_1 = 3$$

The above element values can be impedance scaled (Section 7.7) to yield practical element values, by multiplying the resistors by 10^9 and dividing the capacitors by this same factor. Then

$$C_2 = 0.001\ \mu\text{F} \qquad C_1 = 0.017\ \mu\text{F} \qquad R_2 = 46\ \text{k}\Omega \qquad R_1 = 7.96\ \text{k}\Omega$$

and the potential divider resistors r_1 and r_2 can be scaled by 1000 to yield

$$r_2 = 1\ \text{k}\Omega \qquad r_1 = 3\ \text{k}\Omega$$

The sensitivities of ω_p, Q_p, and K to the element values can be evaluated from Table 8.1, with $Q_p = 10$. These sensitivities are listed in Table 8.2.

Observations

From Table 8.2 the component sensitivities to ω_p and K are seen to be the same for the three designs. However, the component sensitivities to Q_p are quite different. In particular, Design 1 has the lowest component sensitivity. To establish the relative sensitivities of the resulting circuits, we must relate these component sensitivities to the corresponding deviations in gain. As we mentioned in Chapter 5, the gain deviation is a function of the component sensitivities, the biquadratic parameter sensitivities ($\mathcal{S}^G_{\omega_p}$, $\mathcal{S}^G_{Q_p}$, S^G_{K}), and the element variabilities, V_x. The computation of gain deviation for the three designs is illustrated by the following example.

Table 8.2 Sensitivities for Sallen and Key Circuit ($Q_p = 10$)

	Design 1	Design 2	Design 3 (Saraga)
$S^{\omega_p}_{R_1, R_2, C_1, C_2}$	$-\frac{1}{2}$	$-\frac{1}{2}$	$-\frac{1}{2}$
$S^{\omega_p}_{k}$	0	0	0
$S^{Q_p}_{R_1}$	0	9.5	5.3
$S^{Q_p}_{R_2}$	0	−9.5	−5.3
$S^{Q_p}_{C_1}$	$\frac{1}{2}$	19.5	6.3
$S^{Q_p}_{C_2}$	$-\frac{1}{2}$	−19.5	−6.3
$S^{Q_p}_{r_2} = -S^{Q_p}_{r_1}$	0	19	5.8
$S^{K}_{R_1, R_2, C_1, C_2}$	−1	−1	−1
$S^{K}_{r_2} = -S^{K}_{r_1}$	0	0.66	$\frac{1}{4}$

■

Example 8.2

For the *LP* filter described in Example 8.1, compare the gain deviations of Designs 1, 2, and 3 at 2.1 kHz (the upper 3 dB passband edge frequency) due to the manufacturing tolerances of resistors and capacitors. Assume the components to have a ± 1 percent tolerance about their nominal values, where the tolerance limits represent the 3σ points of a Gaussian distribution.

Solution

The gain deviation is given by an equation similar to Equation 5.38:

$$\Delta G = \sum_i [\mathcal{S}^G_{Q_p} S^{Q_p}_{x_i} V_{x_i} + \mathcal{S}^G_{\omega_p} S^{\omega_p}_{x_i} V_{x_i} + \mathcal{S}^G_K S^K_{x_i} V_{x_i}] \tag{8.13}$$

where x_i represents the resistors and capacitors. The component sensitivity terms ($S^{Q_p}_{x_i}$, $S^{\omega_p}_{x_i}$, $S^K_{x_i}$) were evaluated in Example 8.1 and are listed in Table 8.2.

From the given information, the variabilities of the elements are defined by

$$\mu(V_R) = 0 \qquad \mu(V_C) = 0$$
$$3\sigma(V_R) = 0.01 \qquad 3\sigma(V_C) = 0.01$$

Using Equations 5.43, 5.44, and 5.45 the biquadratic parameter sensitivity terms are evaluated to be

$$\mathcal{S}^G_K = 8.686 \text{ dB} \qquad \mathcal{S}^G_{\omega_p} = 78.25 \text{ dB} \qquad \mathcal{S}^G_{Q_p} = 4.45 \text{ dB}$$

With the above information the mean and standard deviation of ΔG can be evaluated using the following equations, which were developed in Section 5.4.3 (Equation 5.58 and 5.59):

$$\mu(\Delta G) = \sum_i [\mathcal{S}^G_{Q_p} S^{Q_p}_{x_i} \mu(V_{x_i}) + \mathcal{S}^G_{\omega_p} S^{\omega_p}_{x_i} \mu(V_{x_i}) + \mathcal{S}^G_K S^K_{x_i} \mu(V_{x_i})] \tag{8.14}$$

$$[3\sigma(\Delta G)]^2 = \sum_i [(\mathcal{S}^G_{Q_p})^2 (S^{Q_p}_{x_i})^2 (3\sigma(V_{x_i}))^2 + (\mathcal{S}^G_{\omega_p})^2 (S^{\omega_p}_{x_i})^2 (3\sigma(V_{x_i}))^2 + (\mathcal{S}^G_K)^2 (S^K_{x_i})^2 (3\sigma(V_{x_i}))^2] \tag{8.15}$$

From Equation 8.14 it can easily be seen that in this problem, since all the variabilities are given to have a mean value of zero, $\mu(\Delta G) = 0$.

The standard deviations of ΔG for the three designs are evaluated below:

Design 1:

$$\begin{aligned}[3\sigma(\Delta G)]^2 &= (4.45)^2(\tfrac{1}{4} + \tfrac{1}{4})(0.01)^2 && \rightarrow Q_p \text{ term}\\ &\quad + (78.25)^2(\tfrac{1}{4} + \tfrac{1}{4} + \tfrac{1}{4} + \tfrac{1}{4})(0.01)^2 && \rightarrow \omega_p \text{ term}\\ &\quad + (8.686)^2(1 + 1 + 1 + 1)(0.01)^2 && \rightarrow K \text{ term}\\ &= \underset{(Q_p \text{ term})}{9.9(10)^{-4}} + \underset{(\omega_p \text{ term})}{0.61} + \underset{(K \text{ term})}{0.03}\end{aligned}$$

$$[3\sigma(\Delta G)]^2 = 0.64$$

or

$$3\sigma(\Delta G) = 0.80 \text{ dB}$$

Thus, the gain deviation has a Gaussian distribution described by

$$\Delta G = 0 \pm 0.80 \text{ dB} \quad \text{for Design 1}$$

The gain deviations for Designs 2 and 3 can be evaluated in a similar way. The results of the computations are:

Design 2

$$[3\sigma(\Delta G)]^2 = \underset{(Q_p \text{ term})}{3.29} + \underset{(\omega_p \text{ term})}{0.61} + \underset{(K \text{ term})}{0.033} = 3.93$$

$$3\sigma(\Delta G) = 1.98 \text{ dB}$$

Design 3 (Saraga)

$$[3\sigma(\Delta G)]^2 = \underset{(Q_p \text{ term})}{0.40} + \underset{(\omega_p \text{ term})}{0.61} + \underset{(K \text{ term})}{0.03} = 1.04$$

$$3\sigma(\Delta G) = 1.02 \text{ dB}$$

Observations

Based on the above results, it would appear that Design 1 yields the smallest gain deviation. These results suggest that k should be chosen to be unity. However, so far we have only considered the sensitivity of the network to the passive elements. To complete the comparison of the three designs it is also necessary to study the gain deviation due to the finite gain of the op amp. This is done next. ■

To evaluate the sensitivities to the op amp gain, we must first find an expression for the transfer function of the Sallen and Key circuit, assuming a finite gain amplifier. From Equation 7.16, the transfer function for a general positive feedback topology is

$$T_V = \frac{kN_{FF}}{D - kN_{FB} + \dfrac{kD}{A(s)}} \tag{8.16}$$

The expressions for N_{FF}, N_{FB}, and D for the Sallen and Key circuit were evaluated in Example 7.1. Substituting these in (8.16), we get

$$T_V = \frac{k/R_1R_2C_1C_2}{s^2 + s\left(\dfrac{1}{R_1C_1} + \dfrac{1}{R_2C_1} + \dfrac{1}{R_2C_2}\right) + \dfrac{1}{R_1R_2C_1C_2} - \dfrac{ks}{R_2C_2} + \dfrac{k}{A(s)}\left[s^2 + s\left(\dfrac{1}{R_1C_1} + \dfrac{1}{R_2C_1} + \dfrac{1}{R_2C_2}\right) + \dfrac{1}{R_1R_2C_1C_2}\right]} \tag{8.17}$$

Dividing numerator and denominator by $1 + k/A(s)$ and recognizing that for large amplifier gain $[1 + k/A(s)]^{-1} \cong 1 - k/A(s)$, the above expression reduces to

$$T_V \cong \frac{[1 - k/A(s)]k/R_1R_2C_1C_2}{s^2 + s\left[\dfrac{1}{R_1C_1} + \dfrac{1}{R_2C_1} + \dfrac{1}{R_2C_2}\left(1 - k + \dfrac{k^2}{A(s)}\right)\right] + \dfrac{1}{R_1R_2C_1C_2}} \tag{8.18}$$

Many op amps have a gain characteristic that can be approximated by a single pole,* as

$$A(s) = \frac{A_0\alpha}{s + \alpha} \tag{8.19a}$$

where α is a low frequency pole (typically around $2\pi 10$ rad/sec); A_0 is the *dc* gain (typically $A_0 = 10^5$); and the term $A_0\alpha$ is the frequency at which the gain of the amplifier is unity, and is known as the *gain-bandwidth product*. For active filter applications above approximately 100 Hz, the gain characteristic can be approximated by a more convenient expression that has a pole at the origin, as

$$A(s) = \frac{A_0\alpha}{s} \tag{8.19b}$$

Substituting this expression in (8.18):

$$T_V \cong \frac{(1 - ks/A_0\alpha)k/R_1R_2C_1C_2}{s^2\left(1 + \dfrac{k^2}{R_2C_2A_0\alpha}\right) + s\left(\dfrac{1}{R_1C_1} + \dfrac{1}{R_2C_1} + \dfrac{1}{R_2C_2}(1 - k)\right) + \dfrac{1}{R_1R_2C_1C_2}}$$

* One example is the popular 741 op amp [4], for which $\alpha \cong 2\pi 10$ rad/sec and $A_0\alpha \cong 2\pi 10^6$ rad/sec. The analysis of active RC circuits with more complex gain characteristics will be studied in Chapter 12.

Dividing by $1 + k^2/R_2C_2A_0\alpha \cong 1/(1 - k^2/R_2C_2A_0\alpha)$, we get

$$T_V \cong \frac{(1 - ks/A_0\alpha)(1 - k^2/R_2C_2A_0\alpha)k/R_1R_2C_1C_2}{s^2 + s\left(\frac{1}{R_1C_1} + \frac{1}{R_2C_1} + \frac{1}{R_2C_2}(1-k)\right)\left(1 - \frac{k^2}{R_2C_2A_0\alpha}\right) + \frac{1}{R_1R_2C_1C_2}\left(1 - \frac{k^2}{R_2C_2A_0\alpha}\right)} \tag{8.20}$$

$$T_V = \frac{N(s)}{s^2 + \frac{\omega_p}{Q_p}s + \omega_p^2} \tag{8.21}$$

where

$$\omega_p = \sqrt{\frac{1}{R_1R_2C_1C_2}\left(1 - \frac{k^2}{R_2C_2A_0\alpha}\right)} \tag{8.22}$$

$$(\text{bw})_p = \frac{\omega_p}{Q_p} = \left(\frac{1}{R_1C_1} + \frac{1}{R_2C_1} + \frac{1}{R_2C_2}(1-k)\right)\left(1 - \frac{k^2}{R_2C_2A_0\alpha}\right) \tag{8.23}$$

and $N(s)$ is the numerator of Equation 8.20. From these expressions, the sensitivities of ω_p and Q_p to the op amp parameter $A_0\alpha$ can now be derived, as follows:

$$S_{A_0\alpha}^{\omega_p} = \frac{\frac{1}{2}\frac{1}{R_1R_2C_1C_2}\frac{-k^2(-1)}{R_2C_2A_0\alpha}}{\omega_p^2}$$

Since ω_p^2 is approximately equal to $1/R_1R_2C_1C_2$, this expression reduces to

$$S_{A_0\alpha}^{\omega_p} \cong \frac{1}{2}\frac{k^2}{A_0\alpha}\frac{1}{R_2C_2} = \frac{k^2}{2}\frac{\omega_p}{A_0\alpha}\sqrt{\frac{R_1C_1}{R_2C_2}} \tag{8.24}$$

In a similar way, it can readily be seen that

$$S_{A_0\alpha}^{Q_p} = S_{A_0\alpha}^{\omega_p} - S_{A_0\alpha}^{(\text{bw})_p} = -\frac{k^2}{2}\frac{\omega_p}{A_0\alpha}\sqrt{\frac{R_1C_1}{R_2C_2}} \tag{8.25}$$

These equations indicate that the sensitivity of ω_p and Q_p to the active element increases with the pole frequency, and decreases with the gain-bandwidth product. Most present-day op amps have gain-bandwidth products around 1 MHz. This limits the maximum pole frequency that can be achieved by active *RC* filters. For op amps characterized by a single pole, the sensitivity of the active filter becomes intolerably high above approximately 10 kHz. More sophisticated shaping of the op amp gain-phase characteristics, which will be described in Chapter 12, permit active filters to be used up to approximately 30 kHz.

The sensitivity of ω_p and Q_p to the active parameter can be related to the corresponding deviation in the gain of the filter, as illustrated in the following example.

Example 8.3

Compute the deviations in gain for Designs 1 to 3 for the Sallen and Key filter of Example 8.1, assuming a finite gain op amp. It is given that the op amp can be modeled by a single pole at the origin (as in Equation 8.19b) and that the gain-bandwidth product of the op amp can deviate by ± 50 percent about its nominal value of $2\pi 10^6$ rad/sec. Once again, the tolerance limits refer to the 3σ points of a Gaussian distribution.

Solution

The expression for the total gain deviation in terms of the variations in the active and passive components is obtained from Appendix C (Equation C.12 and C.13):

$$\mu(\Delta G)_{\text{total}} = \mu(\Delta G)_{\text{active}} + \mu(\Delta G)_{\text{passive}} \tag{8.26a}$$

$$[3\sigma(\Delta G)]^2_{\text{total}} = [3\sigma(\Delta G)]^2_{\text{active}} + [3\sigma(\Delta G)]^2_{\text{passive}} \tag{8.26b}$$

The contributions of the passive elements were calculated in Example 8.1.* In the following the contribution of the active element is computed.

Since $\mu(V_{A_0\alpha}) = 0$, the mean change in ΔG is zero for all three designs, that is

$$\mu(\Delta G)_{\text{active}} = 0 \text{ dB} \tag{8.27}$$

It can be verified that the contribution of the numerator term $N(s)$ to the standard deviation of ΔG is negligible compared with that due to the ω_p term. With this assumption, the standard deviation of ΔG is given by:

$$[3\sigma(\Delta G)]^2_{\text{active}} = (\mathscr{S}^G_{Q_p})^2(S^{Q_p}_{A_0\alpha})^2[3\sigma(V_{A_0\alpha})]^2 + (\mathscr{S}^G_{\omega_p})^2(S^{\omega_p}_{A_0\alpha})^2[3\sigma(V_{A_0\alpha})]^2 \tag{8.28}$$

The values of $\mathscr{S}^G_{Q_p}$ and $\mathscr{S}^G_{\omega_p}$ at 2.1 kHz were calculated in Example 8.1 to be

$$\mathscr{S}^G_{Q_p} = 4.45 \text{ dB} \qquad \mathscr{S}^G_{\omega_p} = 78.25 \text{ dB}$$

From the information given in the problem, $3\sigma(V_{A_0\alpha}) = 0.5$. The terms $S^{\omega_p}_{A_0\alpha}$ and $S^{Q_p}_{A_0\alpha}$ are evaluated by substituting $\omega_p = 2\pi(2000)$ and $A_0\alpha = 2\pi 10^6$ in Equation 8.24 and 8.25:

$$S^{\omega_p}_{A_0\alpha} = -S^{Q_p}_{A_0\alpha} = 0.001k^2\sqrt{\frac{R_1C_1}{R_2C_2}} \tag{8.29}$$

* Strictly speaking, the passive sensitivities should be recomputed from the ω_p and Q_p expressions assuming a finite gain op amp. However, for filter applications in the audio band, the change is negligibly small; thus it is reasonable to assume the contribution of the passive terms to be the same as in Example 8.1.

Using the above computations, $3\sigma(\Delta G)$ can be evaluated for the three designs, as follows.

Design 1

For this design $k = 1$ and $\sqrt{R_1C_1/R_2C_2} = 2Q_p = 20$. Thus, from Equation 8.28

$$[3\sigma(\Delta G)]^2_{\text{active}} = \underset{(Q_p \text{ term})}{(4.45)^2(0.001 \times 20)^2(0.5)^2} + \underset{(\omega_p \text{ term})}{(78.25)^2(0.001 \times 20)^2(0.5)^2}$$
$$= 0.002 + 0.612 \cong 0.61$$

From Example 8.2, $[3\sigma(\Delta G)]^2_{\text{passive}} = 0.64$; therefore, using Equation 8.26*b*,

$$[3\sigma(\Delta G)]_{\text{total}} = 1.12 \text{ dB}$$

Design 2

In this case $k \cong 3$ and $\sqrt{R_1C_1/R_2C_2} = 1$. Substituting, we get

$$[3\sigma(\Delta G)]^2_{\text{active}} = \underset{(Q_p \text{ term})}{0.0004} + \underset{(\omega_p \text{ term})}{0.124} \cong 0.124$$

And since, $[3\sigma(\Delta G)]^2_{\text{passive}} = 3.93$

$$3\sigma(\Delta G)_{\text{total}} = 2.01 \text{ dB}$$

Design 3

In this case, $k = \frac{4}{3}$ and $\sqrt{R_1C_1/R_2C_2} = \sqrt{3}$. Thus:

$$[3\sigma(\Delta G)]^2_{\text{active}} = \underset{(Q_p \text{ term})}{5(10)^{-5}} + \underset{(\omega_p \text{ term})}{0.015} \cong 0.015$$

From Example 8.2, $[3\sigma(\Delta G)]^2_{\text{passive}} = 1.04$, so

$$3\sigma(\Delta G)_{\text{total}} = 1.027 \text{ dB}$$

Observations

1. The above calculations show that Design 3, the one due to Saraga, yields the smallest deviation in gain. Comparing this with Example 8.2, it is seen that if only the passive components are considered, Design 1 is the best; however, if the variations of the active element are also considered, it is found to be inferior to Design 3. This example brings out the important fact that it is not sufficient to analyze only the deviation due to the passive elements. In fact, ignoring the active term will often result in the wrong conclusion.
2. For a different choice of element tolerances, Design 1 might be better than Design 3 (see Problem 8.14). Thus, conclusions regarding the gain variation of a circuit depend on the tolerances assumed.

3. The frequency chosen for the comparison of gain deviations was $\omega = 2\pi(2100)$, which corresponds to a normalized frequency of $\Omega = 1 + 1/2Q_p$. The reason for choosing this frequency becomes apparent from the sensitivity curves of Figure 5.3*a*. It is seen that the magnitude of the $\mathcal{S}^G_{\omega_p}$ term reaches a maximum value of approximately $8.686Q_p$ at the 3 dB passband edge frequencies ($\Omega = 1 \pm 1/2Q_p$); accordingly, the gain deviation will also be a maximum at these frequencies (since in the majority of designs the ω_p-terms contribute the most to the gain deviation).* Thus the gain deviation at either of these frequencies will, in practice, determine whether or not the passband filter requirements will be satisfied. Consequently, for purposes of *comparing* circuits, it is adequate to study the gain deviation at the upper (or lower) 3 dB passband edge frequency alone. To obtain a more complete picture, one would have to calculate $\sigma(\Delta G)$ at several frequencies—a job that is usually reserved for the computer.
4. A compact form for the deviation in gain at $1 \pm 1/2Q_p$ can be obtained by using the following approximate relationships derived in Section 5.4.1 (page 162):

$$\mathcal{S}^G_{\omega_p} \cong \pm 8.686Q_p \qquad \mathcal{S}^G_{Q_p} \cong \tfrac{1}{2}(8.686) \tag{8.30}$$

In particular for the Saraga design, from Table 8.1:

$$\sum_i [(S^{\omega_p}_{R_i})^2 + (S^{\omega_p}_{C_i})^2] = 1 \tag{8.31}$$

$$\sum_i [(S^{Q_p}_{R_i})^2 + (S^{Q_p}_{C_i})^2] \cong 6(0.58Q_p)^2 = 2.02Q_p^2 \tag{8.32}$$

and from Equation 8.25:

$$(S^{Q_p}_{A_0\alpha})^2 = (S^{\omega_p}_{A_0\alpha})^2 = \left[\frac{\sqrt{3}}{2}\left(\frac{4}{3}\right)^2 \frac{\omega_p}{A_0\alpha}\right]^2 = 2.37\left(\frac{\omega_p}{A_0\alpha}\right)^2 \tag{8.33}$$

Substituting the above relationships in Equation 8.15, for $Q_p \gg 1$, we get:

Saraga Design ($k = 4/3$)

$$[3\sigma(\Delta G)]^2_{\text{passive}} \cong 1.5\{8.686Q_p[3\sigma(V_{R,C})]\}^2 \tag{8.34a}$$

$$[3\sigma(\Delta G)]^2_{\text{active}} \cong 2.37\left\{8.686Q_p\,\frac{\omega_p}{A_0\alpha}\,[3\sigma(V_{A_0\alpha})]\right\}^2 \tag{8.34b}$$

where $V_{R,C} = V_R = V_C$. These equations give the approximate deviation in gain at the upper (or lower) 3 dB passband edge frequency for the Saraga design. As an illustration, let us evaluate the gain change for the filter described in the example for which $Q_p = 10$, $\omega_p = 2\pi 2000$, $A_0\alpha = 2\pi 10^6$,

* An exception is Design 2, in which case the Q_p-passive term is the most significant. However, it is reasonable to ignore this design in our discussions, since it results in an extremely sensitive and, therefore, impractical circuit.

$3\sigma(V_{R,C}) = 0.01$, and $3\sigma(V_{A_0\alpha}) = 0.5$. For this case, from (8.34a) and (8.34b)

$$[3\sigma(\Delta G)]^2_{\text{passive}} \cong 1.13 \qquad [3\sigma(\Delta G)]^2_{\text{active}} \cong 0.018$$

These values are seen to be only slightly higher than the exact deviations computed using the expressions given by Equation 5.44 and 5.45 for $\mathscr{S}^G_{\omega_p}$ and $\mathscr{S}^G_{Q_p}$. This demonstrates that for high pole Q's these approximations are very good. Similar approximate expressions can be derived for Designs 1 and 2 (see Problem 8.8 and 8.9).

5. The deviations computed in this example reflect the change in gain at the *time of manufacture*, assuming ± 1 percent resistors and capacitors, and op amps whose gain-bandwidth products can vary by ± 50 percent. Further deviations in the gain are expected to occur due to changes in *temperature*, *humidity*, and *aging*. The so-called end-of-life deviation for the filter must take all these effects into account (just as was done in Example 5.5). In computing these changes we should consider that components of the same type, which are manufactured under the same conditions, have a tendency to *track*. As a result, although the absolute value of a particular resistance may change by, say, 0.5 percent due to environmental effects, the *ratios* of resistors will tend to change by much less (typically, 0.1 percent). This correlation of the element changes effectively makes the gain deviation much less than if the resistor changes were independent. For example, Equation 8.7, which gives the pole Q, can be rewritten in terms of ratios of resistors and capacitors as*

$$Q_p = \frac{1}{\sqrt{\frac{R_1}{R_2}\frac{C_2}{C_1}} + \sqrt{\frac{R_2}{R_1}\frac{C_2}{C_1}} - \frac{r_2}{r_1}\sqrt{\frac{R_1}{R_2}\frac{C_1}{C_2}}} \tag{8.35}$$

Thus tracking will make the gain deviation due to the Q_p-passive term much less than if the component changes were independent. Notice that the ω_p-terms and the Q_p-active terms do not benefit from tracking, since these terms are not dimensionless and therefore cannot be expressed in terms of resistor and capacitor ratios. ■

8.3 HIGH-PASS CIRCUIT USING RC → CR TRANSFORMATIONS

A second-order high-pass filter function could be synthesized using the *RC* circuit of Figure 8.2*a*, following the procedure of the last section. In this section, however, we develop an alternate synthesis technique based on a transformation of the elements of the low-pass realization.

* Since Q_p is a dimensionless quantity [$Q_p = \omega_p/(\text{bw})_p$], it can always be expressed as a function of resistor and capacitor ratios.

Recall from Section 4.7.1 that, given a normalized *LP* filter function $T_{LP_N}(S)$ with $\omega_P = 1$, the equivalent *HP* filter function with passband edge at $\omega = \omega_P$ is obtained by the frequency transformation (Equation 4.53):

$$T_{HP}(s) = T_{LP_N}(S)|_{S=\omega_P/s} \tag{8.36a}$$

A simple extension of this equation, to transform a *LP* function $T_{LP}(S)$ with passband edge at $\omega = \omega_P$ to a *HP* function with passband edge at $\omega = \omega_P$, is given by

$$T_{HP}(s) = T_{LP}(S)|_{S=\omega_P \times \omega_P/s} \tag{8.36b}$$

For instance, suppose the *LP* filter requirement of Figure 8.4*a* is realized by the second-order function

$$T_{LP}(S) = \frac{\omega_p^2}{S^2 + \dfrac{\omega_p}{Q_p} S + \omega_p^2} \tag{8.37}$$

From Equation 8.36b, the transformed *HP* function is

$$T_{HP}(s) = \frac{s^2}{s^2 + \dfrac{\omega_p}{Q_p} s + \omega_p^2} \tag{8.38}$$

This *HP* function satisfies the requirements shown in Figure 8.4*b*. Comparing Figure 8.4*a* and 8.4*b*, it is seen that this transformation makes the cutoff frequencies of both filters the same; moreover, the transition band ratio is the same,

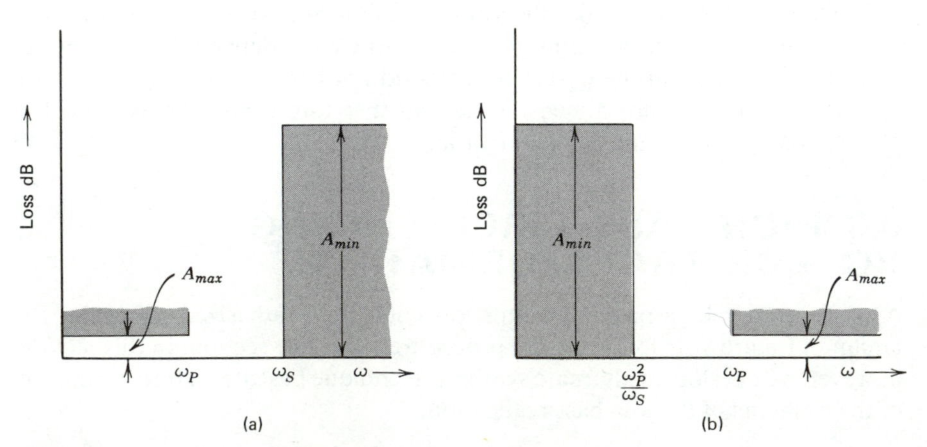

Figure 8.4 *RC* → *CR* Transformation: (*a*) *LP* requirements. (*b*) *HP* requirements.

that is

$$\left(\frac{\omega_{stop}}{\omega_{pass}}\right)_{LP} = \left(\frac{\omega_{pass}}{\omega_{stop}}\right)_{HP} \tag{8.39}$$

In this section we describe a way of obtaining the high-pass filter realization directly from the low-pass circuit, by effectively using the *LP* to *HP* transformation.

Equation 8.37 may be written in the form

$$T_{LP}(S) = \frac{1}{\left(\frac{1}{\omega_p^2}\right)S^2 + \left(\frac{1}{\omega_p Q_p}\right)S + 1} \tag{8.40}$$

In *RC* circuits the dimension of frequency is

$$\dim(\omega) = [R]^{-1}[C]^{-1} \tag{8.41}$$

Since each of the denominator terms of (8.40) is dimensionless, the dimensions of ω_p^2 and $\omega_p Q_p$ must be

$$\dim(\omega_p)^2 = [R]^{-2}[C]^{-2} \qquad \dim(\omega_p Q_p) = [R]^{-1}[C]^{-1} \tag{8.42}$$

respectively. Therefore, a dimensionless representation of Equation 8.40 is

$$\begin{aligned} T_{LP}(S) &= \frac{1}{([R]^2[C]^2)S^2 + ([R][C])S + 1} \\ &= \frac{1}{[R]^2([C]S)^2 + [R]([C]S) + 1} \end{aligned} \tag{8.43}$$

From Equation 8.36b, the $LP \to HP$ transformation requires that S be replaced by ω_p^2/s. Considering (8.43), this may be achieved by replacing

$$R \quad \text{by} \quad \frac{R\omega_p}{S} \tag{8.44}$$

and

$$C \quad \text{by} \quad \frac{C\omega_p}{S} \tag{8.45}$$

Thus the $LP \to HP$ frequency transformation of Equation 8.36 can be effected by replacing the resistors, R, in the low-pass circuit by capacitors of value $1/R\omega_p$; and capacitors, C, in the low-pass circuit by resistors of value $1/C\omega_p$. This is known as the $RC \to CR$ transformation technique for obtaining an *HP* filter circuit directly from a known *LP* circuit. As we mentioned earlier, the *HP* filter will have the same cutoff frequency and transition band ratio as the *LP* filter.

Example 8.4
Synthesize the following *HP* filter function using the $RC \rightarrow CR$ transformation

$$T_{HP}(s) = K \frac{s^2}{s^2 + s + 25}$$

where K is an arbitrary constant.

Solution
The corresponding *LP* function that needs to be realized is

$$T_{LP}(S) = K \frac{25}{S^2 + S + 25}$$

The biquadratic parameters describing this function are $\omega_p = 5$ and $Q_p = 5$. Using the Saraga design, the element values for the *LP* circuit of Figure 8.3 are (Equation 8.12):

$$C_2 = 1 \qquad C_1 = \sqrt{3}\,Q_p = \sqrt{3}(5)$$

$$R_2 = \frac{1}{\sqrt{3}\,\omega_p} = \frac{1}{\sqrt{3}(5)} \qquad R_1 = \frac{1}{Q_p \omega_p} = \frac{1}{5(5)}$$

$$k = \tfrac{4}{3}$$

The factor $k = 1 + r_2/r_1$ can be realized by letting $r_1 = 3$ and $r_2 = 1$.
The remaining elements for the *HP* circuit are obtained using the $RC \rightarrow CR$ transformation, as follows:

LP elements	*HP* elements
$C_1 = 5\sqrt{3}$	$R_A = \dfrac{1}{25\sqrt{3}}$
$C_2 = 1$	$R_B = \frac{1}{5}$
$R_1 = \frac{1}{25}$	$C_A = 5$
$R_2 = \dfrac{1}{5\sqrt{3}}$	$C_B = \sqrt{3}$

The desired *HP* circuit is shown in Figure 8.5.

Observations

1. The resistors r_1 and r_2 form a potential divider, appearing only as a ratio in the transfer function. Therefore, these two resistors need not be transformed to capacitors to effect the $RC \rightarrow CR$ transformation.

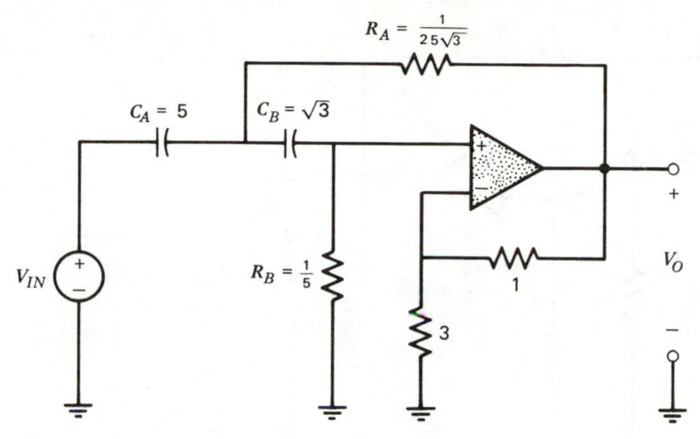

Figure 8.5 High-pass circuit for Example 8.4.

2. Practical voltage sources have a small but finite output resistance, r_s. If such a voltage were being fed into a *LP* filter, the series resistor R_1 (Figure 8.3) could be used to absorb r_s, by using the resistance $(R_1 - r_s)$ in place of r_s. Then the effective series resistance is the desired value R_1 and the voltage source may be assumed to have zero output impedance. In the *HP* case, however, the voltage source is fed into the circuit through the capacitor C_A, so the output resistance r_s cannot be absorbed into the circuit and, therefore, it will introduce an error term in the transfer function.
3. The inverse $CR \rightarrow RC$ transformation can be used to derive a *LP* filter from a given *HP* filter circuit. ■

8.4 SALLEN AND KEY BAND-PASS CIRCUIT

The *RC* circuit of Figure 8.2*b* can be used to generate a band-pass transfer function. This circuit, also developed by Sallen and Key, is shown in Figure 8.6.

The circuit could be analyzed by considering the feedforward and feedback transfer functions, as was done for the *LP* filter. An alternate approach is to write the nodal equations of the active *RC* networks as follows. From Figure 8.6, the nodal equations for nodes 1 and 2 are:

node 1:

$$\frac{V_0}{k}\left(\frac{1}{R_3} + sC_2\right) - V_2(sC_2) = 0 \tag{8.46}$$

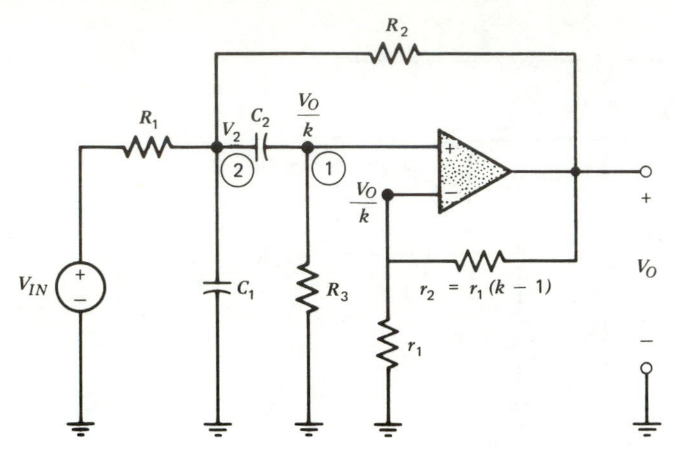

Figure 8.6 Sallen and Key band-pass circuit.

node 2:

$$V_2\left(sC_1 + sC_2 + \frac{1}{R_1} + \frac{1}{R_2}\right) - \frac{V_0}{k}(sC_2) - V_0\left(\frac{1}{R_2}\right) = \frac{V_{IN}}{R_1} \tag{8.47}$$

Solving

$$\frac{V_O}{V_{IN}} = \frac{ks/R_1C_1}{s^2 + s\left(\frac{1}{R_1C_1} + \frac{1}{R_3C_2} + \frac{1}{R_3C_1} + \frac{1-k}{R_2C_1}\right) + \frac{R_1+R_2}{R_1R_2R_3C_1C_2}} \tag{8.48}$$

Consider the synthesis of the second-order *BP* function

$$K\frac{s}{s^2 + \frac{\omega_p}{Q_p}s + \omega_p^2} \tag{8.49}$$

Comparing (8.48) and (8.49), it is seen that there are six elements (R_1, R_2, R_3, C_1, C_2, and k) to satisfy the two constraints imposed by ω_p and Q_p (assuming that K is an arbitrary constant). Therefore, four of the elements can be fixed. One simple solution is achieved by letting

$$C_1 = C_2 = 1 \qquad R_1 = R_2 = R_3 = R$$

Then the remaining elements are given by

$$R = R_1 = R_2 = R_3 = \frac{\sqrt{2}}{\omega_p} \tag{8.50a}$$

$$k = 1 + \frac{r_2}{r_1} = 4 - \frac{\sqrt{2}}{Q_p} \tag{8.50b}$$

Note that the ratio r_2/r_1 is positive for $Q_p > \sqrt{2}/3$.

The above solution results in a gain constant of

$$K = \frac{k}{R_1 C_1} = \omega_p\left(2\sqrt{2} - \frac{1}{Q_p}\right) \tag{8.50c}$$

This gain constant can be changed by using the techniques of input attenuation or gain enhancement, as mentioned in Section 7.6.

Example 8.5

Synthesize a second-order *BP* filter with a center frequency at 1000 rad/sec and a pole Q of 10. The gain at the center frequency is required to be 0 dB.

Solution

The desired transfer function is

$$T(s) = \frac{100s}{s^2 + 100s + (1000)^2}$$

From Equation 8.50, the required element values are

$$C_1 = C_2 = 1\text{F}$$

$$R_1 = R_2 = R_3 = \frac{\sqrt{2}}{1000} = 1.414(10)^{-3}\ \Omega$$

$$\frac{r_2}{r_1} = 3 - \frac{\sqrt{2}}{10} = 2.858$$

To obtain practical element values, the elements are impedance scaled by 10^7, to yield

$$C_1 = C_2 = 0.1\ \mu\text{F}$$
$$R_1 = R_2 = R_3 = 14.14\ \text{k}\Omega$$

The ratio $r_2/r_1 = 2.858$ can be realized by making $r_1 = 1\ \text{k}\Omega$ and $r_2 = 2.858\ \text{k}\Omega$. With the above choice of elements the gain constant that is realized, from Equation 8.50c, is

$$K = 2728$$

Since the desired scale factor is 100, it is necessary to attenuate the input by a factor of 27.28. Using the input attenuation scheme described in Section 7.6 the input resistance R is replaced by the resistors R_4 and R_5, as shown in Figure 8.7, where:

$$\frac{R_5}{R_4 + R_5} = \frac{1}{27.28} \qquad \frac{R_4 R_5}{R_4 + R_5} = 14.14\ \text{k}\Omega$$

Solving, we get $R_4 = 386\ \text{k}\Omega$ and $R_5 = 14.7\ \text{k}\Omega$. The complete circuit is shown in Figure 8.7. ∎

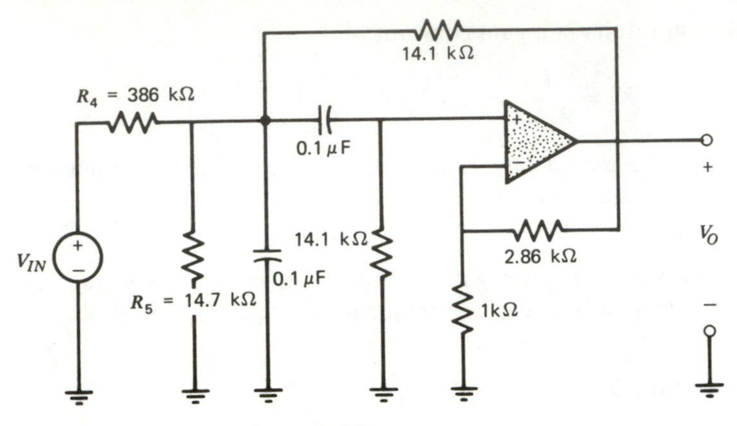

Figure 8.7 Circuit for Example 8.5.

8.5 TWIN-T NETWORKS FOR REALIZATION OF COMPLEX ZEROS

In this section we present a qualitative description of a circuit for the realization of a second-order function with complex zeros, namely,

$$T(s) = \frac{s^2 + \omega_z^2}{s^2 + \dfrac{\omega_p}{Q_p} s + \omega_p^2} \tag{8.51}$$

As mentioned in Chapter 3, this function is used in the realization of band-reject, low-pass-notch, and high-pass-notch filters.

From the basic properties of the positive feedback topology, we know that the zeros of the transfer function are the zeros of the feedforward function of the *RC* network. Thus, we need an *RC* structure that exhibits zeros on the $j\omega$ axis. This requirement rules out *RC* ladder networks because they can only realize zeros on the real axis [10]. This fundamental property of *RC* ladders is a consequence of the fact that in a ladder circuit the zeros of transmission can only be realized when a series branch is an open circuit or a shunt branch is a short circuit. Put differently, the zeros of transmission are the impedance poles of the series branches and the impedance zeros of the shunt branches. However, it is known that *RC* impedances have all their poles and zeros on the negative real axis (Chapter 2, page 42); hence, the transmission zeros of an *RC* ladder are constrained to lie on the negative real axis. Therefore, Equation 8.51 cannot be implemented using the class of ladder networks shown in Figure 8.2, and we must consider alternate topologies. One such topology, is the so-called **Twin-T** [6] shown in Figure 8.8. In this topology there are two paths from input to output. Complex zeros of transmission are formed by choosing the component

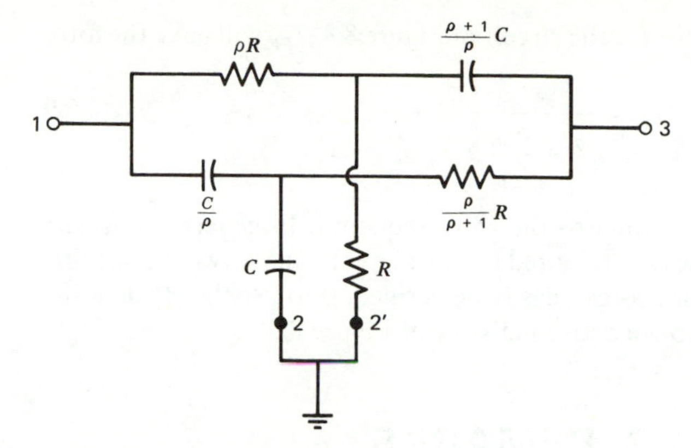

Figure 8.8 Twin-T *RC* network.

values so that the electrical signals arriving at the output via these two paths exactly cancel. Notice that this network has three capacitors and as such it will realize a third-order function. However, if the elements are chosen as shown in the figure, a pole-zero cancellation occurs, resulting in a second-order function with zeros on the $j\omega$ axis, as desired.

In the network of Figure 8.8, the dc gain of the feedforward function T_{FF} is seen to be unity. This is verified by replacing the capacitors with open circuits. Also, the gain at infinite frequency, obtained by shorting the capacitors, can

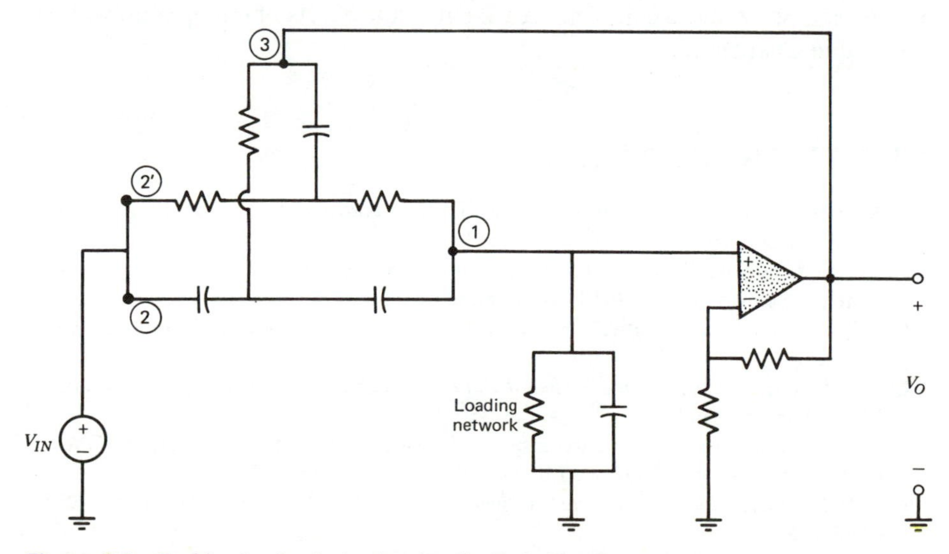

Figure 8.9 Positive feedback circuit using the Twin-T, with a loading network.

be seen to be unity. Therefore, for the circuit of Figure 8.8, T_{FF} will have the form

$$T_{FF} = \frac{s^2 + \omega_{p_1}^2}{s^2 + \dfrac{\omega_{p_1}}{Q_{p_1}} s + \omega_{p_1}^2} \tag{8.52}$$

where the pole frequency is equal to the zero frequency. In general, however, the pole and zero frequencies are required to be different. One way of separating the pole and zero frequencies, as can easily be verified, is to use the *RC* loading network shown in the complete active network of Figure 8.9.

8.6 CONCLUDING REMARKS

In this chapter we discussed the synthesis, sensitivity, and some practical design aspects of positive feedback biquad structures. These biquad circuits can be cascaded to synthesize a more complex filter function, as was mentioned in Chapter 7. We only covered a sampling of the numerous *RC* circuits that are commonly used—a more complete selection can be found in [6], [7], and [9].

We showed that the choice of the fixed elements in the solution of the synthesis equations greatly influenced the sensitivity of the resulting circuit. The analytical techniques presented in this chapter provide a simple yet useful approach to this problem. A more detailed analysis to determine the least sensitive design is best performed using Monte Carlo techniques described in Section 5.5.

The importance of considering the finite gain of the op amp was illustrated by an example. A more complete treatment of the effects of the op amp will be covered in Chapter 12.

FURTHER READING

1. A. Budak, *Passive and Active Network Analysis and Synthesis*, Houghton Mifflin, Boston, Mass., 1974, Chapter 10.
2. Fairchild Semiconductor, *The Linear Integrated Circuits Data Catalog*, 1973, Fairchild Semiconductor, 464 Ellis Street, Mountain View, Calif.
3. S. S. Haykin, *Synthesis of RC Active Filter Networks*, McGraw-Hill, London, 1969, Chapter 4.
4. L. P. Huelsman, *Theory and Design of Active RC Circuits*, McGraw-Hill, New York, 1968, Chapters 3 and 6.
5. W. J. Kerwin, "Active RC network synthesis using voltage amplifiers," *Active Filters*, L. P. Huelsman, ed., McGraw-Hill, New York, 1970, Chapter 2.
6. G. S. Moschytz, *Linear Integrated Networks Design*, Van Nostrand, New York, 1975, Chapter 3.
7. R. P. Sallen and E. L. Key, "A practical method of designing RC active filters," *IRE Trans. Circuit Theory*, *CT-2*, May 1955, pp. 74–85.

8. W. Saraga, "Sensitivity of 2nd-order Sallen-Key-type active RC filters," *Electronics Letters*, *3*, 10, October 1967, pp. 442–444.
9. A. S. Sedra, "Generation and classification of single amplifier filters," *Intl. J. Circuit Theory and Appl.* *2*, 1, March 1974, pp. 51–67.
10. H. H. Sun, *Synthesis of RC Networks*, Hayden, N.Y., 1967, Chapter 3.

PROBLEMS

8.1 *Low-pass synthesis.* Synthesize the low-pass transfer function

$$\frac{K}{s^2 + 100s + 25(10)^4}$$

using the Saraga design, with practical element values. Show how the circuit can be adapted to:

(a) Change the cutoff frequency to 200 rad/sec.
(b) Achieve a gain of 0 dB at *dc*.

8.2 Synthesize the low-pass function

$$\frac{20{,}000}{(s^2 + 2s + 100)(s^2 + 5s + 200)}$$

8.3 *Low-pass Chebyshev filter.* A low-pass filter is required to meet the following specifications:

$$f_P = 1000 \text{ Hz} \qquad f_S = 3000 \text{ Hz}$$
$$A_{\max} = 0.5 \text{ dB} \qquad A_{\min} = 25 \text{ dB} \qquad dc \text{ gain} = 0 \text{ dB}$$

Find the Chebyshev approximation function for these requirements and realize it using the Saraga design. You should not need extra elements for adjusting the gain constant. (*Hint*: first synthesize the normalized *LP* filter.)

8.4 *Low-pass Butterworth filter.* Repeat Problem 8.4 using a Butterworth approximation. You may use one extra element for adjusting the gain constant.

8.5 *Low-pass design.* The *RC* circuit shown is to be used in the positive feedback topology to realize a low-pass transfer function.

(a) Sketch the active *RC* circuit.
(b) Obtain the transfer function and one set of design equations.
(c) Compute the sensitivities of ω_p and Q_p to the passive elements R_1 and C_1 for $Q_p = \sqrt{3}$.

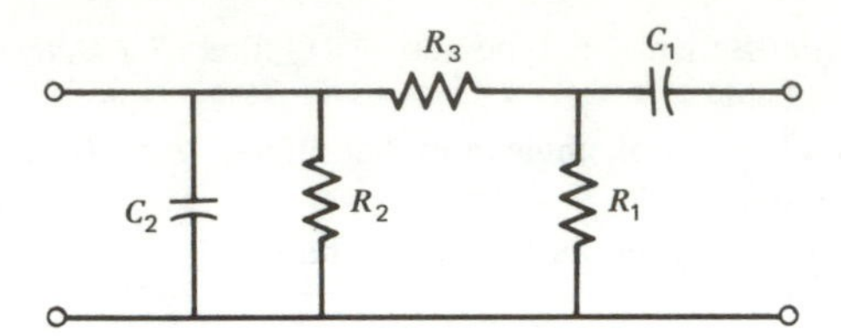

Figure P8.5

8.6 *Alternate design for Sallen and Key LP circuit.* Three designs for the Sallen and Key *LP* circuit were evaluated in Section 8.2. Consider yet another design (Design 4) based on the following choice for the fixed elements:

$$k = 2 \qquad R_1C_1 = R_2C_2 \qquad C_1 = 1$$

(a) Find the synthesis equations for this design. Show that the resistor and capacitor ratios are equal to Q_p.
(b) Compute the sensitivities of ω_p and Q_p to the passive elements and to the gain-bandwidth product $A_0\alpha$. Assume the op amp gain is $A(s) = A_0\alpha/s$.
(c) Determine the statistics (μ and σ) of the gain deviation at the upper 3 dB passband edge frequency for the filter described in Examples 8.1, 8.2, and 8.3.

8.7 *General expressions for ΔG.* For purposes of comparing designs it is convenient to develop expressions for the standard deviation of the gain change $\sigma(\Delta G)$ at the 3 dB passband edge frequencies, as was done for the Saraga design in Equation 8.34. Derive a similar expression for Design 1 ($k = 1$) of the Sallen and Key low-pass circuit. Assume $Q_p \gg 1$, and make reasonable approximations.

Answer: $[3\sigma(\Delta G)]^2_{\text{passive}} \cong \alpha_P\{8.686Q_p[3\sigma(V_{R,C})]\}^2$

$$[3\sigma(\Delta G)]^2_{\text{active}} \cong \alpha_A\left\{8.686Q_p\frac{\omega_p}{A_0\alpha}[3\sigma(V_{A_0\alpha})]\right\}^2$$

where $\alpha_P = 1$, $\alpha_A = Q_p^2$.

8.8 Repeat Problem 8.7 for Design 2 ($k = 3 - 1/Q_p$) of the Sallen and Key low-pass circuit.

Answer: $\alpha_P = 5.5$, $\alpha_A = 20.25$

8.9 Repeat Problem 8.7 for Design 4 (Problem 8.6) of the Sallen and Key low-pass circuit.

Answer: $\alpha_P = 2$, $\alpha_A = 4$.

8.10 *Optimum k for Sallen and Key LP circuit.* For the filter described in Examples 8.2 and 8.3, determine the value of k that yields the smallest gain deviation at the 3 dB passband edge frequency by sketching $3\sigma(\Delta G)_{\text{total}}$ versus k for Designs 1, 2, 3, and 4.

Comparison of Sallen and Key LP designs. In problems 8.11 *to* 8.15 *assume the component statistics to be those given in Examples* 8.2 *and* 8.3, *unless otherwise stated.*

8.11 Compare the statistics of the gain deviation due to the changes in the passive and active components for Design 1 ($k = 1$) and Design 3 (Saraga $k = 4/3$), at the upper 3 dB passband edge frequency, for a second-order low-pass filter with a cutoff frequency at 2000 Hz and a pole Q of 40. (*Hint*: use the general expressions derived in Equation 8.34 and Problem 8.7.)

8.12 Repeat Problem 8.11 for a low-pass filter with a cutoff frequency at 10,000 Hz and a pole Q of 10.

8.13 Show that of the three designs considered in Examples 8.2 and 8.3, Design 1 ($k = 1$) has the smallest gain deviation at the cutoff frequency $\omega_p = 2\pi 2000$ rad/sec.

8.14 The Saraga design ($k = 4/3$) was found to be superior to Design 1 ($k = 1$) for the component tolerances specified in Examples 8.2 and 8.3.

(a) Determine the passive element tolerance (assumed the same for all the components) above which Design 1 has a lower gain deviation than the Saraga design at 2100 Hz. As in the examples in the text assume $f_p = 2000$ Hz, $Q_p = 10$, $3\sigma(V_{A_0\alpha}) = 0.5$.

(b) Determine the *LP* filter cutoff frequency below which Design 1 has a lower gain deviation than the Saraga design. Assume $3\sigma(V_R) = 3\sigma(V_C) = 0.01$, $3\sigma(V_{A_0\alpha}) = 0.5$, and $Q_p = 10$.

8.15 Suppose the filter of Examples 8.2 and 8.3 is built using a technology in which the ratios of resistors and capacitors track to within 0.1 percent (i.e., $3\sigma(V_{R_2/R_1}) = 3\sigma(V_{C_2/C_1}) = 0.001$). Compute the statistics of the deviation in gain for Design 1 ($k = 1$) and Design 3 (Saraga):

(a) At the cutoff frequency.

(b) At the upper 3 dB passband edge frequency (*Hint*: you will need to determine $S^{Q_p}_{R_2/R_1}$, $S^{Q_p}_{C_2/C_1}$, and $S^{Q_p}_{r_2/r_1}$.)

8.16 *RC → CR transformation.* Synthesize the low-pass function

$$\frac{120{,}000}{s^2 + 100s + 90{,}000}$$

with practical element values. Use the design requiring the lowest number of elements. Transform the elements to realize a high-pass function with a cutoff frequency at (a) 300 rad/sec; (b) 1200 rad/sec.

8.17 *Band-pass synthesis.* Synthesize the band-pass function

$$\frac{400s}{s^2 + 4s + 500}$$

Use the element-splitting method of gain enhancement (described in Problem 7.21). This method is particularly attractive for positive feedback band-pass circuits, since it does not require additional capacitors.

8.18 *HP and BP synthesis.* Synthesize the function

$$\frac{Ks^3}{(s^2 + s + 100)(s^2 + 2s + 81)}$$

using the positive feedback circuits developed in this chapter, so as to obtain as large a gain constant as possible (without using gain enhancement).

8.19 *Band-pass Chebyshev filter.* A fourth-order band-pass filter is required to satisfy the requirements sketched in Figure P8.19. Determine the approximation function using the CHEB program. Synthesize this function using two sections of the circuit of Figure 8.6.

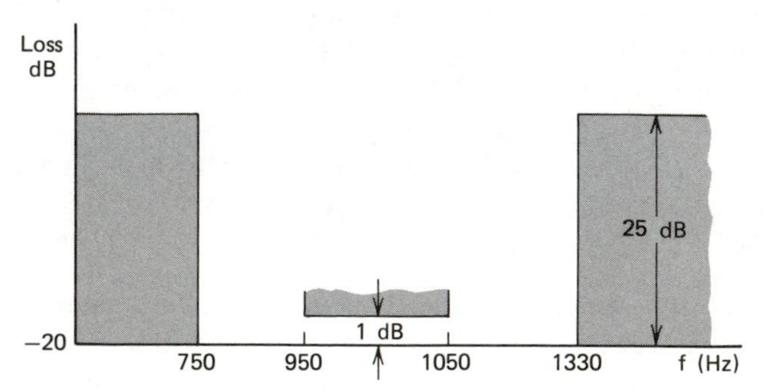

Figure P8.19

8.20 *Band-pass design.* Use the *RC* circuit of Figure P8.20 in the positive feedback topology to synthesize a band-pass filter. Determine the transfer function and a set of design equations.

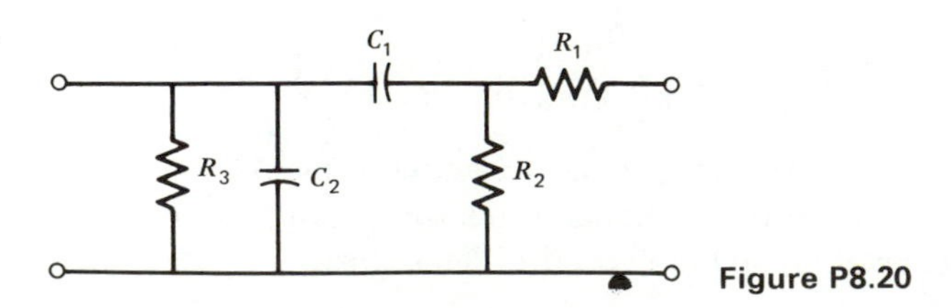

Figure P8.20

8.21 *Band-reject synthesis using* $(1 - T)$ *topology.* Use the band-pass circuit of Figure 8.6 in the $(1 - T)$ topology described in Problem 7.19 to synthesize the band-reject function

$$\frac{s^2 + (1000)^2}{s^2 + 100s + (1000)^2}$$

8.22 *Complete band-pass design.* Consider the band-pass circuit of Figure 8.6.

(a) Determine the transfer function V_O/V_{IN} in terms of the feedforward and feedback transfer functions of the RC network, assuming an ideal op amp.

(b) Using the design equations developed in Section 8.4, find expressions for the sensitivities of ω_p, Q_p, and K to the passive elements, in terms of Q_p.

(c) Determine the transfer function V_O/V_{IN} if $A(s) = A_0\alpha/s$. Reduce the transfer function to a biquadratic (as in Equation 8.20).

(d) Find expressions for the sensitivities of ω_p and Q_p to $A_0\alpha$.

(e) The circuit is used to design a band-pass filter with a pole frequency of 2000 Hz and a pole Q of 10. Compute the statistics of the gain deviation at the upper 3 dB passband edge frequency assuming the active and passive component tolerances are as specified in Examples 8.2 and 8.3.

8.23 *Feedforward zero formation.* In these problems we show how the same RC network can be used in the positive feedback topology to realize more than one type of filter function, by choosing different ports for the input signal.

The RC network shown can be used in the positive feedback topology to realize a high-pass filter by introducing the input signal via the capacitor C_1. Show this:

(a) By analyzing the circuit.

(b) By considering the slopes of the high and low frequency asymptotes of the feedforward and feedback transfer functions. (*Hint*: recognize that the feedback transfer function is a second-order band-pass function.)

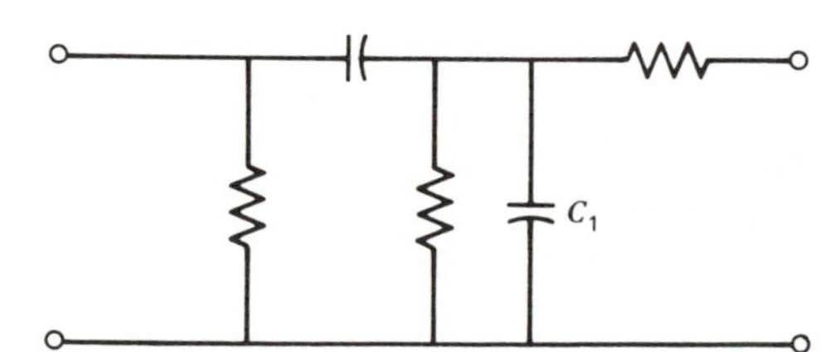

Figure P8.23

8.24 Investigate the types of filter functions that can be realized by using the RC circuit of Figure P8.24 in the positive feedback topology.

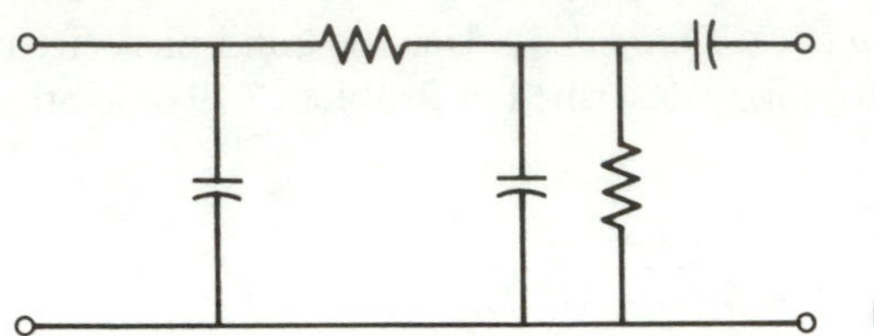

Figure P8.24

8.25 Repeat Problem 8.24 for the *RC* circuit of Figure P8.5.

8 26 Repeat Problem 8.24 for the *RC* circuit of Figure P8.20.

8.27 *Steffen all-pass filter*. In this problem we develop design equations for realizing a biquadratic all-pass network that is often used for the equalization of delay in data transmission systems (Ref. [3], Chapter 9).

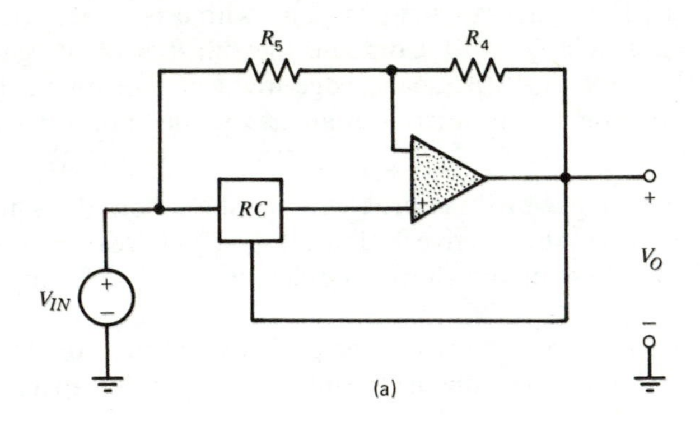

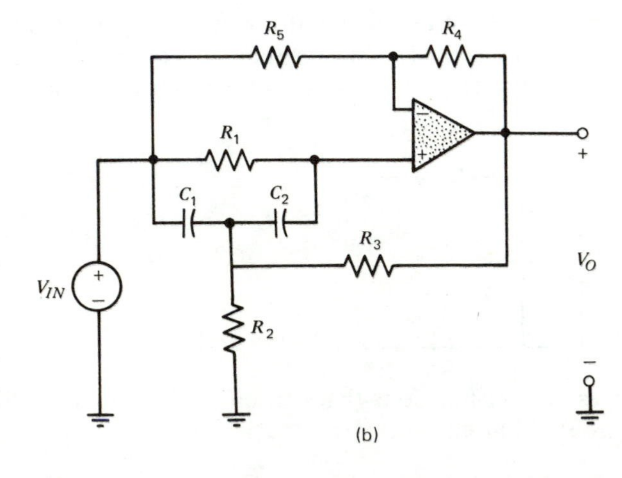

Figure P8.27

(a) Show that the transfer function of the general positive feedback topology of Figure P8.27*a* is

$$\frac{V_O}{V_{IN}} = \frac{\left(1 + \dfrac{R_4}{R_5}\right)T_{FF} - \dfrac{R_4}{R_5}}{1 - \left(1 + \dfrac{R_4}{R_5}\right)T_{FB}}$$

where T_{FF} and T_{FB} are the feedforward and feedback transfer functions of the RC network.

(b) Using part (a) show that the transfer function for the circuit of Figure P8.27*b* is

$$\frac{s^2 + s\left[\dfrac{1}{R_1C_2} + \dfrac{1}{R_1C_1} - \dfrac{R_4}{R_5}\left(\dfrac{1}{R_2} + \dfrac{1}{R_3}\right)\dfrac{1}{C_1}\right] + \dfrac{1}{R_1C_1C_2}\left(\dfrac{1}{R_2} + \dfrac{1}{R_3}\right)}{s^2 + s\left[\dfrac{1}{R_1C_2} + \dfrac{1}{R_1C_1} + \left(\dfrac{1}{R_2} + \dfrac{1}{R_3}\right)\dfrac{1}{C_1} - \dfrac{1}{R_3C_1}\left(1 + \dfrac{R_4}{R_5}\right)\right] + \dfrac{1}{R_1C_1C_2}\left(\dfrac{1}{R_2} + \dfrac{1}{R_3}\right)}$$

(c) Synthesize the all-pass function

$$\frac{s^2 - s + 4}{s^2 + s + 4}$$

Using $C_1 = C_2 = 1$ and $R_4/R_5 = 1$ for the fixed elements.

8.28 Use the results of Problem 8.27 to synthesize the delay equalizer function needed to equalize the delay of a fourth-order Chebyshev filter for which $\omega_P = 1$ rad/sec, $A_{\max} = 0.25$ dB. Refer to Section 4.6 for the transfer function.

9

NEGATIVE FEEDBACK BIQUAD CIRCUITS

The design procedures for negative feedback biquad circuits are quite similar to those for the positive feedback structures studied in Chapter 8. In this chapter we concentrate on techniques for generalizing the basic negative feedback circuit to realize a variety of filter functions. One circuit, due to Friend [3], is shown to be capable of realizing the high-pass, band-pass, band-reject, and delay equalizer filter functions. In addition, this circuit can be designed to provide a sensitivity which is among the lowest of all active *RC* biquad circuits. The step-by-step development of this rather complex circuit will serve to illustrate many design techniques associated with the synthesis of practical active filters.

9.1 PASSIVE RC CIRCUITS USED IN THE NEGATIVE FEEDBACK TOPOLOGY

The basic negative feedback topology (Figure 9.1) was introduced in Chapter 7. There it was shown that under the assumption of an ideal op amp, the transfer function is given by (Equation (7.14)

$$T_V = -\frac{T_{FF}}{T_{FB}} \tag{9.1}$$

where T_{FF} is the feedforward transfer function and T_{FB} is the feedback transfer function of the *RC* circuit. Since the poles of T_{FF} are the same as the poles of T_{FB}, Equation 9.1 can be written as

$$T_V = -\frac{N_{FF}}{N_{FB}} \tag{9.2}$$

From this equation it is seen that complex poles are realized when the zeros of the feedback transfer function are complex. This may be achieved by using a class of *RC* networks, called **bridged-T** networks* [5], examples of which are

* As mentioned in Section 8.5, the Twin-T networks also yield complex zeros; however, we restrict our discussions in this chapter to bridged-T networks.

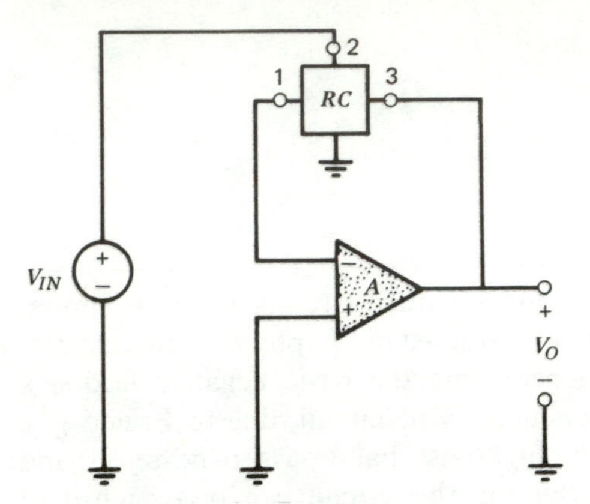

Figure 9.1 Negative feedback topology.

shown in Figure 9.2. For these circuits, the zeros of the feedback transfer function can be made complex by properly choosing the element values.

As in the positive feedback circuits, the zeros of T_V are formed by introducing the input signal at appropriate terminals in the RC network (marked with a numeral 2 in Figure 9.2).

In the following sections we consider the synthesis of some negative feedback biquads using these bridged-T RC networks.

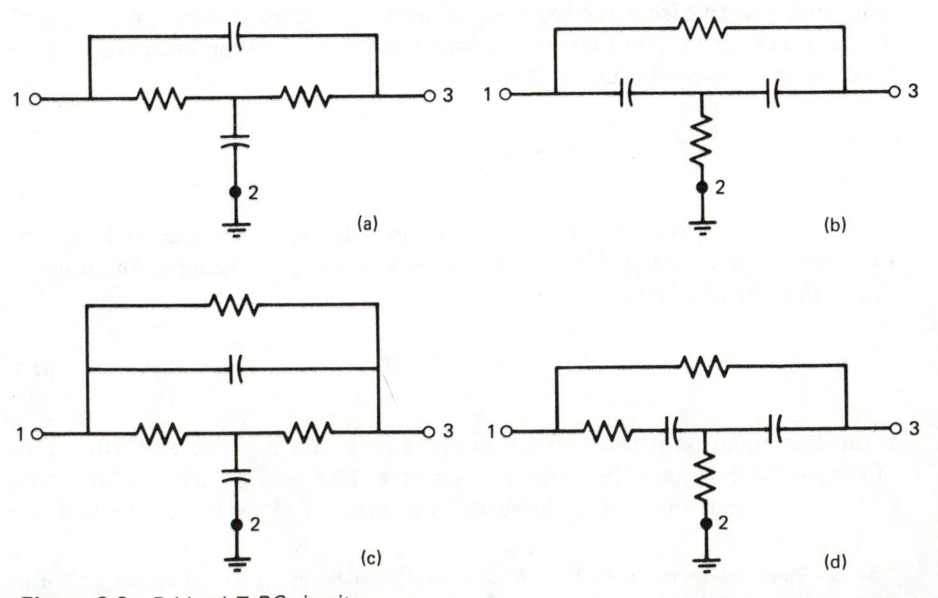

Figure 9.2 Bridged-T *RC* circuits.

9.2 A BAND-PASS CIRCUIT

In this section we will develop a band-pass circuit based on the negative feedback topology. The circuit, shown in Figure 9.3*a*, uses the bridged-T *RC* network of Figure 9.2*b*. The feedback and feedforward transfer functions of the *RC* network are obtained from the following nodal equations:

$$\begin{bmatrix} sC_1 + sC_2 + \dfrac{1}{R_1} & -sC_1 \\ -sC_1 & sC_1 + \dfrac{1}{R_2} \end{bmatrix} \begin{bmatrix} V_x \\ V_1 \end{bmatrix} = \begin{bmatrix} \dfrac{1}{R_1} & sC_2 \\ 0 & \dfrac{1}{R_2} \end{bmatrix} \begin{bmatrix} V_2 \\ V_3 \end{bmatrix} \tag{9.3}$$

Solving, we get:

$$T_{FB} = \left.\frac{V_1}{V_3}\right|_{V_2=0} = \frac{s^2 + s\left(\dfrac{1}{R_2C_1} + \dfrac{1}{R_2C_2}\right) + \dfrac{1}{R_1R_2C_1C_2}}{s^2 + s\left(\dfrac{1}{R_2C_1} + \dfrac{1}{R_2C_2} + \dfrac{1}{R_1C_2}\right) + \dfrac{1}{R_1R_2C_1C_2}} \tag{9.4}$$

and

$$T_{FF} = \left.\frac{V_1}{V_2}\right|_{V_3=0} = \frac{s/R_1C_2}{D} \tag{9.5}$$

where D is the denominator of (9.4). From Equation 9.1, assuming an ideal op amp, the transfer function of the active *RC* circuit is

$$T_{BP} = \frac{-s/R_1C_2}{s^2 + s\left(\dfrac{1}{R_2C_1} + \dfrac{1}{R_2C_2}\right) + \dfrac{1}{R_1R_2C_1C_2}} \tag{9.6}$$

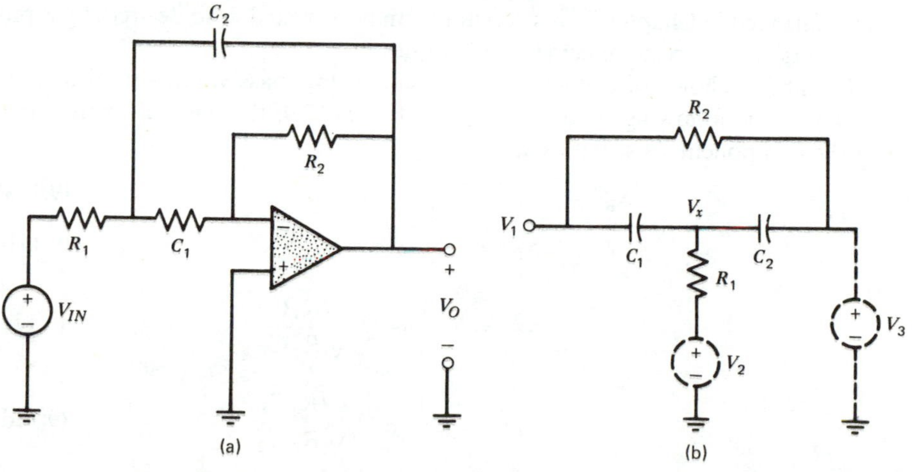

Figure 9.3 (*a*) Band-pass circuit. (*b*) *RC* circuit.

which has the form of the second-order band-pass filter function

$$K_1 \frac{s}{s^2 + \dfrac{\omega_p}{Q_p} s + \omega_p^2} \tag{9.7}$$

where the pole frequency and pole Q are given by

$$\omega_p = \sqrt{\frac{1}{R_1 R_2 C_1 C_2}} \tag{9.8}$$

$$Q_p = \frac{\sqrt{\dfrac{1}{R_1 R_2 C_1 C_2}}}{\dfrac{1}{R_2 C_1} + \dfrac{1}{R_2 C_2}} = \frac{\sqrt{\dfrac{R_2}{R_1}}}{\sqrt{\dfrac{C_2}{C_1}} + \sqrt{\dfrac{C_1}{C_2}}} \tag{9.9}$$

The band-pass function of Equation 9.7 can be synthesized (assume K_1 is arbitrary) by letting

$$C_1 = C_2 = 1 \tag{9.10a}$$

Then the remaining elements, obtained from Equation 9.8 and 9.9, are

$$R_2 = 2\frac{Q_p}{\omega_p} \qquad R_1 = \frac{1}{2\omega_p Q_p} \tag{9.10b}$$

The gain constant attained with this design is

$$K_1 = \frac{-1}{R_1 C_2} = -2\omega_p Q_p$$

As mentioned in Chapter 7, this constant can be adjusted, if so desired, by input attenuation or gain enhancement techniques.

The above choice of elements does result in low passive sensitivities, as is shown in the following. From Equation 9.8 and 9.9, the general expressions for the component sensitivities are:

$$S^{\omega_p}_{R_1, R_2, C_1, C_2} = -\tfrac{1}{2} \tag{9.11a}$$

$$S^{Q_p}_{R_1} = -\tfrac{1}{2}, \ S^{Q_p}_{R_2} = +\tfrac{1}{2} \tag{9.11b}$$

$$S^{Q_p}_{C_1} = -\frac{1}{2} + Q_p \sqrt{\frac{R_1 C_2}{R_2 C_1}} \tag{9.11c}$$

$$S^{Q_p}_{C_2} = -\frac{1}{2} + Q_p \sqrt{\frac{R_1 C_1}{R_2 C_2}} \tag{9.11d}$$

$$S^{K_1}_{R_1, C_2} = -1 \tag{9.11e}$$

Particularly, for the design equations given by (9.10), since $\sqrt{R_2/R_1} = 2Q_p$ and $C_1 = C_2 = 1$

$$S_{C_1}^{Q_p} = S_{C_2}^{Q_p} = 0 \tag{9.11f}$$

Besides resulting in low sensitivities, the choice $C_1 = C_2$ offers yet another advantage, related to a practical consideration in the manufacture of circuits, namely, that the maximum resistance ratio attainable is limited. Referring to Equation 9.9, it can be seen that limiting R_2/R_1 effectively limits the pole Q that can be realized. The choice $C_1 = C_2$ minimizes the denominator of Equation 9.9, and hence results in the maximum pole Q attainable for a given resistance ratio. For example, if R_2/R_1 is limited to 100, the maximum pole Q that can be realized is

$$Q_{p\max} = \frac{1}{2}\sqrt{\frac{R_2}{R_1}} = 5 \tag{9.12}$$

This is a rather severe limitation, considering that active filters are often required to realize much higher pole Q's. A way of alleviating this problem will be discussed in Section 9.4.

Example 9.1

Design an active filter to meet the band-pass requirements sketched in Figure 9.4, using a Chebyshev approximation.

Solution

The required *BP* function, obtained by using the CHEB approximation program (Chapter 4), is

$$T(s) = \frac{5252.26s}{s^2 + 3184.42s + 66320656} \cdot \frac{3126.49s}{s^2 + 1895.58s + 23500096}$$

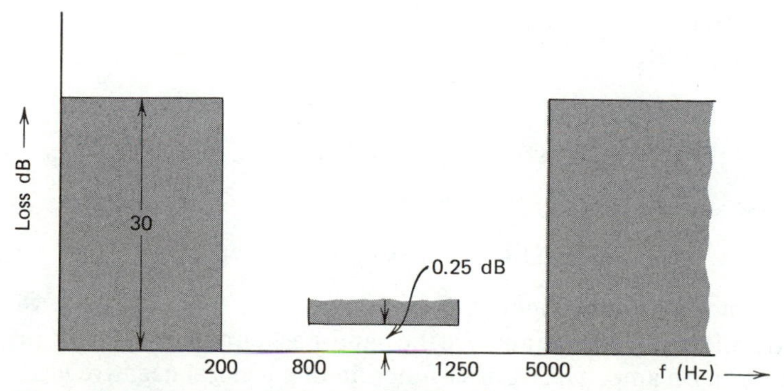

Figure 9.4 Requirements for Example 9.1.

From Equation 9.10, the element values for the *first* biquadratic (identified by the second suffix) are

$$C_{11} = C_{21} = 1 \text{ F} \qquad R_{11} = 2.4(10)^{-5}\ \Omega \qquad R_{21} = 6.28(10)^{-4}\ \Omega$$

and for the second biquadratic

$$C_{12} = C_{22} = 1 \text{ F} \qquad R_{12} = 4.03(10)^{-5}\ \Omega \qquad R_{22} = 1.06(10)^{-3}\ \Omega$$

To get practical values, the elements are impedance scaled by 10^8, to yield

$$C_{11} = C_{21} = 0.01\ \mu\text{F} \qquad R_{11} = 2.4\ \text{k}\Omega \qquad R_{21} = 62.8\ \text{k}\Omega$$
$$C_{12} = C_{22} = 0.01\ \mu\text{F} \qquad R_{12} = 4.03\ \text{k}\Omega \qquad R_{22} = 106\ \text{k}\Omega$$

The gain constant obtained by this realization is

$$K_{obt} = \frac{1}{R_{11}C_{21}} \cdot \frac{1}{R_{12}C_{22}} = 1.0339(10)^9$$

whereas the desired constant is

$$K_{des} = 5252.26 \times 3126.49 = 1.6421(10)^7$$

Since the desired constant is lower than that obtained, the input will need to be attenuated by the factor

$$\frac{1.0339(10)^9}{1.6421(10)^7} = 62.96$$

The distribution of this attenuation factor between the two sections is usually based on dynamic range considerations (this relates to the maximum signal level the filter can accept without clipping the output voltage, and is discussed further in Chapter 12). However, for simplicity, we will portion the attenuation factor equally, thereby requiring each section to attenuate the signal by:

$$\sqrt{62.96} = 7.93$$

The equations used to determine the input potential divider for the first section are

$$\frac{R_{41}R_{51}}{R_{41} + R_{51}} = 2400 \qquad \frac{R_{51}}{R_{41} + R_{51}} = \frac{1}{7.93}$$

Solving, we get

$$R_{41} = 19.0\ \text{k}\Omega \qquad R_{51} = 2.75\ \text{k}\Omega$$

Similarly, for the second section,

$$R_{42} = 31.9\ \text{k}\Omega \qquad R_{52} = 4.61\ \text{k}\Omega$$

The complete circuit is shown in Figure 9.5. ■

In the remainder of this section we analyze the band-pass circuit of Figure 9.3*a*, assuming a finite gain op amp. The transfer function of a general negative feedback structure (Figure 9.1) with a finite gain $A(s)$ was derived in Chapter 7,

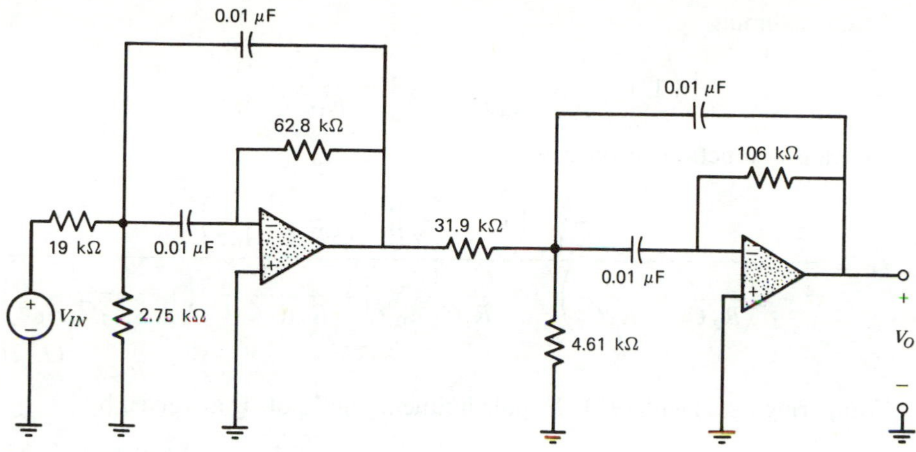

Figure 9.5 Circuit for Example 9.1.

to be (Equation 7.13):

$$T_V = -\frac{T_{FF}}{T_{FB} + \dfrac{1}{A(s)}} = -\frac{N_{FF}}{N_{FB} + \dfrac{D}{A(s)}} \tag{9.13}$$

For the band-pass circuit being considered N_{FF}, N_{FB}, and D are given by Equation 9.4 and 9.5. Substituting these expressions in (9.13), we get

$$T_V = \frac{-s/R_1C_2}{s^2 + s\left(\dfrac{1}{R_2C_1} + \dfrac{1}{R_2C_2}\right) + \dfrac{1}{R_1R_2C_1C_2} + \dfrac{1}{A(s)}\left[s^2 + s\left(\dfrac{1}{R_2C_1} + \dfrac{1}{R_2C_2} + \dfrac{1}{R_1C_2}\right) + \dfrac{1}{R_1R_2C_1C_2}\right]} \tag{9.14}$$

This function is simplified by proceeding as in Section 8.2 (Equation 8.17 to 8.20). First, assuming $[1 + 1/A(s)]^{-1} \cong 1 - 1/A(s)$, the transfer function is approximated as:

$$T_V \cong \frac{-[1 - 1/A(s)]s/R_1C_2}{s^2 + s\left[\dfrac{1}{R_2C_1} + \dfrac{1}{R_2C_2} + \dfrac{1}{R_1C_2}\cdot\dfrac{1}{A(s)}\left(1 - \dfrac{1}{A(s)}\right)\right] + \dfrac{1}{R_1R_2C_1C_2}} \tag{9.15}$$

For an op amp whose gain is characterized by a single pole at the origin, as $A(s) = A_0\alpha/s$, the above transfer function reduces to*

$$T_V \cong \frac{-(1 - s/A_0\alpha)s/R_1C_2}{s^2\left(1 + \dfrac{1}{R_1C_2}\cdot\dfrac{1}{A_0\alpha}\right) + s\left(\dfrac{1}{R_2C_1} + \dfrac{1}{R_2C_2}\right) + \dfrac{1}{R_1R_2C_1C_2}} \tag{9.16}$$

* Assuming $\dfrac{s}{A_0\alpha}\left(1 - \dfrac{s}{A_0\alpha}\right) \cong \dfrac{s}{A_0\alpha}$.

Again, assuming

$$\left(1 + \frac{1}{R_1 C_2} \cdot \frac{1}{A_0 \alpha}\right)^{-1} \cong 1 - \frac{1}{R_1 C_2} \cdot \frac{1}{A_0 \alpha}$$

the transfer function becomes

$$T_V \cong \frac{-\dfrac{s}{R_1 C_2}\left(1 - \dfrac{s}{A_0 \alpha}\right)\left(1 - \dfrac{1}{R_1 C_2 A_0 \alpha}\right)}{s^2 + s\left(\dfrac{1}{R_2 C_1} + \dfrac{1}{R_2 C_2}\right)\left(1 - \dfrac{1}{R_1 C_2 A_0 \alpha}\right) + \dfrac{1}{R_1 R_2 C_1 C_2}\left(1 - \dfrac{1}{R_1 C_2 A_0 \alpha}\right)} \tag{9.17}$$

Comparing (9.17) with (9.7), the pole frequency and pole Q are given by

$$\omega_p = \frac{1}{\sqrt{R_1 R_2 C_1 C_2}} \sqrt{1 - \frac{1}{R_1 C_2 A_0 \alpha}} \tag{9.18}$$

$$Q_p = \frac{\dfrac{1}{\sqrt{R_1 R_2 C_1 C_2}}}{\left(\dfrac{1}{R_2 C_1} + \dfrac{1}{R_2 C_2}\right)\sqrt{1 - \dfrac{1}{R_1 C_2 A_0 \alpha}}} \tag{9.19}$$

The sensitivities of ω_p and Q_p to the parameter $A_0 \alpha$ are computed to be

$$S_{A_0 \alpha}^{\omega_p} = \frac{1}{2} \cdot \frac{1}{R_1 R_2 C_1 C_2} \cdot \frac{\dfrac{1}{R_1 C_2 A_0 \alpha}}{\omega_p^2}$$

and since $\omega_p^2 \cong 1/R_1 R_2 C_1 C_2$, this expression simplifies to

$$S_{A_0 \alpha}^{\omega_p} \cong \frac{1}{2 R_1 C_2 A_0 \alpha} \cong \frac{1}{2} \frac{\omega_p}{A_0 \alpha} \sqrt{\frac{R_2 C_1}{R_1 C_2}} \tag{9.20a}$$

Similarly

$$S_{A_0 \alpha}^{Q_p} \cong -\frac{1}{2} \frac{\omega_p}{A_0 \alpha} \sqrt{\frac{R_2 C_1}{R_1 C_2}} \tag{9.20b}$$

For the design equations given by Equations 9.10, these sensitivities reduce to

$$S_{A_0 \alpha}^{\omega_p} = -S_{A_0 \alpha}^{Q_p} = \frac{\omega_p Q_p}{A_0 \alpha} \tag{9.21}$$

As in the positive feedback circuits, these sensitivities increase with the pole frequency and decrease with the gain-bandwidth product $A_0 \alpha$. The contribution of the active and passive sensitivity terms to deviations in the gain are computed as in the following example.

Example 9.2

A second-order band-pass filter having a pole frequency of 2 kHz and a pole Q of 10, is designed using the circuit of Figure 9.3*a*. Compute the deviation in gain at the upper 3 dB passband edge frequency, 2.1 kHz, assuming the passive elements to have a tolerance of ± 1 percent, and the tolerance on $A_0\alpha$ to be ± 50 percent about its nominal value of $2\pi 10^6$ rad/sec. The tolerance limits represent the 3σ points of a Gaussian distribution.

Solution

The gain deviation is computed as in Example 8.2 and 8.3, using the expressions derived there for $\mu(\Delta G)$ and $\sigma(\Delta G)$ (Equation 8.14 and 8.15).

The sensitivities of ω_p, Q_p, and K to the passive elements are given in Equation 9.11. The sensitivities to the parameter $A_0\alpha$, from Equation 9.21 are

$$S_{A_0\alpha}^{\omega_p} = -S_{A_0\alpha}^{Q_p} = \frac{\omega_p Q_p}{A_0\alpha} = \frac{2\pi 2000 \times 10}{2\pi 10^6} = 0.02$$

The biquadratic parameter sensitivities $\mathcal{S}_{\omega_p}^G$, $\mathcal{S}_{Q_p}^G$, and $\mathcal{S}_K^G$ will have the same values as those computed in Example 8.2, since the normalized frequency ($\Omega = 21/20 = 1.05$) and pole Q are the same as for this example:

$$\mathcal{S}_{\omega_p}^G = 78.25 \text{ dB} \qquad \mathcal{S}_{Q_p}^G = 4.45 \text{ dB} \qquad \mathcal{S}_K^G = 8.686 \text{ dB}$$

From the given information, the variabilities of the elements are described by

$$\mu(V_R) = \mu(V_C) = \mu(A_0\alpha) = 0$$
$$3\sigma(V_R) = 3\sigma(V_C) = 0.01 \qquad 3\sigma(V_{A_0\alpha}) = 0.5$$

Substituting the numerical values derived above in Equation 8.14 and 8.15 we see that $\mu(\Delta G) = 0$ dB, and $\sigma(\Delta G)$ is given by:*

$$\begin{aligned}
[3\sigma(\Delta G)]^2 &= (4.45)^2(\tfrac{1}{4} + \tfrac{1}{4})(0.01)^2 && \rightarrow Q_p\text{-passive} \\
&\quad + (78.25)^2(\tfrac{1}{4} + \tfrac{1}{4} + \tfrac{1}{4} + \tfrac{1}{4})(0.01)^2 && \rightarrow \omega_p\text{-passive} \\
&\quad + (8.686)^2(1 + 1)(0.01)^2 && \rightarrow K\text{-passive} \\
&\quad + (4.45)^2(0.02)^2(0.5)^2 && \rightarrow Q_p\text{-active} \\
&\quad + (78.25)^2(0.02)^2(0.5)^2 && \rightarrow \omega_p\text{-active} \\
&= \underbrace{\underset{Q_p}{0.002} + \underset{\omega_p}{0.61} + \underset{K}{0.015}}_{\text{passive terms}} + \underbrace{\underset{Q_p}{0.002} + \underset{\omega_p}{0.61}}_{\text{active terms}} \\
&= 1.24
\end{aligned}$$

Thus

$$3\sigma(\Delta G) = 1.11 \text{ dB}$$

* As in Example 8.2, the contribution of the numerator-active term is assumed to be negligible.

Observations

1. A comparison with Example 8.3 shows that the low-pass positive feedback circuit with $k = 1$ (Design 1) has almost the same gain deviation as the band-pass circuit of this example. This is expected because, for both filters, the expressions defining the contributions of the ω_p-passive and ω_p-active terms are the same, and these two terms are seen to dominate in the expression for $\sigma(\Delta G)$. In particular, $S_{R,C}^{\omega_p} = -1/2, S_{A_0\alpha}^{\omega_p} = \omega_p Q_p / A_0 \alpha$, and $\mathscr{S}_{\omega_p}^{G} = 78.25$ for both circuits. Therefore, for the same normalized frequency Ω_p, and the same pole Q, these two circuits will always exhibit almost equal gain deviations (assuming, of course, that both circuits use the same types of components). It is interesting to note that the numerators, which define whether the function is a low-pass or a band-pass, contribute insignificantly to $\sigma(\Delta G)$. Therefore, the gain deviation becomes independent of the filter type. This will be found to be true in general for gain deviations in the passband.
2. In the positive feedback *LP* circuit, we were able to improve Design 1 by choosing a different value for k, as in the Saraga design where $k = 4/3$. In comparison, the negative feedback band-pass circuit does not have any such free parameter. In other words, the negative feedback topology has one less degree of freedom than the positive feedback topology. This deficiency will be remedied later in the chapter, by introducing a controlled amount of positive feedback in the general negative feedback topology.

■

9.3 FORMATION OF ZEROS

In this section we show how the band-pass circuit of Figure 9.3*a* can be generalized to realize the general biquadratic function:

$$K \frac{ms^2 + cs + d}{ns^2 + as + b} \qquad \begin{pmatrix} m = 1 \text{ or } 0 \\ n = 1 \text{ or } 0 \end{pmatrix} \tag{9.22}$$

As mentioned in Section 7.1, this general function allows the realization of second-order low-pass, high-pass, band-pass, band-reject, and equalizer filter characteristics. Thus, a biquad realization of this function will provide these various filter functions, all with the same circuit by simply changing the element values to obtain the desired coefficients. Such a circuit topology will greatly reduce the cost of manufacturing filter circuits. Let us see how this biquad realization is achieved.

Referring to Figure 9.3*a*, it is seen that the positive input terminal of the op amp is connected to ground. This node, therefore, can serve as an input port, as shown in Figure 9.6. The transfer function from this input port to the output can be obtained by equating the voltages at the negative and positive input

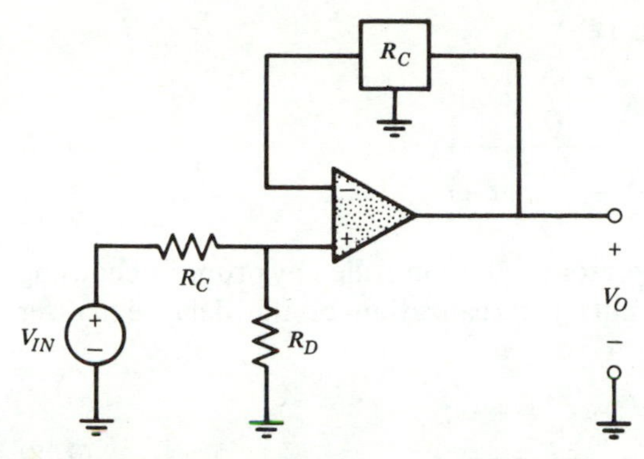

Figure 9.6 Negative feedback circuit with input at positive terminal.

terminals of the op amp:

$$K_2 V_{IN} = T_{FB} V_O \tag{9.23}$$

where

$$K_2 = \frac{R_D}{R_C + R_D} \tag{9.24}$$

Substituting for T_{FB} from Equation 9.4, the resulting transfer function is

$$T_{POS} = \frac{V_O}{V_{IN}} = K_2 \frac{s^2 + s\left(\frac{1}{R_2 C_1} + \frac{1}{R_2 C_2} + \frac{1}{R_1 C_2}\right) + \frac{1}{R_1 R_2 C_1 C_2}}{s^2 + s\left(\frac{1}{R_2 C_1} + \frac{1}{R_2 C_2}\right) + \frac{1}{R_1 R_2 C_1 C_2}} \tag{9.25}$$

If the same input voltage, V_{IN}, is applied to both the positive input and the bandpass input ports, the resulting transfer function is given by the sum of T_{POS} and T_{BP}, as

$$T_{POS} + T_{BP} = \frac{K_2\left[s^2 + s\left(\frac{1}{R_2 C_1} + \frac{1}{R_2 C_2} + \frac{1}{R_1 C_2}\right) + \frac{1}{R_1 R_2 C_1 C_2}\right] - K_1 \frac{s}{R_1 C_2}}{s^2 + s\left(\frac{1}{R_2 C_1} + \frac{1}{R_2 C_2}\right) + \frac{1}{R_1 R_2 C_1 C_2}} \tag{9.26}$$

where K_1 and K_2 are the gain constants associated with T_{POS} and T_{BP}, respectively. A comparison of (9.26) with the general biquadratic function expressed in the form ($m = 1, n = 1$)

$$K \frac{s^2 + \dfrac{\omega_z}{Q_z} + \omega_z^2}{s^2 + \dfrac{\omega_p}{Q_p} s + \omega_p^2} \tag{9.27}$$

shows that the Q_z term in the numerator can be controlled by properly choosing K_1 and K_2. This flexibility permits the realization of the delay equalizer function:

$$K \frac{s^2 - \dfrac{\omega_p}{Q_p} s + \omega_p^2}{s^2 + \dfrac{\omega_p}{Q_p} s + \omega_p^2} \tag{9.28}$$

However, a limitation of the transfer function of Equation 9.26 is that the pole frequency is equal to the zero frequency ($\omega_p = \omega_z = \sqrt{1/R_1R_2C_1C_2}$). It is this restriction that precludes it from realizing the remaining filter functions. The fundamental reason for this limitation is explained in the following. An inspection of Equation 9.26 shows that the pole frequency ω_p and the zero frequency ω_z of the active circuit are, respectively, the zero and pole frequencies of the feedback transfer function of the *RC* circuit. From Figure 9.3*a*, it can easily be seen that the *dc* gain and the infinite frequency gain of T_{FB} are both equal to unity. Therefore, ω_z *must* equal ω_p. The reader will recall a similar situation in the discussion of the Twin-T network, in Section 8.5. The separation of the pole and zero frequencies is effected just as for the Twin-T network, by adding a loading network (in this case the resistor R_3) to the passive circuit as shown in Figure 9.7*a*. The resulting negative feedback circuit is sketched in Figure 9.7*b*.

On analysis of this new active circuit, the feedback transfer function T_{FB} of the *RC* network can be shown to be

$$T'_{FB} = \frac{s^2 + s\left(\dfrac{1}{R_2C_1} + \dfrac{1}{R_2C_2}\right) + \dfrac{1}{R_1R_2C_1C_2}}{s^2 + s\left(\dfrac{1}{R_1C_2} + \dfrac{1}{R_2C_2} + \dfrac{1}{R_3C_2} + \dfrac{1}{R_2C_1} + \dfrac{1}{R_3C_1}\right) + \dfrac{1}{R_1R_2C_1C_2} + \dfrac{1}{R_1R_3C_1C_2}} \tag{9.29}$$

The feedforward transfer function, due to an input signal K_1V_{IN} at node 2 alone, is

$$T'_{FF2} = \frac{\dfrac{s}{R_1C_2}}{D'} \tag{9.30}$$

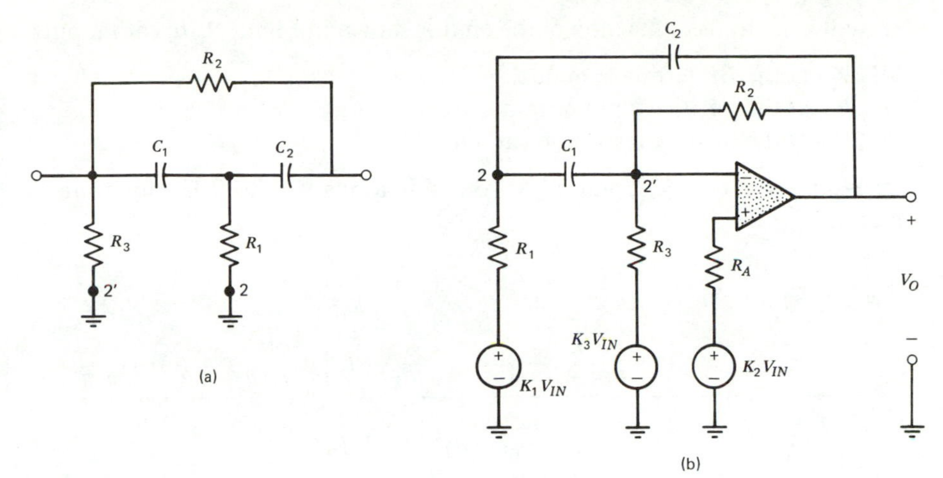

Figure 9.7 (*a*) Bridged-T circuit with a loading resistor R_3. (*b*) Resulting negative feedback circuit showing zero forming inputs.

where D' is the denominator of (9.29). Thus, the transfer function with this input is the band-pass function:

$$T'_{BP} = K_1 \frac{-\dfrac{s}{R_1C_2}}{s^2 + s\left(\dfrac{1}{R_2C_1} + \dfrac{1}{R_2C_2}\right) + \dfrac{1}{R_1R_2C_1C_2}} \tag{9.31}$$

If the input signal K_3V_{IN} is introduced at node 2′, via the loading resistor R_3, the resulting transfer function due to this input alone, can be shown to be

$$T'_L = -K_3 \frac{s\left(\dfrac{1}{R_3C_1} + \dfrac{1}{R_3C_2}\right) + \dfrac{1}{R_1R_3C_1C_2}}{s^2 + s\left(\dfrac{1}{R_2C_1} + \dfrac{1}{R_2C_2}\right) + \dfrac{1}{R_1R_2C_1C_2}} \tag{9.32}$$

The transfer function due to an input K_2V_{IN} at the positive terminal of the op amp, obtained by using Equation 9.23, is given by

$$T'_{POS} = \frac{K_2}{T'_{FB}} = K_2 \frac{s^2 + s\left(\dfrac{1}{R_1C_2} + \dfrac{1}{R_2C_2} + \dfrac{1}{R_3C_2} + \dfrac{1}{R_2C_1} + \dfrac{1}{R_3C_1}\right) + \dfrac{1}{R_1R_2C_1C_2} + \dfrac{1}{R_1R_3C_1C_2}}{s^2 + s\left(\dfrac{1}{R_2C_1} + \dfrac{1}{R_2C_2}\right) + \dfrac{1}{R_1R_2C_1C_2}} \tag{9.33}$$

Finally, the transfer function of the circuit shown in Figure 9.7*b*, with inputs:

$K_1 V_{IN}$ at the *BP* terminal, node 2
$K_2 V_{IN}$ at the positive input terminal
$K_3 V_{IN}$ at the loading resistor, terminal 2′

is the sum of T'_{BP}, T'_{POS}, and T'_L. From Equations 9.31 to 9.33, this transfer function is

$$T' = \frac{K_2\left[s^2 + s\left(\frac{1}{R_1C_2} + \frac{1}{R_2C_2} + \frac{1}{R_3C_2} + \frac{1}{R_2C_1} + \frac{1}{R_3C_1}\right) + \frac{1}{R_1R_2C_1C_2} + \frac{1}{R_1R_3C_1C_2}\right] - K_1\frac{s}{R_1C_2} - K_3\left[s\left(\frac{1}{R_3C_1} + \frac{1}{R_3C_2}\right) + \frac{1}{R_1R_3C_1C_2}\right]}{s^2 + s\left(\frac{1}{R_2C_1} + \frac{1}{R_2C_2}\right) + \frac{1}{R_1R_2C_1C_2}} \tag{9.34}$$

which may be written in the general form

$$T' = \frac{K_2(s^2 + cs + d) - (K_1e)s - K_3(fs + g)}{s^2 + as + b} \tag{9.35}$$

Observe that the denominator is the same as for the *BP* circuit without the loading resistor (Equation 9.6). Therefore, the sensitivity in the passband, which is predominantly controlled by the denominator, will be essentially the same as for the *BP* circuit of Figure 9.3*a*.

An inspection of this general transfer function shows that the following classes of second-order filter functions can be generated by properly choosing the gain constants K_1, K_2, and K_3:

Filter function	Numerator
Band-pass	s
High-pass	s^2
Band-reject	$s^2 + \omega_z^2 \quad (\omega_z = \omega_p)$
High-pass-notch	$s^2 + \omega_z^2 \quad (\omega_z < \omega_p)$
Low-pass-notch	$s^2 + \omega_z^2 \quad (\omega_z > \omega_p)$
Delay equalizer	$s^2 - \frac{\omega_z}{Q_z}s + \omega_z^2 \quad (\omega_z = \omega_p, Q_z = Q_p)$

The circuit of Figure 9.7*b* is therefore seen to be capable of providing all the standard-filter functions (except for the low-pass, in which case the numerator is a constant). This feature is extremely attractive from a manufacturing standpoint, because it allows one circuit layout to be used for the production of most filters. Where many different filter functions are needed in production, this singular approach results in a considerable savings in cost.

9.4 THE USE OF POSITIVE FEEDBACK IN NEGATIVE FEEDBACK TOPOLOGIES

In Section 9.2 it was pointed out that the maximum pole Q attainable by the band-pass circuit of Figure 9.3*a* is limited by the maximum manufacturable ratio of resistors (R_2/R_1). From Equation 9.12, this maximum pole Q is

$$Q_{p_{\max}} = \frac{1}{2}\sqrt{\frac{R_2}{R_1}} \tag{9.36}$$

Thus, in the negative feedback topology, high pole Q's are obtained at the expense of large spreads in the element values. In contrast, recall that in the positive feedback topology the pole Q's could be increased by subtracting a term from the s term in the denominator (Equation 7.19). This subtraction can also be achieved in the negative feedback topology in an analogous manner, by providing some positive feedback via a potential divider (as illustrated in Figure 9.8). In this section we study the effects of this modification on the basic negative feedback topology.

Consider, for instance, the basic band-pass circuit of Figure 9.3*a* where the positive feedback is provided via resistors R_A and R_B (Figure 9.9). This circuit was originally proposed by Delyiannis [1]. The transfer function for this

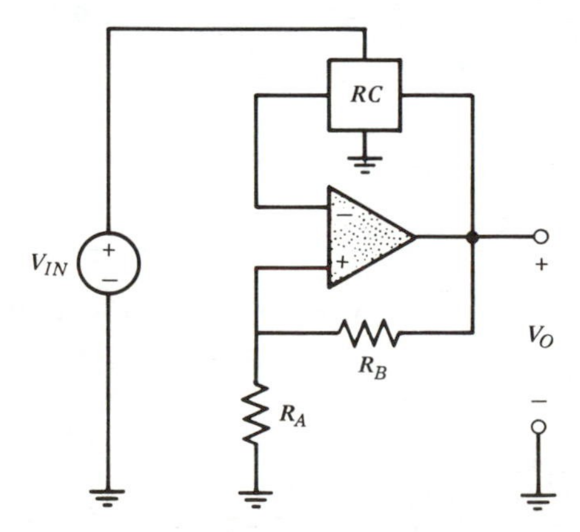

Figure 9.8 The negative feedback topology with positive feedback.

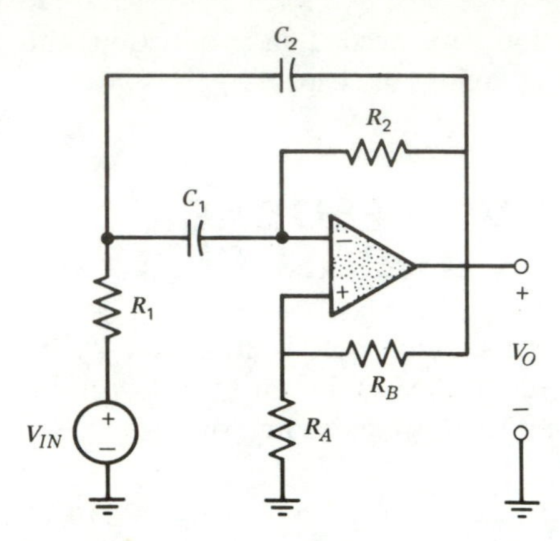

Figure 9.9 Delyiannis band-pass circuit.

circuit can be obtained by equating the voltages at the negative and positive input terminals (assuming an ideal op amp):

$$\frac{V_O}{k} = V_{IN} T_{FF} + V_O T_{FB} \tag{9.37}$$

where

$$k = 1 + \frac{R_B}{R_A} \tag{9.38}$$

From (9.37)

$$\frac{V_O}{V_{IN}} = -\frac{T_{FF}}{T_{FB} - \dfrac{1}{k}} = -\frac{N_{FF}}{N_{FB} - \dfrac{D}{k}} \tag{9.39}$$

Substituting for N_{FF}, N_{FB}, and D from Equation 9.4 and 9.5, we get

$$\frac{V_O}{V_{IN}} = \frac{-s/R_1C_2}{s^2 + s\left(\dfrac{1}{R_2C_1} + \dfrac{1}{R_2C_2}\right) + \dfrac{1}{R_1R_2C_1C_2} - \dfrac{1}{k}\left[s^2 + s\left(\dfrac{1}{R_2C_1} + \dfrac{1}{R_2C_2} + \dfrac{1}{R_1C_2}\right) + \dfrac{1}{R_1R_2C_1C_2}\right]}$$

Dividing numerator and denominator by $(1 - 1/k)$

$$\frac{V_O}{V_{IN}} = \frac{-\dfrac{s}{R_1C_2(1 - 1/k)}}{s^2 + s\left(\dfrac{1}{R_2C_1} + \dfrac{1}{R_2C_2} - \dfrac{1}{k-1}\dfrac{1}{R_1C_2}\right) + \dfrac{1}{R_1R_2C_1C_2}} \tag{9.40}$$

Comparing this expression with the general band-pass function of Equation 9.7, we get

$$\omega_p = \sqrt{\frac{1}{R_1 R_2 C_1 C_2}} \tag{9.41}$$

$$Q_p = \frac{\dfrac{1}{\sqrt{R_1 R_2 C_1 C_2}}}{\dfrac{1}{R_2 C_1} + \dfrac{1}{R_2 C_2} - \dfrac{1}{k-1}\dfrac{1}{R_1 C_2}}$$

$$= \frac{\sqrt{\dfrac{R_2}{R_1}}}{\sqrt{\dfrac{C_2}{C_1}} + \sqrt{\dfrac{C_1}{C_2}} - \dfrac{1}{k-1}\dfrac{R_2}{R_1}\sqrt{\dfrac{C_1}{C_2}}} \tag{9.42}$$

$$K = \frac{-1}{R_1 C_2\left(1 - \dfrac{1}{k}\right)} \tag{9.43}$$

From (9.42) it can be seen that the subtractive term, which occurs due to the introduction of positive feedback, allows the realization of high pole Q's even when R_2/R_1 is limited. This technique is frequently referred to as **Q enhancement.**

Let us next consider the synthesis of a band-pass function in the form of Equation 9.7, where K is arbitrary. In this case there are five elements (R_1, R_2, C_1, C_2, k) and two constraints. Therefore, three of the elements can be fixed. As in the basic band-pass circuit of Figure 9.3*a*, the capacitors are chosen to be equal

$$C_1 = C_2 = 1 \tag{9.44a}$$

For the third fixed parameter, we choose the ratio of the resistors

$$\frac{R_2}{R_1} = \beta \tag{9.44b}$$

where β is some constant. With the above choice, the remaining elements can be evaluated from (9.41) and (9.42) to be

$$R_1 = \frac{1}{\sqrt{\beta}\,\omega_p} \qquad R_2 = \frac{\sqrt{\beta}}{\omega_p} \tag{9.45a}$$

$$k = \frac{Q_p(\beta + 2) - \sqrt{\beta}}{2Q_p - \sqrt{\beta}} \tag{9.45b}$$

Observe that the addition of positive feedback results in one extra degree of freedom, in the form of the parameter β. This parameter can be chosen so as to minimize the sensitivity of the network, in much the same way as in the positive feedback topology, where k was chosen to minimize sensitivity. To properly choose β, we must first evaluate the sensitivities of the biquadratic parameters ω_p, Q_p, and K to the passive and active elements. These passive sensitivities are readily obtained from Equation 9.41, 9.42, and 9.43:

$$S^{\omega_p}_{R_1, R_2, C_1, C_2} = -\tfrac{1}{2} \tag{9.46a}$$

$$S^{Q_p}_{R_2} = -S^{Q_p}_{R_1} = -\frac{1}{2} + \frac{Q_p}{\sqrt{\beta}}\left(\sqrt{\frac{C_2}{C_1}} + \sqrt{\frac{C_1}{C_2}}\right) \tag{9.46b}$$

$$S^{Q_p}_{C_1} = -S^{Q_p}_{C_2} = -\frac{1}{2} + \frac{Q_p}{\sqrt{\beta}}\sqrt{\frac{C_2}{C_1}} \tag{9.46c}$$

$$S^{Q_p}_{R_A} = -S^{Q_p}_{R_B} = \left(\frac{2Q_p}{\sqrt{\beta}} - 1\right)\sqrt{\frac{C_1}{C_2}} \tag{9.46d}$$

$$S^{K}_{R_1, C_2} = -1 \tag{9.46e}$$

$$S^{K}_{R_A} = -S^{K}_{R_B} = \frac{1}{k} \tag{9.46f}$$

To obtain the active sensitivities, the circuit of Figure 9.9 must first be analyzed with a finite $A(s)$. The output voltage in this figure is given by

$$\left(\frac{V_O}{k} - T_{FB}V_O - T_{FF}V_{IN}\right)A(s) = V_O \tag{9.47}$$

Thus

$$\frac{V_O}{V_{IN}} - \frac{-N_{FF}}{N_{FB} - \dfrac{D}{k} + \dfrac{D}{A(s)}} \tag{9.48}$$

Substituting the expressions for N_{FF}, N_{FB}, and D, from Equations 9.4 and 9.5, and simplifying

$$\frac{V_O}{V_{IN}} = \frac{-\dfrac{s}{R_1C_2[1 + 1/A(s) - 1/k]}}{s^2 + s\left(\dfrac{1}{R_2C_1} + \dfrac{1}{R_2C_2} + \dfrac{1}{R_1C_2}\dfrac{1/A(s) - 1/k}{1 + 1/A(s) - 1/k}\right) + \dfrac{1}{R_1R_2C_1C_2}} \tag{9.49}$$

As before, let us assume the op amp gain to be characterized by $A(s) = A_0\alpha/s$.

Then the above expression can be approximated by proceeding as in the derivation of Equation 9.17. This will yield

$$\frac{V_O}{V_{IN}} \cong \frac{-\dfrac{s}{R_1C_2}\dfrac{k}{k-1}\left(1-\dfrac{s}{A_0\alpha}\dfrac{k}{k-1}-\dfrac{k_2}{A_0\alpha(k-1)^2R_1C_2}\right)}{s^2+s\left(\dfrac{1}{R_2C_1}+\dfrac{1}{R_2C_2}-\dfrac{1}{R_1C_2(k-1)}\right)\left(1-\dfrac{k^2}{A_0\alpha(k-1)^2R_1C_2}\right)+\dfrac{1}{R_1R_2C_1C_2}\left(1-\dfrac{k^2}{A_0\alpha(k-1)^2R_1C_2}\right)} \tag{9.50}$$

Thus

$$\omega_p = \frac{1}{\sqrt{R_1R_2C_1C_2}}\sqrt{\left(1-\frac{k^2}{A_0\alpha(k-1)^2R_1C_2}\right)} \tag{9.51}$$

$$(\mathrm{bw})_p = \left(\frac{1}{R_2C_1}+\frac{1}{R_2C_2}-\frac{1}{R_1C_2(k-1)}\right)\left(1-\frac{k^2}{A_0\alpha(k-1)^2R_1R_2}\right) \tag{9.52}$$

From these expressions, the sensitivity of ω_p and Q_p to $A_0\alpha$ can be shown to be

$$S_{A_0\alpha}^{\omega_p} \cong -S_{A_0\alpha}^{Q_p} \cong \frac{1}{2}\frac{\omega_p}{A_0\alpha}\frac{k^2}{(k-1)^2}\sqrt{\frac{R_2C_1}{R_1C_2}} = \frac{1}{2}\frac{\omega_p}{A_0\alpha}\frac{k^2}{(k-1)^2}\sqrt{\beta}\sqrt{\frac{C_1}{C_2}} \tag{9.53}$$

An inspection of Equation 9.46 for the passive component sensitivities and the above expression for the active component sensitivities shows that some of the sensitivities increase with β, while others decrease with β. This apparent disparity is resolved by choosing β so that the total gain deviation due to the active and passive elements is minimized, as illustrated by the following example.

Example 9.3

Synthesize the second-order *BP* function of Example 9.2 using the Delyiannis circuit. Find the value of β that minimizes the gain deviation at the upper 3 dB passband edge frequency 2.1 kHz, given the manufacturing constraint that the ratios of resistors cannot exceed 100. The component tolerances are as specified in Example 9.2.

Solution

The value of β that minimizes ΔG is obtained by plotting ΔG versus β (for β in the prescribed range $0.01 \le \beta \le 100$). This optimum value of β is then used in Equation 9.45 to compute the element values.

The sensitivity terms $\mathcal{S}_{\omega_p}^G$, $\mathcal{S}_{Q_p}^G$, and $\mathcal{S}_K^G$ and the variabilities are the same as in Example 9.2. The sensitivities to the passive elements are obtained from Equation 9.46. For example, if $\beta = 100$, from Equation 9.45b

$$k = 101$$

and from Equation 9.46, for $C_1 = C_2 = 1$:

$$S^{\omega_p}_{R_1, R_2, C_1, C_2} = -\tfrac{1}{2} \qquad S^{Q_p}_{R_2} = -S^{Q_p}_{R_1} = \tfrac{3}{2}$$

$$S^{Q_p}_{C_1} = -S^{Q_p}_{C_2} = \tfrac{1}{2} \qquad S^{Q_p}_{R_A} = -S^{Q_p}_{R_B} = 1$$

$$S^{K}_{R_1, C_2} = -1 \qquad S^{K}_{R_A} = -S^{K}_{R_B} = \tfrac{1}{101}$$

The sensitivities to the active term $A_0\alpha$, obtained from Equation 9.53, are

$$S^{\omega_p}_{A_0\alpha} = -S^{Q_p}_{A_0\alpha} = 0.01$$

Substituting the above values in Equation 8.14 and 8.15, the μ and σ of ΔG are given by

$$\mu(\Delta G) = 0 \text{ dB}$$

$$\begin{aligned}
[3\sigma(\Delta G)]^2 &= (4.45)^2(7)(0.01)^2 && \rightarrow Q_p\text{-passive} \\
&\quad + (78.25)^2(1)(0.01)^2 && \rightarrow \omega_p\text{-passive} \\
&\quad + (8.686)^2(2)(0.01)^2 && \rightarrow K\text{-passive} \\
&\quad + (4.45)^2(0.01)^2(0.5)^2 && \rightarrow Q_p\text{-active} \\
&\quad + (78.25)^2(0.01)^2(0.5)^2 && \rightarrow \omega_p\text{-active} \\
&= \underbrace{\underset{Q_p}{0.014} + \underset{\omega_p}{0.612} + \underset{K}{0.015}}_{\text{passive terms}} + \underbrace{\underset{Q_p}{5(10)^{-4}} + \underset{\omega_p}{0.153}}_{\text{active terms}} \\
&= 0.794
\end{aligned}$$

Therefore, $3\sigma(\Delta G) = 0.891$ dB

In a similar manner, $3\sigma(\Delta G)$ can be computed for other values of β in the prescribed range $0.01 \leq \beta \leq 100$. The results of these computations are plotted in Figure 9.10, from which the optimum value of β is seen to be

$$\beta = 42$$

For this value of β the individual contributions to $[3\sigma(\Delta G)]^2$ are:

$$\begin{aligned}
[3\sigma(\Delta G)]^2 &= \underbrace{\underset{Q_p}{0.048} + \underset{\omega_p}{0.612} + \underset{K}{0.015}}_{\text{passive terms}} + \underbrace{\underset{Q_p}{2.3(10)^{-4}} + \underset{\omega_p}{0.073}}_{\text{active terms}} \\
&= 0.75
\end{aligned}$$

Thus

$$3\sigma(\Delta G) = 0.866 \text{ dB}$$

The element values for the Delyiannis *BP* circuit with $\beta = 42$, from Equation 9.45, are

$$C_1 = 1 \text{ F} \qquad C_2 = 1 \text{ F} \qquad R_1 = 1.23(10)^{-5}$$

$$R_2 = 5.16(10)^{-4} \qquad k = 32.06$$

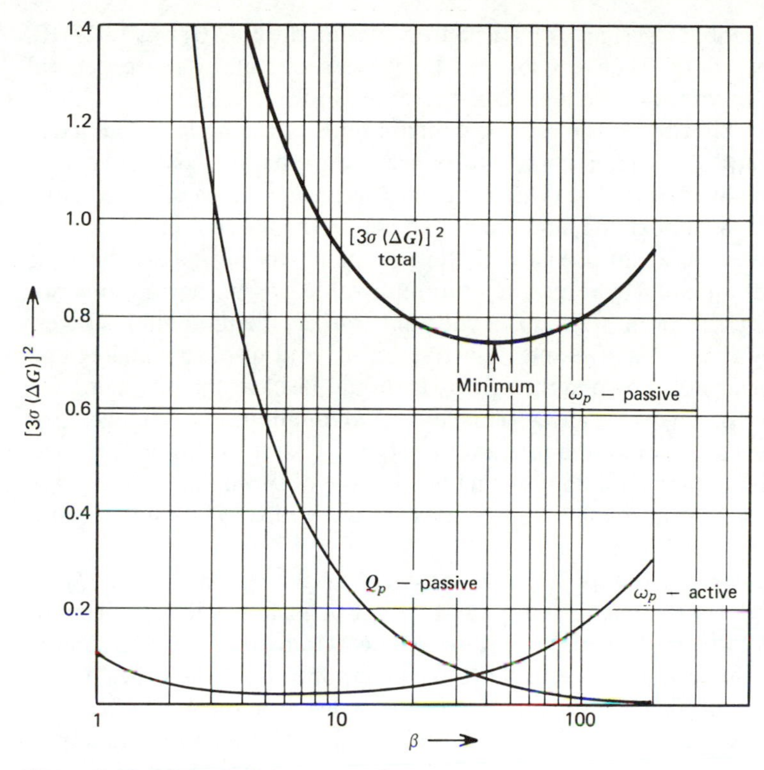

Figure 9.10 Gain deviation versus β for Example 9.3.

Impedance scaling by 10^8 to obtain practical elements, we get:

$$C_1 = 0.01\ \mu\text{F} \qquad C_2 = 0.01\ \mu\text{F} \qquad R_1 = 1.23\ \text{k}\Omega \qquad R_2 = 51.6\ \text{k}\Omega$$

The factor $k = 1 + R_B/R_A$ can be realized using

$$R_A = 1\ \text{k}\Omega \qquad R_B = 31.06\ \text{k}\Omega$$

Observations

1. This example demonstrates the impressive benefits attained by introducing positive feedback in the negative feedback topology. The resistance ratio is decreased from an intolerably high value of 400 to an easily achievable 42. Moreover, the extra degree of freedom provided by the factor β allows a significant reduction in the gain deviation. Recall that without positive feedback $3\sigma(\Delta G)$ was 1.11 dB (Example 9.2), which is 0.24 dB more than attained by the above design with positive feedback.
2. A comparison with the positive feedback LP circuit of Example 8.3, which has similar requirements, shows that this Delyiannis circuit is less sensitive than the Saraga ($k = 4/3$) design (which was the best of the three

positive feedback designs). For the Saraga design the $3\sigma(\Delta G)$ was 1.027 dB, as compared to 0.866 dB for the Delyiannis circuit. A more general comparison of these two circuits is presented in Section 9.6.

3. Figure 9.10 illustrates the relative contributions of the passive and active terms. The three major contributors are the ω_p-passive, ω_p-active, and the Q_p-passive terms. The other terms, Q_p-active and K-passive, are quite negligible in comparison, and are not shown in the figure. For the passive and active tolerances specified in this problem, the ω_p-passive term is seen to dominate. However, the minimum value of gain deviation is not affected by this term, since it is a constant; rather, the minimum is determined by the Q_p-passive and ω_p-active terms (and also, to some extent, by the terms not shown in the figure). It should be clear that changing the active or passive tolerances will cause the minimum to change. For instance, a decrease in the tolerance of $A_0\alpha$ will shift the ω_p-active curve down, resulting in a minimum at a higher value of β. Similarly, decreasing the passive tolerances shifts the Q_p-passive curve up, resulting in a minimum at a lower value of β.
4. An inspection of Figure 9.10 shows that near the optimum value of β, the gain deviation changes very gradually. For instance, if the resistance ratio were dropped to 10, $3\sigma(\Delta G)$ would increase by only 0.01 dB. Therefore, it is often quite reasonable to use a suboptimum β, if other design considerations (such as dynamic range, noise level) so dictate.
5. Fleischer [2] has shown that the optimum value of β is approximately

$$\beta \cong 4\frac{A_0\alpha}{\omega_p}\frac{\sigma_{R,C}}{\sigma_{A_0\alpha}} \tag{9.54}$$

where $\sigma_{R,C}$ and $\sigma_{A_0\alpha}$ represent the standard deviations of the passive elements (assumed equal) and the gain-bandwidth product, respectively. Substituting the values given in this example, we get

$$\beta \cong 4\frac{2\pi(10)^6}{2\pi(2000)}\frac{0.01}{0.5} = 40$$

which is reasonably close to the optimum value of $\beta = 42$ obtained above. Note that the form of Equation 9.54 confirms our comments on the dependence of β on $\sigma_{R,C}$ and $\sigma_{A_0\alpha}$. ■

9.5 THE FRIEND BIQUAD

In this section we consider a negative feedback circuit due to Friend [3] for the realization of the general biquadratic function. This circuit, shown in Figure 9.11*a*, is a generalization of the Delyiannis band-pass circuit with input ports provided for the formation of zeros, just as was done in Section 9.3.

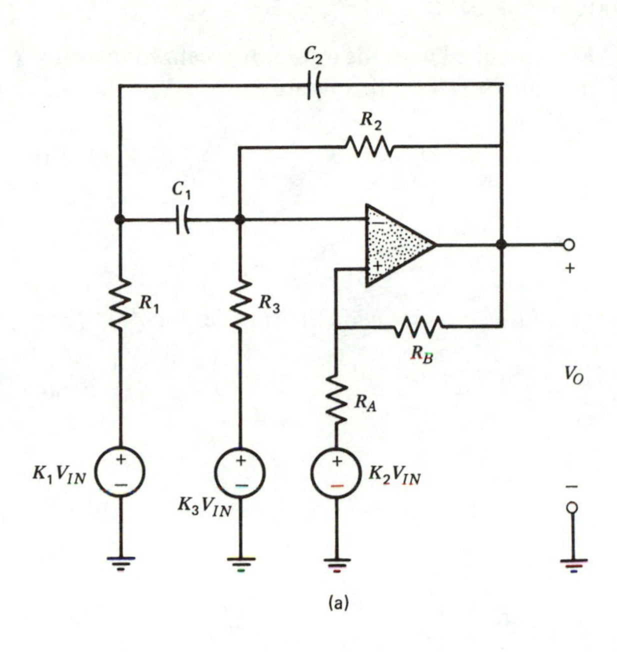

(a)

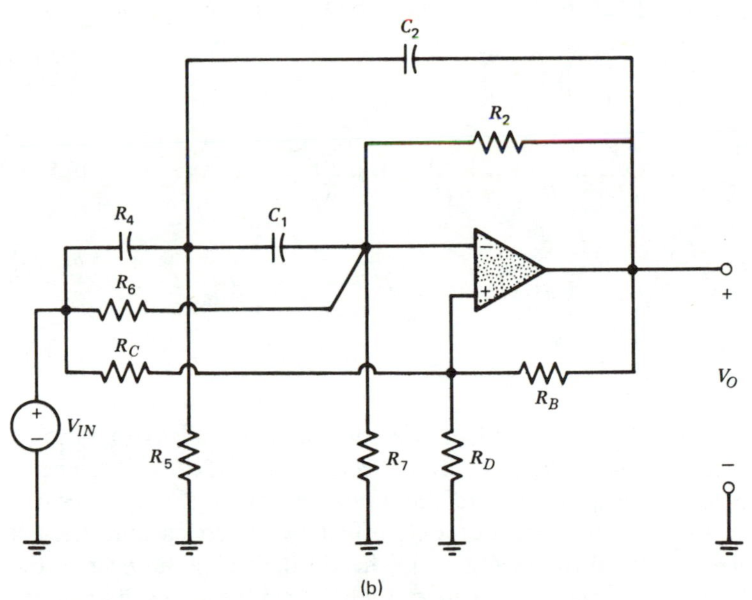

(b)

Figure 9.11 The Friend biquad circuit: (*a*) Showing voltage sources. (*b*) Showing realization of K_1, K_2, and K_3.

The constants K_1, K_2, and K_3 associated with the inputs are realized by using potential dividers, as shown in Figure 9.11*b*. In this figure

$$K_1 = \frac{R_5}{R_4 + R_5} \qquad K_2 = \frac{R_D}{R_C + R_D} \qquad K_3 = \frac{R_7}{R_6 + R_7} \tag{9.55a}$$

$$R_1 = \frac{R_4 R_5}{R_4 + R_5} \qquad R_A = \frac{R_C R_D}{R_C + R_D} \qquad R_3 = \frac{R_6 R_7}{R_6 + R_7} \tag{9.55b}$$

where $K_1, K_2, K_3 \leqq 1$. Analysis of this circuit yields the transfer function [3]:

$$\frac{V_O}{V_{IN}} = K\frac{ms^2 + cs + d}{s^2 + as + b} \tag{9.56}$$

where

$$K = 1 \tag{9.57a}$$

$$m = K_2 \tag{9.57b}$$

$$c = \frac{K_2}{C_2}\left(\frac{1}{R_1} + \frac{1}{R_2} + \frac{1}{R_3}\right) + \frac{K_2}{C_1}\left(\frac{1}{R_2} + \frac{1}{R_3}\right) - \frac{K_1}{R_1 C_2}\left(1 + \frac{R_A}{R_B}\right)$$

$$- \frac{K_3}{R_3}\left(\frac{1}{C_1} + \frac{1}{C_2}\right)\left(1 + \frac{R_A}{R_B}\right) \tag{9.57c}$$

$$d = \frac{1}{C_1 C_2}\left[\frac{K_2}{R_1}\left(\frac{1}{R_2} + \frac{1}{R_3}\right) - \frac{K_3}{R_1 R_3}\left(1 + \frac{R_A}{R_B}\right)\right] \tag{9.57d}$$

$$a = \frac{C_1 + C_2}{C_1 C_2}\left(\frac{1}{R_2} - \frac{R_A}{R_B R_3}\right) - \frac{R_A}{R_B R_1 C_2} \tag{9.57e}$$

$$b = \frac{1}{R_1 C_1 C_2}\left(\frac{1}{R_2} - \frac{R_A}{R_B R_3}\right) \tag{9.57f}$$

Comparing Equations 9.56 and 9.57, it is seen that there are five unknown coefficients and nine variables ($C_1, C_2, R_1, R_2, R_3, K_1, K_2, K_3, R_A/R_B$). Therefore, four of these variables can be fixed, and one resistor in the ratio term is arbitrary. As in the previous sections, for the first two fixed parameters, the capacitors are made equal ($C_1 = C_2 = 1$). The third fixed parameter is the scale factor K_3, which is chosen to ensure that the synthesis results in non-negative element values. The fourth fixed parameter is the ratio R_A/R_B, which is selected to yield a low circuit sensitivity and reasonable element spreads, just as was done in Section 9.4. In particular, the optimum value of R_A/R_B can be derived from the synthesis relations for the Delyiannis *band-pass* circuit, as follows. First, Equation 9.54 is used to obtain an approximate value for β, which is the optimum ratio (R_2/R_1) for the band-pass circuit:

$$\beta = \left(\frac{R_2}{R_1}\right)_{BP} \cong 4\frac{A_0\alpha}{\omega_p}\frac{\sigma_{R,C}}{\sigma_{A_0\alpha}} \tag{9.58}$$

This ratio is related to $(R_A/R_B)_{BP}$ by Equations 9.38 and 9.45b:

$$\left(\frac{R_A}{R_B}\right)_{BP} = \frac{1}{k-1} = \frac{2Q_p - \sqrt{\beta}}{\beta Q_p} \tag{9.59}$$

where $Q_p = \sqrt{b}/a$. This value of $(R_A/R_B)_{BP}$ (henceforth designated as γ) will result in a band-pass circuit that has a gain deviation that is close to the minimum. Based on the assumption that the zeros do not appreciably affect the passband gain deviation, we may assume that this ratio $(R_A/R_B)_{BP} = \gamma$ will be approximately equal to the optimum R_A/R_B for the realization of the general biquadratic of Equation 9.56.

With the above choice of fixed parameters:

$$C_1 = C_2 = 1 \qquad \frac{R_A}{R_B} = \gamma \qquad 0 \le K_3 \le 1$$

the expressions for the remaining elements can be evaluated from Equation 9.57. After much algebraic manipulation, these elements are found to be [3]:

$$R_1 = \frac{2\gamma}{(-a + \sqrt{a^2 + 8\gamma b})} \tag{9.60a}$$

$$K_1 = \frac{m + 2dR_1^2 - cR_1}{1 + \gamma} \tag{9.60b}$$

$$R_3 = \frac{(1+\gamma)(m - K_3)}{R_1 b(d/b - m)} \tag{9.60c}$$

$$R_2 = \frac{R_3}{R_1 R_3 b + \gamma} \tag{9.60d}$$

Finally, from Equations 9.55:

$$R_4 = \frac{R_1}{K_1} \qquad R_5 = \frac{R_4 R_1}{R_4 - R_1} \tag{9.61a}$$

$$R_6 = \frac{R_3}{K_3} \qquad R_7 = \frac{R_6 R_3}{R_6 - R_3} \tag{9.61b}$$

$$R_C = \frac{R_A}{m} \qquad R_D = \frac{R_C R_A}{R_C - R_A} \tag{9.61c}$$

Equations 9.60 and 9.61 yield all the element values for the general circuit of Figure 9.11. In some cases, however, a direct application of these equations will result in unrealizable element values. These situations can be circumvented by using the following artifices:

1. The fixed parameter K_3 should be chosen so that the resulting value of R_3, as given by (9.60c), is nonnegative. From Equation 9.55, the permitted range of values for K_3 is between 0 and 1. However, it can be shown that choosing K_3 at its extreme values ($K_3 = 0$ or $K_3 = 1$) results in a minimum circuit sensitivity [3].
2. To realize a gain constant K that is greater than 1, it is necessary to use gain enhancement at the output, as explained in Section 7.6.
3. If $d < b$ and $m = 1$, then R_3 becomes zero. In this case decrease the numerator coefficients m, c, and d by a small factor (say 1.1), and use gain enhancement at the output to increase the gain constant K of the transfer function.
4. If in the solution of (9.60b) K_1 is found to be greater than one, scale the numerator coefficients m, c, and d to make K_1 equal to unity, and increase K accordingly.
5. In the band-pass case $m = d = 0$, so K_1 becomes negative. In this case change the sign of the numerator coefficient c. Note that such a sign change does not affect $|T(s)|$.

With these artifices, it is always possible to attain nonnegative element values, with only one exception.* This exception occurs when the complex zeros are much to left of the complex poles, in the s plane. In most filter applications, however, the zeros are on the $j\omega$ axis, and the problem does not arise.

Example 9.4

Synthesize the following low-pass notch filter function using the Friend biquad circuit

$$T(s) = 2\frac{[s^2 + (2000)^2]}{s^2 + 100s + (1000)^2} \tag{9.62}$$

The elements for the circuit realization are described by

$$3\sigma(V_R) = 3\sigma(V_C) = 0.001$$

$$A_0\alpha = 2(10)^6 \qquad 3\sigma(V_{A_0\alpha}) = 0.5$$

Solution

Comparing (9.56) and (9.62), the biquadratic parameters are seen to be

$$K = 2 \qquad m = 1 \qquad c = 0 \qquad d = (2000)^2 \qquad a = 100 \qquad b = (1000)^2$$

* As in the circuit of Figure 9.7*b*, the Friend circuit cannot realize the low-pass function. An alternate negative feedback circuit for the realization of the *LP* function can be found in [3] (see Problem 9.15).

from which $\omega_p = 1000$ and $Q_p = 10$. The design parameter γ is obtained using 9.58 and 9.59:

$$\beta = 4\,\frac{2(10)^6(0.001)}{(10)^3(0.5)} = 16$$

$$\gamma = \frac{2(10) - \sqrt{16}}{16(10)} = 0.1$$

The parameter K_3 is chosen to be zero, to make R_3 nonnegative. Thus the fixed design parameters for the synthesis are

$$C_1 = C_2 = 1 \qquad \gamma = 0.1 \qquad K_3 = 0$$

The remaining elements can be obtained from (9.60) and (9.61). From (9.60a)

$$R_1 = 2.5(10)^{-4}$$

Again from (9.60b)

$$K_1 = 1.3636$$

But from Equation 9.55a we know that K_1 cannot exceed unity. Using the artifice (4) mentioned above, this problem is circumvented by decreasing all the numerator coefficients by the factor 1.3636, and correspondingly increasing K by this same factor. Then

$$m' = 0.7333 \qquad c' = 0 \qquad d' = 2.9333(10)^6 \qquad K' = 2.7272$$

With this modified set of coefficients the synthesis equations can be solved directly to yield the remaining element values. From Equation 9.60:

$$R_1 = 2.5(10)^{-4} \qquad K_1 = 1 \qquad R_3 = 1.467(10)^{-3} \qquad R_2 = 3.143(10)^{-3}$$

Again from (9.61a) and (9.61b),

$$R_4 = 2.5(10)^{-4} \qquad R_5 = \infty \qquad R_6 = \infty \qquad R_7 = 1.467(10)^{-3}$$

Impedance scaling by 10^8 to obtain practical element values, we get

$$C_1 = C_2 = 0.01\ \mu\text{F} \qquad R_2 = 314\ \text{k}\Omega \qquad R_4 = 25\ \text{k}\Omega$$

$$R_5 = R_6 = \infty \qquad R_7 = 147\ \text{k}\Omega$$

The resistors R_A and R_B must be chosen to realize the factor $\gamma = R_A/R_B = 0.1$. A practical choice is

$$R_A = 1\ \text{k}\Omega \qquad R_B = 10\ \text{k}\Omega$$

With this choice R_C and R_D are evaluated from Equation 9.61c to be

$$R_C = 1.36\ \text{k}\Omega \qquad R_D = 3.78\ \text{k}\Omega$$

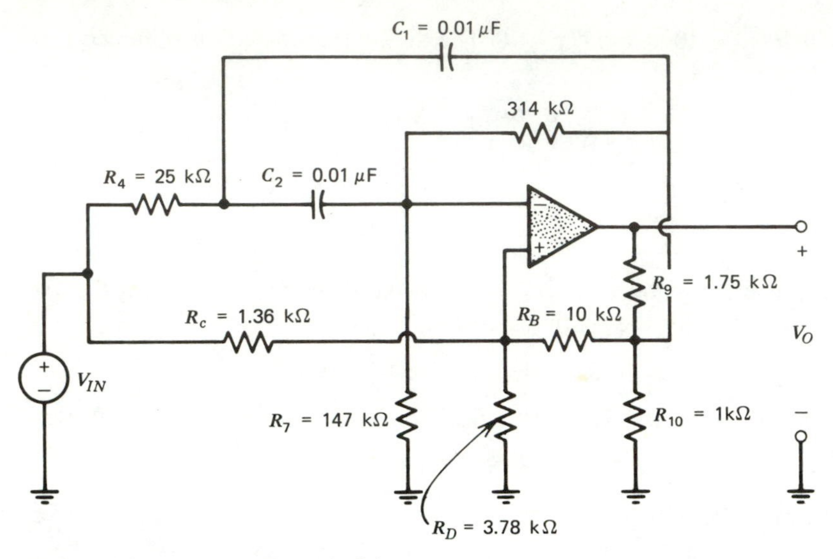

Figure 9.12 Circuit for Example 9.4.

Finally, the scale factor $K' = 2.7272$ is realized using the gain enhancement potential divider resistors R_9 and R_{10} where

$$K' = 1 + \frac{R_9}{R_{10}} = 2.7272$$

A practical choice for R_9 and R_{10} is

$$R_9 = 1.73 \text{ k}\Omega \qquad R_{10} = 1 \text{ k}\Omega$$

This completes the synthesis, and the required circuit is shown in Figure 9.12.

Observations

1. The ratio R_2/R_1 realized is 12.6. Comparing this with the approximate optimum $\beta = 16$ given by Equation 9.58, and considering the shallowness of the gain deviation versus β curve near the optimum, it is seen that the gain deviation of this circuit will be close enough to the minimum, for all practical purposes.
2. The resistive gain enhancement technique does introduce some error, as was mentioned in Section 7.6, since it assumes the impedance of the potential divider resistors to be negligible compared to the input impedance of the RC circuit. An alternate, and exact, scheme which does not rely on such an assumption (but does need one extra capacitor) is described in [3].* ■

* This is referred to as element-splitting gain enhancement (see Problem 7.21).

9.6 COMPARISON OF SENSITIVITIES OF NEGATIVE AND POSITIVE FEEDBACK CIRCUITS

In this section we compare the sensitivities of the negative feedback circuits described in this chapter with the positive feedback circuits discussed in Chapter 8. By considering the contributions of the passive and active terms to the total gain deviations, we are able to determine the most suitable circuit for a given filter application.

The circuits we compare are:

Design A. Positive feedback Sallen and Key circuit for $k = 1$ (Example 8.1, Design 1).
Design B. Saraga's positive feedback design with $k = 4/3$ (Example 8.1, Design 3).
Design C. The basic negative feedback circuit (Example 9.2).
Design D. The Delyiannis negative feedback circuit (Example 9.3).

CASE 1 PASSIVE TERMS DOMINATE

Let us first consider the case when the pole frequency is low enough so that the gain deviation is essentially determined by the passive terms. Designs A and C yield passive element spreads that are of the order of $4Q_p^2$—therefore, these designs are immediately ruled out in all but those applications requiring very low pole Q's. The remaining two designs, due to Saraga and Delyiannis, are next compared.

At the frequency of maximum gain deviation ($\Omega = 1 \pm 1/2Q_p$), the contribution of the passive terms to the total gain deviation for the Saraga design was shown to be (Equation 8.34a):

$$\text{Saraga: } [3\sigma(\Delta G)]^2_{\text{passive}} \cong 1.5\{8.686Q_p[3\sigma(V_{R,C})]\}^2 \tag{9.63}$$

In the derivation of this equation, we used the approximate relationships

$$\mathcal{S}^G_{\omega_p} \cong \pm 8.686Q_p \qquad \mathcal{S}^G_{Q_p} = \tfrac{1}{2}(8.686) \tag{9.64}$$

at $\Omega = 1 \pm 1/2Q_p$. Moreover, it was assumed that the contribution of the numerator terms to the total gain deviation was negligible. The passive term contribution for the Delyiannis circuit is obtained as in Example 8.3 (Observation 4). From (9.46), assuming $C_1 = C_2 = 1$ and $Q_p \gg 1$, we have

$$\sum_i [(S^{\omega_p}_{R_i})^2 + (S^{\omega_p}_{C_i})^2] = 1 \tag{9.65a}$$

$$\sum_i [(S^{Q_p}_{R_i})^2 + (S^{Q_p}_{C_i})^2] \cong \frac{18}{\beta} Q_p \tag{9.65b}$$

Substituting in Equation 8.15, the contribution of the passive terms for the Delyiannis circuit is

$$\text{Delyiannis: } [3\sigma(\Delta G)]^2_{\text{passive}} = \left(1 + \frac{4.5}{\beta}\right)\{8.686Q_p[3\sigma(V_{R,C})]\}^2 \quad (9.66)$$

To minimize this expression, $\beta = R_2/R_1$ should be chosen as large as possible, within the constraints of manufacturability. A comparison of (9.66) with (9.63) shows that the passive term for the Delyiannis design will be lower than for the Saraga design if

$$1 + \frac{4.5}{\beta} < 1.5$$

or

$$\beta > 9$$

Since this inequality can always be satisfied, it can be concluded that at low frequencies the Delyiannis design is the best choice.

In the above discussion Designs A and C were not considered because they result in large element spreads (of the order of $4Q_p^2$). However, for low pole Q's (say, $Q_p < 5$) the element spreads may be attainable, thereby making these designs viable alternates. In both these designs, the element sensitivities satisfy the relationships:

$$\sum_i [(S_{R_i}^{\omega_p})^2 + (S_{C_i}^{\omega_p})^2] = 1 \qquad \sum_i [(S_{R_i}^{Q_p})^2 + (S_{C_i}^{Q_p})^2] = \tfrac{1}{2} \quad (9.67)$$

Thus the contribution of the passive term is

$$\text{Designs A and C: } [3\sigma(\Delta G)]^2 \cong 1\{8.686Q_p[3\sigma(V_{R,C})]\}^2 \quad (9.68)$$

which is seen to be lower than the passive term contributions of the Delyiannis and Saraga designs. Thus, if the pole frequency is low enough that the passive terms dominate, and if the pole Q's are low enough that the element spreads can be attained, the best choices are the $k = 1$ positive feedback design and the basic negative feedback circuit.

CASE 2 ACTIVE TERMS DOMINATE

Next, let us consider the case when the pole frequency is high enough so that the passive terms are negligible in comparison with the active term. Designs A and C are quickly dispensed with on the basis of the results of Example 8.3, 9.2, and 9.3 from which it is seen that these designs yield a much higher active term than do the Saraga and Delyiannis designs. We therefore restrict our discussions to the latter two designs.

From Equation 8.34b, the active term for the Saraga design is

$$\text{Saraga: } [3\sigma(\Delta G)]^2_{\text{active}} = 2.37\left\{8.686Q_p \frac{\omega_p}{A_0\alpha}[3\sigma(V_{A_0\alpha})]\right\}^2 \tag{9.69}$$

Considering the Delyiannis design for $C_1 = C_2 = 1$, the sensitivities of ω_p and Q_p to $A_0\alpha$ are (Equation 9.53)

$$S^{\omega_p}_{A_0\alpha} = -S^{Q_p}_{A_0\alpha} = \frac{1}{2}\frac{\omega_p}{A_0\alpha}\frac{k^2}{(k-1)^2}\sqrt{\beta} \tag{9.70a}$$

where

$$k = \frac{Q_p(\beta + 2) - \sqrt{\beta}}{2Q_p - \sqrt{\beta}} \tag{9.70b}$$

For $Q_p \gg 1$, the term $k/(k-1)$ is approximately

$$\frac{k}{k-1} \cong 1 + \frac{2}{\beta} \tag{9.71}$$

Substituting this in (9.70a), we get

$$S^{\omega_p}_{A_0\alpha} = -S^{Q_p}_{A_0\alpha} = \frac{1}{2}\frac{\omega_p}{A_0\alpha}\left[\sqrt{\beta}\left(1 + \frac{2}{\beta}\right)^2\right] \tag{9.72}$$

It can easily be shown that this expression has a minimum value at $\beta = 6$. For this value of β

$$S^{\omega_p}_{A_0\alpha} = -S^{Q_p}_{A_0\alpha} = 2.177\frac{\omega_p}{A_0\alpha} \tag{9.73}$$

and the corresponding contribution of the active terms to the gain deviation, from Equation 8.15, is

$$\text{Delyiannis: } [3\sigma(\Delta G)]^2_{\text{active}} \cong 4.74\left\{8.686Q_p \frac{\omega_p}{A_0\alpha}[3\sigma(V_{A_0\alpha})]\right\}^2 \tag{9.74}$$

Comparing this equation with (9.69), we conclude that when the pole frequency is high and the active terms dominate, the Saraga design is the best choice.*

In summary, the choice of the least sensitive design depends on the pole Q and pole frequency. It was shown that:

- For low pole frequencies, when the passive terms dominate, the Delyiannis design is the least sensitive. Furthermore, if the pole Q is low enough that the element spreads can be attained, then the $k = 1$ positive feedback

* The above equations apply to op amps characterized by a single pole at the origin. Similar results may be derived for other op amp gain characteristics, using the more general analytical techniques derived in Chapter 12.

design and the basic negative feedback designs provide the least sensitive circuits.

- For high pole frequencies, when the active terms dominate, the Saraga ($k = 4/3$) design is the least sensitive.
- If neither the passive nor the active terms dominate, all the designs will need to be analyzed to determine the least sensitive one.

Before concluding this section, it should be mentioned that while sensitivity is a very important consideration, it is not the only criterion used in choosing between designs. Depending on the application, other factors that may need to be considered are (a) element spreads, (b) the number of elements used, (c) the classes of filter functions provided, and (d) tunability (i.e., how difficult it is to adjust the elements to achieve the nominal design).

9.7 CONCLUDING REMARKS

In this chapter it was demonstrated that the Friend negative feedback circuit is capable of realizing a large variety of filter functions. Moreover this circuit is canonic, in the sense that it uses the minimum number of capacitors (two) and only one op amp to realize the general biquadratic function. These features have made the circuit economically attractive from a manufacturing standpoint [3].

The positive feedback option in this circuit allows the designer to find the best compromise considering element spreads and sensitivities to active and passive components. A comparison with other biquad circuits shows that this circuit is among the least sensitive. In particular, for low pole frequencies it was shown to be the best choice of the biquad circuits considered thus far.

FURTHER READING

1. T. Delyiannis, "High-Q factor circuit with reduced sensitivity," *Electronics Letters*, 4, December 1968, p. 577.
2. P. E. Fleischer, "Sensitivity Minimization in a Single Amplifier Biquad Circuit," *IEEE Trans. Circuits and Systems*, *CAS-23*, No. 1, January 1976, pp. 45–55.
3. J. J. Friend, C. A. Harris and D. Hilberman, "STAR: An active biquadratic filter section," *IEEE Trans. Circuits and Systems*, *CAS-22*, No. 2, February 1975, pp. 115–121.
4. G. S. Moschytz, *Linear Integrated Networks Design*, Van Nostrand, New York, 1975, Chapter 2.
5. A. S. Sedra, "Generation and classification of single amplifier filters," *Int. J. of Circuit Theory and Applications*, *2*, March 1974, pp. 1–57.
6. B. A. Shenoi, "Optimum variability design and comparative evaluation of thin-film RC active filters," *IEEE Trans. Circuits and Systems*, *CAS-21*, 1970, pp. 263–267.

PROBLEMS

9.1 *Band-pass, basic negative feedback.* Synthesize the band-pass function

$$\frac{400s}{s^2 + 400s + 1.024(10)^7}$$

using the basic negative feedback circuit of Figure 9.3a, with practical element values. Determine the maximum resistance spread.

9.2 Synthesize the function

$$\frac{400s^2}{(s^2 + 400s + 1.024(10)^7)(s + 1000)}$$

without using extra elements for adjusting the gain constant.

9.3 *ΔG expression, basic negative feedback BP.* Derive a general expression for the standard deviation of the gain change $3\sigma(\Delta G)$ for the basic negative feedback band-pass circuit (Figure 9.3*a*) at the 3 dB passband edge frequencies. Use the design formula given by Equations 9.10 and make reasonable approximations, assuming $Q_p \gg 1$. The answer should be the same as that for the Sallen and Key *LP* circuit for $k = 1$ (Design 1, Problem 8.7). The result shows that these two circuits have similar sensitivities.

9.4 *Delay equalizer, using positive terminal input.* Synthesize the delay equalizer biquadratic

$$\frac{1}{2}\left(\frac{s^2 - 5s + 100}{s^2 + 5s + 100}\right)$$

using the basic negative feedback circuit with the V_{POS} input (Figure 9.6).

9.5 *Band-pass, Delyiannis circuit.* Synthesize the band-pass function of Problem 9.1 with the Delyiannis circuit using the maximum resistance ratio, which is given to be 100.

9.6 Synthesize the band-pass function of Problem 9.1 using the optimum β as given by Fleischer's formula (Equation 9.54), assuming component tolerances as in Example 9.2.

9.7 Synthesize a second-order band-pass filter to have a center frequency at 100 Hz, pole Q of 10, and a gain of 10 at the center of the passband. Use the Delyiannis circuit, and assume the component tolerances of Example 9.2.

9.8 *Band-pass design, Delyiannis.* Synthesize the band-pass function of Example 9.2 using the Delyiannis circuit, assuming the manufacturing tolerances for the resistors and capacitors are ± 0.1 percent (the limits refer to the 3σ points of a Gaussian distribution). Pick β according to

Fleischer's formula. Also compute the statistics of the gain deviation at 2.1 kHz.

9.9 *Root locus, Delyiannis.* The basic negative feedback circuit of Figure 9.3*a* is used to synthesize a band-pass filter with a pole Q of 2 and a pole frequency of 10. Now, positive feedback is introduced in the circuit using a potential divider, as in Figure 9.9. Plot the locus of the poles of the band-pass function versus $k = 1 + R_B/R_A$. For what value of k will the poles be on the imaginary axis?

9.10 *Band-pass design equations.* Determine the transfer function for the circuit shown, and derive a set of design equations to realize the band-pass function

$$\frac{Ks}{s^2 + as + b}$$

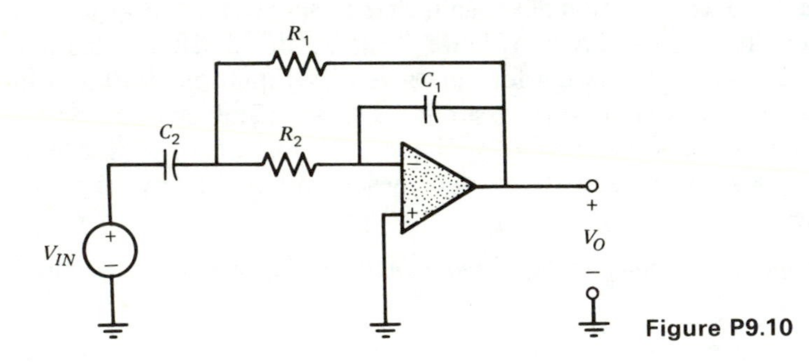

Figure P9.10

9.11 By introducing an additional input to the positive terminal of Figure P9.10, the delay equalizer function

$$K\frac{s^2 - as + b}{s^2 + as + b}$$

can be realized. Sketch the complete circuit and derive the design equations.

9.12 *High-pass notch, Friend.* Synthesize the high-pass notch filter function

$$\frac{0.5s^2 + (2000)^2}{s^2 + 500s + (4000)^2}$$

using the Friend negative feedback circuit, assuming $R_A/R_B = 0.1$.

9.13 *Friend circuit, no positive feedback.* For low-pole Q circuits, the Friend circuit without positive feedback is often used. Derive the design equations for this case from the design equations for the Friend circuit with positive feedback.

9.14 *Delay Equalizer.* Synthesize the two section delay equalizer function described in Section 4.6, using the Friend circuit with no positive feedback.

9.15 *Design equations for Friend low-pass circuit* [3]. Show that the low-pass transfer function realized by using the RC circuit shown in Figure P9.15 in the modified negative feedback topology of Figure 9.8a is

$$\frac{-\dfrac{1}{R_2R_3C_1C_2}\left(1+\dfrac{R_A}{R_B}\right)}{s^2+s\left(\dfrac{1}{R_1C_1}+\dfrac{1}{R_2C_1}+\dfrac{1}{R_3C_1}-\dfrac{R_A}{R_B}\dfrac{1}{R_2C_2}\right)+\left(\dfrac{1}{R_1R_2C_1C_2}-\dfrac{R_A}{R_B}\dfrac{1}{R_2R_3C_1C_2}\right)}$$

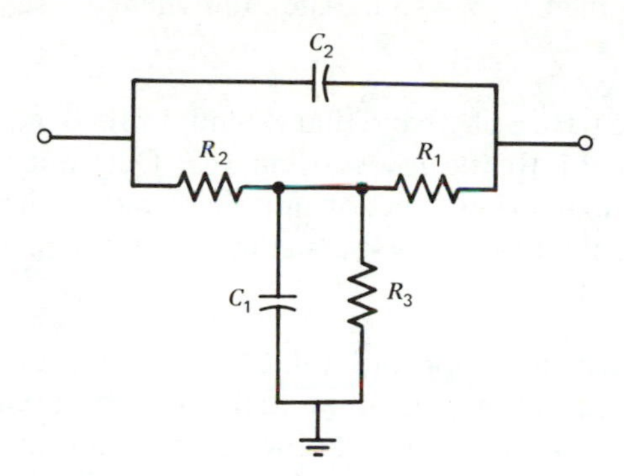

Figure P9.15

If the fixed elements are chosen as $C_1 = 1$, $C_2 = 0.1$, and $R_A/R_B = \gamma$, show that the design equations for synthesizing the low-pass biquadratic

$$\frac{-d}{s^2+as+b}$$

are given by

$$R_2 = \frac{2(1-10\gamma)}{a \pm \sqrt{a^2 - 0.4(b+d)(1-10\gamma)}}$$

$$R_1 = \frac{10}{R_2\left(b + d\dfrac{\gamma}{1+\gamma}\right)}$$

$$R_3 = \frac{10(1+\gamma)}{dR_2}$$

As in the general Friend circuit, the optimum value for γ depends on the component tolerances. Equation 9.59 and 9.54 will yield a γ value that is close to the optimum.

9.16 Synthesize the low-pass function

$$-\frac{10^6}{s^2 + 200s + 10^6}$$

using the Friend low-pass circuit of Problem 9.15 (assume $R_A/R_B = 0.2$).

9.17 Synthesize the low-pass function

$$\frac{10^6}{s^2 + 50s + 10^6}$$

using the Friend circuit of Problem 9.15. Assume the component tolerances are as in Example 9.4.

9.18 A low-pass filter is required to have a passband that extends to 1000 rad/sec with a maximum attenuation of 1 dB, the *dc* loss being 0 dB. Determine the third-order Chebyshev approximation function for these requirements. Synthesize the function using the Friend low-pass circuit of Problem 9.15, with no positive feedback.

9.19 *Tone-separation filters.* In the discussion of TOUCH–TONE® dialing in Section 3.1.3, we described a low-pass filter that could be used to separate the low-band tones from the high-band tones. Suppose the requirements for the low-pass filter can be met by a fourth-order Chebyshev approximation function which has a passband ripple of $A_{max} = 1$ dB, the minimum loss in the passband being 0 dB. Synthesize the required low-pass function using the Friend circuit of Problem 9.15, with $R_A/R_B = 0.2$. Determine the attenuation on the closest high-band tone, which is at 1209 Hz.

9.20 *Voice-frequency low-pass filter.* In many voice-communication systems a low-pass filter is needed to isolate the voice band of frequencies (from *dc* to 4 kHz) from higher frequency tones and noise. One such filter, designed

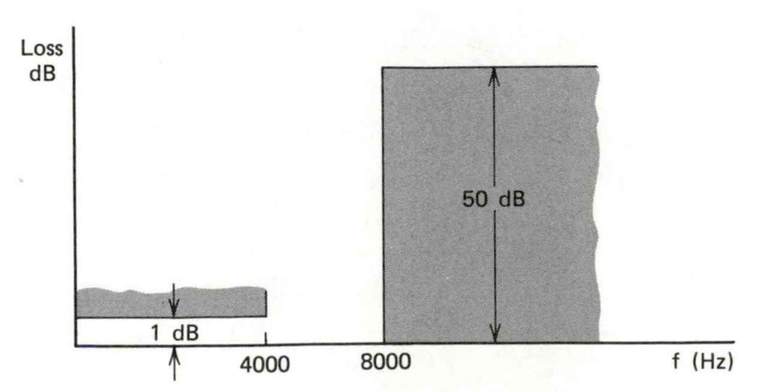

Figure P9.20

to meet the requirements sketched in Figure P9.20, has the transfer function

$$\frac{s^2 + 2.9(10)^9}{s^2 + 6.1(10)^3 s + 6.25(10)^8} \cdot \frac{s^2 + 1.5(10)^{10}}{s^2 + 1.65(10)^4 s + 2.0(10)^8}$$

Synthesize this function using the Friend circuit without positive feedback.

9.21 *Band-reject filter for 60 Hz hum.* A band-reject filter is required to remove an objectionable 60 Hz hum associated with the power supply in an audio application. The filter must pass frequencies below 55 Hz and above 65 Hz with at most 3 dB attenuation, and the *dc* loss must be 0 dB. Synthesize the required function using the Friend circuit, assuming the passive component tolerances have Gaussian distributions with $3\sigma(V_R) = 3\sigma(V_C) = 0.001$. Assume the gain of the op amp is $A(s) = A_0\alpha/s$ where $A_0\alpha = 10^6$ and $3\sigma(V_{A_0\alpha}) = 0.5$.

9.22 *Delay equalizer.* Synthesize the delay equalizer function needed to equalize the passband delay of a fourth-order Chebyshev *LP* filter which has a cutoff frequency of 1000 rad/sec and a passband ripple of 0.25 dB. The required transfer function can be derived from the function given in Section 4.6. Use the Friend circuit with $\gamma = .1$ for the synthesis.

9.23 *First order all-pass.* A first-order all-pass section characterized by

$$-K\frac{s-a}{s+a}$$

can be realized using the circuit shown. Develop the design equations. Use element-splitting gain enhancement to realize a gain constant of unity.

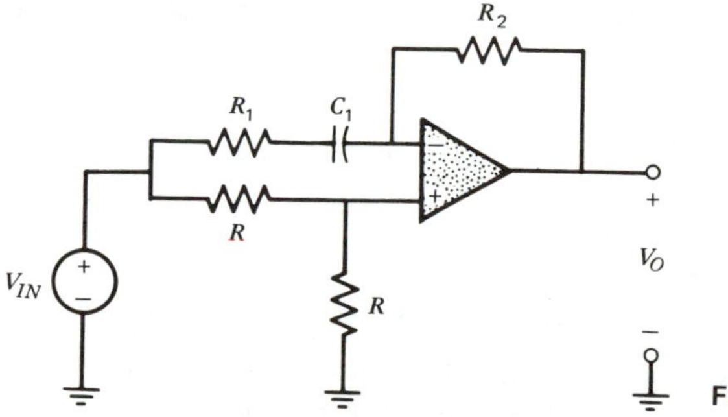

Figure P9.23

9.24 *Complete low-pass design.* The *RC* circuit of Figure P9.15 is to be used in the basic negative feedback topology of Figure 9.1 to realize a low-pass transfer function:

(a) Derive an expression for the transfer function V_O/V_{IN}, assuming an ideal op amp.

(b) Determine the synthesis equations given the choice $R_1 = R_2 = 0.01R_3 = 1$ for the fixed elements.

(c) Find expressions for the sensitivities of the pole frequency, pole Q, and gain constant to the passive elements. Use the design formula from part (b) to express these sensitivities in terms of the pole Q.

(d) Find expressions for the pole Q and pole frequency if the gain of the op amp is $A(s) = A_0\alpha/s$. Show that for $Q_p \gg 1$

$$S_{A_0\alpha}^{\omega_p} \cong -S_{A_0\alpha}^{Q_p} = \frac{\omega_p Q_p}{A_0\alpha}$$

(e) Use the sensitivity relationships from (b) and (d) to obtain a general expression for the statistics of the gain deviation at the 3 dB passband edge frequencies.

9.25 *Comparison of single amplifier biquads.* Determine which of the four amplifier designs discussed in Section 9.6 results in the lowest passband sensitivity for the realization of a biquadratic characterized by the denominator:

$$s^2 + (1000\pi)s + (2\pi)^2(10)^8$$

The passive components are characterized by $3\sigma(V_R) = 3\sigma(V_C) = 0.005$, and the maximum capacitor and resistor ratios may not exceed 100. The op amp gain is $A(s) = A_0\alpha/s$ where $A_0\alpha = 2\pi 10^6$ rad/sec, and $3\sigma(V_{A_0\alpha}) = 0.5$.
You may use the approximate expressions for gain deviation which were developed in Section 9.6 for the Saraga and Delyiannis designs, in Problem 8.7 for the $k = 1$ positive feedback design, and in Problem 9.3 for the basic negative feedback design.

9.26 Repeat Problem 9.25 for the denominator functions:
(a) $s^2 + (100\pi)s + (2\pi 10^3)^2$
(b) $s^2 + (1000\pi)s + (\pi 10^4)^2$
(c) $s^2 + (500\pi)s + (\pi 10^3)^2$

9.27 *Transfer functions by inspection.* Determine the types of biquadratic functions that can be realized by using the RC circuit shown in Figure P9.27 in the basic negative feedback topology. The answer can be obtained by considering the low and high frequency asymptotes of the feedforward and feedback transfer functions of the RC network.

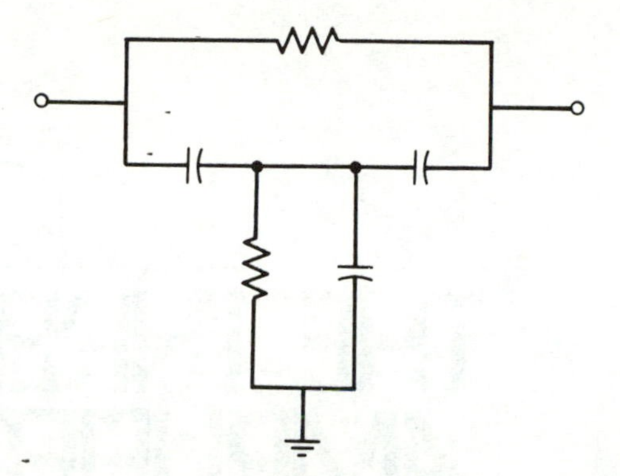

Figure P9.27

9.28 Repeat Problem 9.27 for the *RC* circuit shown in Figure P9.28.

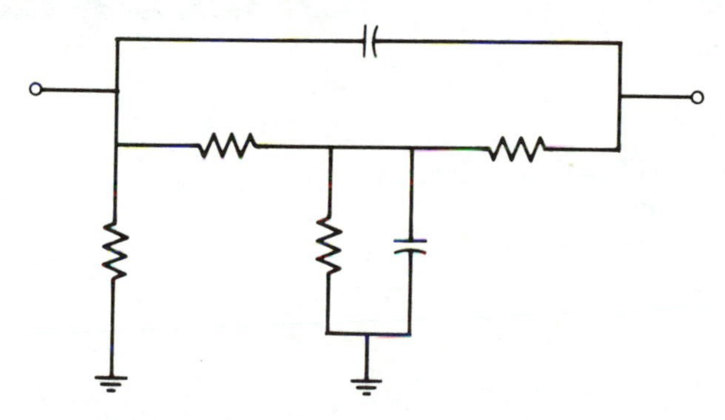

Figure P9.28

10.

THE THREE AMPLIFIER BIQUAD

In this chapter we present a topology that uses three op amps for the realization of the biquadratic function. The topology is known as the three amplifier biquad, the state-variable-biquad [4], or just the BIQUAD [7]. Sometimes, for convenience, a fourth amplifier is added.

In spite of its requiring more op amps than the single amplifier structures, this topology is quite popular, because of several desirable features it has to offer. In particular, the topology can realize the general biquadratic function

$$T(s) = K\frac{ms^2 + cs + d}{ns^2 + as + b} \qquad \begin{pmatrix} m = 1 \text{ or } 0 \\ n = 1 \text{ or } 0 \end{pmatrix} \tag{10.1}$$

with no exceptions. Therefore it allows the realization of *all* the filter functions—*LP*, *HP*, *BP*, *BR*, and delay equalizers. As mentioned in the last chapter, the manufacture of filters becomes more economical with such a universal circuit. Another salient feature is that, as part of the manufacture, the circuit can easily be tuned to match the nominal requirements. In addition, the sensitivity of the circuit is reasonably low. The ease of tuning and low sensitivity allow the three amplifier biquad to realize high Q filters with stringent requirements. One other distinguishing feature, useful in some applications, is that it permits the simultaneous realization of a variety of filter functions with the addition of a minimal number of components. In these so-called multifilter applications, the three amplifier biquad is the only contender.

10.1 THE BASIC LOW-PASS AND BAND-PASS CIRCUIT

In this section we describe a three amplifier biquad [7]*, which is used to realize the low-pass and band-pass filter functions. The realization of the general biquadratic function of Equation 10.1 will be covered in the next section.

Consider first the realization of the low-pass function

$$T_{LP}(s) = \frac{V_O}{V_{IN}} = \frac{-d}{s^2 + as + b} \tag{10.2}$$

The approach used in the synthesis is to rearrange the given transfer function so that it lends itself to a realization using elementary op amp circuit blocks,

* Similar three amplifier biquad structures are described in [1], [4], and [6].

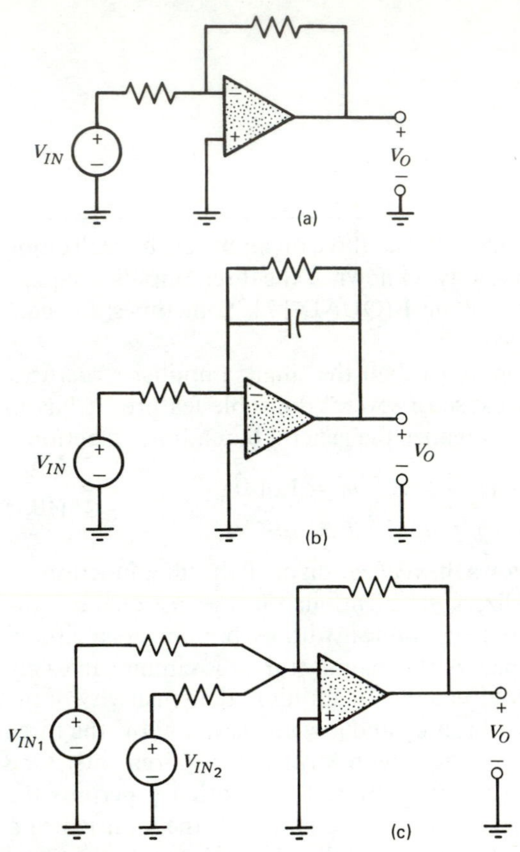

Figure 10.1 Circuit blocks for basic three amplifier biquad: (*a*) Inverter. (*b*) Leaky integrator. (*c*) Summer.

such as the inverter, the leaky integrator, and the summer (Figure 10.1). Note that these blocks are also used in analog computer simulations of dynamic systems. As in analog computer simulations, Equation 10.2 is first rearranged in the following way:

$$(s^2 + as + b)V_O = -dV_{IN}$$

$$\left(s + a + \frac{b}{s}\right)V_O = -\frac{dV_{IN}}{s}$$

$$\left(1 + \frac{b}{s(s+a)}\right)V_O = -\frac{d}{s(s+a)}V_{IN}$$

$$V_O = -\frac{b}{s(s+a)}V_O - \frac{d}{s(s+a)}V_{IN} \tag{10.3}$$

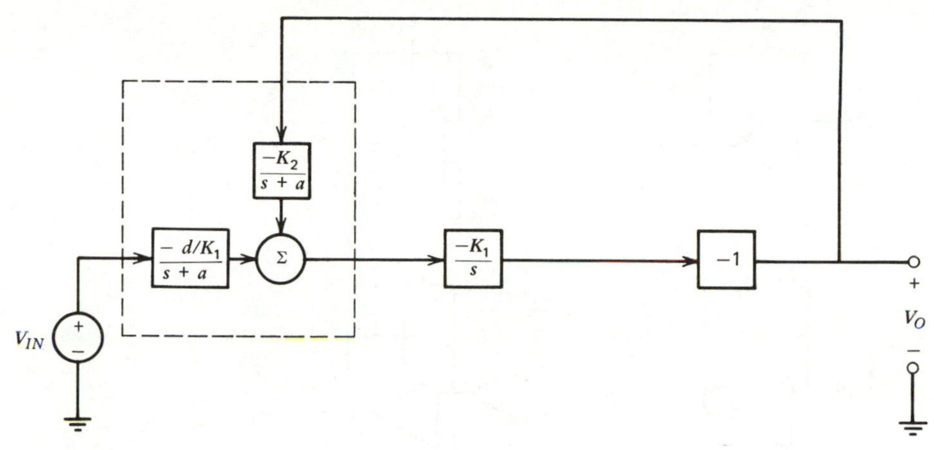

Figure 10.2 Block diagram representation of Equation 10.4.

This equation may be written in the form

$$V_O = (-1)\left(-\frac{K_1}{s}\right)\left[\left(-\frac{K_2}{s+a}\right)V_O + \left(-\frac{d/K_1}{s+a}\right)V_{IN}\right] \qquad (10.4)$$

where $b = K_1 K_2$. A block diagram representation of this last equation is shown in Figure 10.2. From this figure it is seen that the functional blocks that need to be realized are:

(a) -1 (10.5a)

(b) $-\dfrac{K_1}{s}$ (10.5b)

(c) $-\dfrac{K_2}{s+a}$ and $-\dfrac{d/K_1}{s+a}$ (10.5c)

Also, a means must be provided for summing the two blocks in (10.5c). The function (10.5a) is realized by the simple inverter circuit of Figure 10.3*a*, where $R_5 = R_6 = R$. The function (10.5b), representing an integration, is realized using the circuit shown in Figure 10.3*b*. This circuit has the desired transfer function:

$$T(s) = -\frac{\dfrac{1}{R_2 C_2}}{s} \qquad (10.6)$$

with $K_1 = 1/R_2 C_2$. The functions (10.5c) are realized by using the leaky

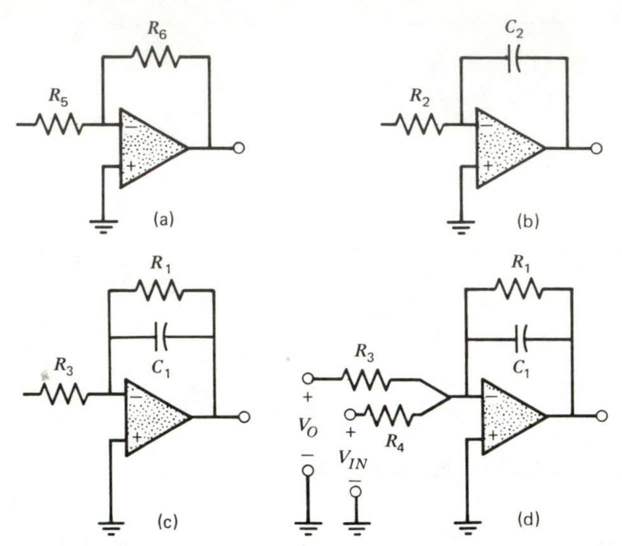

Figure 10.3 Realization of circuit blocks for three amplifier biquad: (*a*) Inverter. (*b*) Integrator. (*c*) Leaky integrator. (*d*) Summing-leaky-integrator.

integrator shown in Figure 10.3*c*, which can easily be seen to have the transfer function:

$$T(s) = \frac{-\dfrac{1}{R_3 C_1}}{s + \dfrac{1}{R_1 C_1}} \tag{10.7}$$

Finally, the required summation is readily attained by connecting another resistor to the op amp negative input terminal, as shown in Figure 10.3*d*. The resulting output voltage of this circuit is easily seen to be

$$V_1 = \frac{-\dfrac{1}{R_3 C_1}}{s + \dfrac{1}{R_1 C_1}} V_O + \frac{-\dfrac{1}{R_4 C_1}}{s + \dfrac{1}{R_1 C_1}} V_{IN} \tag{10.8}$$

Substituting the op amp circuit realizations of Equations 10.5 in the block diagram of Figure 10.2, we get the circuit shown in Figure 10.4. The output voltage in this circuit is:

$$V_3 = (-1)\left(\frac{-\dfrac{1}{R_2 C_2}}{s}\right)\left[\frac{-\dfrac{1}{R_3 C_1}}{s + \dfrac{1}{R_1 C_1}} V_3 + \frac{-\dfrac{1}{R_4 C_1}}{s + \dfrac{1}{R_1 C_1}} V_{IN}\right] \tag{10.9}$$

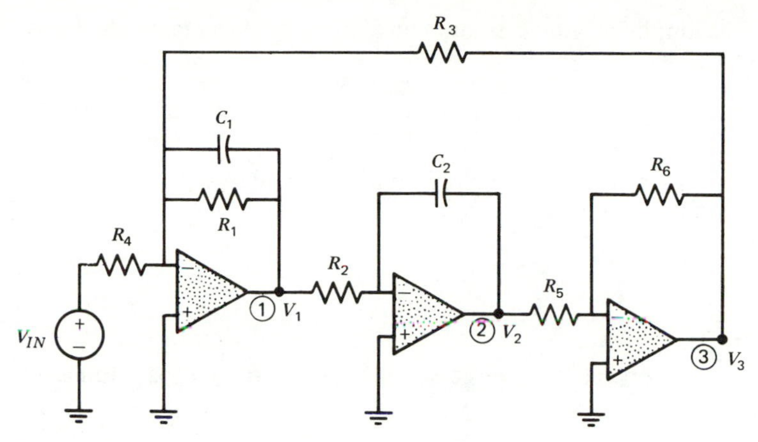

Figure 10.4 Circuit realization of *LP* and *BP* functions.

This equation easily simplifies to the desired low-pass function of Equation 10.2:

$$\frac{V_3}{V_{IN}} = \frac{-\dfrac{1}{R_2 R_4 C_1 C_2}}{s^2 + \dfrac{1}{R_1 C_1} s + \dfrac{1}{R_2 R_3 C_1 C_2}} \tag{10.10}$$

The element values of the low-pass circuit are obtained by matching the coefficients of Equation 10.2 and 10.10. From these two equations

$$b = \frac{1}{R_2 R_3 C_1 C_2} \tag{10.11a}$$

$$d = \frac{1}{R_2 R_4 C_1 C_2} \tag{10.11b}$$

$$a = \frac{1}{R_1 C_1} \tag{10.11c}$$

If the fixed elements are chosen as

$$C_1 = 1 \qquad C_2 = 1 \qquad R_2 = R_3 = R \tag{10.12a}$$

the remaining elements, from Equation 10.11, are

$$R_1 = \frac{1}{a} \qquad R_2 = R_3 = \frac{1}{\sqrt{b}} \qquad R_4 = \frac{\sqrt{b}}{d} \tag{10.12b}$$

This completes the synthesis of the low-pass function.

The basic three amplifier biquad circuit can also be used to realize the band-pass function:

$$T_{BP}(s) = \frac{-cs}{s^2 + as + b} \tag{10.13}$$

This is achieved by taking the output voltage at node 1, as shown in the following. In Figure 10.4, V_1 is related to V_3 by

$$V_3 = \frac{\frac{1}{R_2 C_2}}{s} V_1 \tag{10.14}$$

Substituting in Equation 10.10, we directly get the desired band-pass function:

$$\frac{V_1}{V_{IN}} = \frac{-\frac{1}{R_4 C_1} s}{s^2 + \frac{1}{R_1 C_1} s + \frac{1}{R_2 R_3 C_1 C_2}} \tag{10.15}$$

To synthesize the band-pass function, this equation is compared with (10.13) to yield the following relationships:

$$c = \frac{1}{R_4 C_1} \tag{10.16a}$$

$$a = \frac{1}{R_1 C_1} \tag{10.16b}$$

$$b = \frac{1}{R_2 R_3 C_1 C_2} \tag{10.16c}$$

As in the low-pass case, the fixed elements are chosen as

$$C_1 = 1 \qquad C_2 = 1 \qquad R_2 = R_3 = R \tag{10.17a}$$

Then, the remaining elements are given by

$$R_1 = \frac{1}{a} \qquad R_2 = R_3 = \frac{1}{\sqrt{b}} \qquad R_4 = \frac{1}{c} \tag{10.17b}$$

10.2 REALIZATION OF THE GENERAL BIQUADRATIC FUNCTION

In this section we describe two methods for the realization of the general biquadratic function of Equation 10.1. The first is based on the summation of the voltages already available in the basic circuit developed in the last section. This summation requires an extra summing amplifier. The resulting circuit will be referred to as the **summing four amplifier biquad**. The second method uses the feedforward scheme, developed in Chapters 8 and 9, in which the zeros are formed by introducing the input signal at appropriate nodes in the basic three amplifier circuit. This circuit will be called the **feedforward three amplifier biquad**.

10.2.1 THE SUMMING FOUR AMPLIFIER BIQUAD

In the last section it was shown that the voltage at node 1 of the basic three amplifier circuit (Figure 10.4) yields the band-pass function:

$$V_1 = V_{BP} = \frac{-\dfrac{1}{R_4 C_1} s}{s^2 + \dfrac{1}{R_1 C_1} s + \dfrac{1}{R_2 R_3 C_1 C_2}} V_{IN} \tag{10.18}$$

while node 3 exhibits the low-pass function:

$$V_3 = V_{LP} = \frac{-\dfrac{1}{R_2 R_4 C_1 C_2}}{s^2 + \dfrac{1}{R_1 C_1} s + \dfrac{1}{R_2 R_3 C_1 C_2}} V_{IN} \tag{10.19}$$

The voltage at node 2 is the same as that at node 3 with the sign reversed, that is,

$$V_2 = -V_{LP} \tag{10.20}$$

The band-pass, low-pass, and input voltages may be summed, using a fourth amplifier, as shown in Figure 10.5. The output of the summing amplifier is

$$V_O = -\frac{R_{10}}{R_8} V_{LP} - \frac{R_{10}}{R_7} V_{BP} - \frac{R_{10}}{R_9} V_{IN} \tag{10.21}$$

and the resulting transfer function, obtained by substituting Equation 10.18 and 10.19 for V_{BP} and V_{LP}, respectively, is

$$\frac{V_O}{V_{IN}} = \frac{\dfrac{R_{10}}{R_8}\dfrac{1}{R_2 R_4 C_1 C_2} + \dfrac{R_{10}}{R_7}\dfrac{s}{R_4 C_1} - \dfrac{R_{10}}{R_9}\left(s^2 + \dfrac{1}{R_1 C_1} s + \dfrac{1}{R_2 R_3 C_1 C_2}\right)}{s^2 + \dfrac{1}{R_1 C_1} s + \dfrac{1}{R_2 R_3 C_1 C_2}} \tag{10.22}$$

Comparing this with the general biquadratic (for $m = 1, n = 1$):

$$T(s) = -K \frac{s^2 + cs + d}{s^2 + as + b} \tag{10.23}$$

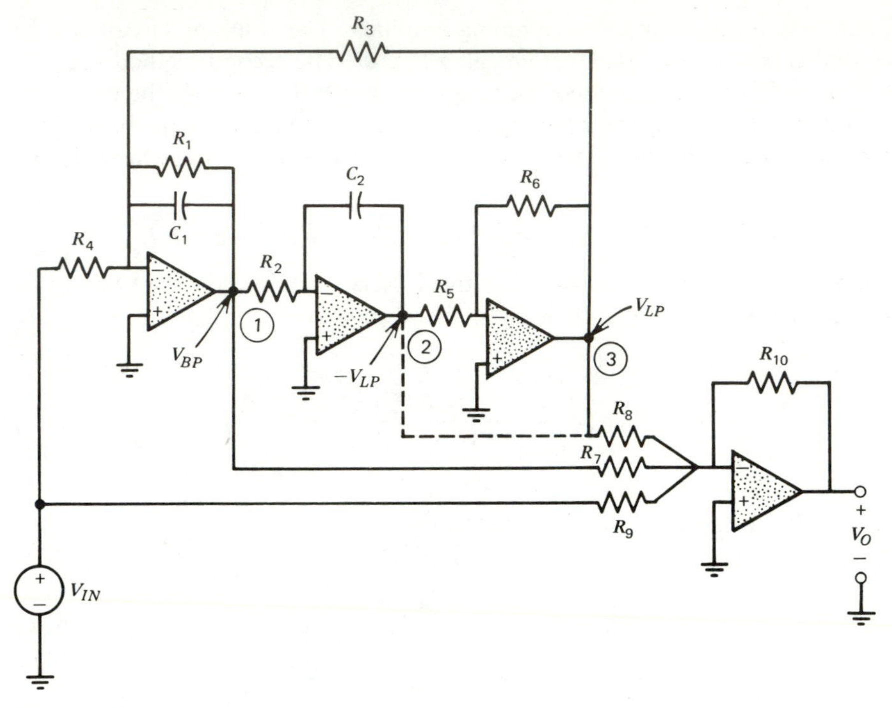

Figure 10.5 Summing four amplifier biquad.

the following relationships are obtained:

$$a = \frac{1}{R_1C_1} \tag{10.24a}$$

$$b = \frac{1}{R_2R_3C_1C_2} \tag{10.24b}$$

$$K = \frac{R_{10}}{R_9} \tag{10.24c}$$

$$c = \frac{1}{R_1C_1} - \frac{R_9}{R_7}\frac{1}{R_4C_1} = a - \frac{R_9}{R_7}\frac{1}{R_4C_1} \tag{10.24d}$$

$$d = \frac{1}{R_2R_3C_1C_2} - \frac{R_9}{R_8}\frac{1}{R_2R_4C_1C_2} = b - \frac{R_9}{R_8}\frac{1}{R_2R_4C_1C_2} \tag{10.24e}$$

We have five equations and ten elements. Therefore, five of the elements can be fixed. One choice for the fixed elements is

$$C_1 = 1 \qquad C_2 = 1 \qquad R_2 = R_3 = R \qquad R_7 = R_{10} = R \tag{10.25a}$$

Then the remaining elements are given by

$$R_1 = \frac{1}{a} \qquad R_2 = R_3 = \frac{1}{\sqrt{b}} \qquad R_4 = \frac{1}{K(a-c)}$$
$$R_7 = R_{10} = \frac{1}{\sqrt{b}} \qquad R_8 = \frac{a-c}{b-d} \qquad R_9 = \frac{1}{K\sqrt{b}} \tag{10.25b}$$

These synthesis equations yield nonegative element values for

$$a \geq c \qquad \text{and} \qquad b \geq d \tag{10.26}$$

The first inequality requires the zero to have a smaller real part than the pole. This condition is satisfied for all the approximation functions described in Chapter 4, since their zeros were constrained to lie on the $j\omega$ axis.* The second inequality requires that the magnitude of the pole frequency be larger than that of the zero frequency. This restriction can be removed by using V_2 instead of V_3 as the input to the summing amplifier (dotted lines in Figure 10.5). Then the output of the summer is

$$+\frac{R_{10}}{R_8} V_{LP} - \frac{R_{10}}{R_7} V_{BP} - \frac{R_{10}}{R_9} V_{IN} \tag{10.27}$$

It can easily be seen that the resulting synthesis equations will be the same as Equation 10.25, except in this case

$$R_8 = \frac{a-c}{d-b} \tag{10.28}$$

Thus, we see that the summing four amplifier biquad can be used to realize the general biquadratic function of Equation 10.1.

Example 10.1

Synthesize the following delay equalizer function using the summing four amplifier biquad:

$$\frac{V_O}{V_{IN}} = -\frac{s^2 - 500s + 25(10)^6}{s^2 + 500s + 25(10)^6}$$

Solution

The coefficients of the biquadratic are

$$K = 1 \qquad c = -500 \qquad d = 25(10)^6 \qquad a = 500 \qquad b = 25(10)^6$$

Since $b = d$, the resistance $R_8 = \infty$ and the low-pass voltage is not needed.

* The circuit can easily be adapted to also realize the case with $a < c$ [9].

From Equation 10.25, the element values for the circuit are

$$C_1 = 1 \qquad C_2 = 1 \qquad R_1 = \frac{1}{a} = 2(10)^{-3} \qquad R_2 = R_3 = \frac{1}{\sqrt{b}} = 2(10)^{-4}$$

$$R_4 = \frac{1}{K(a-c)} = (10)^{-3} \qquad R_7 = R_{10} = 2(10)^{-4} \qquad R_9 = \frac{1}{K\sqrt{b}} = 2(10)^{-4}$$

Impedance scaling by 10^7, we get the following practical element values:

$$C_1 = C_2 = 0.1\ \mu\text{F} \qquad R_1 = 20\ \text{k}\Omega \qquad R_2 = R_3 = 2\ \text{k}\Omega$$

$$R_4 = 10\ \text{k}\Omega \qquad R_7 = R_{10} = 2\ \text{k}\Omega \qquad R_8 = \infty \qquad R_9 = 2\ \text{k}\Omega$$

The complete circuit is drawn in Figure 10.6.

Observations

1. The circuit also provides a low-pass function at node 3 and a band-pass function at node 1. This special feature of simultaneously realizing more than one filter function is unique to three amplifier biquads.

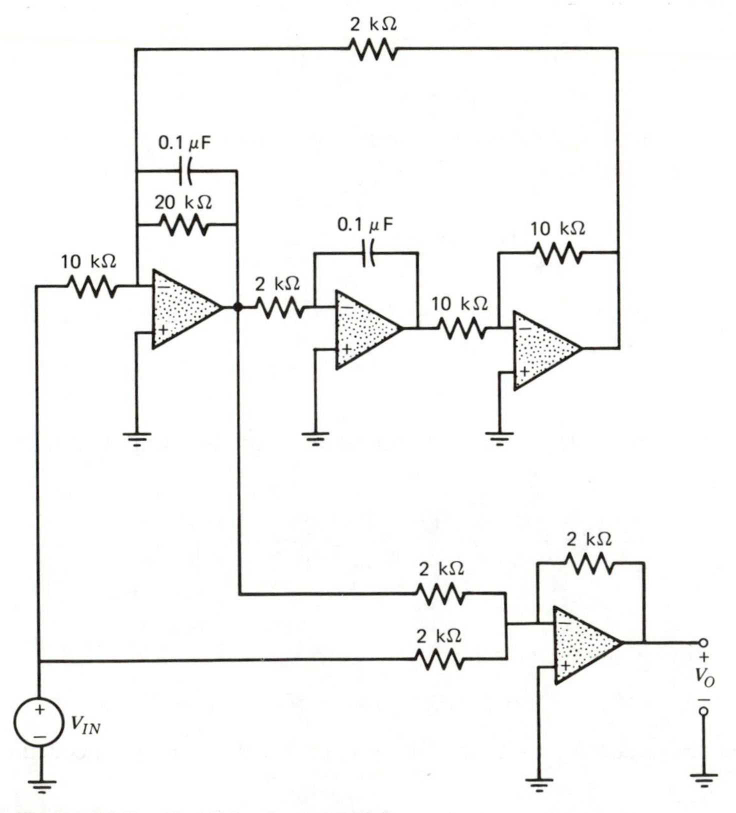

Figure 10.6 Circuit for Example 10.1.

2. The maximum ratio of resistors is 10 to 1 for the required pole Q of 10. This compares favorably with some single amplifier realizations where the element spreads were as high as $4Q_p^2$. ■

10.2.2 THE FEEDFORWARD THREE AMPLIFIER BIQUAD

In the circuit of Figure 10.5 an extra summing amplifier was used for forming the complex zeros. An alternate way of realizing the general biquadratic function, which requires only three operational amplifiers, is described in the following [2]. This approach is based on the feedforward scheme, in which the zeros are formed by introducing the input signal at those nodes of the circuit that are at ground potential. In the basic three amplifier biquad circuit, the input may be introduced at the negative input terminals of the three op amps, as shown in Figure 10.7.* Analyzing this circuit by writing node equations for nodes A, B, and C, we get:

node A:

$$-\left(\frac{1}{R_1} + sC_1\right)V_1 - \frac{1}{R_3}V_3 = \frac{1}{R_4}V_{IN} \tag{10.29a}$$

node B:

$$-\frac{1}{R_7}V_1 - \frac{1}{R_8}V_2 = \frac{1}{R_6}V_{IN} \tag{10.29b}$$

node C:

$$-\frac{1}{R_2}V_2 - sC_2V_3 = \frac{1}{R_5}V_{IN} \tag{10.29c}$$

These equations can be easily solved to yield the transfer function:

$$\frac{V_2}{V_{IN}} = -\frac{R_8}{R_6}\,\frac{s^2 + s\left(\dfrac{1}{R_1C_1} - \dfrac{1}{R_4C_1}\dfrac{R_6}{R_7}\right) + \dfrac{R_6}{R_7}\dfrac{1}{R_3R_5C_1C_2}}{s^2 + s\left(\dfrac{1}{R_1C_1}\right) + \dfrac{R_8}{R_7}\dfrac{1}{R_2R_3C_1C_2}} \tag{10.30}$$

One set of synthesis equations [2] for realizing the general biquadratic of Equation 10.23 is obtained by choosing the fixed elements as

$$C_1 = C_2 = 1 \qquad R_2 = R_3 = R \qquad R_7 = R_8 = R \tag{10.31a}$$

* The positions of the inverter and integrator are interchanged (as compared to Figure 10.4) to provide a convenient solution to the synthesis equations.

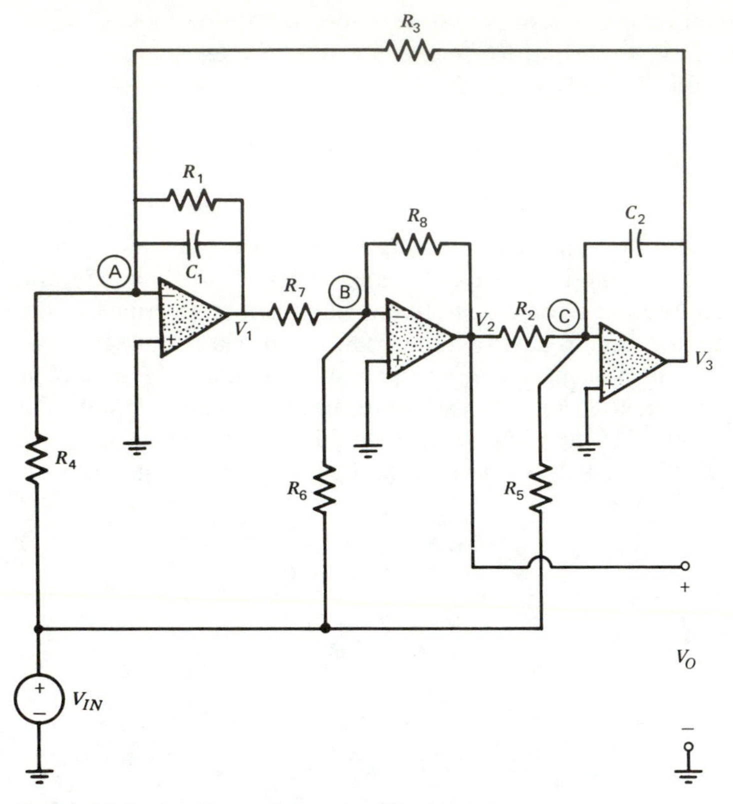

Figure 10.7 Feedforward three amplifier biquad.

Then the remaining element values are given by

$$R_1 = \frac{1}{a} \qquad R_2 = R_3 = \frac{1}{\sqrt{b}} \qquad R_4 = \frac{1}{K(a-c)}$$

$$R_5 = \frac{\sqrt{b}}{Kd} \qquad R_6 = \frac{1}{K\sqrt{b}} \qquad R_7 = R_8 = \frac{1}{\sqrt{b}} \tag{10.31b}$$

10.3 SENSITIVITY

In this section we consider the sensitivity of the three amplifier biquad, with particular reference to the low-pass filter. Insofar as the passband is concerned, the gain deviation for the other filter types will be essentially the same as for the *LP* case, because the contribution of the numerator terms to the passband deviation is quite negligible.

Let us first evaluate the sensitivity of the basic three amplifier low-pass circuit to the passive elements. The transfer function for this circuit, assuming ideal op amps, is given by Equation 10.19. Notice that the inverter resistors do not appear in the transfer function, since their ratio is unity when the components are assumed to be ideal. However, to account for the deviation in these two elements, they must be included in the transfer function. It is easy to verify that the transfer function with R_5 and R_6 included is

$$T_{LP} = \frac{-\dfrac{1}{R_2 R_4 C_1 C_2}}{s^2 + \dfrac{1}{R_1 C_1} s + \dfrac{1}{R_2 R_3 C_1 C_2} \dfrac{R_6}{R_5}} \tag{10.32}$$

From this equation the biquadratic parameters ω_p, Q_p, and K are identified as

$$\omega_p = \sqrt{\frac{1}{R_2 R_3 C_1 C_2} \frac{R_6}{R_5}} \tag{10.33a}$$

$$Q_p = \sqrt{\frac{R_1^2 C_1}{R_2 R_3 C_2} \frac{R_6}{R_5}} \tag{10.33b}$$

$$K = -\frac{1}{R_2 R_4 C_1 C_2} \tag{10.33c}$$

The sensitivities of these parameters to the passive elements are:

$$S^{\omega_p}_{R_2, R_3, C_1, C_2, R_5} = -\tfrac{1}{2} \qquad S^{\omega_p}_{R_6} = \tfrac{1}{2} \tag{10.34a}$$

$$S^{Q_p}_{R_1} = 1 \qquad S^{Q_p}_{R_2, R_3} = -\tfrac{1}{2} \tag{10.34b}$$

$$S^{Q_p}_{R_5} = -S^{Q_p}_{R_6} = -\tfrac{1}{2} \qquad S^{Q_p}_{C_1} = -S^{Q_p}_{C_2} = \tfrac{1}{2} \tag{10.34c}$$

$$S^{K}_{R_2, R_4, C_1, C_2} = -1 \tag{10.34d}$$

The computation of the sensitivities to the active elements is long and arduous. In this discussion we only present the results, as derived by Åkerberg and Mossberg [1]. Assuming the three operational amplifiers to be identical [$A_1(s) = A_2(s) = A_3(s)$] and that the amplifier gain can be modeled by a single pole at the origin, as in Equation 8.19b [$A(s) = A_0 \alpha / s$], the ω_p and Q_p sensitivities are

$$S^{\omega_p}_{A_0 \alpha} = \frac{3}{2} \frac{\omega_p}{A_0 \alpha} \qquad S^{Q_p}_{A_0 \alpha} = -4 Q_p \frac{\omega_p}{A_0 \alpha} \tag{10.35}$$

Observe that for high pole Q's the ω_p-active sensitivity term is much less than the Q_p sensitivity term. In contrast, recall that these two sensitivities had the same magnitude in the single amplifier biquad circuits. The relative importance of these active and passive sensitivities are gauged by evaluating their respective contributions to the gain deviation, as illustrated by the following example.

Example 10.2

Evaluate the gain deviation at the upper 3 dB passband edge frequency, 2.1 kHz, for the three amplifier biquad realization of the *LP* filter of Examples 8.2 and 8.3.

Solution

From Examples 8.2 and 8.3, $\omega_p = 2\pi(2000)$, $Q_p = 10$, and $A_0\alpha = 2\pi 10^6$. Also at $\omega = 2\pi(2100)$

$$\mathcal{S}^G_{\omega_p} = 78.25 \text{ dB} \qquad \mathcal{S}^G_{Q_p} = 4.45 \text{ dB} \qquad \mathcal{S}^G_K = 8.686 \text{ dB}$$

and the variabilities of the active and passive elements are described by

$$\mu(V_R) = \mu(V_C) = \mu(V_{A_0\alpha}) = 0$$
$$3\sigma(V_R) = 3\sigma(V_C) = 0.01 \qquad 3\sigma(V_{A_0\alpha}) = 0.5$$

For the three amplifier biquad the sensitivities of K, ω_p, and Q_p to the passive elements are given by Equation 10.34. The active sensitivities of ω_p and Q_p are obtained by substituting the values of ω_p, Q_p, and $A_0\alpha$ in Equation 10.35:

$$S^{\omega_p}_{A_0\alpha} = 0.003 \qquad S^{Q_p}_{A_0\alpha} = -0.08$$

As in Example 8.3, the contribution of the variations in the numerator is assumed to be negligible.

From Equation 8.14 and 8.15, the deviation in gain is given by:

$$\mu(\Delta G) = 0 \text{ dB}$$

$$\begin{aligned}
[3\sigma(\Delta G)]^2 &= (4.45)^2(2.5)(0.01)^2 && \rightarrow Q_p\text{-passive} \\
&\quad + (78.25)^2(1.5)(0.01)^2 && \rightarrow \omega_p\text{-passive} \\
&\quad + (8.686)^2(4)(0.01)^2 && \rightarrow K\text{-passive} \\
&\quad + (4.45)^2(0.08)^2(0.5)^2 && \rightarrow Q_p\text{-active} \\
&\quad + (78.25)^2(0.003)^2(0.5)^2 && \rightarrow \omega_p\text{-active} \\
&= \underbrace{\underset{Q_p}{0.005} + \underset{\omega_p}{0.918} + \underset{K}{0.03}}_{\text{passive terms}} + \underbrace{\underset{Q_p}{0.032} + \underset{\omega_p}{0.014}}_{\text{active terms}} \\
&= 0.99.
\end{aligned}$$

Thus

$$3\sigma(\Delta G) = 0.99 \text{ dB}$$

Observation

A comparison with Example 8.3 and 9.3 shows that this three amplifier biquad has a gain deviation that is somewhat lower than the Saraga positive feedback design but is larger than the Delyiannis negative feedback design. A general comparison of the sensitivities of these circuits is presented in the next section. ■

10.4 COMPARISON OF SENSITIVITIES OF THREE AMPLIFIER AND SINGLE AMPLIFIER BIQUADS

In this section we compare the sensitivities of the three amplifier biquad with two representative single amplifier designs, namely, the Saraga positive feedback and the Delyiannis negative feedback. To make this comparison we must first derive a general expression for the sensitivity of the three amplifier biquad.

Let us first consider the contribution of the passive terms to the gain deviation. From Equations 10.34, we have

$$\sum_i [(S_{R_i}^{\omega_p})^2 + (S_{C_i}^{\omega_p})^2] = 1.5 \tag{10.36}$$

$$\sum_i [(S_{R_i}^{Q_p})^2 + (S_{C_i}^{Q_p})^2] = 2.5 \tag{10.37}$$

At the frequency of maximum gain deviation ($\Omega = 1 \pm 1/2Q_p$),

$$\mathscr{S}_{\omega_p}^{G} \cong \pm 8.686 Q_p \qquad \mathscr{S}_{Q_p}^{G} = \tfrac{1}{2}(8.686) \tag{10.38}$$

Substituting the above expressions in Equation 8.15, the contribution of the passive terms to the total gain deviation is:

$$\text{Three Amplifier:}\ [3\sigma(\Delta G)]_{\text{passive}}^2 \cong 1.5\left\{8.686 Q_p[3\sigma(V_{R,C})]\right\}^2 \tag{10.39}$$

Comparing this with Equation 9.63 and 9.66, it is seen that the passive term for the three amplifier biquad is equal to that for the Saraga design but is larger than that for the Delyiannis design. To attain some insight into the reason for this difference, consider the expressions for pole frequency in the respective designs. In both single amplifier designs the pole frequency is described by an expression containing four elements of the form

$$\omega_p = \sqrt{\frac{1}{R_1 R_2 C_1 C_2}} \tag{10.40}$$

whereas in the three amplifier biquad, the pole frequency is described by an expression containing six elements:*

$$\omega_p = \sqrt{\frac{1}{R_2 R_3 C_1 C_2} \cdot \frac{R_6}{R_5}} \tag{10.41}$$

* Since the dimensions of ω_p^2 are $[R]^{-2}[C]^{-2}$, it takes *at least* two resistors and two capacitors to describe ω_p^2. Thus, (10.40) represents a canonic realization, while (10.41) is noncanonic.

It is this increase in number of elements that makes the passive term for the three amplifier biquad larger than that for the Delyiannis circuit. The fact that two of the elements occur as a ratio in (10.41) is quite significant, in that the tracking of these resistors with changes in temperature and aging will effectively diminish their contribution to the deviation in gain (refer to Section 8.2, page 281, for a discussion on tracking).

Next, let us consider the contribution of the active terms to the gain deviation. Using the sensitivity expressions given by Equation 10.35, it can be readily seen that the active term for the three amplifier biquad is

$$\text{Three Amplifier: } [3\sigma(\Delta G)]^2 = 6.25\left\{8.686 Q_p \frac{\omega_p}{A_0 \alpha}[3\sigma(V_{A_0 \alpha})]\right\}^2 \tag{10.42}$$

Comparing this with Equation (9.69) and (9.74), it is seen that the active term is somewhat larger for the three amplifier biquad than for either the Delyiannis or the Saraga designs. This is heuristically explained by the fact that three amplifiers are used instead of one.

In summary, the three amplifier biquad has a slightly higher sensitivity than the best of the single amplifier biquads; the main reason for this being that it needs more components for the circuit realization.

10.5 TUNING

In high precision filters it is often necessary to adjust the values of the components at the time of manufacture to correct for deviations in the gain from the nominal. This process of component adjustment, also known as **tuning**, is the subject of this section.

The synthesis equations lead to a set of passive element values that represent the nominal or paper design of the filter. Since the equations are derived *assuming the op amps to be ideal*, the performance of the network built in the laboratory, with real op amps, is expected to differ from the nominal design. Another factor that makes the laboratory model differ from the nominal is the *initial manufacturing tolerances* associated with the resistors and capacitors. Tuning corrects for these two sources of gain deviation. A tuned filter will exhibit the desired nominal performance at the time of manufacture (within the measurement accuracy). However, tuning cannot account for the deviations due to temperature and aging, which occur after the filter is manufactured. In some very critical and, of course, very expensive systems, the filter may be tuned every few years to correct for these environmental changes.

Although the discussion in this section refers to the summing four amplifier biquad, similar techniques may be adapted to the feedforward three amplifier biquad.

Consider the tuning of a second-order biquadratic with a pair of complex zeros, represented by

$$T(s) = K\frac{s^2 + \omega_z^2}{s^2 + (\mathrm{bw})_p s + \omega_p^2} \tag{10.43}$$

where K, ω_z, $(\mathrm{bw})_p$, and ω_p are nominal design parameters (Note that Q_z is infinite). From Equation 10.22, the summing four amplifier biquad realization of this function has the transfer function:

$$T'(s) = -\frac{R_{10}}{R_9}\frac{s^2 + s\left(\dfrac{1}{R_1C_1} - \dfrac{R_9}{R_7}\dfrac{1}{R_4C_1}\right) + \left(\dfrac{1}{R_2R_3C_1C_2} - \dfrac{R_9}{R_8}\dfrac{1}{R_2R_4C_1C_2}\right)}{s^2 + \dfrac{1}{R_1C_1}s + \dfrac{1}{R_2R_3C_1C_2}} \tag{10.44}$$

The tuning operation consists of adjusting the resistors (capacitors being difficult to adjust) to obtain the desired performance, as defined by the five parameters K, ω_z, ω_p, $(\mathrm{bw})_p$, and Q_z. From a cost standpoint it is desirable to accomplish the tuning in the minimum possible number of steps, which in this case is five, assuming one step for each biquadratic parameter. To achieve this ideal goal it is also necessary that each tuning step remain unaffected by the following steps. Such a minimal tuning algorithm is indeed attainable for the summing four amplifier biquad, as explained in the following.

In the algorithm the parameters describing the poles, namely, ω_p and $(\mathrm{bw})_p$, are adjusted by observing the output at the *BP* node (V_1 in Figure 10.5); and the parameters describing the zeros, namely, ω_z, Q_z, and K, are adjusted by observing the output of the summing amplifier.

1. *Pole frequency* ω_p
   The gain of a band-pass function is maximum at the pole frequency. Thus, the network can be tuned to give the correct ω_p by measuring the gain at the *BP* node, and adjusting R_2 to obtain maximum gain at ω_p.
2. *Pole bandwidth* $(\mathrm{bw})_p$
   For a *BP* function, the pole bandwidth defines the two frequencies at which the gain is 3 dB below the maximum gain. Thus, the correct bandwidth $(\mathrm{bw})_p$ can be attained by tuning the resistor R_1 so that the gain, as observed at the *BP* node, is 3 dB down from the maximum at the appropriate frequencies ($\omega \cong \omega_p \pm \omega_p/2Q_p$). Observe that this adjustment of R_1 does not affect the pole frequency ω_p.
3. *Zero frequency* ω_z
   For zeros on the $j\omega$ axis, the gain is a minimum at the zero frequency. Thus, the correct ω_z can be achieved by tuning the resistor R_8 so that the gain minimum, as measured at the output of the summer, occurs at ω_z. This adjustment does not affect the two preceding tuning steps.

4. *Zero Q*
   The zero Q determines the depth of the null at ω_z. For the given function $Q_z = \infty$, so the null depth is infinite. In practice, R_7 is tuned to obtain as deep a null as is measurable. Since R_7 does not appear in the expressions for ω_p, $(\text{bw})_p$, and ω_z, this step leaves the previous tuning steps unaffected.
5. *Scale factor K*
   K determines the gain at very high frequencies. Thus, it can be tuned by measuring the gain at the output of the summer and adjusting R_{10} to obtain the desired gain at some high frequency.

The above tuning algorithm is seen to be minimal in that it requires just five steps, each of which is unaffected by the following steps. Such a tuning algorithm is said to be *orthogonal.* The summing and feedforward biquads are the only topologies discussed in this text that lend themselves to an orthogonal tuning algorithm.

The ease of tuning of these biquads often justifies the cost of the extra op amps. An additional advantage of the easy tunability is that it allows this structure to be used in stringent filter applications with pole Q's as high as 100*, which would be quite unattainable without tuning. These features of the three amplifier biquad explain why it has found such favor with many filter manufacturers.

10.6 SPECIAL APPLICATIONS

The flexibility of the three amplifier biquad also allows it to be used in some special applications. A few of these applications are described in this section.

In the circuit of Figure 10.4, the band-pass function is obtained at node 1, and is given by

$$T_{BP} = \frac{-\dfrac{1}{R_4 C_1} s}{s^2 + \dfrac{1}{R_1 C_1} s + \dfrac{1}{R_2 R_3 C_1 C_2}} \tag{10.45}$$

Comparing this with the general band-pass function

$$T'_{BP} = -K \frac{s}{s^2 + (\text{bw})_p s + \omega_p^2} \tag{10.46}$$

* To achieve such high pole Q's we need to use op amps whose gain characteristics are superior to the single pole characteristic discussed so far. Ways of achieving different gain-phase characteristics are described in Chapter 12.

it is seen that each of the parameters K, $(\mathrm{bw})_p$, and ω_p can be controlled by an independent resistor. Specifically,

R_2 affects ω_p, but no other parameter
R_1 affects $(\mathrm{bw})_p$, but no other parameter
R_4 affects K, but no other parameter

Thus, we can realize a family of band-pass filters with different passband center frequencies ω_p, but with the same bandwidth and K, simply by replacing R_3 by a bank of switched resistors (or a potentiometer), as illustrated in Figure 10.8. These variable frequency band-pass filters find application in circuit testing equipments. In a similar way, it is possible to use the circuit of Figure 10.4 to realize filters with varying bandwidths by switching the resistor R_1; and the scale factor K can be varied by R_4. Also, considering the circuit of Figure 10.5, filters with varying ω_z and Q_z can be attained by adjusting R_8 and R_7, respectively.

The easy controllability of the three amplifier biquad is a direct consequence of the buffering (or isolation) provided by the three op amps between the stages of the circuit. It is this same buffering action that makes the orthogonal tuning of the filter possible. In contrast, the single amplifier biquad circuits do not have this one-to-one relationship between each biquadratic parameter and a circuit

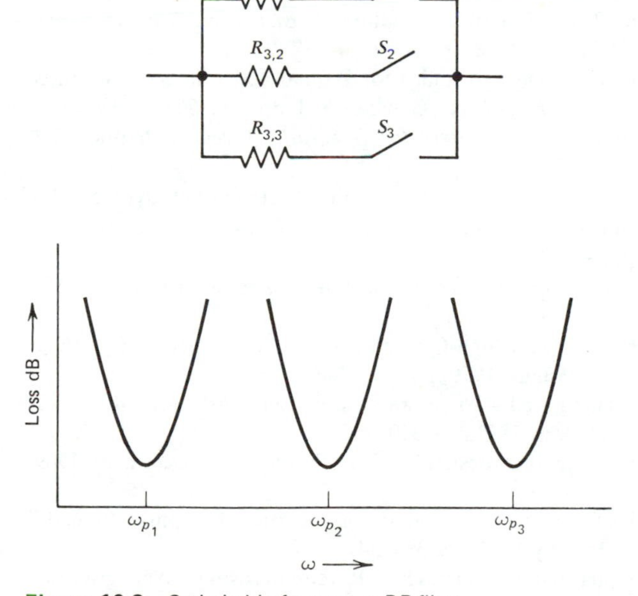

Figure 10.8 Switchable frequency *BP* filter.

element. For this reason, they cannot be tuned as easily and they do not provide the switchable-filter options. This capability of the three amplifier biquad is one of the more impressive advantages of active filters over their passive counterparts. Using passive *RLC* circuits, it would be necessary to switch many more components, or else one would have to use a separate circuit for each filtering function.

10.7 CONCLUDING REMARKS

The advantages of the three amplifier biquad are:

- It realizes the general biquadratic, so the same topology can be used for generating all the filter functions.
- Tuning is easier and more economical than in the single amplifier topologies.
- The one-to-one relationship between the biquadratic parameters and circuit resistors allows its use in switchable-filter applications.

In stringent filter applications requiring tuning, the advantages of the three amplifier biquad will often more than offset the cost of the extra op amps. In such applications this is the most versatile and economical realization.

FURTHER READING

1. D. Åkerberg and K. Mossberg, "A versatile active RC building block with inherent compensation for the finite bandwidth of the amplifier," *IEEE Trans. Circuit Theory*, *CAS-21*, No. 1, January 1974, pp. 75–78.
2. P. E. Fleischer and J. Tow, "Design formulas for Biquad active filters using three operational amplifiers," *Proc. IEEE*, *61*, No. 5, May 1973, pp. 662–663.
3. W. Heinlen and H. Holmes, *Active Filters for Integrated Circuits*, Prentice-Hall International, London, 1974, Chapter 8.
4. W. J. Kerwin, L. P. Huelsman, and R. W. Newcomb, "State-variable synthesis for insensitive integrated circuit transfer functions," *IEEE J. Solid-State Circuits*, *SC-2*, No. 3, September 1967, pp. 87–92.
5. G. S. Moschytz, *Linear Integrated Networks Design*, Van Nostrand, New York, 1975, Chapter 3.
6. R. Tarmy and M. S. Ghausi, "Very high-Q insensitive active RC networks," *IEEE Trans. Circuit Theory*, *CT-17*, August 1970, pp. 358–366.
7. L. C. Thomas, "The Biquad: Part I—Some practical design considerations," *IEEE Trans. Circuit Theory*, *CT-18*, May 1971, pp. 350–357.
8. J. Tow, "A step-by-step active-filter design," *IEEE Spectrum*, *6*, December 1969, pp. 64–68.
9. J. Tow, "Design formulas for active RC filters using operational-amplifier biquad," *Electronics Letters*, *5*, No. 15, July 1969, pp. 339–341.
10. P. W. Vogel, "Method for phase correction in active RC circuits using two integrators," *Electronics Letters*, *10*, May 20, 1971, pp. 273–275.

PROBLEMS

10.1 *Summing biquad synthesis.* Synthesize the following transfer functions using the summing four amplifier biquad:

(a) $-\dfrac{10(s^2 + 9000)}{s^2 + 5s + 4000}$

(b) $-\dfrac{10s^2}{s^2 + 5s + 4000}$

(c) $-\dfrac{10(s^2 - 20s + 4000)}{(s^2 + 20s + 4000)}$

(d) $-\dfrac{20s}{(s^2 + s + 16)(s^2 + 2s + 25)}$

10.2 *Feedforward biquad synthesis.* Synthesize the transfer functions in Problem 10.1 using the feedforward three amplifier biquad.

10.3 *Three amplifier biquad filter.* The requirements for a low-pass filter are

$$A_{max} = 0.5 \text{ dB} \qquad A_{min} = 30 \text{ dB}$$
$$f_P = 1000 \text{ Hz} \qquad f_S = 2000 \text{ Hz} \qquad dc \text{ gain} = 0 \text{ dB}$$

Synthesize the Chebyshev approximation function for these requirements using the three amplifier biquad.

10.4 *Feedforward biquad filter.* Use the feedforward biquad to synthesize the elliptic function approximation for the low-pass requirements of Problem 10.3. The transfer function can be derived from Table 4.3.

10.5 The band-pass requirements shown in Figure P10.5 can be approximated by following elliptic approximation function:

$$T(s) = \frac{597s}{s^2 + 597s + 3.106(10)^8} \cdot \frac{0.36[s^2 + 2.834(10)^8]}{s^2 + 233s + 2.987(10)^8} \cdot \frac{0.56[s^2 + 3.404(10)^8]}{s^2 + 242s + 3.229(10)^8}$$

(a) Synthesize this function using three-amplifier biquads, with 0.02 μF capacitors.
(b) Suppose that, due to dynamic range considerations (see Section 13.2), it is necessary to transfer 6 dB of gain from the second section to the third section. Indicate the changes needed in the circuit.

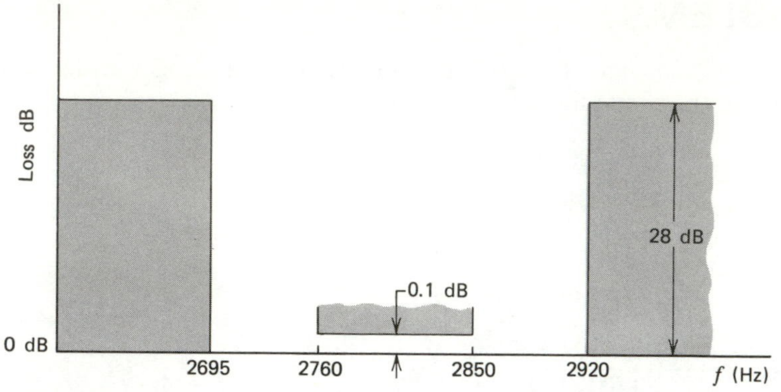

Figure P10.5

10.6 *Feeaforward biquad, sensitivity.* Derive expressions for the sensitivities of the biquad parameters ($K, \omega_p, \omega_z, Q_p, Q_z$) of the feedforward biquad to the passive elements. Express the sensitivities in terms of the biquad parameters.

10.7 The feedforward biquad is used to synthesize the low-pass notch function

$$\frac{s^2 + (500)^2}{s^2 + 50s + (250)^2}$$

using one percent passive elements [i.e., $3\sigma(V_R) = 3\sigma(V_C) = 0.01$]. Compute the statistics of the gain deviation due to the passive elements at

(a) The pole frequency.
(b) The passband edge frequencies $\omega = \omega_p(1 \pm 1/2Q_p)$.
(c) 450 rad/sec.

Use these results to sketch the $\pm 3\sigma$ limits of the gain deviation versus frequency (as in Figure 5.4*b*).

10.8 *Comparison of Saraga, Delyiannis, and three amplifier biquads.* The Saraga positive feedback, Delyiannis negative feedback, and the three amplifier biquads are used to realize a second function whose denominator is

$$s^2 + (2\pi 50)s + (2\pi 1000)^2$$

Show that, of the three biquads, the Delyiannis design has the smallest gain deviation at the lower 3 dB passband edge frequency, given that the components are characterized as in Example 8.2 and 8.3.

10.9 Repeat Problem 10.8 for the denominator function

$$s^2 + (2\pi 500)s + (2\pi 10{,}000)^2$$

and show that the Saraga design has the lowest gain deviation.

10.10 In the thin film technology, ratios of resistors can be matched very closely. Suppose that the ratio R_5/R_6 associated with the inverter in the three amplifier biquad, the ratio r_2/r_1 in the Saraga design, and R_A/R_B in the Delyiannis circuit, are each trimmed so that they are within ± 0.1 percent of their nominal values [i.e., $3\sigma(V_{R5/R6}) = 3\sigma(V_{r2/r1}) = 3\sigma(V_{RA/RB}) = 0.001$]. Repeat Problem 10.8 with this additional stipulation, and show that the three amplifier biquad has a smaller gain deviation than both the Delyiannis and the Saraga designs.

10.11 *Tuning the feedforward biquad.* Describe an orthogonal tuning algorithm for the feedforward biquad realization of a general biquadratic function.

10.12 *Tuning a low-pass biquad.* Devise a tuning procedure for an active RC circuit which has the following low-pass function (assuming $Q_p > 10$):

$$\frac{\dfrac{1}{R_1 R_4 C_1 C_2}}{s^2 + s\left(\dfrac{1}{R_3 C_1} + \dfrac{1}{R_1 C_2}\right) + \dfrac{1}{R_1 R_2 C_1 C_2}}$$

10.13 *Tuning a three amplifier biquad.* A second-order band-pass filter is required to have a center frequency at 1000 Hz, 3 dB bandwidth of 100 Hz, and a gain of 6 dB at the center frequency. Determine the nominal element values for the three amplifier biquad realization, using $C_1 = C_2 = 0.1\ \mu\text{F}$, $R_2 = 1\ \text{k}\Omega$. A laboratory model of this circuit, built using 2 percent components, measures to have a center frequency at 1005 Hz, 3 dB bandwidth of 98 Hz, and center-frequency gain of 6.1 dB. Determine the percentage changes needed in the resistors R_1, R_3, and R_4 to tune the circuit.

10.14 *Phase tuning of three amplifier biquad.* Describe an algorithm for tuning the pole frequency and pole Q of the band-pass three amplifier biquad based on phase measurements at the center frequency and at the upper 3 dB passband edge frequency. (*Hint*: use the results of Problem 2.34c.)

10.15 *Switchable tone detector.* A circuit for detecting the presence of one of two tones, at 1000 Hz and 1300 Hz, requires the two second-order band-pass characteristics shown in Figure P10.15. Show how the two filter characteristics can be realized using one three-amplifier biquad with one additional resistor and a switch.

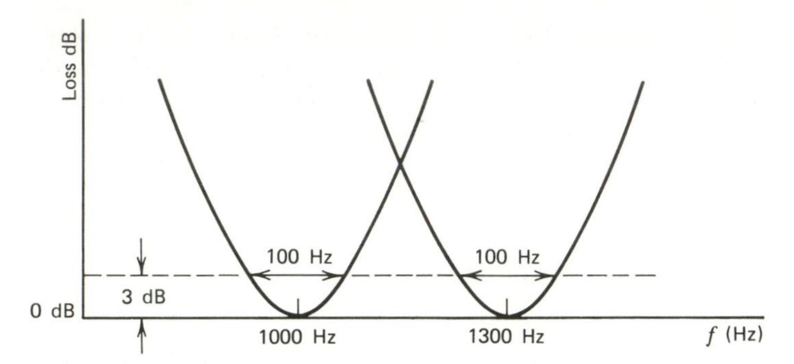

Figure P10.15

10.16 *Switchable frequency filter*. Design a second-order switchable band-pass filter in which the switch positions can be selected to pass any one of the eight frequencies ω_0, $\sqrt{2}\omega_0$, $\sqrt{3}\omega_0$, $\sqrt{4}\omega_0$, $\sqrt{5}\omega_0$, $\sqrt{6}\omega_0$, $\sqrt{7}\omega_0$, or $\sqrt{8}\omega_0$. In each case the gain at the center frequency must be 0 dB and the 3 dB passband width must be $\omega_0/5$. Use one three-amplifier biquad with three additional resistors and three switches.

10.17 *Tone separation*. A signal contains three tones at $F1$, $F2$, and $F3$. Show how a feedforward three-amplifier biquad can be used to eliminate any one of the three tones from the signal (i.e., depending on the switch position the output should be $F1$ and $F2$, $F1$ and $F3$, or $F2$ and $F3$). Use two additional resistors and two switches.

10.18 *Switchable gain equalizer*.* Gain equalizers are often needed to introduce bumps or dips in the gain characteristics of a digital signal to compensate for distortions introduced by the transmission system.

(a) Verify that the biquadratic function

$$\frac{s^2 + kas + b}{s^2 + as + b}$$

introduces a bump at the pole frequency for $k > 1$, and a dip for $k < 1$.

(b) Show how the structure of Figure 10.7, when used as a *summing* four amplifier biquad (without the inputs at nodes B and C), can be used to realize the above gain equalizer function for $k > 1$ and $k < 1$.

(c) Synthesize a second-order function with a 1 dB bump at 1000 rad/sec.

(d) Indicate how the sharpness of the bump can be controlled.

* P. E. Fleischer, "Active adjustable loss and delay equalizers," *IEEE Trans. Communications*, *COM-22*, No. 7, July 1974.

10.19 *Analysis of three amplifier biquad with finite gain op amps.* If the op amps for the three amplifier biquad are contained in the same integrated circuit chip (Section 13.3.1), their characteristics will be very similar. Verify that if the op amps are each assumed to have the same gain $A(s)$, the low-pass transfer function is given by Equation 12.19.

10.20 If the op amp gains (assumed to be the same) in the three amplifier biquad can be approximated by a single pole at the origin, as $A(s) = A_0\alpha/s$, show that the low-pass transfer function of Equation 12.9 reduces to

$$T(s) \cong \frac{N(s)}{s^2 + \dfrac{\omega_{po}}{Q_{po}}\left(1 - 4Q_{po}\dfrac{\omega_{po}}{A_0\alpha}\right)s + \omega_{po}^2\left(1 - \dfrac{3\omega_{po}}{A_0\alpha}\right)}$$

where $\omega_{po}^2 = 1/R_2R_3C_1C_2$, $\omega_{po}/Q_{po} = 1/R_1C_1$. Use the design formula of Equations 10.12 and assume $d = b$. [*Hint*: follow the analysis steps used to derive Equation 8.20, using the approximations $1/A \ll 1/A^2$, $[1+(1/A)]^{-1} \cong 1 - (1/A)$.

10.21 *Q enhancement in three amplifier biquad.* Use the transfer function given in Problem 10.20 to find an expression for the pole Q. Plot the pole Q for increasing pole frequency ω_{po}, given that the gain-bandwidth product $A_0\alpha = 2\pi 10^6$ rad/sec and the low frequency pole Q is $Q_{po} = 20$. Observe the Q enhancement at high frequencies—a phenomena that brings the active circuit closer to instability. Determine the pole frequency ω_{po} at which the circuit becomes unstable.

10.22 *Kerwin, Huelsman, Newcomb biquad.* One of the first three amplifier biquads, proposed by Kerwin, Huelsman, and Newcomb [4], is shown in Figure P10.22. Show that this circuit provides a high-pass function at node 1, a band-pass function at node 2, and a low-pass function at node 3.

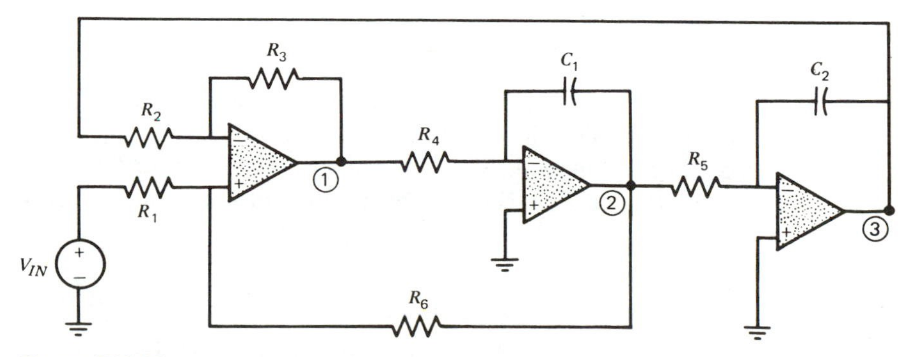

Figure P10.22

Verify that the band-pass transfer function is

$$\frac{-\left(\dfrac{R_6}{R_2R_4C_1}\cdot\dfrac{R_2+R_3}{R_1+R_6}\right)s}{s^2+\dfrac{R_1}{R_2R_4C_1}\left(\dfrac{R_2+R_3}{R_1+R_6}\right)s+\dfrac{R_3}{R_2R_4R_5C_1C_2}}$$

Determine design equations for the band-pass circuit, assuming the choice $R_1 = R_2 = R_3 = R_4 = R_5 = 1$ and $C_1 = C_2 = C$ for the fixed elements.

10.23 Find expressions for the sensitivities of the pole Q and pole frequency to the passive elements for the band-pass transfer function given in Problem 10.22. Express the sensitivities in terms of the pole Q, assuming the design equations are based on $R_1 = R_2 = R_3 = R_4 = R_5 = 1$ and $C_1 = C_2 = C$.

10.24 The biquad of Figure P10.22 can be used to obtain complex zeros, by summing the voltages at nodes 1, 2, and 3. Use the general summing circuit shown in Figure P10.24 to synthesize the high-pass notch filter

$$-\frac{s^2+9}{s^2+s+16}$$

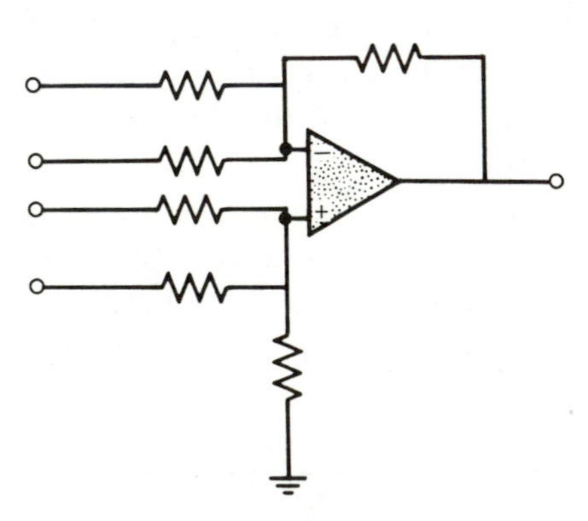

Figure P10.24

10.25 *Vogel biquad* [10]. Show that the Vogel biquad shown in Figure P10.25, when used *without* the summing inputs via R_5 and R_6, provides a low-pass transfer function. Show that inclusion of the feedforward input resistors R_5 and R_6 yields the general biquadratic transfer function:

$$-\frac{R_8}{R_6}\,\frac{s^2+s\left(\dfrac{1}{R_1C_1}-\dfrac{R_6}{R_5C_2}\right)+\left(\dfrac{R_6}{R_2R_4R_7}-\dfrac{R_6}{R_1R_5R_7}\right)\dfrac{1}{C_1C_2}}{s^2+s\left(\dfrac{1}{R_1C_1}\right)+\dfrac{1}{R_2R_3C_1C_2}\dfrac{R_8}{R_7}}$$

Derive design equations to synthesize the function

$$-K\frac{s^2+\omega_z^2}{s^2+\dfrac{\omega_p}{Q_p}s+\omega_p^2}$$

given the fixed element choice $C_1 = C_2 = 1$, $R_2 = R_3 = R_7 = R_8 = R$.

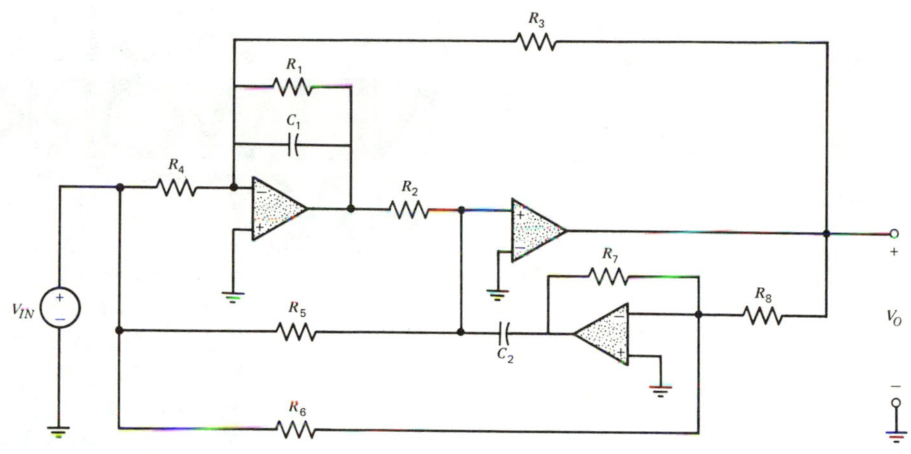

Figure P10.25

10.26 If the op amp gains (assumed to be the same) in the low-pass Vogel circuit can be approximated by a single pole at the origin as $A(s) = A_0\alpha/s$, show that the transfer function reduces to

$$T(s) \cong \frac{N(s)}{s^2+\dfrac{\omega_{po}}{Q_{po}}\left(1-\dfrac{3\omega_{po}}{A_0\alpha}\right)s+\omega_{po}^2\left(1-\dfrac{3\omega_{po}}{A_0\alpha}\right)}$$

where $\omega_{po}^2 = 1/R_2R_3C_1C_2$ and $\omega_{po}/Q_{po} = 1/R_1C_1$. Use the design equations based on the fixed element choice $C_1 = C_2 = 1$, $R_2 = R_3 = R_7 = R_8 = R$ and assume the dc gain is 0 dB.

10.27 Use the transfer function given in Problem 10.26 to plot the pole Q of the Vogel circuit for increasing pole frequency ω_{po}. Assume that the gain-bandwidth product $A_0\alpha$ is $2\pi 10^6$ rad/sec and the low frequency pole Q is $Q_{po} = 20$. Observe that the *Q-enhancement* for this circuit is much less than for the three amplifier biquad analyzed in Problem 10.21. Determine the pole frequency at which the circuit becomes unstable.

11.

ACTIVE NETWORKS BASED ON PASSIVE LADDER STRUCTURES

In Chapter 6 it is shown that double-terminated passive *LC* ladder networks can be designed to have very low sensitivity in the passband. This chapter deals with some active networks, which are derived from the passive ladder structure, in an attempt to emulate this low sensitivity property. These active networks belong to the family of **coupled ladder structures**, which will be shown to have a much lower sensitivity than the cascade topologies discussed in Chapters 7 to 10.

Two approaches will be used for obtaining these coupled active realizations. The first is based on the replacement of some elements in the passive network by their active *RC* equivalents. The active building blocks used in this approach are the gyrator or the *FDNR* (frequency dependent negative resistance). The second method is based on a direct simulation of the equations of the passive network, which leads to an active *RC* coupled topology.

11.1 PASSIVE LADDER STRUCTURES

In the chapter on passive synthesis (Section 6.3) we presented Orchard's argument to explain the low sensitivity of *LC* ladder networks terminated in resistors. There it was shown that if the ladder network is designed to have maximum power transfer, the sensitivity is very low in the passband. To verify this statement, let us compare the sensitivity of a passive realization versus that of the equivalent cascaded active realization by means of an example [11]. Consider a sixth-order elliptic *BP* filter with center frequency 2805 Hz, 0.1 dB passband ripple, passband width of 90 Hz, with a minimum attenuation of 30 dB in the stopbands that extend below 2694.8 Hz and above 2919.8 Hz (Figure 11.1). From [11], the elliptic approximation for these filter requirements is

$$T(s) = \frac{597s}{s^2 + 597s + 3.106172(10)^8} \cdot \frac{0.36[s^2 + 2.834248(10)^8]}{s^2 + 233.33s + 2.987363(10)^8} \cdot \frac{0.562[s^2 + 3.404184(10)^8]}{s^2 + 242.61s + 3.229704(10)^8} \tag{11.1}$$

The passive, doubly terminated ladder realization for this filter function, obtained by using the zero shifting technique developed in Chapter 6, is shown in Figure 11.2. This circuit has four inductors, four capacitors, and the two terminating resistors. The cascaded realization using the feedforward three

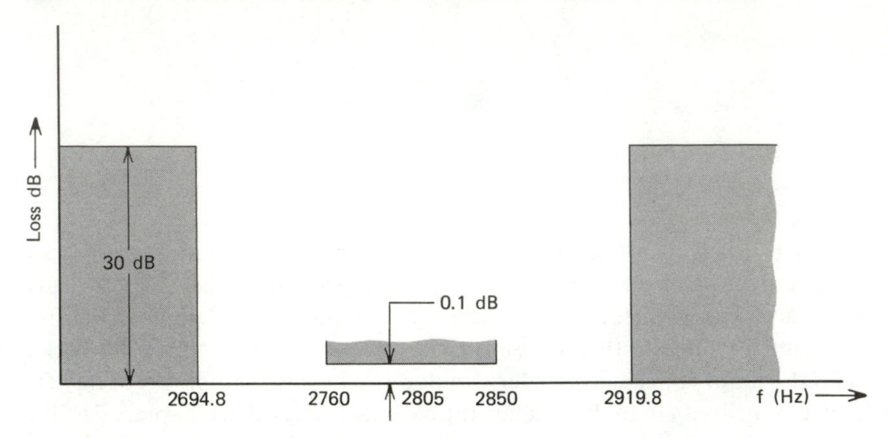

Figure 11.1 Requirements for band-pass example.

amplifier biquad can easily be derived by using the synthesis equations developed in Section 10.2.2 (page 350). The resulting circuit is shown in Figure 11.3. The sensitivities of these two circuits were compared by evaluating the statistical change in gain using a Monte Carlo computer analysis. In the simulation, the resistors, capacitors, and inductors were all assumed to have a uniform distribution with a ± 0.25 percent tolerance. The op amps were modeled by the typical gain characteristics sketched in Figure 1.10, and the tolerance on the gain was assumed to be ± 50 percent. The results of the two simulations are shown in Figure 11.4. It is seen that in the passband, which is usually the critical band for most filters, the standard deviation for the passive circuit is much lower than for the active cascaded circuit. For instance, at the two edges of the passband, the standard deviation σ of the active network is approximately eight

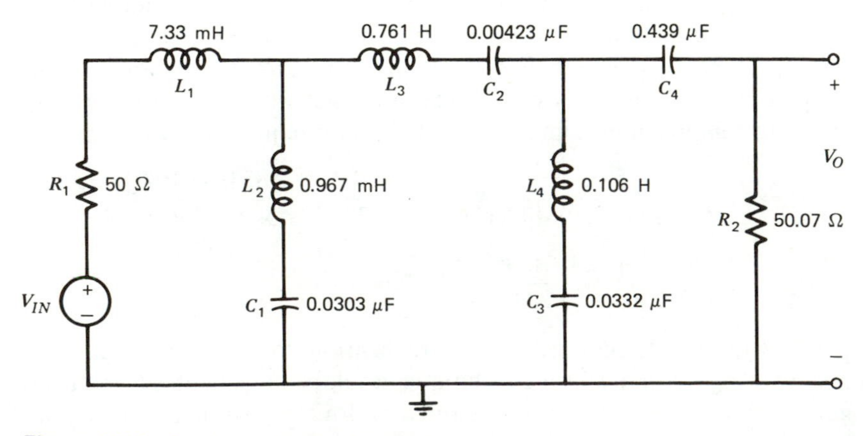

Figure 11.2 Passive realization for band-pass example.

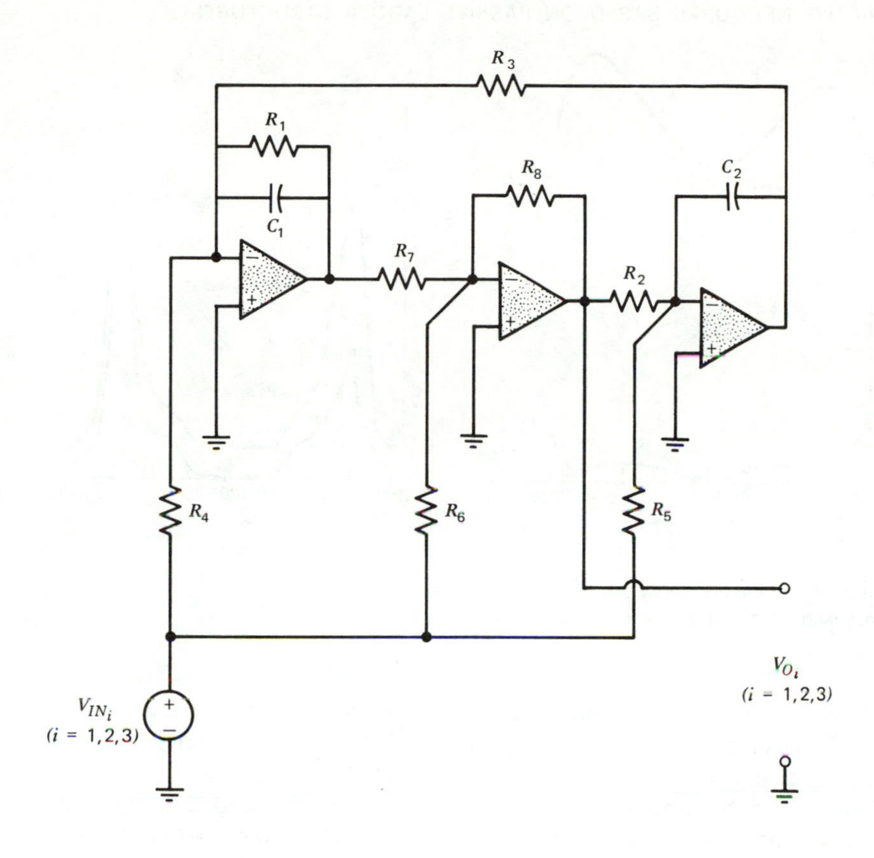

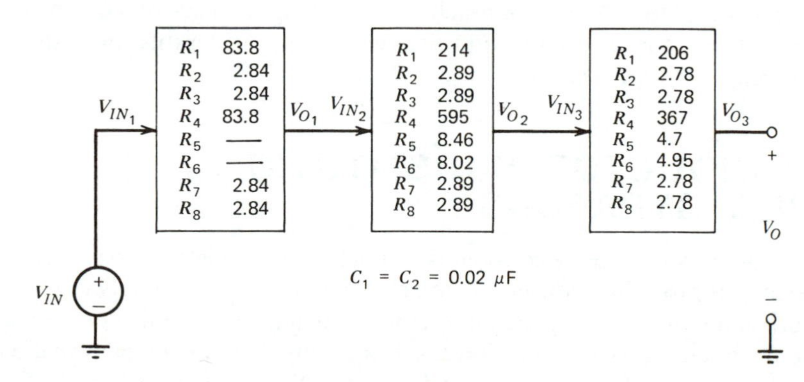

Figure 11.3 Three amplifier biquad cascaded realization of example *BP* filter (resistors in kΩ).

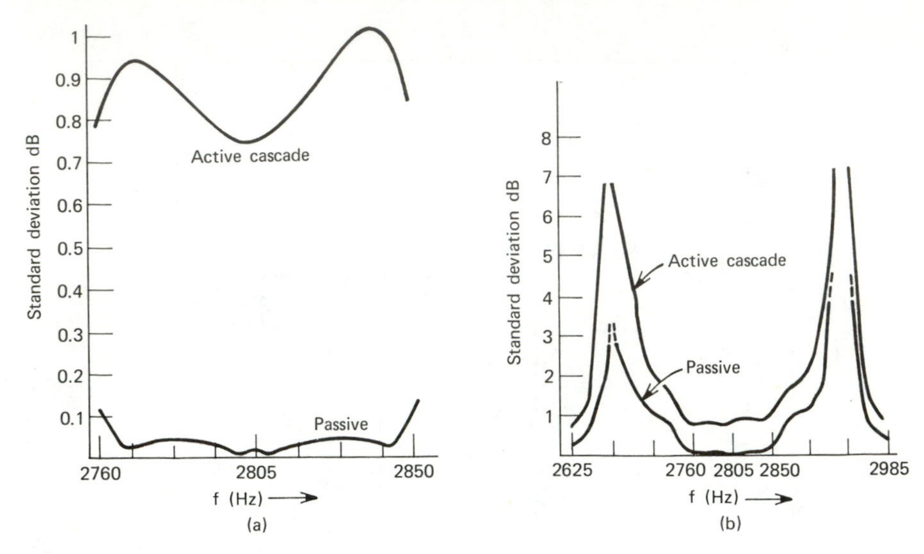

Figure 11.4 Standard deviations of band-pass example: (*a*) Passband. (*b*) Stopbands and passband.

times greater than that of the passive.* In the stopband the standard deviations of the two circuits are not very different (recall that Orchard's argument on the sensitivity of such networks does not apply to the stopband).

Similar results have been obtained in comparing other cascaded designs with their passive equivalents. This large difference in performance motivates us to search for active *RC* realizations based on passive ladder structures. In this chapter we develop different approaches for deriving such active *RC* topologies and in each case the sensitivity will be shown to be considerably lower than for the cascaded active realizations.

11.2 INDUCTOR SUBSTITUTION USING GYRATORS

The first active structure we consider is a direct imitation of the passive network. The synthesis procedure consists of first obtaining the *LC* resistively terminated network and then replacing each inductor by an active RC equivalent. Such an active *RC* circuit is expected to have as low a sensitivity as the passive circuit, except for the imperfections in the realization of the active inductor.

* In this example the gain deviations of the active filter can be shown to be mostly due to the variations in the passive elements. However, if the passband were at a higher frequency, or if the op amp used had a lower gain in the passband of the filter, the contribution of the active terms would become more significant. This would result in an even larger gain deviation for the active filter.

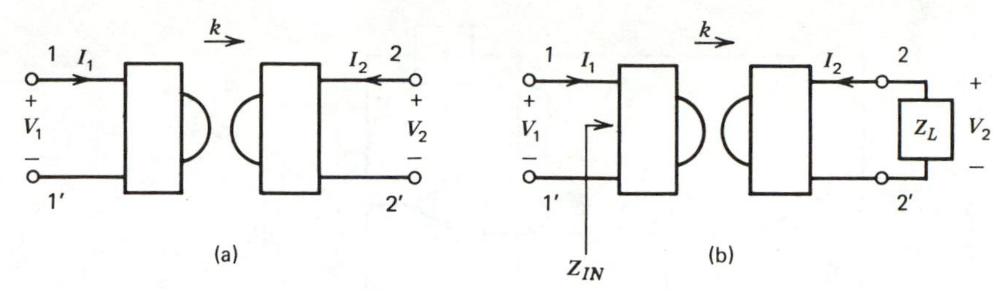

Figure 11.5 The gyrator: (*a*) Symbolic representation. (*b*) Terminated by a load at port 2 − 2′.

The most commonly used active circuit for the realization of the inductor is the **gyrator**, shown symbolically in Figure 11.5*a*. The gyrator is a two port network that has the z matrix representation:

$$\begin{bmatrix} V_1 \\ V_2 \end{bmatrix} = \begin{bmatrix} 0 & -k \\ k & 0 \end{bmatrix} \begin{bmatrix} I_1 \\ I_2 \end{bmatrix} \tag{11.2}$$

where k is known as the gyration resistance. A useful property of the gyrator is that the input impedance at either port is proportional to the reciprocal of the impedance terminating the other port. To show this, suppose port 2 is terminated in the impedance Z_L (Figure 11.5*b*). Then

$$V_2 = -I_2 Z_L \tag{11.3}$$

Substituting this relation in (11.2), the input impedance at port 1-1′ is

$$Z_{IN} = \frac{V_1}{I_1} = \frac{k^2}{Z_L} \tag{11.4}$$

In particular, if port 2-2′ is terminated in a capacitor C, the input impedance seen at port 1-1′ is

$$Z_{IN} = k^2 sC \tag{11.5}$$

which corresponds to an inductor of value k^2C henries. Therefore, the gyrator, terminated in a capacitor, can be used to realize an inductor.

Let us next consider the circuit realization of the gyrator.* One such circuit, due to Riordan [8], is shown in Figure 11.6. The node equations for this circuit are

node A:

$$(Y_1 + Y_2)V_{IN} - Y_2 V_2 = 0 \tag{11.6a}$$

* For our purposes, it is only necessary that circuit invert the terminating impedance at *one* port, and a general gyrator is not needed. Nevertheless we will refer to the one-port-impedance-inverter as a gyrator, as is customary in much of the literature.

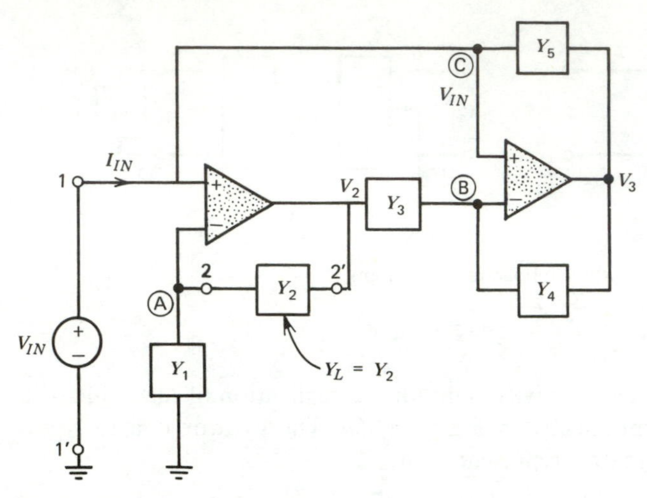

Figure 11.6 Riordan gyrator terminated in a load $Y_L = Y_2$.

node B:

$$(Y_3 + Y_4)V_{IN} - Y_3 V_2 - Y_4 V_3 = 0 \tag{11.6b}$$

node C:

$$Y_5 V_{IN} - Y_5 V_3 = I_{IN} \tag{11.6c}$$

Solving these equations, the input impedance is found to be

$$Z_{IN} = \frac{V_{IN}}{I_{IN}} = \frac{Y_2 Y_4}{Y_1 Y_3 Y_5} = \frac{Z_1 Z_3 Z_5}{Z_2 Z_4} \tag{11.7}$$

In particular, if Z_2 represents a capacitor C and all the other elements are resistors R, the input impedance becomes

$$Z_{IN} = sR^2C \tag{11.8a}$$

which is the impedance of an inductor

$$L = R^2C \text{ henries} \tag{11.8b}$$

Thus, Riordan's circuit allows the realization of an active RC inductor with a gyration resistance $k = R$. Notice that one terminal of the input port is ground; thus, the circuit can only realize grounded inductors. The realization of a floating inductor requires two gyrators connected back-to-back, as shown in Figure 11.7. An analysis of this circuit will show that the input impedance at port 1-1′ is sR^2C.

The *inductor-substitution* technique described above leads to a realization that has the same topology as the parent passive ladder network. The difference is that each grounded inductor is replaced by a circuit using two op amps, four resistors, and one capacitor; while each floating inductor requires a circuit using four op amps, seven resistors, and two capacitors. We would expect the sensitivity of the gyrator-RC realizations to be equivalent to that of the passive ladder, except for this increase in the total number of components required to realize the inductor. Moreover, the fact that the op amp is not ideal will also increase the sensitivity of the gyrator-RC circuit.

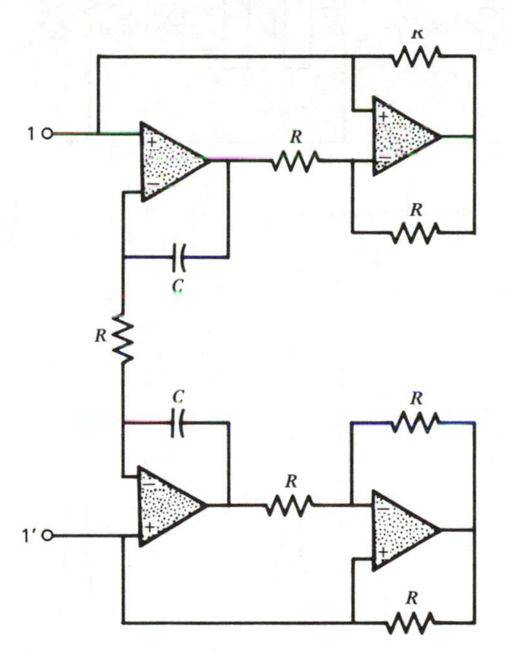

Figure 11.7 Floating inductor realization.

As an example, let us consider the active gyrator-RC realization of the passive circuit of Figure 11.2. By replacing the inductors by their gyrator-RC equivalents (the gyration resistance $k = R = 1\ \mathrm{k}\Omega$), we get the circuit shown in Figure 11.8. The results of a Monte Carlo statistical analysis of this circuit, using the same element types as in the cascaded realization of Equation 11.1, results in the standard deviation curve depicted in Figure 11.9. It is seen that the gyrator-RC realization shows a remarkable improvement over the cascaded biquad circuit. However, the sensitivity is not quite as low as for the passive circuit, for reasons already mentioned.

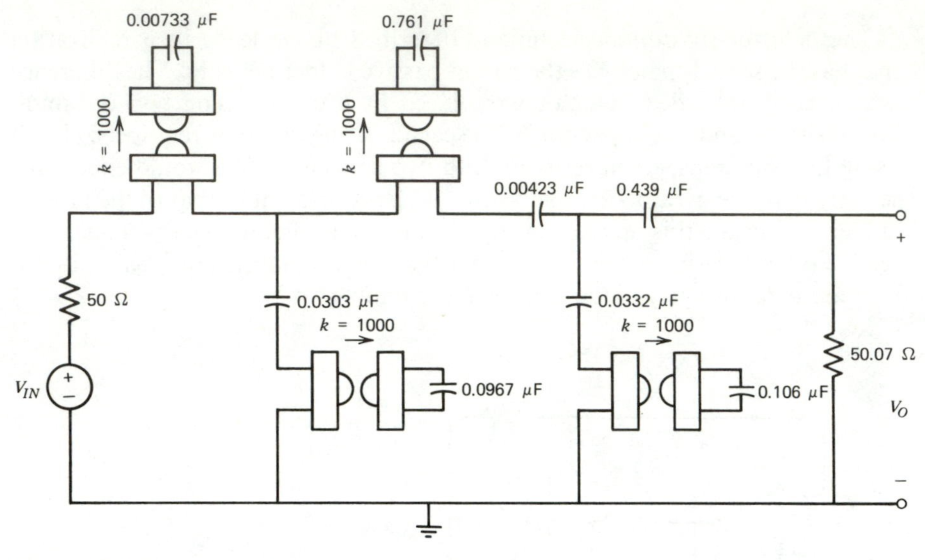

Figure 11.8 Gyrator-*RC* equivalent circuit for passive circuit of Figure 11.2.

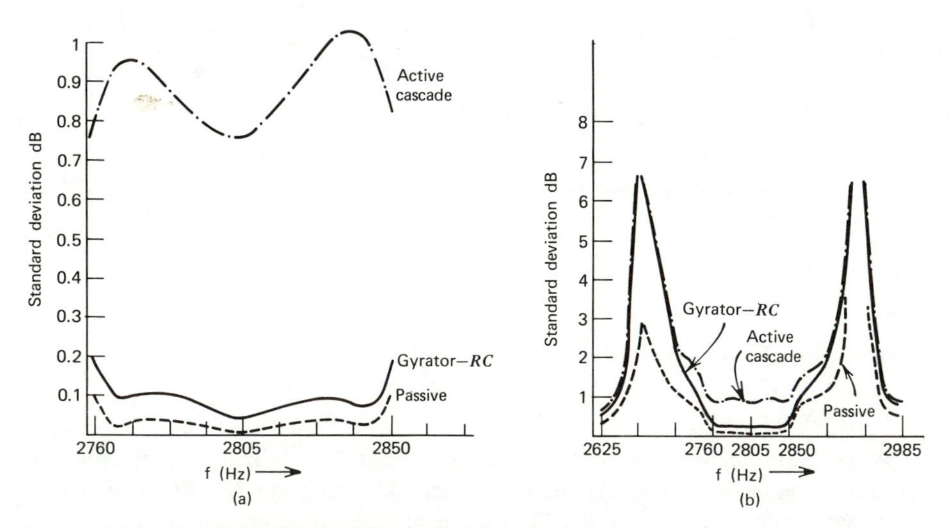

Figure 11.9 Standard deviation of gyrator-*RC* realization: (*a*) Passband. (*b*) Stopbands and passband.

11.3 TRANSFORMATION OF ELEMENTS USING THE FDNR

In this section we present an alternate way of obtaining an active *RC* equivalent of the *LC* ladder network in which the inductors, capacitors, and resistors are transformed to a different set of elements. As will be seen, the resistors transform to capacitors, the inductors to resistors, and the capacitors to a new element known as the **frequency-dependent-negative-resistance** (*FDNR*), which can be realized using an active *RC* circuit.

The principle of element transformations is related to the impedance scaling of a network, and is explained as follows. Consider the ladder network in Figure 11.10*a* where the voltage transfer function is $T = V_O/V_{IN}$, and the input impedance is $Z_{IN} = V_{IN}/I_{IN}$. If the impedance of each branch is multiplied by α,

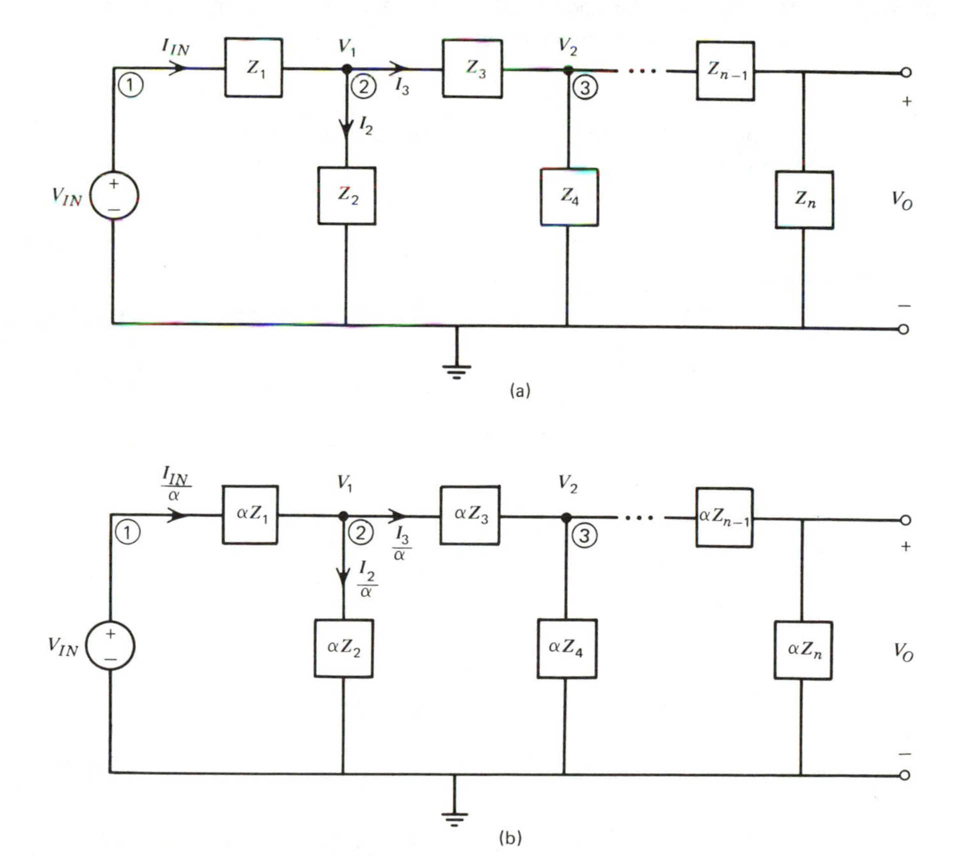

Figure 11.10 Impedance scaling: (*a*) Original ladder network. (*b*) Scaled network.

the input impedance of the scaled network, shown in Figure 11.10*b*, is αZ_{IN} and the input current is I_{IN}/α. The voltage drop across the first branch, αZ_1, is therefore

$$(\alpha Z_1)\left(\frac{I_{IN}}{\alpha}\right) = I_{IN} Z_1 \tag{11.9}$$

which is the same as the voltage drop across Z_1 in the original network. Thus, the voltage at node 2 in the scaled network must be the same as that in the original network. By extending this argument to the rest of the ladder, the scaling is seen to divide all the branch currents by a factor α, while all the node voltages remain unchanged. Therefore, the voltage transfer function of the network is unaffected by impedance scaling.

If, instead of scaling by α, the impedances were scaled by the factor K/s, it should be evident that then also the voltage transfer function would remain unchanged. It can easily be seen that such a scaling by K/s results in the following transformations on the elements:

Impedances in the Passive Network			Impedances in the Network Scaled by K/s		
—⩘⩘— (resistor)	R	→	—||— (capacitor)	$\frac{RK}{s}$	(11.10a)
—ooo— (inductor)	sL	→	—⩘⩘— (resistor)	LK	(11.10b)
—||— (capacitor)	$\frac{1}{sC}$	→	?	$\frac{K}{s^2C}$	(11.10c)

Resistors and inductors in the passive network transform to capacitors and resistors, respectively. However, each capacitor in the passive network transforms to a new element, which has an impedance given by

$$Z(s) = \frac{1}{s^2 D} \tag{11.11}$$

where $D = C/K$. For frequencies on the $j\omega$ axis, $s = j\omega$, we have

$$Z(j\omega) = \left.\frac{1}{s^2 D}\right|_{s=j\omega} = -\frac{1}{\omega^2 D} \tag{11.12}$$

Thus the impedance is real, but frequency dependent; hence its name, the frequency-dependent-negative-resistance (abbreviated as *FDNR*). The symbolic representation of the *FDNR* element is shown in Figure 11.11.

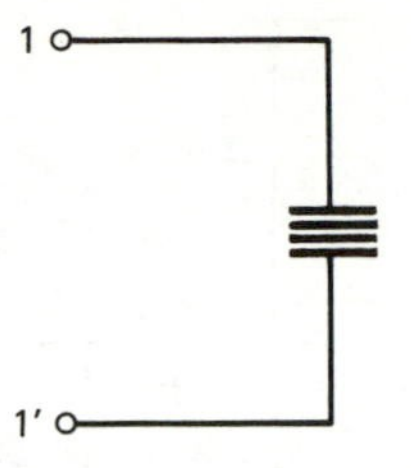

Figure 11.11 Symbolic representation of the *FDNR*.

An active *RC* realization of the *FDNR*, due to Bruton [2], is shown in Figure 11.12. Analyzing this circuit, the node voltages are

node A:

$$Y_1 V_{IN} - Y_1 V_2 = I_{IN}$$

node B:

$$(Y_2 + Y_3)V_{IN} - Y_2 V_2 - Y_3 V_3 = 0$$

node C:

$$(Y_4 + Y_5)V_{IN} - Y_4 V_3 = 0$$

Solving, we get

$$Z_{IN} = \frac{Y_2 Y_4}{Y_1 Y_3 Y_5} = \frac{Z_1 Z_3 Z_5}{Z_2 Z_4} \tag{11.13}$$

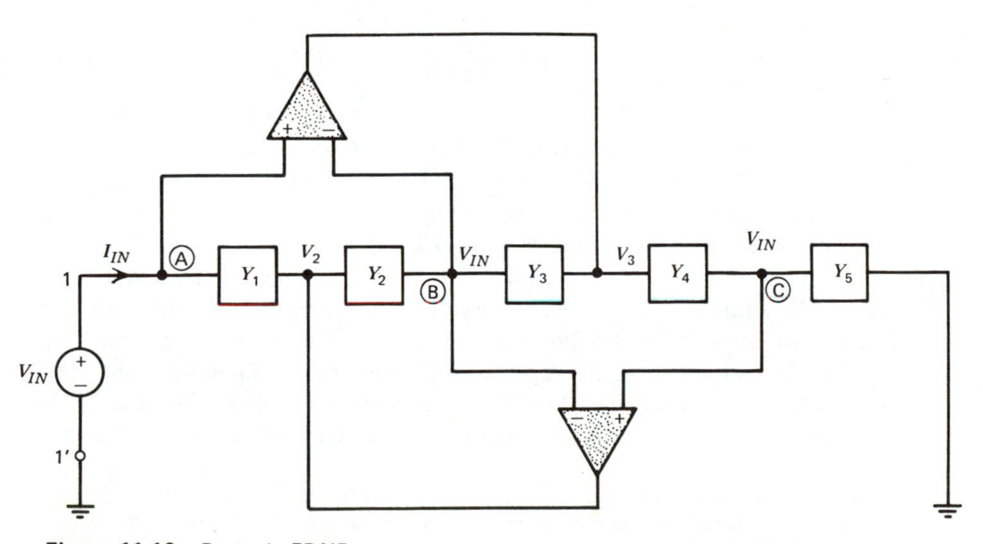

Figure 11.12 Bruton's *FDNR*.

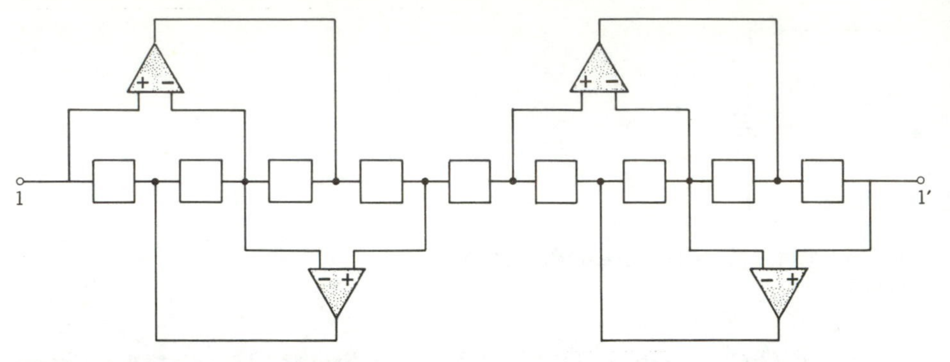

Figure 11.13 The floating *FDNR*.

The realization of the impedance given by Equation 11.12 is achieved by making any two of the impedances Z_1, Z_3, or Z_5 capacitors, and the remaining impedances resistors.* In particular, Bruton and Lim [3] have shown that the specific choice:

$$Z_1 = Z_3 = \frac{1}{sC_D} \qquad Z_2 = Z_4 = Z_5 = R_D \tag{11.14}$$

results in the lowest sensitivity to the active elements. With this choice, the impedance realized is

$$Z_{IN} = \frac{1}{s^2 R_D C_D^2} = \frac{1}{s^2 D} \tag{11.15}$$

where

$$D = R_D C_D^2 \tag{11.16}$$

The circuit of Figure 11.12 has one of the input terminals connected to ground, and as such it can only realize a grounded *FDNR*. The floating *FDNR* is realized by placing two of these circuits back-to-back (Figure 11.13), just as was done in the realization of the floating active inductor.

In review, the active *RC* synthesis using *FDNR*'s goes as follows. The given approximation function is first realized by an *LC* double-terminated ladder network. The equivalent active *RC* network is then obtained by replacing the resistors R by capacitors of value $1/RK$ (where K is any convenient positive constant); the inductors L are replaced by resistors of value LK; and, the capacitors C are replaced by FDNR's whose D value is C/K. The *FDNR* is realized using Bruton's circuit, following Equation 11.14, where the resistors R_D

* A comparison of Equation 11.7 and 11.13 shows that this circuit can also be used to realize the active inductor. In fact, the circuit was first proposed by Antoniou [1] for the gyrator-*RC* realization of an inductor.

can be chosen arbitrarily and the capacitors C_D, from Equation 11.16, are given by

$$C_D = \sqrt{\frac{D}{R_D}} = \sqrt{\frac{C}{KR_D}} \tag{11.17}$$

Applying the above synthesis technique to the passive circuit of Figure 11.2, we get the FDNR-*RC* circuit shown in Figure 11.14 (in this circuit $K = 10^6$, $R_D = 1000$). The transformation K/s does not affect the topology of the ladder circuit, nor does it affect the transfer function realized. Therefore, we again expect this *FDNR-RC* circuit to have as low a sensitivity as the passive circuit except for the increase in the number of elements and the imperfections in the op amps which realize the *FDNR*. In fact, a Monte Carlo analysis of the circuit of Figure 11.14, assuming the same component types as in the cascaded realization of Equation 11.1, results in a gain deviation which is almost the same as for the gyrator-*RC* circuit (Figure 11.9). The gain deviation is much smaller than for the cascaded circuit, but not quite as small as for the passive circuit.

In the above example the gyrator-*RC* and the *FDNR-RC* realizations required the same number of components and the gain deviations were almost equivalent. Thus, both from a cost and performance standpoint, there is little to choose between the two. In general, however, the number of components needed for the two approaches will be different and this factor will dictate the choice between the two circuits. In particular, for an *FDNR-RC* realization the total number of op amps is equal to twice the number of grounded capacitors plus four times the number of floating capacitors in the passive circuit. In contrast,

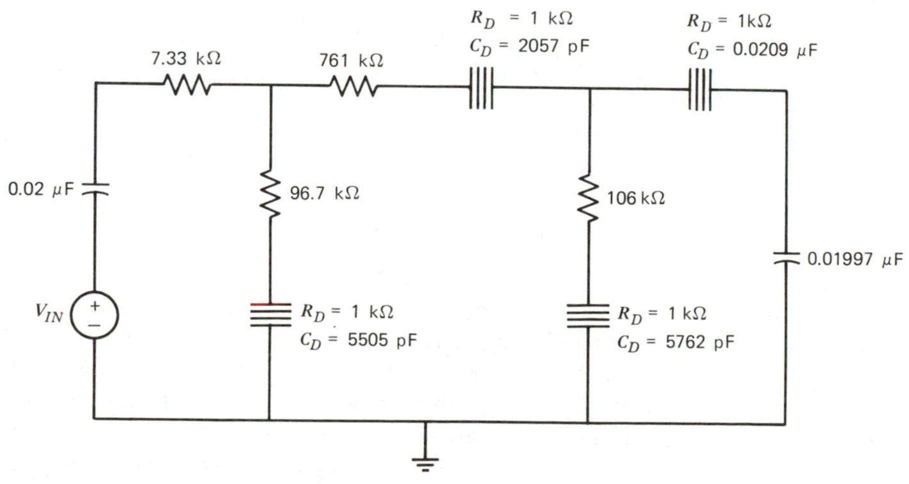

Figure 11.14 *FDNR-RC* realization for band-pass circuit of Figure 11.2.

the number of op amps needed in the gyrator-*RC* realizations is twice the number of grounded inductors and four times the number of floating inductors in the passive circuit. Therefore, in low-pass filters, where each series arm must contain a floating inductor, the gyrator-*RC* realization will invariably need more op amps than the *FDNR-RC* and the latter will be the better approach. However, in high-pass filters, the reverse is true—the gyrator-*RC* realization will require a lower number of op amps and will therefore be the better choice. The band-pass and band-reject realizations will have to be evaluated on a case-by-case basis.

11.4 A COUPLED TOPOLOGY USING BLOCK SUBSTITUTION

In the gyrator-*RC* realization of the passive ladder structure the inductor was replaced by an active *RC* circuit. In the *FDNR-RC* realization, all the elements were transformed to new elements: the inductors to resistors, the resistors to capacitors, and the capacitors to an active *RC* circuit. Thus, in both these realizations one particular element type was realized actively. In this section a different approach is presented, in which the active circuit is obtained by simulating the equations that describe the topology of the passive ladder structure.

To explain the method let us first study the topology of the general ladder network shown in Figure 11.15. Here the series elements are represented by admittances and the shunt elements by impedances. The equations describing this circuit are:

$$\begin{aligned} I_1 &= Y_1(V_{IN} - V_2) \\ V_2 &= Z_2(I_1 - I_3) \\ I_3 &= Y_3(V_2 - V_4) \\ V_4 &= Z_4(I_3 - I_5) \\ I_5 &= Y_5(V_4 - V_6) \\ V_O &= V_6 = Z_6 I_5 \end{aligned} \tag{11.18}$$

The transfer function V_O/V_{IN} can, of course, be obtained from these equations by eliminating the intermediate variables I_1, V_2, I_3, V_4, and I_5. Equation 11.18 can be represented by the block diagram depicted in Figure 11.16. Observe that the output of each block is fed back to the input of the preceding block. In contrast with the cascaded topology, these blocks are not isolated from each other, and any change in one block affects the voltages and currents in all the other blocks. This is a characteristic of the family of **coupled topologies**. One expected effect of this coupling between the blocks is that it makes the tuning of the complete network more difficult. On the other hand, it has been observed

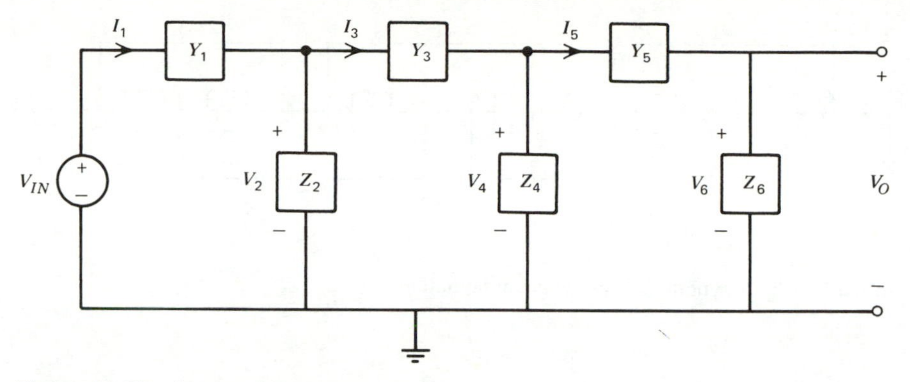

Figure 11.15 A general ladder structure.

that such coupled structures exhibit a much lower passband sensitivity than do the cascaded topologies. The gyrator-*RC* and *FDNR-RC*, being ladder structures, are examples of coupled topologies that bear out this point. Indeed, based on experimental results on these and other [4, 5, 7, 9, 11] coupled circuits, it can be hypothesized that *it is the coupling that makes the sensitivity low.*

In the remainder of this section we describe an active *RC* coupled realization that is obtained by simulating the equations describing the ladder structure. Consider the hypothetical network shown in Figure 11.17, which is defined by equations similar to Equation 11.18, except that the intermediate variables are all voltages. The defining equations of this network are:

$$\begin{aligned} E_1 &= Y_1(E_{IN} - E_2) \\ E_2 &= Z_2(E_1 - E_3) \\ E_3 &= Y_3(E_2 - E_4) \\ E_4 &= Z_4(E_3 - E_5) \\ E_5 &= Y_5(E_4 - E_6) \\ E_O &= E_6 = Z_6 E_5 \end{aligned} \tag{11.19}$$

Observe that since the variables on either side of the equation are voltages, the Y's and the Z's are now voltage transfer functions. If the intermediate variables

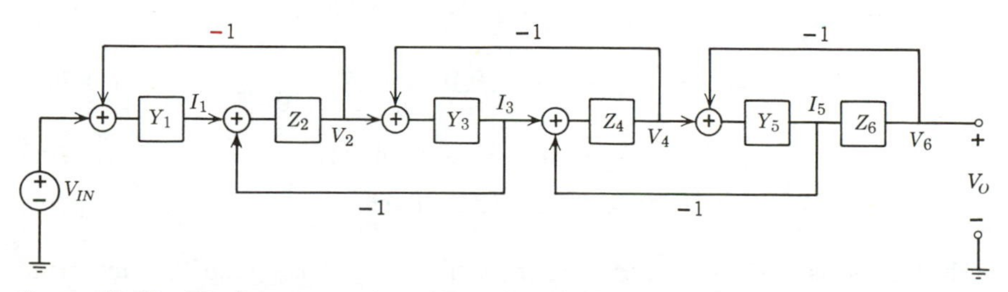

Figure 11.16 Block representation of Equation 11.18.

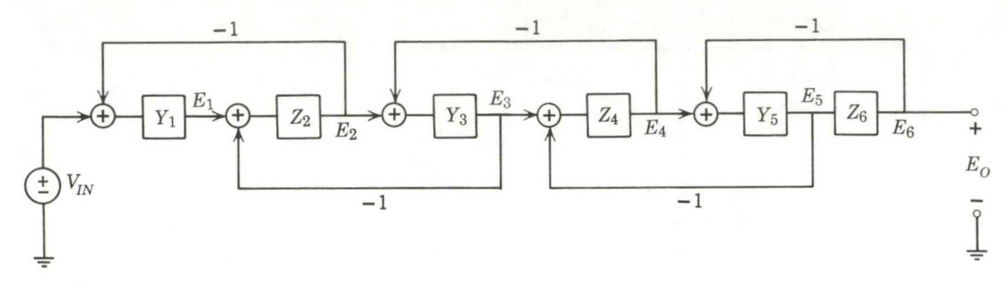

Figure 11.17 A hypothetical network configuration.

E_1, E_2, E_3, E_4, and E_5 are eliminated in this set of equations, the transfer function E_O/E_{IN} obtained will be the same as the transfer function V_O/V_{IN} for Equation 11.18. This simple discussion leads to the important result that the circuit represented in Figure 11.17 has the same transfer function as the parent ladder structure. The active realization of the passive ladder, therefore, reduces to the synthesis of the voltage transfer functions Y_i and Z_i. The poles and zeros of the functions are all in the left half s plane, or on the $j\omega$ axis, so the active RC synthesis techniques developed in the previous chapters can be used for their realization. We will refer to this method of realizing the coupled ladder topology as **block substitution.**

Let us apply this block substitution method to the passive band-pass filter of Figure 11.2. Since there are six branches in this circuit, its block diagram representation is just that shown in Figure 11.17, with

$$T_1 = Y_1 = \frac{1}{R_1 + sL_1} = \frac{1}{50 + 7.33(10)^{-3}s} \tag{11.20a}$$

$$T_2 = Z_2 = sL_2 + \frac{1}{sC_1} = 9.67(10)^{-2}s + \frac{1}{3.03(10)^{-8}s} \tag{11.20b}$$

$$T_3 = Y_3 = \frac{\dfrac{s}{L_3}}{s^2 + \dfrac{1}{L_3C_2}} = \frac{\dfrac{s}{0.761}}{s^2 + \dfrac{1}{3.22(10)^{-9}}} \tag{11.20c}$$

$$T_4 = Z_4 = sL_4 + \frac{1}{sC_3} = 0.106s + \frac{1}{3.32(10)^{-8}s} \tag{11.20d}$$

$$T_5 = Y_5 = sC_4 = 4.39(10)^{-7}s \tag{11.20e}$$

$$T_6 = Z_6 = R_2 = 50.07 \tag{11.20f}$$

The functions $-T_1$, $-T_5$, and $-T_6$ are easily realized using the basic inverting amplifier structure, as shown in Figure 11.18a, e, and f, respectively.

The function T_3 represents a second-order band-pass filter which can be realized using the three amplifier biquad circuit (Figure 10.4).* From Equation 10.17, if the fixed elements are chosen as

$$C_1 = 1\ \mu\text{F} \qquad C_2 = 0.001\ \mu\text{F} \qquad R_2 = R_3 = R$$

the remaining elements are given by

$$R_1 = \infty \qquad R_2 = R_3 = 1.79\ \text{k}\Omega \qquad R_4 = 761\ \text{k}\Omega$$

The circuit realization is shown in Figure 11.18*c*.

Consider next the circuit realization for T_2. This function can be realized as the sum of two transfer functions

$$T_2 = T_{21} + T_{22}$$

where

$$T_{21} = 9.67(10^{-2})s \qquad \text{and} \qquad T_{22} = \frac{1}{3.03(10)^{-8}s}$$

Once again, T_{21} and T_{22} are realized using the inverting amplifier structure; the functions are then summed as in Figure 11.18*b* to realize T_2. The function T_4 is realized in a similar way, to yield the circuit shown in Figure 11.18*d*.

Finally, the separate circuits for T_1, T_2, T_3, T_4, T_5, and T_6 are interconnected in accordance with the structure of Figure 11.17. The complete circuit is shown in Figure 11.19.

Since the block substitution method retains the coupled topology of the passive ladder structure, it is expected that the sensitivity of the circuit of Figure 11.19 will be similar to the gyrator-*RC* and *FDNR-RC* circuits. This is borne out by a Monte-Carlo analysis (using the same tolerances as before), which shows that the gain deviation is almost the same as that for the gyrator-*RC* realization (Figure 11.19).

In the example considered we needed first- and second-order blocks. In general, the complexity of the blocks will depend on the number of elements in the series and shunt branches of the passive ladder realization. In fact, the zero shifting technique can result in third- and even fourth-order circuit blocks, the realization of which is usually quite complex.

Because of these problems the above method is not usually used in the form presented. Szentirmai has shown a method [9] for realizing this coupled topology using biquad circuits. The structure of this so-called multiple-feedback or leapfrog [4] realization is shown in Figure 11.20. Since the method is fairly involved it will not be covered in this book.

An alternate coupled structure that is popular among filter designers, is shown in Figure 11.21. Hurtig's primary resonator block [5], Laker and Ghausi's

* The op amps 6 and 7 have been interchanged in position, to allow the realization of both $+T_3$ and $-T_3$. The synthesis equations remain the same.

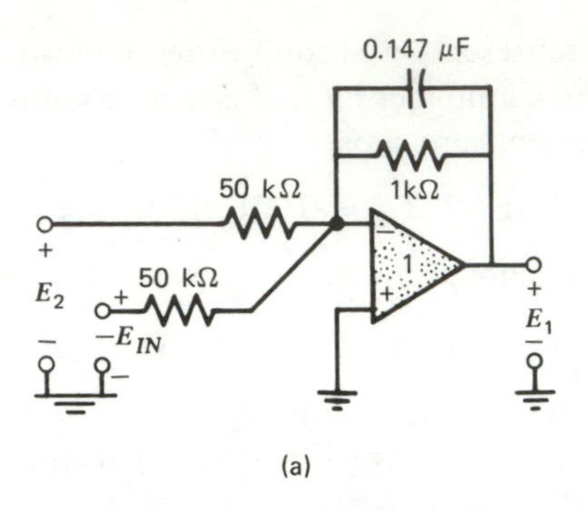
0.147 μF
1kΩ
50 kΩ
50 kΩ
1
+
E2
−
+
−E_IN
−
+
E1
−
(a)

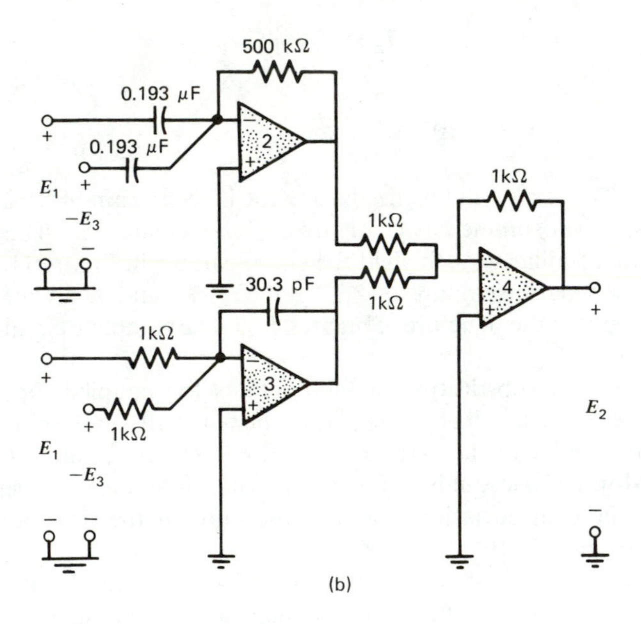
500 kΩ
0.193 μF
0.193 μF
2
E1
−E3
1kΩ
1kΩ
1kΩ
4
30.3 pF
1kΩ
3
1kΩ
1kΩ
E1
−E3
E2
(b)

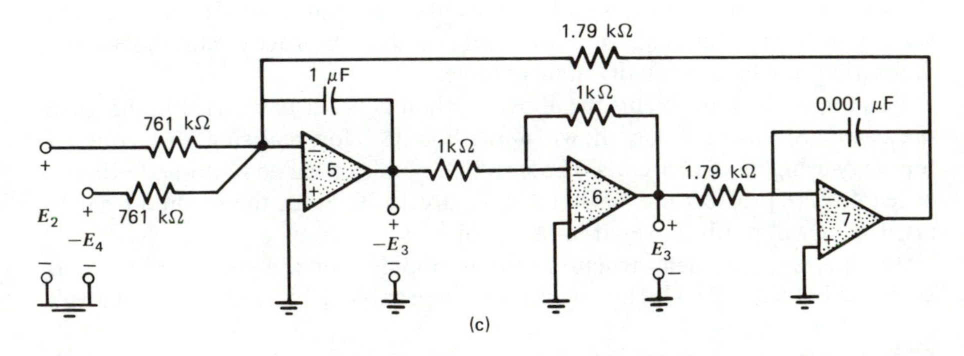
1.79 kΩ
1 μF
1kΩ
0.001 μF
761 kΩ
761 kΩ
5
1kΩ
6
1.79 kΩ
7
E2
−E4
−E3
E3
(c)

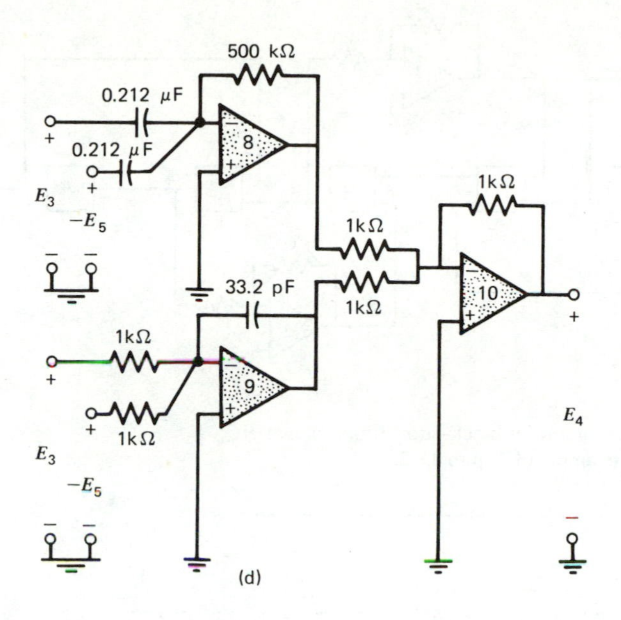

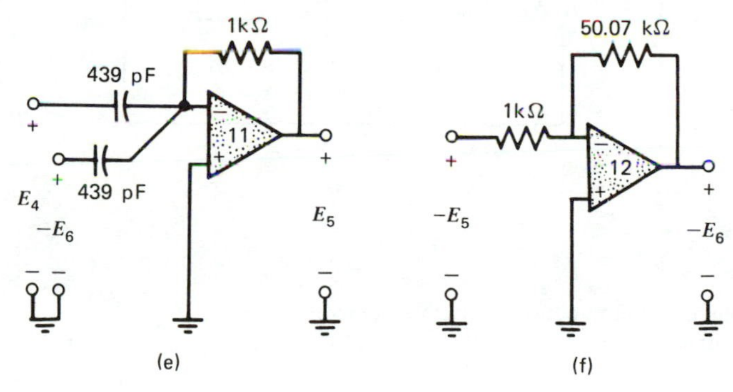

Figure 11.18 Realizations for: (*a*) $-T_1$. (*b*) T_2. (*c*) $\pm T_3$. (*d*) T_4. (*e*) $-T_5$. (*f*) $-T_6$.

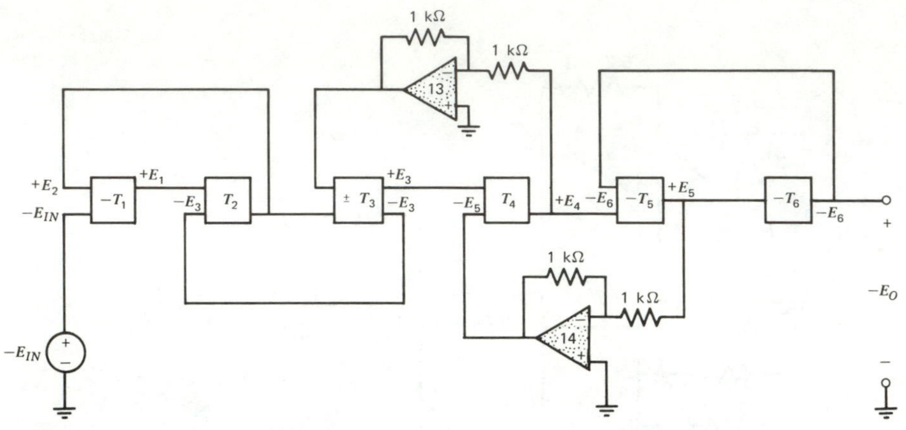

Figure 11.19 Complete circuit for block-substitution active-*RC* realization of band-pass example of Figure 11.2.

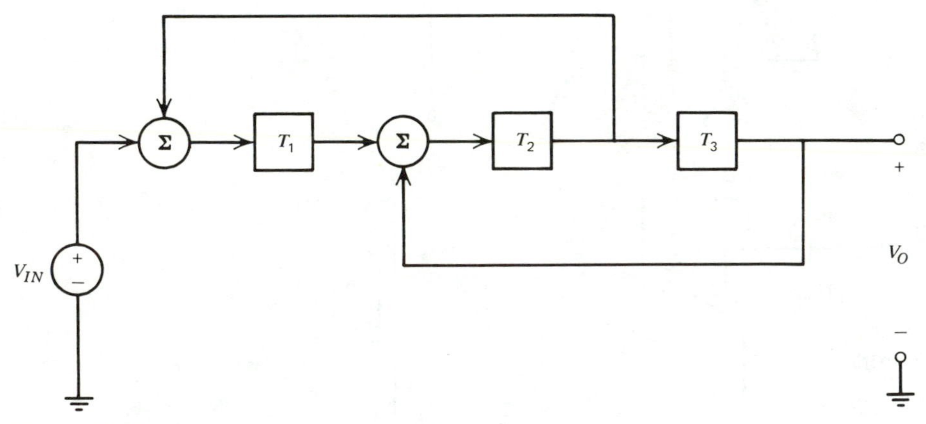

Figure 11.20 Schematic of leapfrog coupled topology, where T_1, T_2, and T_3 are biquadratics.

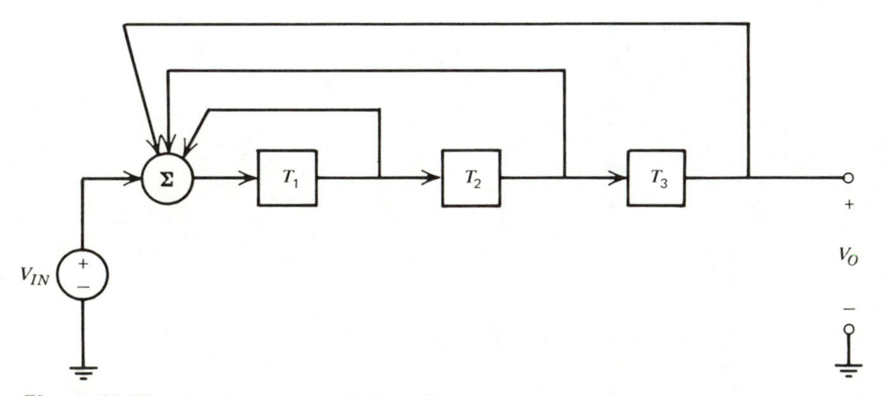

Figure 11.21 An alternate coupled topology.

follow-the-leader feedback structure [6], and Tow's shifted-comparison form [11] all use this basic topology. As shown by Tow [11], all these coupled structures have a significantly lower sensitivity than the equivalent cascaded circuit.

11.5 CONCLUDING REMARKS

In this chapter it was seen that the coupled active realizations derived from a double terminated *LC* network have a much lower passband sensitivity than the equivalent cascaded realizations. The sensitivity is not quite as low as for the equivalent passive ladder circuit because the active circuit requires many more components and the op amps are not ideal.

One common characteristic of the coupled circuit is the lack of isolation between the sections. This makes the synthesis and the tuning of such circuits more difficult than in the cascaded case. Also, these circuits will usually require more op amps than the equivalent cascaded structure.

These drawbacks need to be weighed against the sensitivity advantage, in deciding between a coupled and a cascaded realization. In most high-order filters with stringent requirements, the sensitivity advantage usually prevails, making it desirable to use a coupled structure.

FURTHER READING

1. A. Antoniou, "Realization of gyrators using operational amplifiers and their use in RC-active network synthesis," *Proc. IEE* (London), *116*, November 1969, pp. 1838–1850.
2. L. T. Bruton, "Network transfer functions using the concept of frequency-dependent negative resistance," *IEEE Trans. Circuit Theory*, *CT-16*, No. 3, August 1969, pp. 406–408.
3. L. T. Bruton and J. T. Lim, "High-frequency comparison of GIC-Simulated inductance circuits," *Int. J. Circuit Theory and Applications*, *2*, No. 4, 1974, pp. 401–404.
4. F. E. J. Girling and E. F. Good, "The leapfrog or active ladder synthesis," Part 12, *Wireless World*, July 1970, pp. 341–345; "Applications of the active ladder synthesis," Part 13, September 1970, pp. 445–450.
5. G. Hurtig III, "The primary resonator block technique of filter synthesis," *Int. Filter Symp.*, Santa Monica, Calif., April 15–18, 1972, p. 84.
6. K. R. Laker and M. S. Ghausi, "A comparison of active multi-loop feedback techniques for realizing high-order bandpass filters," *IEEE Trans. Circuits and Systems*, *CAS-21*, No. 6, November 1974, pp. 774–783.
7. K. R. Laker and M. S. Ghausi, "Synthesis of a low sensitivity multiloop feedback active RC filter," *IEEE Trans. Circuits and Systems*, *CAS-21*, No. 2, March 1974, pp. 252–259.
8. R. H. S. Riordan, "Simulated inductors using differential amplifiers," *Electronic Letters*, *3*, 1967, pp. 50–51.

9. G. Szentirmai, "Synthesis of multiple-feedback active filters," *Bell System Tech. J.*, *52*, No. 4, April 1973, pp. 527–555.
10. J. Tow and Y. L. Kuo, "Coupled biquad active filters," *IEEE Proc. Int. Symp. Circuit Theory*, April 1972, pp. 164–168.
11. J. Tow, "Design and evaluation of shifted-companion-form active filters," *Bell System Tech. J.*, *54*, No. 3, March 1975, pp. 545–567.

PROBLEMS

11.1 *Gyrator-RC filter.* The *LC* filter shown realizes a third-order low-pass Chebyshev approximation function for which $A_{max} = 0.5$ dB and $\omega_P = 1$ rad/sec.
(a) Find a gyrator-*RC* realization for the network.
(b) Scale the active circuit so that the passband edge frequency is at 1000 Hz.

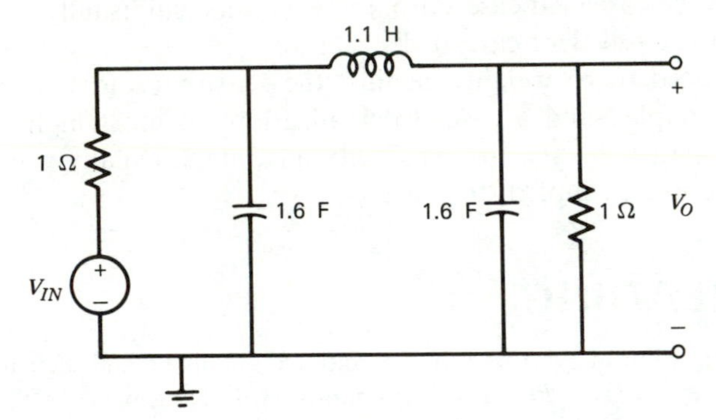

Figure P11.1

11.2 Find a gyrator-*RC* realization for the sixth-order band-pass elliptic filter shown in Figure P11.2.

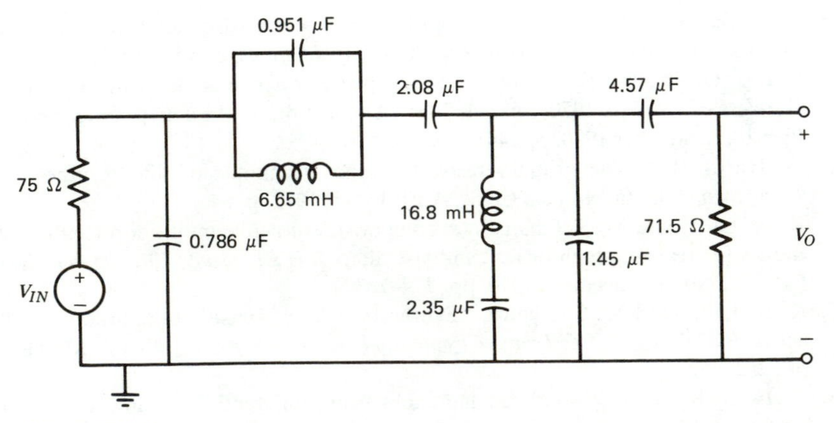

Figure P11.2

11.3 *Riordan gyrator.* Show that the Riordan gyrator circuit of Figure 11.7 realizes a floating inductor.

11.4 *Riordan gyrator, real op amps.* In this problem we investigate the nonideal properties of the Riordan active inductor (Figure 11.6 where $Y_1 = Y_3 = Y_4 = Y_5 = 1/R$ and $Y_2 = sC$) due to the finite gain characteristics of the op amps. If the op amp gains (assumed the same) can be approximated by $A(s) = A_0\alpha/s$, show that the inductance realized is frequency dependent. In particular, show that at the resonant frequency, $\omega = 1/RC$, the inductance realized is

$$L \cong CR^2\left(1 + \frac{2\omega}{A_0\alpha}\right)$$

and the quality factor of the inductor (defined in Equation 3.15) is

$$Q_L \cong -\frac{A_0\alpha}{4\omega}$$

What is the effect of using a negative quality factor inductor in a gyrator-*RC* circuit?

11.5 *Antoniou gyrator.* Consider the active inductor realization using Antoniou's circuit of Figure 11.12, where $Y_1 = Y_2 = Y_3 = Y_5 = 1/R$ and $Y_4 = sC$. If the op amp gains are characterized by $A_1(s) = s/A_1\alpha_1$ and $A_2(s) = s/A_2\alpha_2$, show that the inductance realized at the resonant frequency $\omega = 1/RC$ is

$$L = CR^2\left(1 + \frac{2\omega}{A_1\alpha_1} + \frac{2\omega}{A_2\alpha_2}\right)$$

and that the quality factor of the inductor is infinite. It is this high quality factor that makes the Antoniou circuit one of the best active inductor realizations.

11.6 *Single op amp gyrator.** Orchard and Wilson's single op amp realization of an active inductor is shown in Figure P11.6. Show that the inductance realized is 4C. (It may be mentioned that the quality factor of this active inductor is considerably lower than that attained by Antoniou's circuit.)

* H. J. Orchard and A. N. Wilson, Jr., "New active gyrator circuit," *Electronics Letters, 10*, No. 13, June 27, 1974, pp. 261–262.

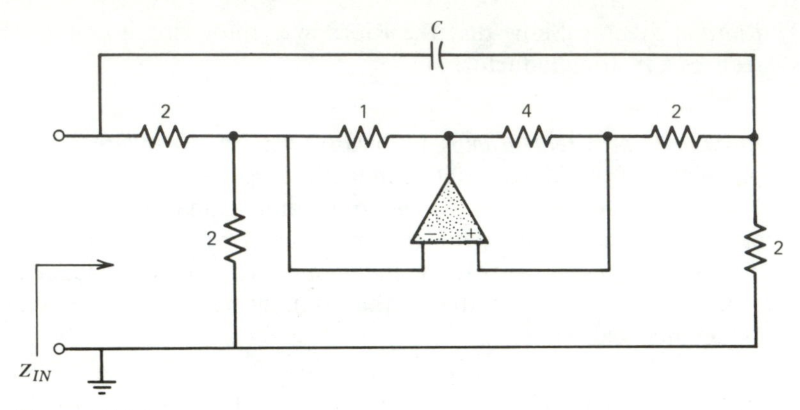

Figure P11.6

11.7 *FDNR-RC filter.* Find an FDNR-*RC* realization for the third-order low-pass circuit of Figure P11.1.

11.8 The dual of the passive circuit will often require less op amps for the equivalent FDNR-*RC* realization. The circuit shown is obtained from the circuit of Figure P11.1 by first replacing the voltage source by its Norton equivalent current source and then taking the dual. Find an FDNR-*RC* realization for the dual network. Observe that this active realization uses less op amps than the active circuits of Problems 11.1 and 11.7.

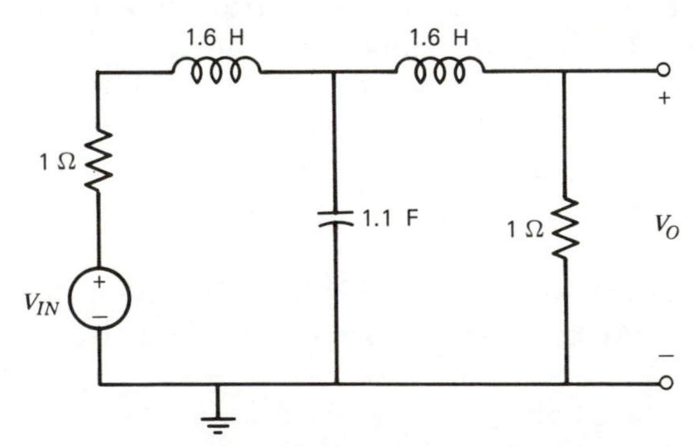

Figure 11.8

11.9 The circuit shown is derived from the circuit of Figure P11.2 by first replacing the voltage source by an equivalent current source, then taking the dual of the network, and finally impedance scaling the dual network to obtain a 75 Ω source termination.

(a) Determine the remaining element values.

(b) Find an FDNR-*RC* equivalent circuit for the passive network.

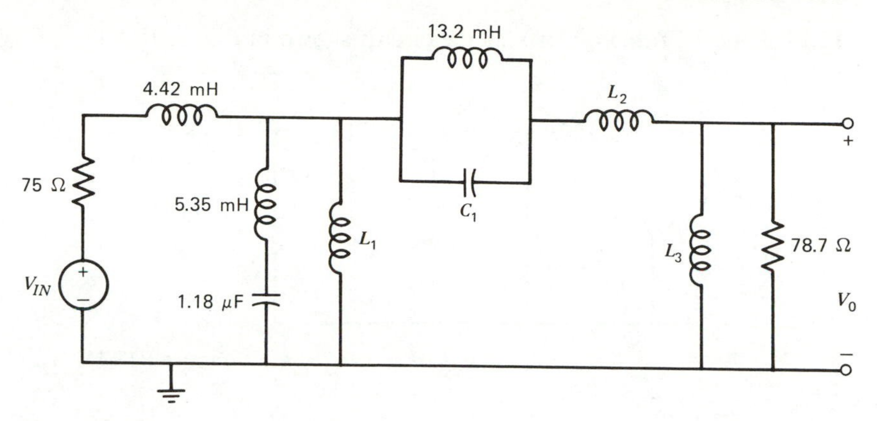

Figure P11.9

11.10 *Gyrator-RC vs FDNR-RC.* Determine the number of op amps needed for the circuit shown and for its dual

(a) Using the FDNR-*RC* realization with the Bruton circuit.

(b) Using the gyrator-*RC* realization with the Riordan circuit.

Determine the least expensive of the four active circuits, if the cost is given by

$$\text{Cost} = N_A + 0.5N_C + 0.1N_R$$

where

N_A is the number of op amps

N_C is the number of capacitors

N_R is the number of resistors

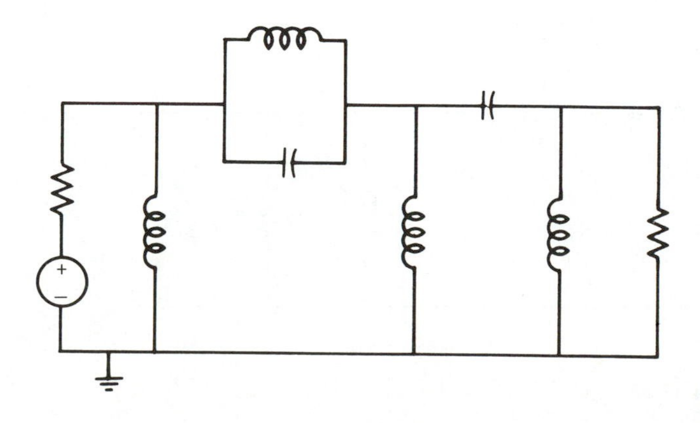

Figure P11.10

11.11 Repeat Problem 11.10 for the circuit shown in Figure P11.11.

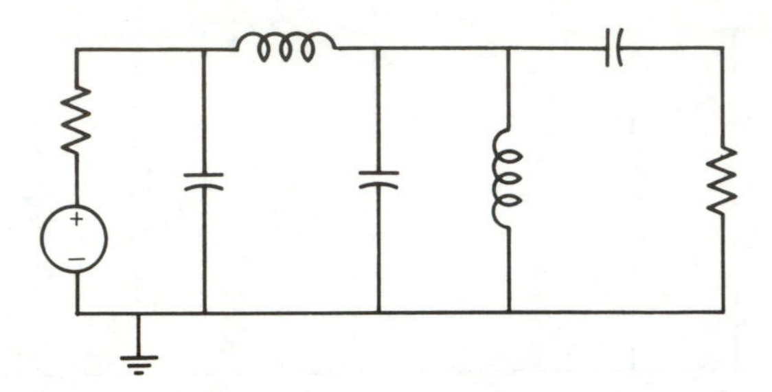

Figure P11.11

11.12 Repeat Problem 11.10 for the circuit of Figure P11.12.

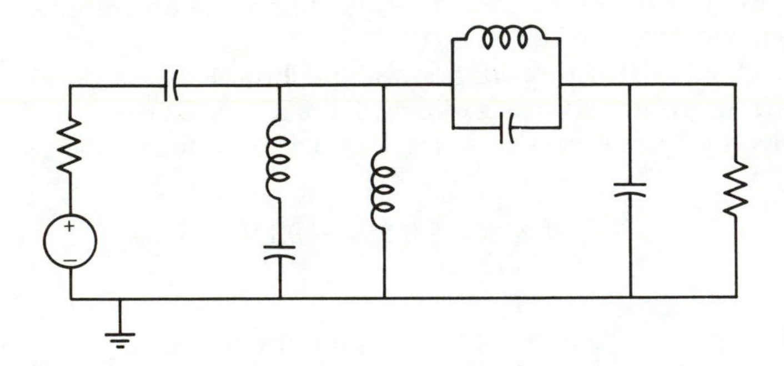

Figure P11.12

11.13 *Block substitution coupled network.* Use the block substitution method to find an active-*RC* realization for the passive circuit of Figure P11.13. (*Hint*: use the three amplifier band-pass circuit of Figure 11.18*c*).

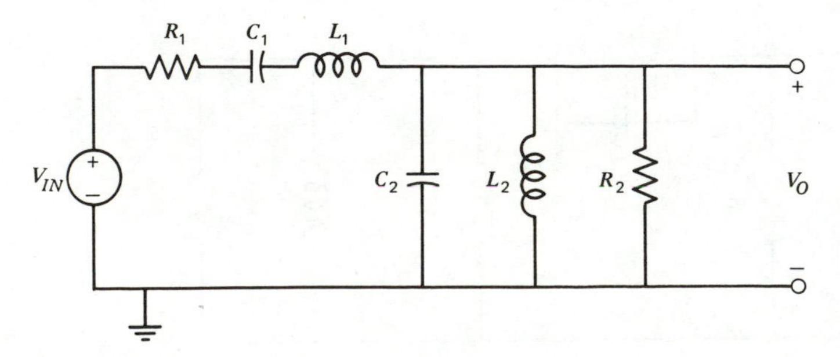

Figure P11.13

11.14 Repeat Problem 11.13 for the circuit of Figure P11.14.

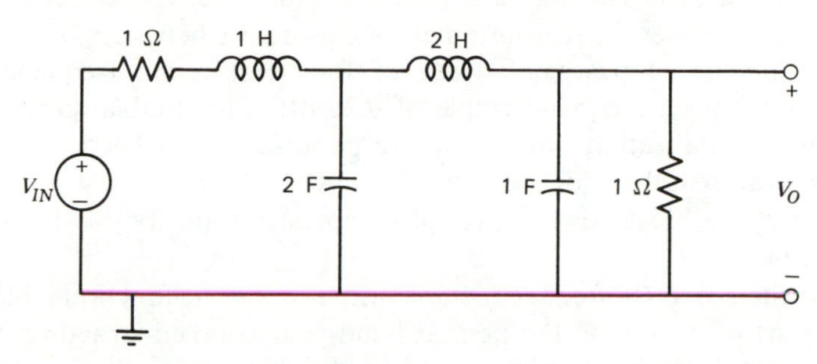

Figure P11.14

11.15 Repeat Problem 11.13 for the circuit of Figure P11.15.

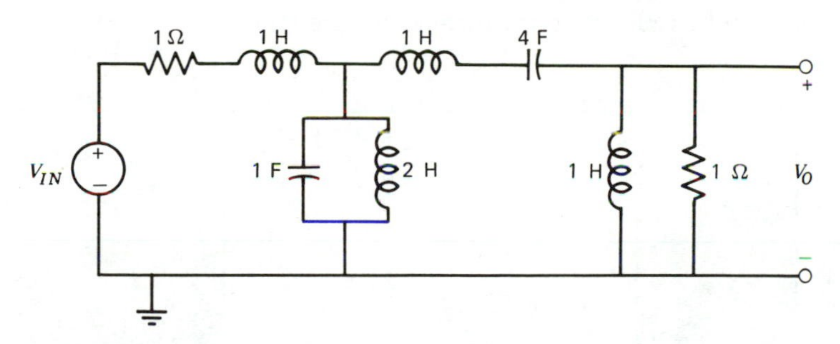

Figure P11.15

11.16 Repeat Problem 11.13 for the circuit of Figure P11.16.

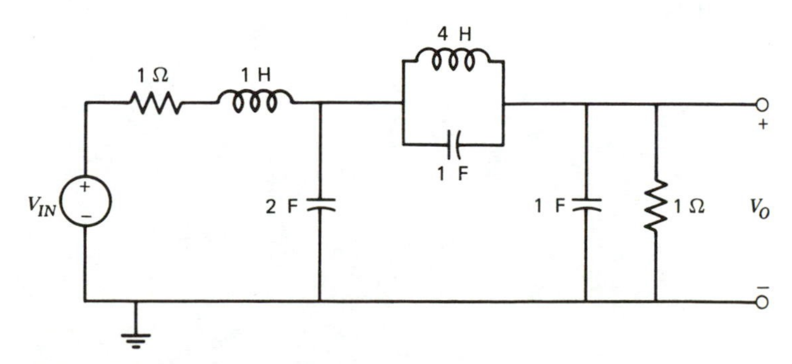

Figure P11.16

11.17 *Chebyshev band-pass coupled filter.* This problem describes a coupled biquad realization for all-pole (Butterworth, Chebyshev) band-pass functions. Consider the realization of a sixth-order Chebyshev band-pass filter which has a passband width of 1000 Hz, center frequency at 10,000 Hz, and a passband ripple of 0.25 dB. The double terminated *LC* ladder realization for the corresponding normalized low-pass function can readily be derived, and is shown in Figure P11.8.

(a) Sketch the block diagram coupled representation for the low-pass circuit.

(b) Use the *LP* to *BP* frequency transformation on the functional blocks in part (a) to obtain the desired band-pass coupled structure. Synthesize this structure using three amplifier biquads.

11.18 *Butterworth band-pass coupled filter.* Find a Butterworth approximation function for the band-pass requirements sketched in Figure P11.18. Synthesize this function using a coupled biquad topology, following the procedure outlined in Problem 11.17. (*Hint*: the normalized low-pass circuit can be reduced to the form of Figure P11.8.)

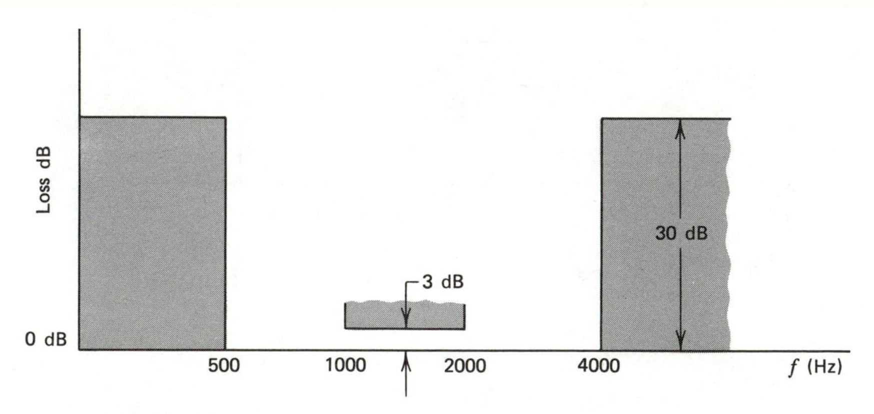

Figure P11.18

12.

EFFECTS OF REAL OPERATIONAL AMPLIFIERS ON ACTIVE FILTERS

In this chapter we investigate the properties of real op amps and their effects on the design and performance of active filters. The characteristic of the op amp that is of greatest concern in filter synthesis is its finite, frequency-dependent gain. One effect of this characteristic is that the transfer function realized is slightly different from that computed using the ideal op amp. A second, more serious, effect is the possible instability of the circuit. These problems will be discussed in some depth in this chapter. Other characteristics of op amps that will be described are dynamic range, slew-rate limiting, offset-voltage, input-bias and input-offset currents, common-mode signals, and noise.

12.1 REVIEW OF FEEDBACK THEORY AND STABILITY

The study of the stability of networks requires some concepts from feedback theory, which are reviewed in this section. A block diagram representation of a basic feedback system is drawn in Figure 12.1. In this figure:

$A(s)$ is called the **forward gain** of the amplifying stage, and
$\beta(s)$ is the **feedback factor**, which is the fraction of the output voltage returned to the input.

From this block diagram the input and output voltages are related by

$$(V_{IN} - \beta V_0)A = V_O$$

Thus, the transfer function V_O/V_{IN}, also known as the **closed loop gain**, is given by

$$\frac{V_O}{V_{IN}} = \frac{A}{1 + A\beta} \tag{12.1}$$

In Chapter 2 it was shown that the stability of a circuit is assured if none of the poles of the transfer function are in the right-half s plane and if the poles on the imaginary axis are simple. This suggests a straightforward way for checking the stability of a circuit—evaluate the transfer function; factor the denominator; and, observe the location of the roots. If none of the roots are in the right half plane and if the roots on the $j\omega$ axis are simple, the circuit is stable; otherwise,

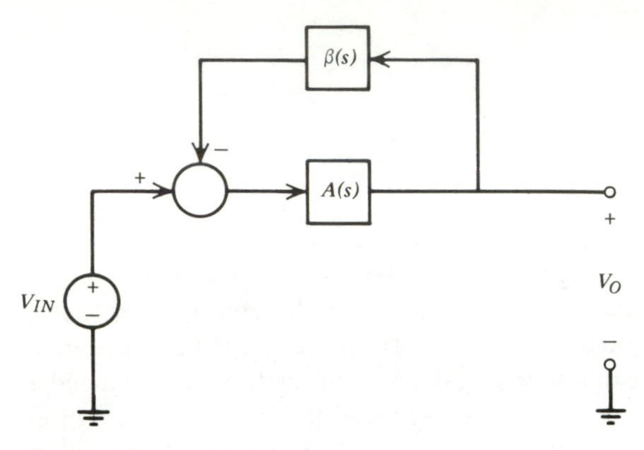

Figure 12.1 Basic feedback system block diagram.

it is unstable. The difficulty with this straightforward approach is that it requires the factoring of a polynomial—a step that increases in complexity with the order of the polynomial.

An alternate procedure, developed by Nyquist, depends on the plotting of $A\beta$ in a special way. The function $A\beta$ can be obtained directly by opening the feedback loop, as shown in Figure 12.2, to measure the transfer function E_O/E_{IN}. The transfer function $E_O/E_{IN} = A\beta$, is also known as the **open loop gain**. The Nyquist procedure consists of plotting the imaginary part of $A\beta$ versus the real part of $A\beta$ (or equivalently the magnitude and phase of $A\beta$) as a function of frequency, for $0 \leq \omega < \infty$. Some sample sketches are shown in Figure 12.3.

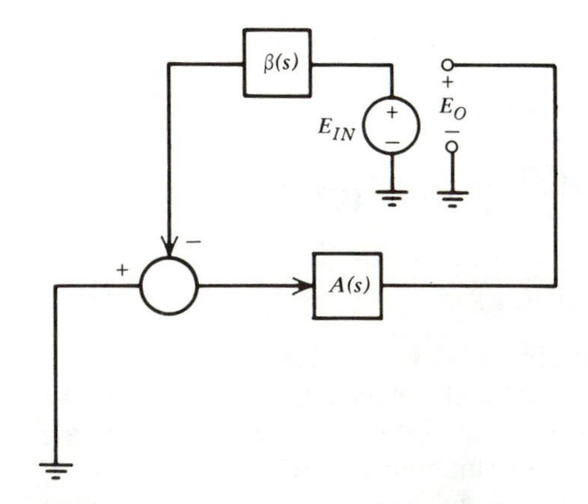

Figure 12.2 Breaking the loop to measure open loop gain $A(s)\beta(s)$.

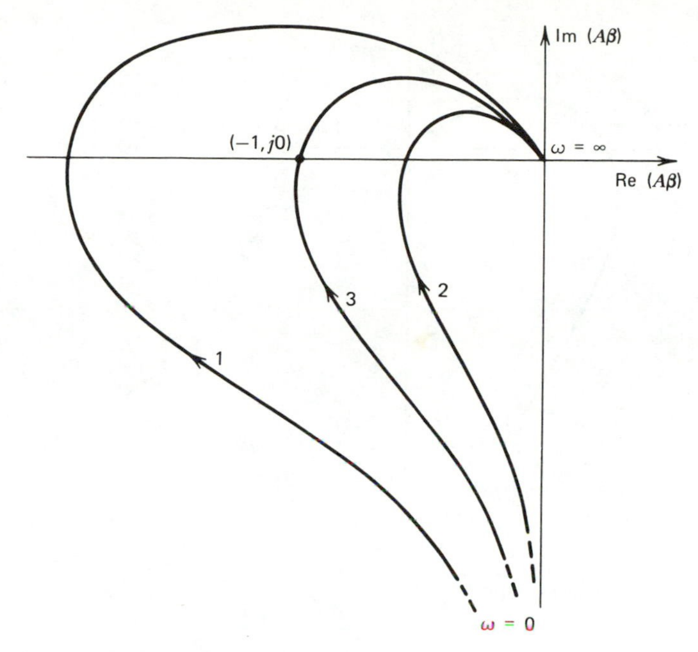

Figure 12.3 Nyquist plots for (1) unstable (2) stable (3) marginally stable systems.

Referring to this figure, the Nyquist stability criterion states that:*

If the curve generated encircles the $(-1, j0)$ point, the system is unstable (curve 1).

If the curve does not encircle the $(-1, j0)$ point, the system is stable (curve 2).

If the curve goes through the $(-1, j0)$ point, the system is marginally stable, that is, any small change in the system could make it unstable (curve 3).

The proof of the Nyquist criterion, which can be found in several text books [11], is omitted here.

In practice, the elements comprising a circuit will vary with environmental changes and with time; therefore, some margin must be allowed in the degree of stability, when designing the circuit. Referring to the Nyquist plot of Figure 12.4, this margin of stability is related to the closeness of the plot to the $(-1, j0)$ point. The closer the plot is to $(-1, j0)$, the more apt the circuit is to becoming unstable due to deviations in the components. One way of characterizing the

* The criterion, as stated above, does not apply to functions having zeros in the right-half s plane. A more general treatment of the Nyquist criterion, which includes such functions (called non-minimum phase functions), can be found in any standard text on Control Systems [11].

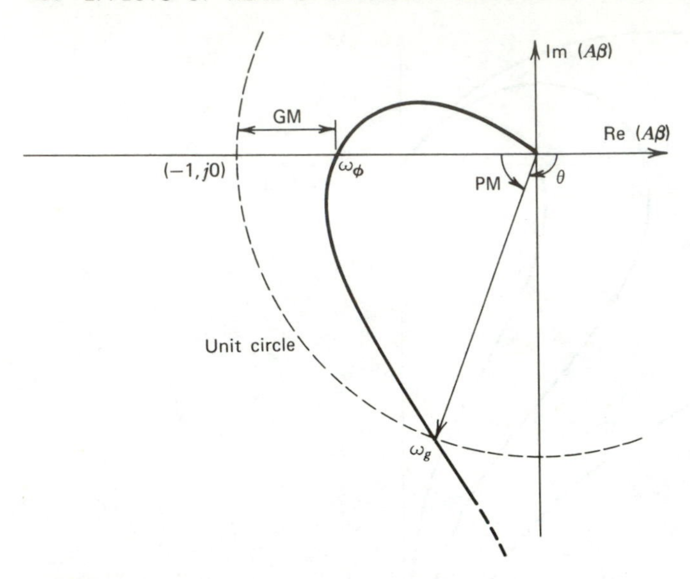

Figure 12.4 Gain margin and phase margin in a Nyquist plot.

margin of stability is by examining the magnitude of $A\beta$ at the frequency ω_ϕ where the phase is—180 degrees.* This so-called **gain margin** (GM) is defined as the additional gain (in dB) permitted before the system becomes unstable. In Figure 12.4

$$\begin{aligned} GM &= 20 \log_{10}|-1 + j0| - 20 \log_{10}|A(j\omega)\beta(j\omega)|_{\omega=\omega_\phi} \\ &= -20 \log_{10}|A(j\omega)\beta(j\omega)|_{\omega=\omega_\phi} \end{aligned} \tag{12.2}$$

A good rule of thumb is that the gain margin should be greater than 6 dB in the nominal design of op amp RC network, to allow for all the variations in the components and amplifier parameters. In some systems the gain margin alone does not adequately describe the margin of stability. This is illustrated in Figure 12.5, which shows a plot of two functions having the same gain margins but with curve 2 obviously being closer to instability than curve 1. In this case the stability is better characterized in terms of the phase at ω_g, which is the frequency† at which $|A\beta| = 1$. This so-called **phase margin** (PM) is defined as the number of degrees of additional phase lag permitted before the system becomes unstable. Referring to Figure 12.4

$$PM = \theta - (-180°) \tag{12.3}$$

* ω_ϕ is called the phase crossover frequency.
† ω_g is called the gain crossover frequency.

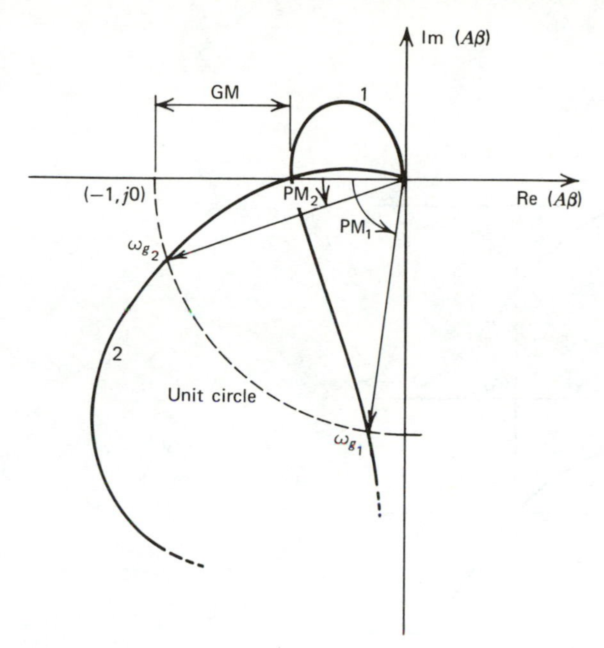

Figure 12.5 Two functions having the same gain margins but different stability margins.

where

$$\theta = \sphericalangle\, A(j\omega)\beta(j\omega)|_{\omega=\omega_g} \tag{12.4}$$

Since θ is negative, PM is a positive angle. In Figure 12.5 the phase margin for curve 2, which is closer to the critical $(-1, j0)$ point, is seen to be smaller than the phase margin for curve 1, whereas the gain margins are the same for both curves. In this case the phase margin does provide additional stability information. In general, both gain and phase margins should be considered in the analysis of stability.

A good rule of thumb for active RC circuits is to provide a phase margin of at least 45 degrees in the nominal design, to allow for the deviations in the components.

A convenient way of computing these stability margins is to plot the magnitude (in dB) and phase (in degrees) of the open loop gain $A\beta$, as illustrated in Figure 12.6. Given the poles and zeros of $A\beta$, it is quite easy to sketch the gain and phase using the Bode asymptotes, as explained in Chapter 2. Then the parameters ω_ϕ, ω_g, GM, and PM can easily be identified, as in Figure 12.6.

The Bode plot technique is very convenient for determining gain and phase margins and will be used in future discussions of the stability of active RC circuits.

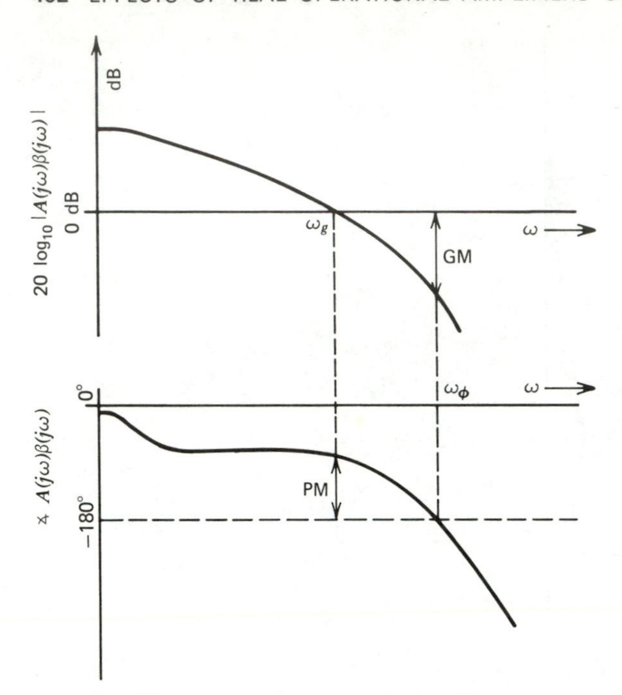

Figure 12.6 Gain and phase margins from Bode plots.

12.2 OPERATIONAL AMPLIFIER FREQUENCY CHARACTERISTICS AND COMPENSATION TECHNIQUES

Consider the simple feedback circuit of a noninverting amplifier shown in Figure 12.7. In this circuit the feedback factor β, which is the fraction of the output voltage returned to the negative summing junction, is

$$\beta = \frac{R_1}{R_1 + R_2} \tag{12.5}$$

The amplifier gain $A(s)$, for a typical op amp, can be approximated by a 3-pole function of the form

$$A(s) = 10^5 \cdot \frac{2\pi 10^3}{(s + 2\pi 10^3)} \cdot \frac{2\pi 10^5}{(s + 2\pi 10^5)} \cdot \frac{2\pi 5(10)^6}{[s + 2\pi 5(10)^6]} \tag{12.6}$$

The *dc* gain of this op amp is 10^5, or 100 dB. The low-frequency poles, at 1000 Hz and 100 kHz, are due to stray capacitors at high impedance points of the transistor circuit comprising the op amp. The additional high frequency roll-off,

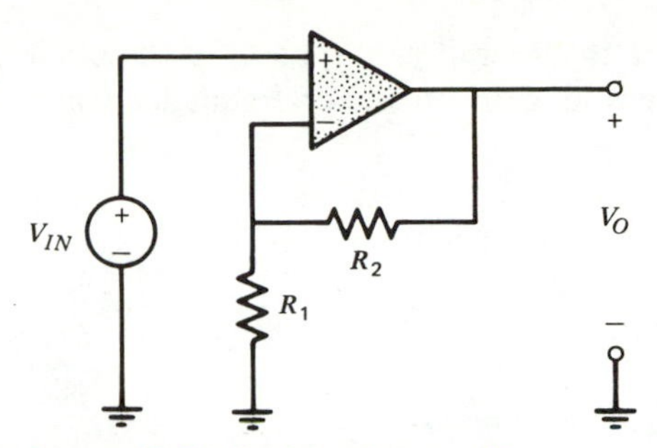

Figure 12.7 Noninverting amplifier.

represented by the pole at 5 MHz, is attributed to the finite time taken by the minority carriers to travel across the bases of the transistors.

To investigate the stability characteristics the Bode plots will be sketched for the open loop gain, $A(s)\beta$. Let us first consider the case with $R_2 = 0$ (i.e., $\beta = 1$), when all of the output voltage is fed back to the input. In this case

$$A(s)\beta = A(s) \tag{12.7}$$

and the open loop gain is given by Equation 12.6. The Bode asymptotes for this function are drawn in Figure 12.8. Here the dotted lines indicate the asymptotes

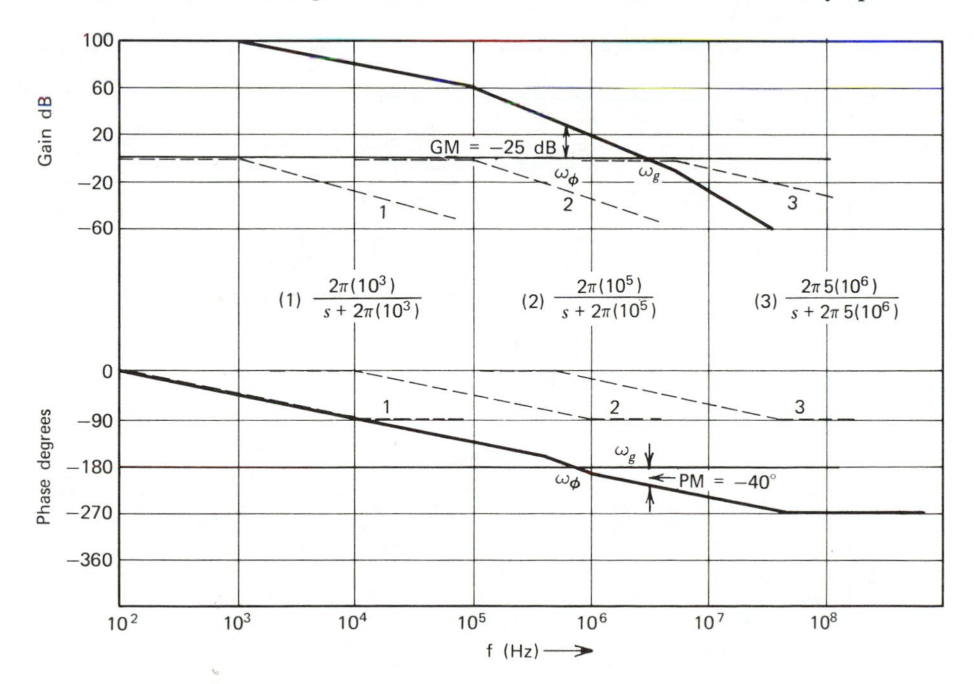

Figure 12.8 Open loop gain and phase characteristics for a typical op amp with $\beta = 1$.

for each of the poles and the solid line is the overall gain (phase) of the open loop gain. From this sketch the phase and gain crossover frequencies are, respectively

$$\omega_\phi = 2\pi 7.5(10)^5 \qquad \omega_g = 2\pi 3(10)^6$$

The corresponding phase and gain margins are

$$PM = -220^\circ - (-180^\circ) = -40^\circ$$
$$GM = -25 \text{ dB}$$

Thus, the circuit is definitely unstable.

The feedback circuit considered above provided the maximum possible feedback factor, with $\beta = 1$. One way of making the circuit stable is by decreasing the feedback factor by increasing R_2, as explained in the following. Since β is a real number, independent of frequency, the phase of the open loop gain $A(s)\beta$ will remain unchanged as β is reduced. However, the open loop gain will decrease as β is reduced. In Figure 12.9, β has been reduced by the amount

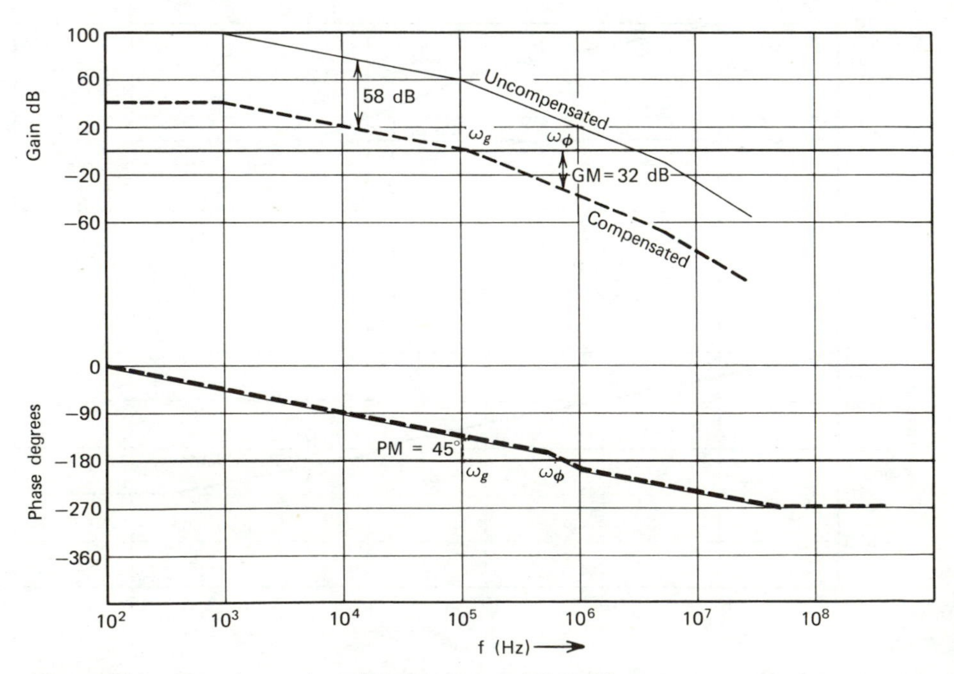

Figure 12.9 Open loop gain and phase characteristics with $\beta = 0.00126$.

necessary to attain a 45 degree phase margin. Here

$$20 \log_{10}(\beta) = -58 \text{ dB}$$

so

$$\beta = 0.00126$$

From Equation 12.5, this value of β can be achieved by making

$$R_1 = 1 \text{ k}\Omega \qquad R_2 = R_1(1 - 1/\beta) = 792.7 \text{ k}\Omega$$

With this new value of β, the parameters defining the stability margins are seen to be (dotted lines, Figure 12.9):

$$\omega_\phi = 2\pi 7.5(10)^5 \qquad \omega_g = 2\pi 1.1(10)^5$$
$$PM = 45^\circ \qquad GM = 32 \text{ dB}$$

which corresponds to a circuit with adequate gain and phase margins.

By generalizing from the above example, it is seen that one way of stabilizing a circuit is by decreasing the feedback ratio. However, we do not often have this option. In particular, the *RC* feedback circuits used in the design of filters are determined by the approximation function; thus, any attempt at reducing $\beta(s)$ by changing the feedback circuit would result in an unwanted change in the transfer function. Therefore, alternate ways of stabilizing op amp circuits, which do not depend on altering the feedback circuit, need to be investigated. One such technique is to use external *RC* circuits that introduce additional poles and/or zeros so as to increase the gain and phase margins. This method, known as **frequency compensation**, is described in the following.

It was mentioned earlier that the low frequency poles were caused by stray capacitors in the transistor circuit comprising the op amp. In particular, the dominant capacitor is a collector to base capacitor (known as the Miller capacitor) associated with a high gain transistor. If this Miller capacitor is increased by a factor α by adding an *external* capacitor* in parallel with it, it can be shown that the first pole frequency at 1 kHz will be decreased by the factor α; the second pole frequency at 100 kHz is increased to some very high frequency; while the third pole at 5 MHz remains unchanged (see Problem 12.9). With a large enough capacitor the high frequency characteristics of the op amp can be modified so as to yield sufficient gain and phase margins, as illustrated in Figure 12.10. In this figure, the unity gain frequency of the compensated characteristic was fixed at 1 MHz, requiring the low frequency pole to be at 10 Hz. This decrease in the first pole frequency from 1000 Hz to 10 Hz was achieved by increasing the Miller capacitor by a factor of 100. Typically the

* In some op amps this capacitor is included in the integrated circuit chip (e.g., Fairchild's $\mu A741$ op amp [3]).

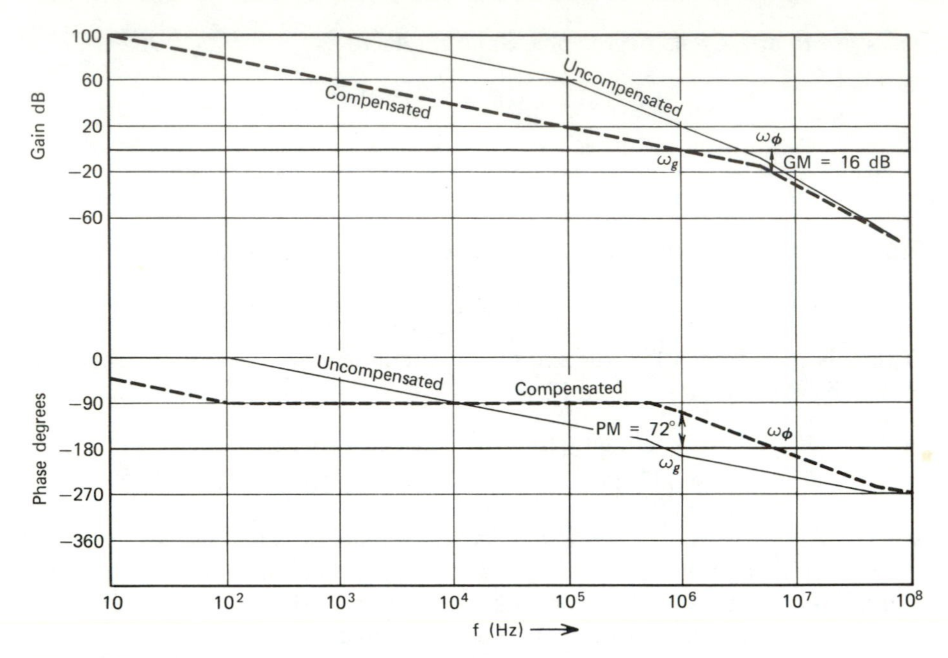

Figure 12.10 Single pole compensated gain and phase characteristics.

Miller capacitor could be 0.25 pF, requiring an external capacitor of 24.75 pF.* The parameters defining the stability margin for the compensated characteristic are:

$$\omega_g = 2\pi(10)^6 \qquad \omega_\phi = 2\pi 6.5(10)^6$$
$$PM = 72^\circ \qquad GM = 16 \text{ dB}$$

The op amp is therefore stable, having sufficient gain and phase margins. This method is referred to as **single pole compensation**. An adverse effect of single pole compensation is the unavoidable reduction in gain at the lower frequencies (note that at 1 kHz the gain has been reduced to 60 dB). As will be seen in the next section, this reduction in the op amp gain in the passband of an active filter can have a significant effect on the filter characteristic.

A method of compensation that results in much higher low-frequency gain uses external *RC* circuits to introduce two poles and a zero in the open loop gain characteristics.† This is illustrated in Figure 12.11, where the unity gain crossover frequency has been fixed at 1 MHz and the phase margin is fixed at

* In practice, the closest available capacitor is used.

† Refer to [9] for details on how this is implemented.

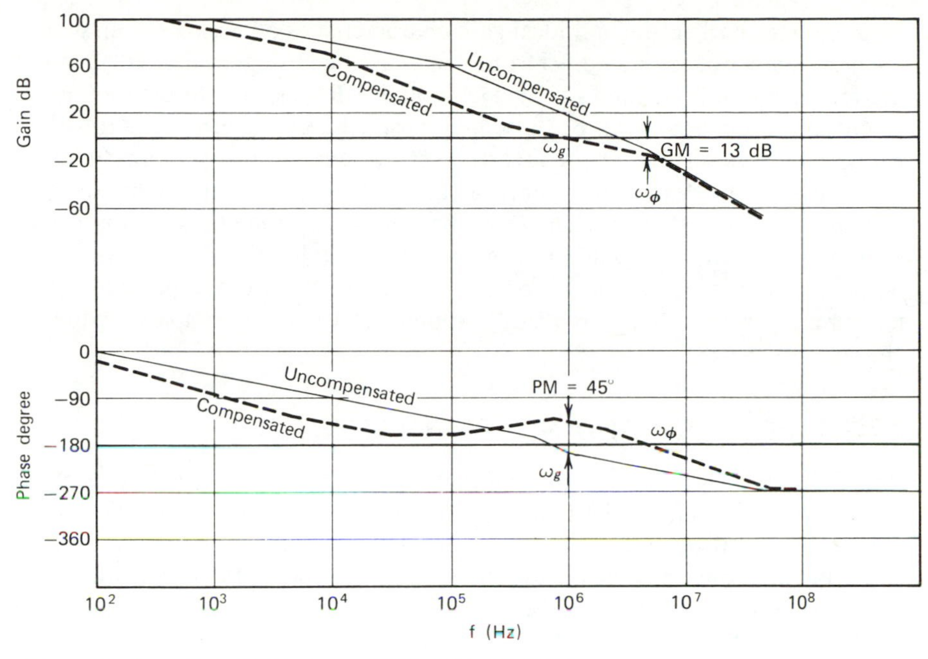

Figure 12.11 Double pole compensated gain and phase characteristics.

45 degrees. The required location of the two poles and zero can be determined graphically. With the pole-zero locations as shown, it can be seen that

$$\omega_g = 2\pi(10)^6 \qquad \omega_\phi = 2\pi 4.7(10)^6$$
$$PM = 45^\circ \qquad GM = 13 \text{ dB}$$

Thus, the compensated open loop gain is sufficiently stable. This method of compensation is often referred to as **double pole compensation**. Observe that double pole compensation provides a significant increase in the low-frequency open loop gain over the single pole compensation. At 1 kHz, for example, this increase is seen to be as much as 30 dB.

12.3 EFFECTS OF OP AMP FREQUENCY CHARACTERISTICS ON FILTER PERFORMANCE

In this section we consider the effects of finite, frequency dependent gain characteristics of the op amp on the filter response. The synthesis equations derived in previous chapters were based on an analysis using ideal op amps.

As mentioned earlier, the non-ideal gain characteristics of a real op amp makes the performance of the filter differ from that computed using the desired approximation function. A method for analyzing active RC circuits to determine the extent of this deviation will be presented in Sec. 12.3.1. Also, a method will be proposed for modifying, or predistorting, the original approximation function to correct for the deviation. In Section 12.3.2 the deviation in the circuit performance due to statistical changes in the op amp characteristics is discussed. This deviation was computed in previous chapters for the special case of an op amp characterized by a single pole at the origin. The analysis presented here is more general in that it is readily applicable to more complex frequency characteristics.

12.3.1 POLE SHIFT AND PREDISTORTION

The first step in the analysis is the computation of the change in the poles of the transfer function due to the finite gain of the real amplifier. The shifts in the zeros are ignored because their major effect is to produce small deviations in the stopband attenuation, which is usually tolerable in most filter applications.

As an aid to understanding the analysis, consider first the simple example of an integrator (Figure 12.12). The nodal equations for this circuit are

$$(0 - V^-)A(s) = V_O \tag{12.8a}$$

$$\frac{V_{IN} - V^-}{R} = (V^- - V_O)sC \tag{12.8b}$$

Thus

$$\frac{V_O}{V_{IN}} = -\frac{1}{sRC + \dfrac{1}{A(s)}(1 + sRC)} \tag{12.8c}$$

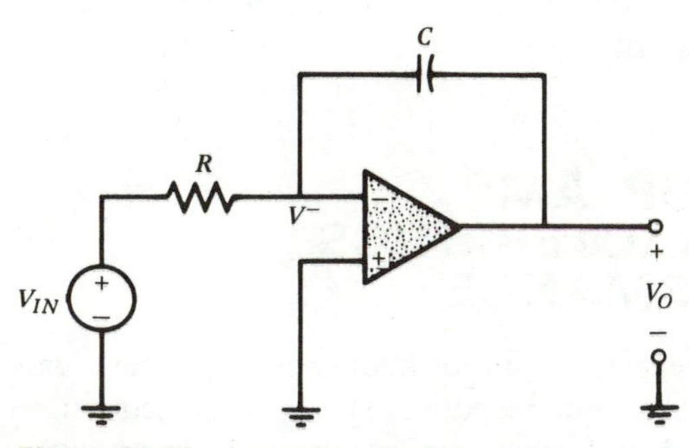

Figure 12.12 Integrator circuit.

If the amplifier were ideal, the transfer function would be

$$\left.\frac{V_O}{V_{IN}}\right|_{A=\infty} = -\frac{1}{sRC} \tag{12.9}$$

and the pole of the transfer function would be at the origin. Assuming that the single pole compensation of Figure 12.10 is used to stabilize the op amp, $A(s)$ can be approximated in the frequency range of interest as

$$A(s) = \frac{A_0\alpha}{s+\alpha} \tag{12.10}$$

where A_0, the *dc* gain, is 10^5; and α, the first pole frequency, is 10 Hz. Substituting (12.10) in (12.8c), we get

$$\frac{V_O}{V_{IN}} = \frac{-1}{sRC + \dfrac{(s+\alpha)}{A_0\alpha}(1+sRC)} \tag{12.11}$$

The poles of this function are given by the roots of

$$D(s) = s^2 + s\left(\frac{1}{RC} + A_0\alpha + \alpha\right) + \frac{\alpha}{RC} \tag{12.12}$$

Since $A_0\alpha$ is much larger than α and $1/RC$

$$D(s) \cong s^2 + sA_0\alpha + \frac{\alpha}{RC} \tag{12.13}$$

the roots of which are

$$s_{1,2} = -\frac{A_0\alpha}{2} \pm \frac{A_0\alpha}{2}\left[1 - \frac{4\alpha}{(A_0\alpha)^2RC}\right]^{1/2}$$

Using the relationship $(1-x)^{1/2} \cong 1 - x/2$ for $x \ll 1$, this expression reduces to

$$s_{1,2} \cong -\frac{A_0\alpha}{2} \pm \frac{A_0\alpha}{2}\left[1 - \frac{2\alpha}{(A_0\alpha)^2RC}\right] \tag{12.14}$$

which finally yields

$$s_1 = -\frac{1}{A_0RC} \tag{12.15}$$

$$s_2 = -A_0\alpha + \frac{1}{A_0RC} \cong -A_0\alpha \tag{12.16}$$

Thus, the pole at the origin is shifted to the left by $1/A_0RC$. Also a new pole is generated at $s = -A_0\alpha$, which is the unity gain cross over frequency (1 MHz in Figure 12.10). This new pole is so far from the dominant pole that its effect on the performance of the integrator will be negligible.

This example illustrates the two effects of using a real op amp, which are also found to occur in more general filter circuits:

1. To shift the dominant pole(s).
2. To create a new pole(s) at very high frequencies.

We next consider the problem of determining the pole shifts for second-order filter functions.* The analysis is applicable to all three biquad structures used in the cascaded topology, namely the negative feedback, the positive feedback, and the three amplifier biquad.

From Equation 8.16, the transfer function of the positive feedback structure is

$$\frac{V_O}{V_{IN}} = \frac{kN_{FF}}{(D - kN_{FB}) + kD/A(s)} \tag{12.17}$$

The negative feedback structure, from Equation 7.13, has the transfer function

$$\frac{V_O}{V_{IN}} = \frac{-N_{FF}}{N_{FB} + D/A(s)} \tag{12.18}$$

Finally, it can be shown that the three amplifier biquad of Figure 10.4 (assuming $A_1 = A_2 = A_3 = A$ and $1/A \gg 1/A^2$) has the LP transfer function:

$$\frac{V_3}{V_{IN}} = \frac{-\dfrac{1}{R_2 R_4 C_1 C_2}}{s^2 + \dfrac{1}{R_1 C_1} s + \dfrac{1}{R_2 R_3 C_1 C_2} + \dfrac{4}{A(s)}\left[s^2 + s\left(\dfrac{1}{R_1 C_1} + \dfrac{1}{4R_2 C_2} + \dfrac{1}{4R_3 C_1} + \dfrac{1}{4R_4 C_1}\right) + \dfrac{1}{4R_1 R_2 C_1 C_2}\right]} \tag{12.19}$$

From the above we see that the transfer functions of the negative feedback topology, the positive feedback topology, and the three amplifier biquad will all have the general form:

$$\frac{V_O}{V_{IN}} = \frac{k_1 D_3}{D_1 + kD_2/A(s)} \tag{12.20}$$

where

D_1 and D_2 are second-order polynomials.
D_3 is at most a second-order polynomial.
k is a positive constant.
k_1 is a positive or a negative constant.

* The analysis presented is based on Fleischer [4].

Also note that the leading coefficients (of the s^2 term) in the polynomials D_1 and D_2 are unity.

Equation 12.20, which applies to all three biquad structures, will next be used to obtain the poles of the transfer function V_O/V_{IN}, which are the roots of the denominator:

$$D(s) = D_1(s) + \frac{kD_2(s)}{A(s)} \tag{12.21}$$

In general $D(s)$ will be a third- or fourth-order polynomial [depending on the number of poles used to describe $A(s)$], and as such is not easily factored. An alternate way of finding the roots is described in the following.

Consider first the nominal case, with $A = \infty$. In this case

$$D(s)|_{A=\infty} = D_1(s) = (s - s_0)(s - s_0^*) \tag{12.22}$$

where s_0 and s_0^* are the complex conjugate roots of the second-order polynomial, $D_1(s)$. As in the example of the integrator, the effects of using a real amplifier are first, to produce a shift in the dominant poles s_0 and s_0^*; and, second, to introduce extra poles at high frequencies.

The high frequency poles can be obtained by approximating the quadratic polynomials $D_1(s)$ and $D_2(s)$ as

$$D_1(s) \cong s^2 \qquad D_2(s) \cong s^2$$

Substituting in Equation 12.21, we get

$$D(s) = s^2\left(1 + \frac{k}{A(s)}\right) \tag{12.23}$$

The high frequency poles of $D(s)$ are given by the roots of the equation

$$A(s) + k = 0 \tag{12.24}$$

Since k is a small number (verify this by inspecting Equations 12.17, 18, and 19), the high frequency roots will lie near the unity gain crossover frequency, which is around 1 MHz for most currently used op amps. Since in most active filter applications the passband is below 30 kHz, the effect of the high frequency poles will, in general, be negligible. Recall that this was also true for the example case of the integrator.

The major effect of the amplifier gain will therefore be the shift in the location of the dominant roots s_0 and s_0^*, which may be estimated as follows. Suppose s_0 and s_0^* move to $s_0 + \Delta s_0$ and $s_0 + \Delta s_0^*$, respectively. Then $D(s)$ is given by

$$D(s) = D_1(s) + \frac{kD_2(s)}{A(s)} = (s - s_0 - \Delta s_0)(s - s_0^* - \Delta s_0^*)M(s) \tag{12.25}$$

where $M(s)$ represents the high frequency factors. Since $s_0 + \Delta s_0$ is a root of $D(s)$ we can write

$$D(s)|_{s=s_0+\Delta s_0} = D_1(s_0 + \Delta s_0) + \frac{k}{A(s_0 + \Delta s_0)} D_2(s_0 + \Delta s_0) = 0 \quad (12.26)$$

But from Equation 12.22

$$D_1(s_0 + \Delta s_0) = \Delta s_0(s_0 + \Delta s_0 - s_0^*)$$

Substituting this in (12.26), the following expression is obtained for Δs_0

$$\Delta s_0 = \frac{-k}{A(s_0 + \Delta s_0)} \frac{D_2(s_0 + \Delta s_0)}{(s_0 + \Delta s_0 - s_0^*)} \quad (12.27)$$

Since none of the terms on the right-hand side of this equation changes rapidly, and since Δs_0 is expected to be small compared to s_0, the change Δs_0 is, approximately

$$\Delta s_0 \cong \frac{-k}{A(s_0)} \frac{D_2(s_0)}{(s_0 - s_0^*)} \quad (12.28)$$

Moreover, if the Q of the pole s_0 is high (so that $\omega_0 \gg \sigma_0$), then

$$s_0 - s_0^* = (-\sigma_0 + j\omega_0) - (-\sigma_0 - j\omega_0) \cong 2j\omega_0$$

Substituting this in (12.28), Δs_0 finally reduces to

$$\boxed{\Delta s_0 \cong j \frac{k}{A(s_0)} \frac{D_2(s_0)}{2\omega_0}} \quad (12.29)$$

This equation gives the shift in the dominant poles of a biquadratic filter function from the nominal case (with an infinite gain op amp), to the real world case (with a finite, frequency dependent gain op amp). The equation is applicable to the negative feedback, positive feedback, and the three amplifier biquads.

Observe that since the pole shift is inversely proportional to the gain of the op amp at the pole frequency, the double pole compensation scheme (Figure 12.11) will introduce a much smaller pole shift than the single pole compensation (Figure 12.10). This is illustrated in the following example.

Example 12.1

A band-pass filter with a pole Q of 5 and a pole frequency of 1 kHz is synthesized using the negative feedback biquad circuit of Figure 9.3*a*. The synthesis equations for this circuit were developed in Section 9.2. Compute the per-unit change in the pole location, $\Delta s_0/s_0$, for (a) the single pole compensated op amp of Figure 12.10, and (b) the double pole compensated op amp of Figure 12.11.

Solution

The transfer function for this circuit, derived in Section 9.2 (Equation 9.14), is:

$$\frac{V_O}{V_{IN}} = \frac{-s/R_1C_2}{s^2 + s\left(\frac{1}{R_2C_1} + \frac{1}{R_2C_2}\right) + \frac{1}{R_1R_2C_1C_2} + \frac{1}{A(s)}\left[s^2 + s\left(\frac{1}{R_2C_1} + \frac{1}{R_2C_2} + \frac{1}{R_1C_2}\right) + \frac{1}{R_1R_2C_1C_2}\right]} \tag{12.30}$$

The nominal pole Q and pole frequency, with $A(s) = \infty$, are given by

$$\omega_0 = \sqrt{\frac{1}{R_1R_2C_1C_2}} \tag{12.31a}$$

$$Q_0 = \frac{\sqrt{\frac{R_2}{R_1}}}{\sqrt{\frac{C_1}{C_2}} + \sqrt{\frac{C_2}{C_1}}} \tag{12.31b}$$

Comparing Equation 12.30 and 12.18, we see that $k = 1$, and $D_2(s_0)$ is

$$D_2(s_0) = \left[s_0^2 + s_0\left(\frac{1}{R_2C_1} + \frac{1}{R_2C_2} + \frac{1}{R_1C_2}\right) + \frac{1}{R_1R_2C_1C_2}\right]$$

For high pole Q's

$$s_0^2 = (\sigma_0 + j\omega_0)^2 \cong (j\omega_0)^2 = -\frac{1}{R_1R_2C_1C_2}$$

thus, $D_2(s_0)$ reduces to

$$D_2(s_0) \cong s_0\left(\frac{1}{R_2C_1} + \frac{1}{R_2C_2} + \frac{1}{R_1C_2}\right)$$

If the circuit is synthesized using $C_1 = C_2 = 1$, as was done in Section 9.2, the remaining elements are given by (Equation 9.10)

$$R_2 = 2\frac{Q_0}{\omega_0} \qquad R_1 = \frac{1}{2\omega_0 Q_0} \tag{12.32}$$

Therefore, for high pole Q's, $R_2 \gg R_1$, and $D_2(s_0)$ can be approximated by

$$D_2(s_0) \cong s_0\left(\frac{1}{R_1C_2}\right) = 2\omega_0 Q_0 s_0 \tag{12.33}$$

Substituting in (12.29), the per-unit change in s_0 is

$$\frac{\Delta s_0}{s_0} = j\frac{Q_0}{A(s_0)} \tag{12.34}$$

The gain of the op amp at the pole frequency, $A(s_0)$, depends on the compensation used.

Consider first the single pole compensation of Figure 12.10. This gain characteristic can be approximated by

$$A(s) = \frac{A_0 \alpha}{s + \alpha}$$

where $A_0 = 10^5$ and $\alpha = 2\pi 10$. Thus,

$$A(s_0) \cong A(-\sigma_0 + j\omega_0) = A\left(-\frac{\omega_0}{2Q_0} + j\omega_0\right) = A(s)|_{s=-\omega_0/2Q_0 + j\omega_0}$$

$$= \frac{2\pi 10^6}{\left(-\dfrac{2\pi 1000}{10} + j2\pi 1000\right) + 2\pi 10}$$

$$= \frac{10^6}{j1000 - 90}$$

Substituting in (12.34)

$$\frac{\Delta s_0}{s_0} = -5(10)^{-3} - j4.5(10)^{-4} = 5.02(10)^{-3} \angle -174.86^\circ$$

Next, consider the double pole compensation of Figure 12.11. This gain characteristic can be approximated by

$$A(s) = (10)^5 \frac{2\pi 400}{s + 2\pi 400} \cdot \frac{2\pi(10)^4}{s + 2\pi(10)^4} \cdot \frac{s + 2\pi 3(10)^5}{2\pi 3(10)^5}$$

At $s = s_0 = (-\omega_0/2Q_0 + j\omega_0)$, the amplifier gain is

$$A(s_0) = (10)^5 \frac{400}{1044.03 \angle 73.33^\circ} \cdot \frac{10^4}{9950.38 \angle 5.77^\circ} \cdot \frac{299901.7 \angle 0.191^\circ}{3(10)^5}$$

$$= 3.8491(10)^4 \angle -78.9^\circ$$

Substituting in (12.34)

$$\frac{\Delta s_0}{s_0} = 1.3(10)^{-4} \angle 168.9^\circ = -1.275(10)^{-4} + j2.5(10)^{-5}$$

Observation

The magnitude of $\Delta s_0/s_0$, using the double pole compensation, is seen to be approximately 39 times less than that using the single pole compensation. This improvement is a direct consequence of the increase in op amp gain (30 dB more at 1 kHz) provided by the double pole compensation. ■

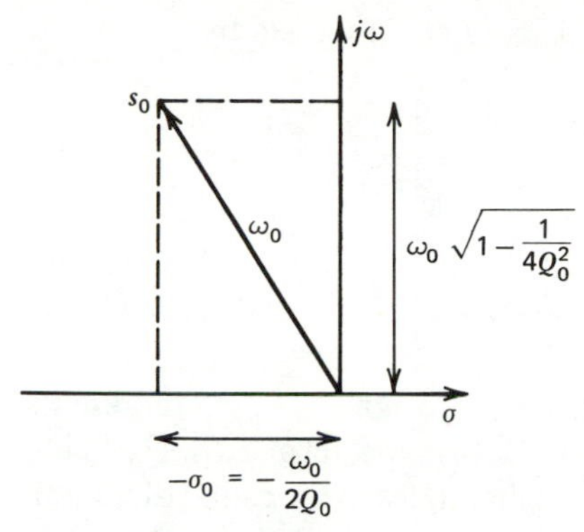

Figure 12.13 Graphical representation of pole s_0.

The change in pole position will, in general, be a complex number. In the following it will be shown that the real and imaginary parts of this complex number are related to the change in pole frequency and pole Q, respectively. From Figure 12.13,

$$s_0 = -\frac{\omega_0}{2Q_0} + j\omega_0\sqrt{1 - \frac{1}{4Q_0^2}} \tag{12.35}$$

Ignoring the second- and higher-order terms, the change Δs_0 is given by the Taylor series expansion:

$$\Delta s_0 = \frac{\partial s_0}{\partial \omega_0}\Delta\omega_0 + \frac{\partial s_0}{\partial Q_0}\Delta Q_0 \tag{12.36}$$

In this expansion, from (12.35)

$$\frac{\partial s_0}{\partial \omega_0} = \left(-\frac{1}{2Q_0} + j\sqrt{1 - \frac{1}{4Q_0^2}}\right) = \frac{s_0}{\omega_0} \tag{12.37a}$$

To evaluate the second term in Equation 12.36, note that for high Q poles $1/4Q_0^2 \ll 1$, so that s_0 can be approximated by

$$s_0 \cong \omega_0\left(\frac{1}{2Q_0} - j\right)$$

Hence

$$\frac{\partial s_0}{\partial Q_0} = \frac{\omega_0}{2Q_0^2} \cong \left(\frac{s_0}{j}\right)\frac{1}{2Q_0^2} = -j\,\frac{s_0}{2Q_0^2} \tag{12.37b}$$

Substituting (12.37a) and (12.37b) in (12.36), we get

$$\frac{\Delta s_0}{s_0} = \frac{\Delta\omega_0}{\omega_0} - \frac{j}{2Q_0}\frac{\Delta Q_0}{Q_0} \tag{12.38}$$

From this equation, the per-unit changes in ω_0 and Q_0 are given by

$$\frac{\Delta\omega_0}{\omega_0} = \text{Re}\left(\frac{\Delta s_0}{s_0}\right) \tag{12.39}$$

$$\frac{\Delta Q_0}{Q_0} = -2Q_0 \, \text{Im}\left(\frac{\Delta s_0}{s_0}\right) \tag{12.40}$$

Since ω_0 is the radial distance to the pole from the origin, $\Delta\omega_0/\omega_0$ represents a change along this radial direction (Figure 12.14); the term $\Delta Q_0/Q_0$ being orthogonal to $\Delta\omega_0/\omega_0$, represents a change at right angles to the radial direction. For the filter of Example 12.1, the per-unit changes in ω_0 and Q_0 for the single pole compensation case are

$$\frac{\Delta\omega_0}{\omega_0} = -5(10)^{-3} \qquad \frac{\Delta Q_0}{Q_0} = 4.5(10)^{-3}$$

and for the double pole compensation case

$$\frac{\Delta\omega_0}{\omega_0} = -1.275(10)^{-4} \qquad \frac{\Delta Q_0}{Q_0} = -2.5(10)^{-4}$$

Thus far we have seen how a biquad circuit can be analyzed to evaluate the shifts in the dominant poles due to the finite gain of the op amp. In the following we show a method for modifying the design to accomodate this shift in the poles; a step that is referred to as **predistortion**.

The finite gain of the op amp effectively changes the pole frequency and pole Q from

$$\omega_0 \to \omega_0 + \Delta\omega_0 \qquad \text{and} \qquad Q_0 \to Q_0 + \Delta Q_0$$

The nominal design equations (which are based on an infinite op amp gain) can be modified to correct for these changes. To elaborate, if the desired param-

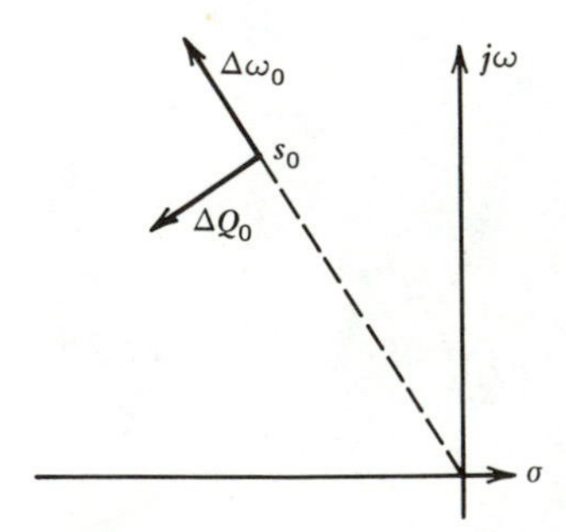

Figure 12.14 Direction of change for $\Delta\omega_0$ and ΔQ_0.

eters are ω_0 and Q_0, we use the nominal design equations with the design parameters changed to

$$\omega_0 - \Delta\omega_0 \qquad \text{and} \qquad Q_0 - \Delta Q_0$$

Then, in the presence of the real op amp, the manufactured circuit will exhibit the desired parameters, namely, ω_0 and Q_0.

For instance, in Example 12.1, for the single pole compensation case, the desired design parameters are $\omega_0 = 2\pi 1000$ and $Q_0 = 5$; these are changed to

$$\omega_0 - \Delta\omega_0 = 2\pi 1005 \qquad Q_0 - \Delta Q_0 = 4.9775$$

for use in the nominal design equations.

From the above it is seen that predistortion compensates for the finite gain of the op amp, thereby allowing the designer to use the ideal [$A(s) = \infty$] synthesis equation.

12.3.2 STATISTICAL DEVIATIONS IN GAIN

In the last section we showed how the approximation function could be predistorted to account for the gain of the op amp, $A(s)$. The resulting nominal filter, with ideal R's and C's and an op amp with gain $A(s)$, will exhibit the desired transfer function. In practice, of course, both the passive elements and the op amp gain will deviate from their nominal values due to manufacturing tolerances and environmental (temperature, humidity, aging) changes. In this section the statistical change in the filter response due to these variations in the elements is investigated. In particular, the mean μ and standard deviation σ of the gain deviation ΔG are evaluated, using the techniques developed in Chapter 5.

If the amplifier gain changes from infinity to $A(s_0)$, the shift in the pole location, Δs_0, is that given by Equation 12.29. By following a similar analysis, it can be shown that if the op amp gain changes from its nominal value $A(s_0)$, to a new value, $A'(s_0)$, (due to manufacturing and environmental effects), the corresponding change in the pole location is

$$\Delta'(s_0) = j\,\frac{kD_2(s_0)}{2\omega_0}\left[\frac{1}{A'(s_0)} - \frac{1}{A(s_0)}\right] \tag{12.41}$$

$$= j\,\frac{kD_2(s_0)}{2\omega_0}\,\Delta\!\left(\frac{1}{A(s_0)}\right) \tag{12.42}$$

where $\Delta(1/A(s_0))$ represents the random shift in amplifier gain due to manufacturing and environmental effects

and $\Delta'(s_0)$ represents the corresponding random change in the pole location.

Observe that $\Delta'(s_0)$, as given by Equation 12.41, reduces to the pole shift Δs_0 as given by Equation 12.29, if the reference value of the op amp gain is infinity, that is

$$\Delta'(s_0) = j\frac{kD_2(s_0)}{2\omega_0}\left[\frac{1}{A(s_0)} - \frac{1}{\infty}\right] = \Delta s_0 \tag{12.43}$$

As in the last section, the per-unit change of ω_0 and Q_0 can be expressed in terms of the pole shift by using Equation 12.39 and 12.40, as

$$\frac{\Delta'\omega_0}{\omega_0} = \mathrm{Re}\left(\frac{\Delta' s_0}{s_0}\right) \tag{12.44}$$

$$\frac{\Delta' Q_0}{Q_0} = -2Q_0\,\mathrm{Im}\left(\frac{\Delta' s_0}{s_0}\right) \tag{12.45}$$

The corresponding deviation in gain, from Equation 5.36, is

$$\Delta G = \mathcal{S}^G_{\omega_0}\left(\frac{\Delta'\omega_0}{\omega_0}\right) + \mathcal{S}^G_{Q_0}\left(\frac{\Delta' Q_0}{Q_0}\right) \tag{12.46}$$

Thus, the mean and standard deviation of ΔG are given by

$$\mu(\Delta G) = \mathcal{S}^G_{\omega_0}\mu\left(\frac{\Delta'\omega_0}{\omega_0}\right) + \mathcal{S}^G_{Q_0}\mu\left(\frac{\Delta' Q_0}{Q_0}\right) \tag{12.47}$$

$$\sigma(\Delta G) = \sqrt{(\mathcal{S}^G_{\omega_0})^2\sigma^2\left(\frac{\Delta'\omega_0}{\omega_0}\right) + (\mathcal{S}^G_{Q_0})^2\sigma^2\left(\frac{\Delta' Q_0}{Q_0}\right)} \tag{12.48}$$

The following example illustrates the use of the above expressions for evaluating the gain deviation.

Example 12.2

The nominal gain of a single pole compensated op amp used in the circuit of Example 12.1 can be modeled as $A(s) = A_0\alpha/(s+\alpha)$. Compute the statistics of the deviation in the gain of the filter at the lower 3 dB passband edge frequency, 900 Hz, due to a manufacturing tolerance in the gain bandwidth product of ± 50 percent. The tolerance limits refer to the $\pm 3\sigma$ points of a Gaussian distribution.

Solution

For the negative feedback circuit of Example 12.1, the per-unit change in s_0 is given by an equation similar to (12.34), namely:

$$\frac{\Delta' s_0}{s_0} = jQ_0\Delta\left[\frac{1}{A(s_0)}\right] \tag{12.49}$$

Using the given single pole approximation for $A(s_0)$, we get

$$\frac{\Delta' s_0}{s_0} = jQ_0 \Delta\left(\frac{s_0 + \alpha}{A_0 \alpha}\right) = jQ_0 \Delta\left(\frac{-\omega_0/2Q_0 + j\omega_0 + \alpha}{A_0 \alpha}\right)$$

$$= jQ_0\left(-\frac{\omega_0}{2Q_0} + j\omega_0 + \alpha\right)\Delta\left(\frac{1}{A_0\alpha}\right)$$

$$= -Q_0\omega_0\Delta\left(\frac{1}{A_0\alpha}\right) - jQ_0\left(\frac{\omega_0}{2Q_0} - \alpha\right)\Delta\left(\frac{1}{A_0\alpha}\right) \tag{12.50}$$

From Equation 12.44 and 12.45, the per-unit changes in ω_0 and Q_0 are related to the real and imaginary parts of 12.50, as

$$\frac{\Delta'\omega_0}{\omega_0} = -Q_0\omega_0\Delta\left(\frac{1}{A_0\alpha}\right) \tag{12.51}$$

$$\frac{\Delta' Q_0}{Q_0} = 2Q_0^2\left(\frac{\omega_0}{2Q_0} - \alpha\right)\Delta\left(\frac{1}{A_0\alpha}\right) \tag{12.52}$$

In this expression $\Delta(1/A_0\alpha)$ is obtained from the relationship

$$\frac{d(1/A_0\alpha)}{d(A_0\alpha)} = -\frac{1}{(A_0\alpha)^2} \tag{12.53}$$

For small changes in $A_0\alpha$, from this equation

$$\Delta\left(\frac{1}{A_0\alpha}\right) \cong -\frac{1}{A_0\alpha}\frac{\Delta(A_0\alpha)}{A_0\alpha} \tag{12.54}$$

In particular, from the given information

$$\mu\left(\frac{\Delta A_0\alpha}{A_0\alpha}\right) = 0 \qquad 3\sigma\left(\frac{\Delta A_0\alpha}{A_0\alpha}\right) = 0.5$$

so

$$\mu\left[\Delta\left(\frac{1}{A_0\alpha}\right)\right] = 0 \qquad 3\sigma\left[\Delta\left(\frac{1}{A_0\alpha}\right)\right] = \frac{-0.5}{2\pi 10^6}$$

Also, from Example 12.1, $\omega_0 = 2\pi 1000$, $Q_0 = 5$, and $\alpha = 2\pi 10$. Substituting in (12.51), we get

$$\mu\left(\frac{\Delta'\omega_0}{\omega_0}\right) = 0 \qquad 3\sigma\left(\frac{\Delta'\omega_0}{\omega_0}\right) = -Q_0\omega_0\left\{3\sigma\left[\Delta\left(\frac{1}{A_0\alpha}\right)\right]\right\} = 0.0025$$

Similarly, from (12.52)

$$\mu\left(\frac{\Delta' Q_0}{Q_0}\right) = 0 \qquad 3\sigma\left(\frac{\Delta' Q_0}{Q_0}\right) = -0.00225$$

To evaluate μ and σ for ΔG, the sensitivity terms $\mathcal{S}^G_{\omega_0}$ and $\mathcal{S}^G_{Q_0}$ are needed. At 900 Hz, the normalized frequency Ω is 900/1000 = 0.9. Substituting for Ω and Q_0 in Equation 5.44 and 5.45, we get

$$\mathcal{S}^G_{\omega_0} = -52.3 \qquad \mathcal{S}^G_{Q_0} = 4.1$$

Finally, from (12.47) and (12.48), the μ and σ for ΔG are:

$$\mu(\Delta G) = 0$$

$$\begin{aligned}[3\sigma(\Delta G)]^2 &= (52.3)^2(0.0025)^2 + (4.1)^2(0.00225)^2 \\ &= 0.0172\end{aligned}$$

So

$$3\sigma(\Delta G) = 0.131 \text{ dB}$$

Observations

1. In Examples 8.3, 9.3, and 10.2 we illustrated a method for computing gain deviation due to changes in $A_0\alpha$, for the special case when the op amp gain could be characterized by a single pole at the origin $[A(s) = A_0\alpha/s]$. The method described in this section, and illustrated by the above example, is more general in that it is applicable to *any* amplifier gain characteristic $A(s)$. As such, the method is equally applicable to the double pole compensated characteristic.
2. The deviation in ΔG due to the passive elements, not considered here, can be computed as in Example 8.2, 9.2, and 10.2. ■

12.4 OTHER OPERATIONAL AMPLIFIER CHARACTERISTICS

In this section we briefly describe some other properties of the op amp that are of interest in the design of practical filters. The characteristics considered are dynamic range, slew-rate limiting, offset-voltage, input-bias and input-offset currents, common-mode signals, and noise. For a detailed discussion of these topics the reader is referred to [1], [5], and [10].

12.4.1 DYNAMIC RANGE

As the input to an op amp active filter is increased, the output continues to increase until eventually the output waveform becomes clipped, as illustrated in Figure 12.15. The op amp is then said to be saturated. The reason for the clipping is that the output voltage cannot swing beyond the dc power supply voltage. For instance, if the power supply voltage is ± 12 volts, the output is

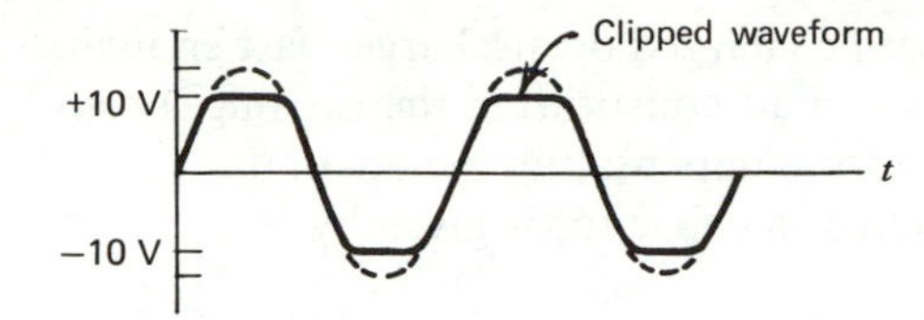

Figure 12.15 Clipping of a sine wave when the op amp saturates.

certainly constrained to lie between +12 and −12 volts.* To prevent this type of a distortion, the maximum input amplitude must be limited. On the other hand, the minimum input signal level should be large enough to maintain the signal levels in the op amp well above the internally generated noise voltages (see Section 12.4.6). The op amp characteristic describing these limitations is known as the **dynamic range**, which is defined as the ratio of the maximum usable output voltage to the noise output voltage. Since the output voltage depends on the filter function, dynamic range depends not only on the noise voltages and the power supply voltage, but also on the frequency characteristics of the active filter. Active filters typically have a dynamic range between 70 and 100 dB.

12.4.2 SLEW-RATE LIMITING

If the frequency of the input signal to an op amp circuit is gradually increased, the output will eventually distort in the manner shown in Figure 12.16. This type of a distortion, called **slew-rate** limiting, is caused by some capacitor

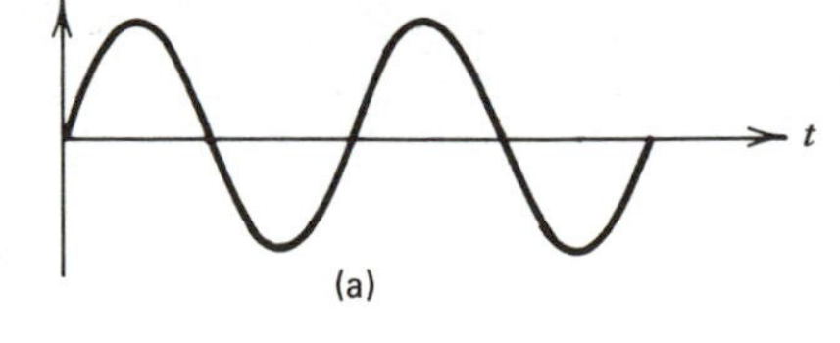

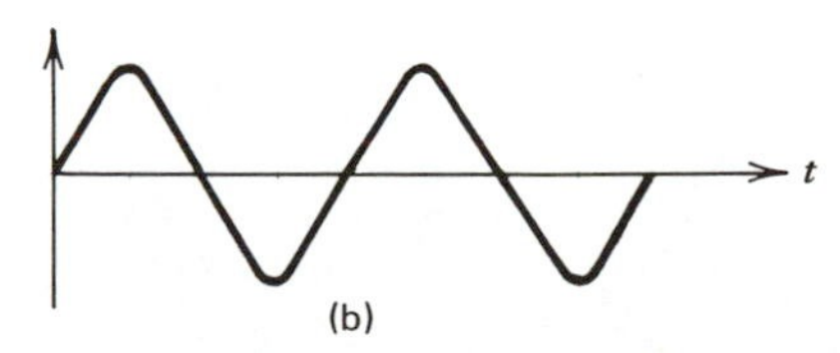

Figure 12.16 Distortion due to slew rate limiting: (*a*) Undistorted sine wave output. (*b*) Slew-rate limited output.

* In practice, the output voltage cannot exceed approximately ±10 volts, because a part of the power supply voltage is needed to bias transistors internal to the op amp.

associated with the op amp that cannot be charged or discharged fast enough. The capacitor could be internal to the circuit constituting the op amp, or an external capacitor such as the one used for compensating the op amp.

The rate at which voltage changes across a capacitor is given by

$$\frac{dv}{dt} = \frac{i}{C} \tag{12.55}$$

In an op amp the available currents are limited. If, for example, the current available to the capacitor in question is limited to $i_{\max}$, the maximum rate at which the voltage across it can change is

$$\left.\frac{dv}{dt}\right|_{\max} = \frac{i_{\max}}{C} = \rho \tag{12.56}$$

If the voltage across the capacitor is required to change any faster, the output waveform will distort in the manner indicated in Figure 12.16. The circuit is then said to be slew-rate limited; the factor ρ being referred to as the slew-rate of the op amp. Typical op amps have a slew-rate of approximately 1 volt/μsec, although higher slew-rates can be achieved by special design.

Example 12.3

The input to the inverter shown in Figure 12.17 is a sine wave of amplitude 5 volts. If the slew-rate of the op amp is 1 V/μsec, find the frequency at which slew-rate limiting occurs.

Solution

The input signal is

$$V_{IN} = 5 \cos \omega t$$

and the corresponding output voltage is

$$V_O = -10 \cos \omega t$$

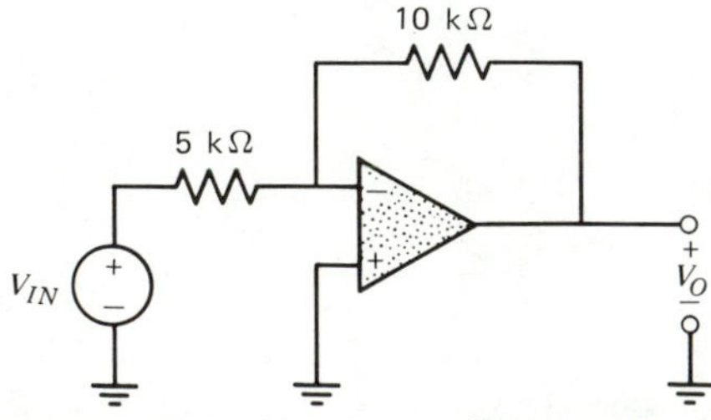

Figure 12.17 Circuit for Example 12.3.

The maximum rate of change of the output signal is

$$\left.\frac{dV_O}{dt}\right|_{\max} = 10\omega$$

Since the output voltage cannot change faster than the slew-rate, ω is limited to

$$\omega_{\max} = \frac{\rho}{10} = 0.1 \text{ rad}/\mu\text{sec}$$

$$= 10^5 \text{ rad/sec}$$ ■

12.4.3 OFFSET VOLTAGE

In an ideal op amp, if the input signal is zero the output will be zero. In real op amps, however, imperfections in the circuit components cause a *dc* voltage to exist at the output even when the input voltage is zero. A convenient way of representing this so-called offset voltage is by an equivalent input voltage source at the positive terminal of the op amp, as shown in Figure 12.18. This input voltage is called the **input-offset voltage,** V_{OS}. Typical values for V_{OS} are less than 5 mV.

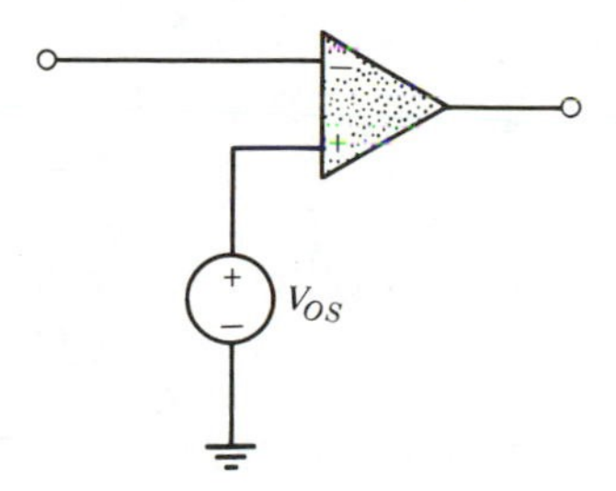

Figure 12.18 Representation of input-offset voltage.

In most filter applications the response at *dc* is not critical, so offset voltages are not of any concern. However, in some applications the filter may have a significant response at *dc* that must be held accurately. In such cases the offset voltage can be reduced by adding a small *dc* voltage at the input of the op amp and adjusting its magnitude and polarity to make the output voltage zero when the input is grounded.* However, even after this nulling, the output-offset will drift due to temperature and aging effects.

* Many op amps provide special terminals for external potentiometers, to be used for nulling the offset voltages.

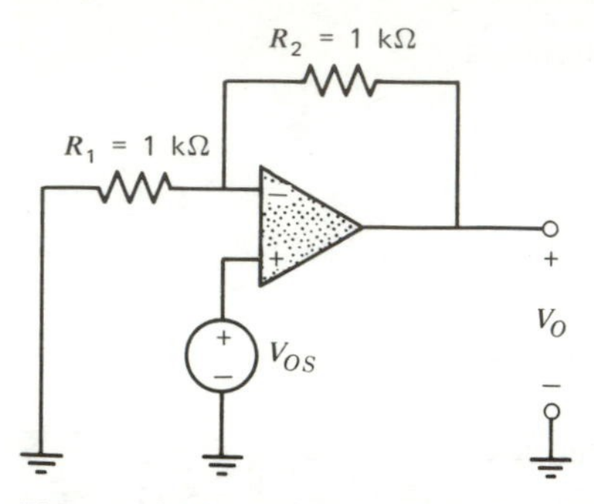

Figure 12.19 Circuit for Example 12.4.

Example 12.4

The output offset voltage measured in the circuit of Figure 12.19 is 10 mV. What is the input-offset voltage, V_{OS}, if $R_2 = R_1 = 10\ \text{k}\Omega$.

Solution

The node equation at the negative terminal is:

$$V_{OS}\left(\frac{1}{R_1} + \frac{1}{R_2}\right) - V_O\left(\frac{1}{R_2}\right) = 0$$

Thus

$$V_{OS} = V_O\left[\frac{1}{1 + R_2/R_1}\right]$$

Since $R_2/R_1 = 1$ and V_O is given to be 10 mV, the input offset voltage is 5 mV. ■

12.4.4 INPUT-BIAS AND INPUT-OFFSET CURRENTS

In an ideal op amp the input impedance is infinite and no current flows into the input terminals. In real amplifiers small currents, needed to bias the input transistors, do flow into these two terminals. These currents are represented by I_{B1} and I_{B2} in Figure 12.20. In most op amp data sheets these currents are given in terms of the **input-bias current**, I_{BS}, which is the average value of I_{B1} and I_{B2}

$$I_{BS} = \frac{I_{B1} + I_{B2}}{2} \tag{12.57}$$

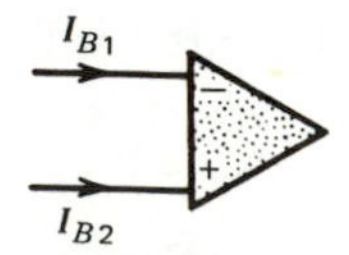

Figure 12.20 Input currents in an op amp.

The currents I_{B1} and I_{B2} tend to track, and a gauge of the degree of tracking is the difference between the currents, known as the **input-offset current**, I_{OS}:

$$I_{OS} = I_{B1} - I_{B2} \tag{12.58}$$

The effect of the input currents is to produce a small output offset voltage, as illustrated by the following example.

Example 12.5

The op amp used in the inverter circuit of Figure 12.21, has an input-bias current of 500 nA and an input-offset current that can range between ± 100 nA. Find the resulting maximum output offset voltage.

Solution

From the given information

$$I_{BS} = \frac{I_{B1} + I_{B2}}{2} = 500 \text{ nA}$$

$$I_{OS} = I_{B1} - I_{B2} = \pm 100 \text{ nA}$$

From these equations, the maximum value of I_{B1} is seen to 550 nA. The *dc* offset voltage at the output is obtained from the node equation

$$V^{-}\left(\frac{1}{R_1} + \frac{1}{R_2}\right) - V_O \frac{1}{R_2} = -I_{B1}$$

However $V^{-} = V^{+} = 0$, thus

$$V_O = I_{B1} R_2$$

from which the output offset voltage is 5.5 mV. ■

In most active filter applications, the offset voltage produced by these currents is negligible. However, if need be, their effect can be nulled by using a corrective *dc* voltage source at the input, as explained in the last section.

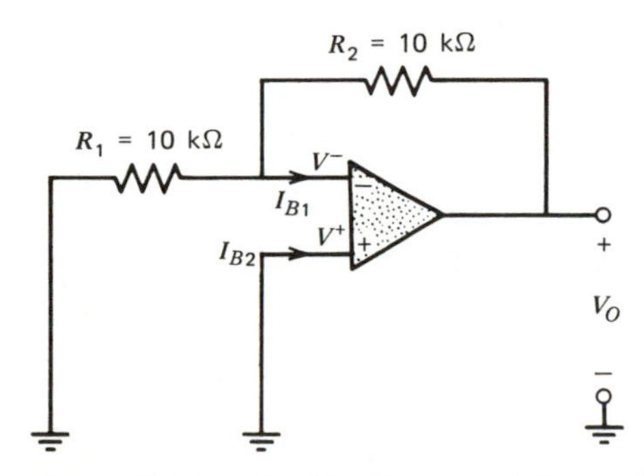

Figure 12.21 Circuit for Example 12.5.

12.4.5 COMMON-MODE SIGNALS

The output of an ideal differential op amp is

$$V_O = (V^+ - V^-)A \tag{12.59}$$

where A is the gain of the op amp:

> V^+ is the voltage at the positive terminal.
> V^- is the voltage at the negative terminal.

If this ideal op amp is connected in the *common-mode* as shown in Figure 12.22*a*, with the same voltage applied to both terminals, the output voltage will be zero. The above equation implies that the ideal op amp is perfectly symmetrical so that the gain from the negative terminal to the output, A^-, is equal to the gain from the positive terminal to the output, A^+. In real op amps there is some degree of asymmetry, so that these gains are not exactly equal, leading to a common-mode output voltage given by (Figure 12.22*a*):

$$V_O = (A^+ - A^-)V_{IN} \tag{12.60}$$

The term $(A^+ - A^-)$ is referred to as the common-mode gain. In the ideal op amp this common-mode gain is zero.

Consider next the op amp connected in the *difference-mode*, as shown in Figure 12.22*b*, where the input voltages are equal in magnitude but opposite

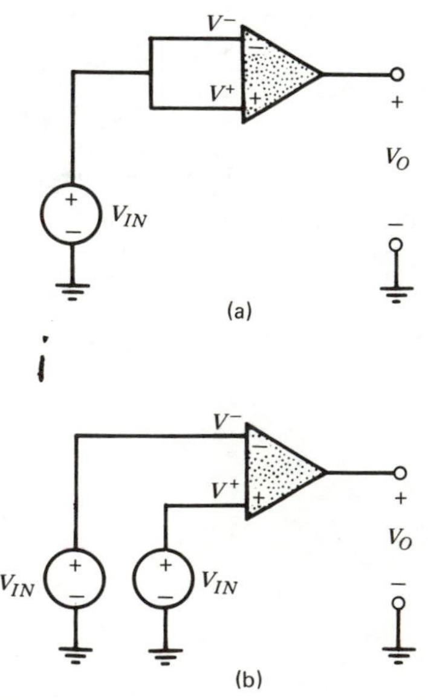

Figure 12.22 (*a*) Common-mode operation. (*b*) Difference-mode operation.

in sign. With an ideal op amp the output voltage is $2AV_{IN}$, whereas with a real op amp the output is

$$V_O = (A^+ + A^-)V_{IN} = 2\left(\frac{A^+ + A^-}{2}\right)V_{IN} \tag{12.61}$$

This difference-mode output is the desired gain in an op amp. The term $(A^+ + A^-)/2$ is called the difference mode gain, which is infinite in an ideal op amp.

A measure of the degree of suppression of the undesired common-mode signal relative to the desired difference-mode signal is expressed by the **common-mode rejection ratio** ($CMRR$), which is defined as

$$CMRR = \frac{\dfrac{A^+ + A^-}{2}}{|(A^+ - A^-)|} \tag{12.62}$$

In typical op amps A^+ and A^- are frequency dependent, with characteristics similar to $A(s)$ (Figure 12.8). The low frequency difference-mode gain is typically 100,000; while the common-mode gain is around 10. With these values the $CMRR$ is 10,000 (or 80 dB). In most active filter applications the common-mode signal is small enough to be ignored.

Example 12.6

The output voltage measured in the circuit of Figure 12.23 is 10 mV when the input voltage is 1 volt. Find the $CMRR$.

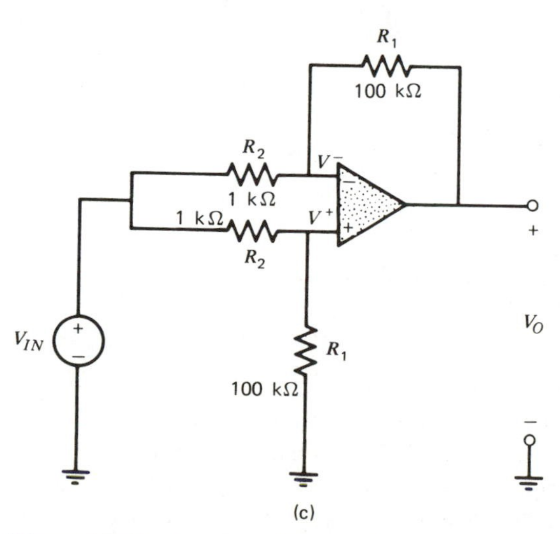

Figure 12.23 Circuit for Example 12.6.

Solution

The voltage at the positive terminal is

$$V^+ = \frac{R_1}{R_1 + R_2} V_{IN}$$

and the voltage at the negative terminal is

$$V^- = \frac{R_2}{R_1 + R_2} V_O + \frac{R_1}{R_1 + R_2} V_{IN}$$

Thus, the output voltage is

$$\begin{aligned} V_O &= A^+V^+ - A^-V^- \\ &= A^+\left(\frac{R_1}{R_1 + R_2}\right)V_{IN} - A^-\left(\frac{R_2}{R_1 + R_2}\right)V_O - A^-\left(\frac{R_1}{R_1 + R_2}\right)V_{IN} \end{aligned}$$

Solving,

$$\frac{V_O}{V_{IN}} = \frac{(A^+ - A^-)\dfrac{R_1}{R_1 + R_2}}{1 + A^-\dfrac{R_2}{R_1 + R_2}}$$

Since $A^-R_2/(R_1 + R_2) \gg 1$, the transfer function is approximately

$$\frac{V_O}{V_{IN}} \cong \frac{R_1}{R_2}\frac{A^+ - A^-}{A^-}$$

Also, A^- is approximately equal to $(A^+ + A^-)/2$; therefore, from Equation 12.62

$$\frac{V_O}{V_{IN}} \cong \frac{R_1}{R_2}\frac{1}{CMRR}$$

Substituting the given values, $R_1 = 100\ \text{k}\Omega$, $R_2 = 1\ \text{k}\Omega$, $V_O = 0.01$ volt, and $V_{IN} = 1$ volt, we get

$$CMRR = \frac{R_1}{R_2}\frac{V_{IN}}{V_O} = 10{,}000 \text{ (or 80 dB)}$$ ■

12.4.6 NOISE

Spurious signals at the output of the op amp that cannot be predicted from a precise knowledge of the input signal and the transfer function are called noise voltages. These signals originate from external sources such as power supply ripple, electromagnetic radiation, and 60 cycle pickup. The op amp itself also has noise sources associated with its internal resistors and transistors.

The noise generated by op amps can be represented by equivalent voltage and current noise sources at the two input terminals. The effect of these noise sources can be evaluated in a manner analogous to offset voltage and current computations. The only difference here is that the noise sources are frequency dependent and random, and need to be characterized accordingly. The reader is referred to [6] and [12] for a detailed discussion of this topic.

In practice, noise effects are quite small and are usually negligible in most filter applications.

12.5 CONCLUDING REMARKS

In this chapter several practical limitations imposed by the op amp on active *RC* circuits were considered. The finite, frequency dependent gain characteristic of the op amp limits the frequency of application for active filters. A practical, but approximate, rule of thumb is that the product of the pole Q and the pole frequency (f_p) should be kept below one tenth of the unity gain crossover frequency (for the double pole compensated op amp) to achieve moderately accurate filter performance. For instance, an op amp with a unity gain crossover frequency of 1 MHz could be used to realize a pole Q of 10 up to approximately 10 kHz. Of course, if the single pole compensated op amp is used, or if the filter requirements are more stringent, as high a $Q_p f_p$ product cannot be realized. It is expected that, with advances in the op amp technology, the future will see better and higher frequency active *RC* filters.

For most voice frequency applications (below 10 kHz) using present day op amps, the performance of active *RC* filters are excellent—in fact, in this frequency range they are the best way of realizing filters.

FURTHER READING

1. A. Barna, *Operational Amplifiers*, Wiley-Interscience, New York, 1971.
2. J. J. D'Azzo and C. H. Houpis, *Feedback Control Systems Analysis and Synthesis*, Second Edition, McGraw-Hill, New York, 1966. Chapter 10.
3. Fairchild Semiconductor, *The Linear Integrated Circuits Data Catalog*, 1973. 464 Ellis Street, Mountain View, Calif., pp. 3–53.
4. P. E. Fleischer, "Sensitivity minimization in a single amplifier biquad circuit," *IEEE Trans. Circuits and Systems*, *CAS-23*, No. 1, January 1976, pp. 45–55.
5. J. G. Graeme, G. E. Tobey, and L. P. Huelsman, *Operational Amplifiers Design and Applications*, McGraw-Hill, New York, 1971.
6. J. W. Haslett, "Noise performance limitations of single amplifier RC active filters," *IEEE Trans. Circuits and Systems*, *CAS-22*, No. 9, September 1975, pp. 743-747.
7. S. S. Haykin, *Active Network Theory*, Addison-Wesley, Reading, Mass., 1970, Chapter 12.

8. J. L. Melsa and D. G. Schultz, *Linear Control Systems*, McGraw-Hill, New York, 1969, Chapter 6.
9. G. S. Moschytz, *Linear Integrated Networks Fundamentals*, Van Nostrand, New York, 1974, Chapter 7.
10. J. K. Roberge, *Operational Amplifiers Theory and Practice*, Wiley, New York, 1975.
11. R. Saucedo and E. Schiring, *Introduction to Continuous and Digital Control Systems*, Macmillan, New York, 1968, Chapter 10.
12. F. N. Tromfimenkoff, D. H. Treleaven, and L. T. Bruton, "Noise performance of RC-active quadratic filter sections," *IEEE Trans. Circuit Theory. CT-20*, No. 5, September 1973, pp. 524–532.
13. B. O. Watkins, *Introduction to Control Systems*, Macmillan, New York, 1969, Chapter 8.

PROBLEMS

12.1 *Stability margins.* The open loop gain of an amplifier circuit is characterized by the function:

$$\frac{A_0 \omega_1 \omega_2 \omega_3}{(s + \omega_1)(s + \omega_2)(s + \omega_3)}$$

where $A_0 = 10^5$, $\omega_1 = 400$ rad/sec, $\omega_2 = 20{,}000$ rad/sec and $\omega_3 = 10^6$ rad/sec. Sketch the Bode gain and phase plots of this function and determine the gain crossover frequency, the phase crossover frequency, the gain margin, and the phase margin. Is the circuit stable?

12.2 *Stability, inverting amplifier.* An inverting amplifier (Figure 1.9*a*) uses an op amp whose gain is given by Equation 12.6. Determine the closed loop gain $(-R_1/R_1)$ above which the circuit provides 30° phase margin.

12.3 An inverter (Figure 1.9*a*) uses an op amp whose open loop gain is characterized by

$$A(s) = 10^5 \frac{2\pi 10}{(s + 2\pi 10)} \cdot \frac{2\pi 10^5}{(s + 2\pi 10^5)}$$

Determine the gain and phase margins and comment on the stability of the inverter if

(a) $R_F = 1\ \text{k}\Omega$, $\quad R_S = 1\ \text{k}\Omega$
(b) $R_F = 10\ \text{k}\Omega$, $\quad R_S = 1\ \text{k}\Omega$

12.4 *Stability, integrator.* The integrator circuit of Figure 12.12 uses the op amp described in Problem 12.3. Determine the gain and phase margins and comment on the stability if $R = 1\ \text{k}\Omega$, $C = 0.01591\ \mu\text{F}$.

12.5 *Stability, negative feedback topology.* The op amp used in the basic negative feedback topology of Figure 9.1 has the gain characteristics described in Problem 12.3. Use the MAG program to determine the stability margins for the active circuit, if the feedback transfer function of the *RC* network is

$$T_{FB} = \frac{(2\pi 500)s}{s^2 + (2\pi 500)s + (2\pi 1000)^2}$$

12.6 Repeat Problem 12.5 for an *RC* network characterized by

$$T_{FB} = \frac{s^2}{s^2 + (2\pi 500)s + (2\pi 1000)^2}$$

12.7 *Stability, noninverting amplifier.* The noninverting amplifier circuit of Figure 12.7 uses an op amp whose gain is characterized by

$$\frac{A_0(2\pi)^3 f_1 f_2 f_3}{(s + 2\pi f_1)(s + 2\pi f_2)(s + 2\pi f_3)}$$

where $A_0 = 10^5$, $f_1 = 20$ kHz, $f_2 = 200$ kHz, and $f_3 = 2$ MHz. Determine the minimum closed loop gain $(1 + R_2/R_1)$ above which:

(a) The circuit is marginally stable.

(b) The circuit provides 45° phase margin.

12.8 *Op amp model.* The small signal model for an op amp is shown in Figure P12.8. Typical element values are

$$R_1 = 1\ \text{M}\Omega \qquad R_2 = 100\ \text{k}\Omega \qquad R_{IN} = 500\ \text{k}\Omega$$
$$C_1 = 5\ \text{pF} \qquad C_2 = 5\ \text{pF}$$

The transconductance g_{m2} is $4(10)^{-4}$ mhos and g_{m1} is given by

$$g_{m1} = \frac{0.0025\omega_3}{(s + \omega_3)}\ \text{mhos}$$

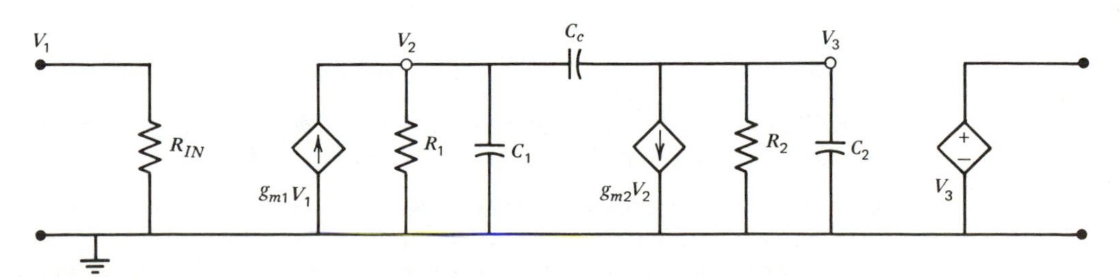

Figure P12.8

where $\omega_3 = 4\pi 10^6$ rad/sec. The intrinsic Miller capacitor is $C_c = 0.25$ pF.
(a) Show that the transfer function V_3/V_1 is

$$\frac{-g_{m1}(g_{m2} - sC_c)}{s^2(C_1C_c + C_2C_c + C_1C_2) + s\left(\frac{C_c}{R_1} + \frac{C_c}{R_2} + \frac{C_2}{R_1} + \frac{C_1}{R_2} + g_{m2}C_c\right) + \frac{1}{R_1R_2}}$$

(b) Using the element values given, write the transfer function in the form

$$\frac{A_0\omega_1\omega_2\omega_3}{(s + \omega_1)(s + \omega_2)(s + \omega_3)}$$

Determine the dc gain A_0, and the pole frequencies ω_1, ω_2, and ω_3. Assume $g_{m2} \gg |sC_c|$.

12.9 *Pole splitting by Miller capacitor.* Use the transfer function given in Problem 12.8 to show that:
(a) If $C_c \ll C_1$ and C_2, then $\omega_1 \cong 1/R_1R_2g_{m2}C_c$ and $\omega_2 \cong C_cg_{m2}/C_1C_2$.
(b) If $C_c \gg C_1$ and C_2, then $\omega_1 \cong 1/R_1R_2g_{m2}C_c$ and $\omega_2 \cong g_{m2}/(C_1 + C_2)$.
Thus, the effect of increasing the Miller capacitor is to decrease the first pole frequency ω_1 and to increase the second pole frequency ω_2. This is known as pole splitting. Observe that the pole at ω_3 does not change.

12.10 Determine the pole frequencies ω_1, ω_2, and ω_3 for the op amp of Problem 12.8, if the Miller capacitor is $C_c = 0.5$ pF:
(a) from the transfer function of Problem 12.8a
(b) using the approximate expressions of Problem 12.9

12.11 *Compensation.* Compute the gain and phase margins for the op amp of Problem 12.8 if the external Miller compensation capacitor is 25 pF. You may use the approximate results of Problem 12.9.

12.12 What value of Miller compensation capacitor will yield a 45° phase margin for the op amp of Problem 12.8. You may use the expressions derived in Problem 12.9.

12.13 *Gain error.* Find an expression for the gain of an inverting amplifier if the op amp gain is

$$A(s) = \frac{A_0\alpha}{(s + \alpha)}$$

where $A_0\alpha = 2\pi 10^6$ rad/sec and $\alpha = 2\pi 10$ rad/sec. If the *dc* gain of the inverter is 40 dB, determine the gain at (a) 4 kHz, and (b) 40 kHz.

12.14 *Integration error.* An integrator (Figure 12.12) uses the op amp of Problem 12.13. If a one volt step is applied to the input at $t = 0$, determine the error in the integration (in volts) at $t = 0.1$ msec, given that the time constant of the integrator is $RC = 0.01$ msec.

12.15 *Inverter bandwidth.* An inverter is built using the op amp of Problem 12.13. If the *dc* gain is 40 dB, determine the 3 dB bandwidth (i.e., the frequency at which the gain is 3 dB down from the *dc* gain).

12.16 Determine the 3 dB bandwidth of an inverter whose *dc* gain is 40 dB, if the gain of the op amp is

$$A(s) = \frac{A_0 \omega_1 \omega_2}{(s + \omega_1)(s + \omega_2)}$$

where

$$A_0 = 10^5, \qquad \omega_1 = 10^4 \text{ rad/sec}, \qquad \omega_2 = 10^5 \text{ rad/sec.}$$

12.17 *Predistortion.* A single pole compensated op amp characterized by $A(s) = 10^5(2\pi 10)/(s + 2\pi 10)$ is used in the negative feedback circuit of Example 12.1 to synthesize a band-pass function with a nominal pole Q of 10 and a nominal pole frequency of 2 kHz. Determine the predistorted pole Q and pole frequency needed to account for the gain of the op amp.

12.18 Repeat Problem 12.17 for a double pole compensated op amp characterized by

$$A(s) = 10^5 \frac{2\pi 10^3}{(s + 2\pi 10^3)} \cdot \frac{2\pi 10^4}{(s + 2\pi 10^4)}$$

12.19 The op amp of Problem 12.13 is used in the negative feedback circuit of Example 12.1 to realize the band-pass function

$$\frac{(2\pi 100)s}{s^2 + (2\pi 100)s + (2\pi 1000)^2}$$

Determine the predistorted denominator function needed to account for the gain of the op amp.

12.20 *Pole deviation, Saraga circuit.* The transfer function for the Saraga design of the Sallen and Key *LP* circuit of Figure 8.3 is given by Equation 8.17. Using the design formula of Equation 8.12, show that for high pole Q's

$$D_2(s_0) \cong 2.3\omega_0 s_0$$

Hence, show that the per-unit change in the pole location is given by

$$\frac{\Delta s_0}{s_0} \cong j\frac{1.53}{A(s_0)}$$

12.21 *Predistortion, Saraga circuit.* A low-pass filter with a pole Q of 5 and pole frequency at 1 kHz is synthesized using the Saraga positive feedback circuit. Assuming the op amp is single pole compensated as in Example 12.1, determine:

(a) The per-unit change in pole position.
(b) The predistorted transfer function needed to account for the gain of the op amp.
(*Hint*: use the results of Example 12.1 and Problem 12.20.)

12.22 *Statistical gain deviation.* A band-pass filter using the circuit of Example 12.1 exhibits a pole Q of 10 and a pole frequency of 10 kHz. The gain of the op amp is $A(s) = A_0\alpha/(s + \alpha)$ where $A_0\alpha = 4\pi 10^6$ rad/sec and $\alpha = 40\pi$ rad/sec, and the variability of $A_0\alpha$ due to environmental changes is $V_{A_0\alpha} = \pm 0.3$ (the tolerance limits refer to the 3σ points of a Gaussian distribution). Determine the statistics of the per-unit change in the pole Q and pole frequency. Hence, determine the statistics of the deviation in gain at the 3 dB passband edge frequencies.

12.23 A low-pass filter based on the Saraga design exhibits a pole Q of 5 and a pole frequency of 1 kHz. Assuming the op amp of Problem 12.22, determine the statistics of the gain deviation at the pole frequency. (*Hint*: use results of Problem 12.20.)

12.24 A low-pass filter based on the Saraga design has the transfer function

$$\frac{K}{s^2 + (2\pi 200)s + (2\pi 2000)^2}$$

The op amp is characterized by $A_0\alpha/(s + \alpha)$ where $A_0\alpha = 10^7$ and $\alpha = 100$. If environmental changes cause the gain bandwidth product to deviate by ± 20 percent (the limits refer to 3σ points of a Gaussian distribution), determine the statistics of the corresponding gain deviation at:
(a) The pole frequency.
(b) The 3 dB passband edge frequencies.
(c) *dc*.
Using these computations, sketch the $\pm 3\sigma$ boundaries of the gain deviation versus frequency.

For Problems 12.25 to 12.37 assume the following typical op amp characteristics (unless otherwise stated):

A_O	open loop dc gain	100 dB
V_{max}	maximum output voltage	± 10 volts
I_{max}	maximum output current	± 15 mA
V_{OS}	input offset voltage	2 mV
I_{BS}	input bias current	150 nA
I_{OS}	input offset current	30 nA
ρ	slew-rate limit	0.5 volts/μsec
$CMRR$	common mode rejection ratio	90 dB

12.25 *Dynamic range.* The op amp described above is used to build an inverter with a gain of 5 (Figure 1.9*a*).

(a) Find the amplitude of the input at which the output voltage just saturates.

(b) If $R_S = 1$ kΩ and $V_{IN} = 1$ volt, show that the current delivered by the op amp is less than I_{max}. Now suppose a resistive load R_L is connected across the output of the inverter. Determine the minimum load resistance below which the output current exceeds I_{max}. The op amp is then said to *current limit*.

12.26 A single op amp biquad realizes the band-pass function

$$\frac{100s}{s^2 + 50s + (100)^2}$$

If the input is a sinusoid at the frequency of maximum gain, determine the amplitude of the input signal above which the output voltage will just saturate.

12.27 Three single amplifier biquads are cascaded to realize the transfer function

$$\frac{V_O}{V_{IN}} = (K_1 T_1)(K_2 T_2)(K_3 T_3)$$

where K_1, K_2, and K_3 are the gain constants associated with the three biquads. The overall gain constant is required to be 25, and each gain constant must exceed 2. Choose the gain constants to maximize the allowable input voltage.

12.28 *Slew-rate limiting.* The input to the inverter of Figure 12.17 is the sinusoid $A_1 \sin(2\pi 10{,}000t)$. What is the maximum amplitude A_1 above which the output will slew-rate limit.

12.29 What is the maximum frequency which can be amplified without slew-rate distortion, with maximum possible output voltage swing.

12.30 The output load across a unity gain inverter is a capacitor $C_F = 0.05\ \mu$F and the input signal is $V_{IN} = A_1 \sin(2\pi 10{,}000t)$. Determine the amplitude A_1 above which the output just distorts, and indicate whether the distortion is caused by voltage saturation, current limiting, or slew-rate limiting.

12.31 Repeat Problem 12.30 for $C_F = 0.01\ \mu$F.

12.32 *Input offsets.* An inverter with a gain of 100 is realized using $R_F = 500$ kΩ and $R_S = 5$ kΩ. Assuming the effects of offset voltage and bias currents add, determine the total output offset voltage.
(*Hint*: first determine I_{B1} and I_{B2}.)

12.33 Suppose the op amp used in an inverter circuit has input bias currents $I_{B1} = I_{B2} = 0.2\ \mu A$, and the input offset voltage is negligible. If $R_S = 10$ kΩ and the closed loop gain is 100:

(a) Determine the output offset voltage.

(b) If the closed loop gain is to remain 100, but the output offset voltage must be less than 30 mV, determine the maximum input resistance this configuration may have.

12.34 Consider the inverter circuit shown in Figure P12.34.

(a) Determine the output offset voltage for $R = 0$.

(b) Find the value of the resistor R for which the output offset voltage is zero.

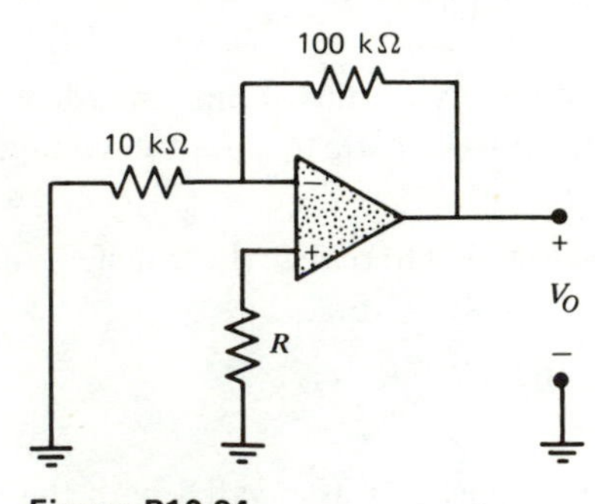

Figure P12.34

12.35 Determine the error of integration due to input bias currents in the circuit of Figure 12.12, for $R = 10$ kΩ and $C = 0.1\ \mu F$ when $t = 1$ sec.

12.36 *Common mode signals.* Determine the common mode output voltage for the circuit of Figure 12.23, if the input voltage is 2 volts.

12.37 Compute the common mode gain for the given op amp.

13.

DESIGN OPTIMIZATION AND MANUFACTURE OF ACTIVE FILTERS

The objective in any practical design is to obtain the most economical product that will meet the given specifications. In this chapter we describe the complete filter design process, emphasizing the factors which are related to cost. The cost considerations in the choice of the approximation function and the choice of the biquad have already been alluded to in the previous chapters. It is further necessary to consider the choice of the physical characteristics and the tolerances of the components used to fabricate the filter. These factors will be discussed in the light of the available ways of manufacturing filters—using discrete, thin-film, thick-film, and the integrated circuit technologies. Computer aids form an integral part of any efficient design, and they will be mentioned whereever applicable.

The discussions in this chapter relate directly to the cascaded topologies; however, the design philosophy is applicable to other filter structures, and indeed to circuits in general.

13.1 REVIEW OF THE NOMINAL DESIGN

To put into perspective those steps of the design process already discussed in previous chapters, we will review the design of the nominal filter. The design of practical filters, with real components, will then be described in the next section.

A flow diagram of the steps leading to the nominal design is shown in Figure 13.1. The starting point is the nominal filter requirements, which must be met by the nominal filter using ideal resistors, ideal capacitors, and a given op amp gain characteristic, $A(s)$. The first step in the design procedure is to find an approximation function to fit these requirements. The commonly used approximation functions are the Butterworth, Chebyshev, elliptic, and Bessel. The choice of the approximation function is based on their relative advantages, as discussed in Chapter 4. The poles, zeros, and gain constant describing the approximation function are obtained using standard tables [2], or from computer programs (such as CHEB, described in Chapter 4). The resulting transfer function can be written in this form

$$T(s) = K \frac{\prod_{i=1}^{N} m_i s^2 + \frac{\omega_{z_i}^2}{Q_{z_i}} s + \omega_{z_i}^2}{\prod_{j=1}^{N} n_j s^2 + \frac{\omega_{p_j}}{Q_{p_j}} s + \omega_{p_j}^2} \qquad \begin{pmatrix} m_i = 1 \text{ or } 0 \\ n_j = 1 \text{ or } 0 \end{pmatrix} \tag{13.1}$$

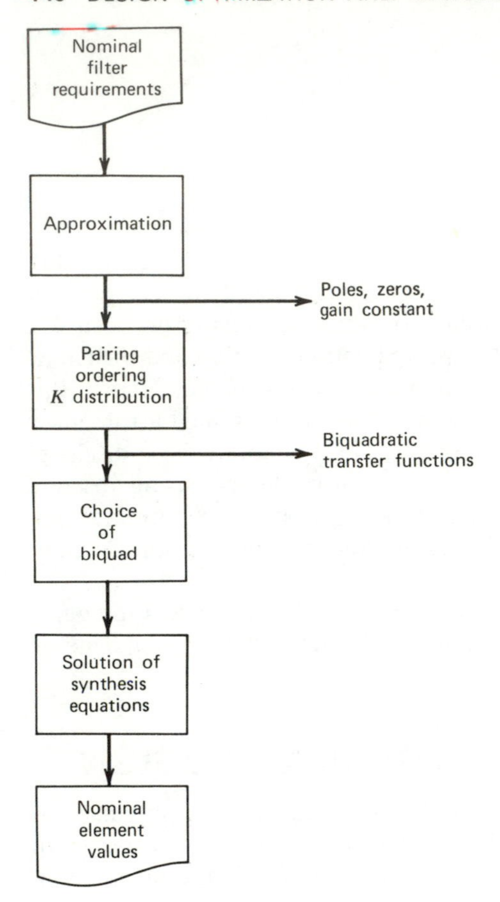

Figure 13.1 Flow diagram of nominal filter design.

Given that a cascaded realization is desired, Equation 13.1 must be expressed as the product of biquadratics. This requires:

(a) *pairing* of the poles and zeros to form biquadratic functions,
(b) determining the *order* in which the biquads will be cascaded,
(c) *distribution of the gain constant* K among the biquadratics.

Each of these steps affects the performance of the resulting filters. In [4], Halfin shows that the pairing of poles and zeros influences the sensitivity of the filter. A useful rule of thumb is to pair the pole closest to the passband edge with the zero closest to the stopband edge—this pairing minimizes the sensitivity near the edge of the passband. The optimal pairing of poles and zeros in a high-order filter would require a computer optimization algorithm [4]. The distribution of the gain constant and the ordering of the biquadratics have a direct bearing on the dynamic range of the filter. As explained in Chapter 12, the dynamic range is related to the maximum signal level the filter can transmit without distortion. In [5, 8] it is shown that the dynamic range is maximum when the maximum output voltage levels of each of the biquads is the same. This objective can be met by a proper choice of the gain constants of each biquadratic, and by an appropriate ordering of the biquads. An algorithm for optimizing the dynamic range is described in [5]. After completing the pairing, ordering, and constant distribution, the approximation function can be expressed as:

$$T(s) = \prod_{i=1}^{N} T_i(s) = \prod_{i=1}^{N} K_i \frac{m_i s^2 + \dfrac{\omega_{z_i}}{Q_{z_i}} s + \omega_{z_i}^2}{n_i s^2 + \dfrac{\omega_{p_i}}{Q_{p_i}} s + \omega_{p_i}^2} \tag{13.2}$$

The next step in the design sequence is the choice of the circuit for realizing the biquadratics. The circuit may be chosen from the negative feedback, positive feedback, and the three amplifier biquad topologies. The relative advantages of these circuits were discussed in Chapters 8 to 10. From an economical standpoint it is usually desirable to use a single amplifier biquad circuit. However, if tuning is required, it may be more convenient and economical to use a three amplifier biquad. Some other situations in which the three amplifier biquad is preferred were discussed in Chapter 10.

The last step in the design of the nominal filter is the synthesis of each $T_i(s)$ to obtain the element values. Recall that the synthesis equations assumed ideal op amps, with $A(s) = \infty$. To account for the finite gain of the op amp $T(s)$ must be predistorted, as explained in Chapter 12 (page 416). Knowing the biquad transfer function and the op amp gain characteristics, the shift in pole frequency ($\Delta\omega_{p_i}$) and pole Q (ΔQ_{p_i}) can be evaluated by using Equation 12.39 and 12.40. The predistorted transfer function, $T_{PD}(s)$, is obtained by shifting the pole frequency to $\omega'_{p_i} = \omega_{p_i} - \Delta\omega_{p_i}$ and the pole Q to $Q'_{p_i} = Q_{p_i} - \Delta Q_{p_i}$:

$$T_{PD}(s) = \prod_{i=1}^{N} K_i \frac{m_i s^2 + \dfrac{\omega_{z_i}}{Q_{z_i}} s + \omega_{z_i}^2}{n_i s^2 + \dfrac{\omega'_{p_i}}{Q'_{p_i}} s + \omega'^2_{p_i}} \tag{13.3}$$

This transfer function is synthesized to obtain the element values, by using the coefficient matching technique. The reader will recall that in the solution of the synthesis equations, the fixed elements are chosen so as to minimize the circuit sensitivity while retaining reasonable element spreads. The synthesis equations yield the desired set of nominal resistor and capacitor values.

13.2 DESIGN OF PRACTICAL FILTERS

Suppose that the nominal filter has been designed so that it *just* meets the filter requirements, with no margin to spare. When this filter is built, it is clear that the response of the filter will deviate from the nominal because of the manufacturing tolerances associated with the elements; consequently, the manufactured filter will not meet the requirements. Therefore, the nominal filter

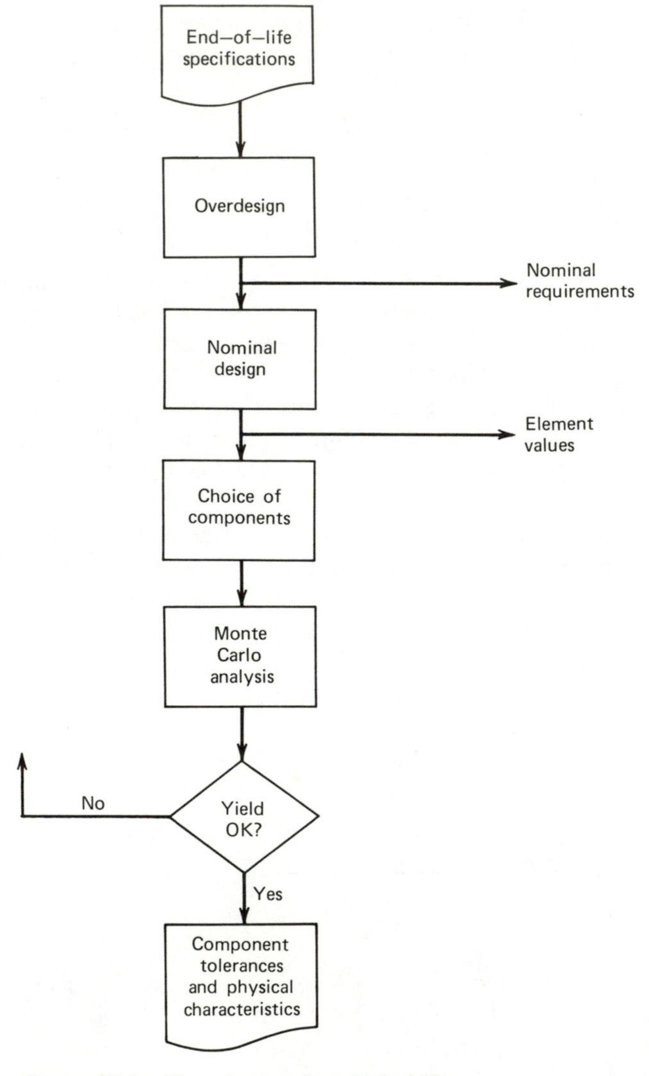

Figure 13.2 Flow diagram for practical filter design.

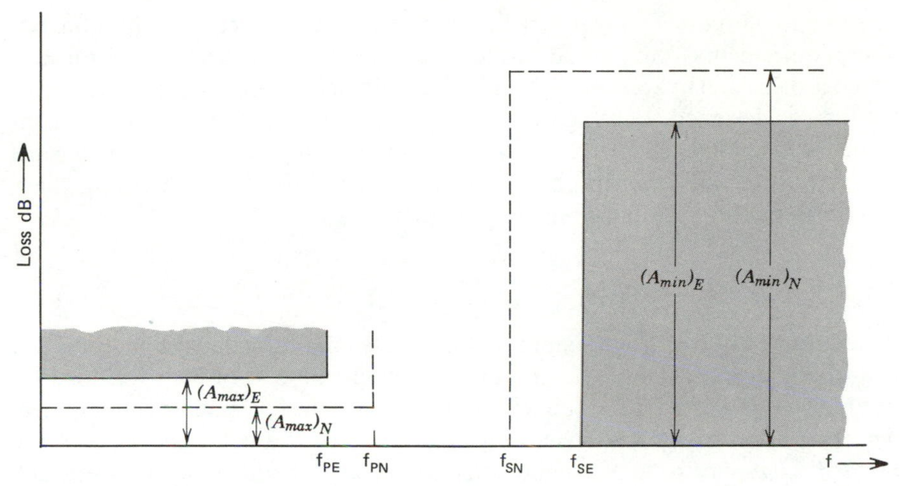

Figure 13.3 Tightening of low-pass requirements.

must be designed to meet the filter requirements with enough of a margin to allow for the component deviations. Again, after the filter is manufactured, we know that the filter response will deviate further due to environmental changes (temperature and humidity) and due to aging. These deviations must also be accounted for by providing a sufficient margin in the nominal design.

A flow diagram of the design steps for practical filters is shown in Figure 13.2. The filter requirements are usually gain versus frequency specifications that must be met over a prescribed temperature range and for a given period of time. These are known as the **end-of-life** specifications. The first step in the design sequence is to obtain the nominal filter requirements. To account for the deviations in the nominal response because of changes in the elements, the nominal requirements will have to be tighter (more stringent) than the end-of-life requirements. Suppose, for instance, that the end-of-life requirements for a LP filter are (Figure 13.3 solid lines):

$(A_{\text{max}})_E$	the passband ripple
$(A_{\text{min}})_E$	the stopband loss
f_{PE}	the passband edge frequency
f_{SE}	the stopband edge frequency

These requirements are tightened by

reducing	$(A_{\text{max}})_E$ to $(A_{\text{max}})_N$
increasing	$(A_{\text{min}})_E$ to $(A_{\text{min}})_N$
increasing	f_{PE} to f_{PN}
decreasing	f_{SE} to f_{SN}

as shown in Figure 13.3 (broken lines). The process of deriving the nominal filter requirements from the end-of-life requirements is referred to as **overdesign**, which is discussed in Section 13.2.1. Having obtained the nominal requirements, the filter is designed as in the previous section to obtain the nominal element values of the biquad circuits. Finally, the manufacturing tolerances and physical characteristics of the elements are selected so as to minimize the cost of the filter, while still meeting the end-of-life requirements.

13.2.1 OVERDESIGN

As mentioned above, the nominal requirements are obtained by tightening the end-of-life specifications. The amount by which the end-of-life specifications need to be tightened depends on the total change in the components anticipated. These are due to the manufacturing *tolerances* of the components, and the *physical characteristics* of components which are identified by temperature, humidity, and aging coefficients. However, at this early stage in the design process it is only possible to *estimate* the quality of the components needed, based on the complexity of the filter. Although an accurate estimate would be desirable, it is possible to continue with the design process with a reasonable guess.

Suppose the resistors considered for the circuit realization are all of the same type and that the per-unit deviation of the resistors is $\Delta R/R$. This random deviation can be characterized by a mean value of $\mu(\Delta R/R)$ and a standard deviation of $\sigma(\Delta R/R)$. Similarly, the per-unit deviations of the capacitors can be characterized by $\mu(\Delta C/C)$ and $\sigma(\Delta C/C)$. The change in the RC product is given by

$$\Delta RC = (R + \Delta R)(C + \Delta C) - RC = R\Delta C + C\Delta R + \Delta R\Delta C$$

For small changes in R and C, the $\Delta R\Delta C$ term is much smaller than the other two terms, so that

$$\frac{\Delta RC}{RC} \cong \frac{\Delta R}{R} + \frac{\Delta C}{C} \tag{13.4}$$

Thus the per-unit deviation in the RC product is a random variable characterized by (see Appendix C, Equations C.12 and C.13):

$$\mu\left(\frac{\Delta RC}{RC}\right) = \mu\left(\frac{\Delta R}{R}\right) + \mu\left(\frac{\Delta C}{C}\right) \tag{13.5}$$

$$\sigma\left(\frac{\Delta RC}{RC}\right) = \sqrt{\sigma^2\left(\frac{\Delta R}{R}\right) + \sigma^2\left(\frac{\Delta C}{C}\right)} \tag{13.6}$$

This deviation in the RC product is related to the required shifts in the passband and stopband edge frequencies, as shown in the following. Recall from Section

8.3 on $RC \rightarrow CR$ transformations that a biquadratic can be represented in the dimensionless form

$$T(s) = \frac{[R]^2[C]^2s^2 + [R][C]s + 1}{[R]^2[C]^2s^2 + [R][C]s + 1} \tag{13.7}$$

where the $[R][C]$ and $[R]^2[C]^2$ terms indicate the dimensions of the coefficients of s and s^2, respectively. Suppose now that all the RC products in the circuit increased by one percent. The deviated transfer function would then be

$$T'(s) = \frac{[R]^2[C]^2(1 + 0.01)^2s^2 + [R][C](1 + 0.01)s + 1}{[R]^2[C]^2(1 + 0.01)^2s^2 + [R][C](1 + 0.01)s + 1} \tag{13.8a}$$

$$= \frac{[R]^2[C]^2(1.01s)^2 + [R][C](1.01s) + 1}{[R]^2[C]^2(1.01s)^2 + [R][C](1.01s) + 1} \tag{13.8b}$$

From this equation it is apparent that the one percent increase in the RC products is equivalent to replacing s by $1.01s$. Considering real frequencies ($s = j\omega$), this represents a one percent shift in the frequency response along the $j\omega$ axis, towards $\omega = 0$ (Figure 13.4). This argument extends to the general active RC realization, in that a fractional change $(\Delta RC)/RC$ results in a similar shift in the fr:quency response. In particular, the change in the passband edge and stopband edge frequencies are

$$\frac{\Delta f_P}{f_P} = -\frac{\Delta RC}{RC} \quad \text{and} \quad \frac{\Delta f_S}{f_S} = -\frac{\Delta RC}{RC} \tag{13.9}$$

To anticipate this change, the nominal requirements must be made more stringent than the end-of-life requirements. This is accomplished by increasing

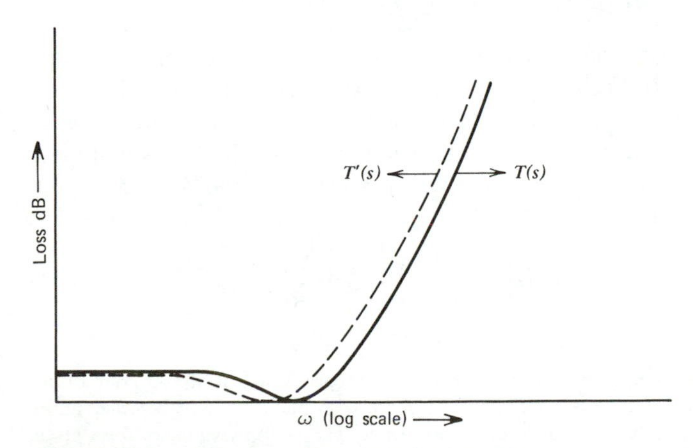

Figure 13.4 Shift of response along the frequency axis.

the passband edge frequency and decreasing the stopband edge frequency by the largest expected deviation in value of $\Delta RC/RC$, that is

$$f_{PN} = f_{PE}\left[1 + \max\left(\frac{\Delta RC}{RC}\right)\right] \tag{13.10}$$

$$f_{SN} = f_{SE}\left[1 + \min\left(\frac{\Delta RC}{RC}\right)\right] \tag{13.11}$$

The range of values for $\Delta RC/RC$ depends not only on the manufacturing tolerances and physical characteristics of the components, but also on the range of environmental conditions over which the filter is required to meet the end-of-life requirements. For example, suppose that the requirements must be met for t years after manufacture, from $-\Delta T$°C to ΔT°C about room temperature. Then the maximum and minimum values for $\Delta RC/RC$ are computed by considering the statistics of the filter response at each of the extreme environmental conditions:

1. 0 years, $-\Delta T$°C
2. 0 years, $+\Delta T$°C
3. t years, $-\Delta T$°C
4. t years, $+\Delta T$°C

If the change in $\Delta RC/RC$ can be characterized as a Gaussian random variable, then considering the $\pm 3\sigma$ limits (Appendix C), for 99.74 percent of the filters

$$\max\left(\frac{\Delta RC}{RC}\right) = \max\left[\mu\left(\frac{\Delta RC}{RC}\right) + 3\sigma\left(\frac{\Delta RC}{RC}\right)\right] \tag{13.12}$$

$$\min\left(\frac{\Delta RC}{RC}\right) = \min\left[\mu\left(\frac{\Delta RC}{RC}\right) - 3\sigma\left(\frac{\Delta RC}{RC}\right)\right] \tag{13.13}$$

where the standard deviation $\sigma(\Delta RC/RC)$ is always a positive number, while the mean change $\mu(\Delta RC/RC)$ depends on the extreme condition being considered, and it may be positive or negative. Substituting these expressions in Equations 3.10, we see that to ensure that 99.74 percent of the filters will meet the passband requirement (i.e., for the so-called **yield** to be 99.74 percent), the nominal passband should be*

$$f_{PN} = f_{PE}\left\{1 + \max\left[\mu\left(\frac{\Delta RC}{RC}\right) + 3\sigma\left(\frac{\Delta RC}{RC}\right)\right]\right\} \tag{13.14}$$

* It is convenient to use 99.74 percent as the yield, because this corresponds to the 3σ point of the Gaussian distribution. In practice, the required end-of-life yield may very well be different, in which case the factor multiplying σ will be other than three. The multiplying factor for a given yield is readily obtained from standard tables of areas under the Gaussian distribution curve (see references in Appendix C).

Similarly, for a 99.74 percent yield, the nominal stopband edge requirement is

$$f_{SN} = f_{SE}\left\{1 + \min\left[\left(\frac{\Delta RC}{RC}\right) - 3\sigma\left(\frac{\Delta RC}{RC}\right)\right]\right\} \tag{13.15}$$

Let us next consider the requirement on the stopband loss. Since the deviation in the stopband loss is not very critical in most filter applications, a rather rough estimate for $(A_{\min})_N$ will be quite satisfactory. A good rule of thumb, developed from experience, is to increase $(A_{\min})_E$ by approximately twice the passband ripple $(A_{\max})_E$, that is

$$(A_{\min})_N = (A_{\min})_E + 2(A_{\max})_E \tag{13.16}$$

The last requirement that needs to be determined is the nominal passband ripple $(A_{\max})_N$. In most filter applications this requirement is very critical. Therefore, it is desirable *to allow as much margin in the passband ripple as is possible*, without having to increase the order of the filter. To achieve this we first calculate the order of the filter defined by

$$f_{SN}, f_{PN}, (A_{\min})_N, \quad \text{and} \quad (A_{\max})_E \tag{13.17}$$

(Note that the end-of-life value is used for the passband ripple.) The order n can be obtained from standard tables [2], or by using the expressions derived in Chapter 4. Typically, the value computed for n will be fractional and the order will have to be the next higher integer. For example, if n were computed to be 3.4, a fourth-order filter should be tried. This fourth-order filter will, of course, meet the filter requirements specified by Equation (13.17), with a margin to spare. This means that the nominal passband ripple $(A_{\max})_N$ can be made less than the end-of-life ripple, $(A_{\max})_E$. To take full advantage of the margin, the nominal ripple should be made as small as possible, within the constraint that the order may not exceed four. This value of $(A_{\max})_N$ can be obtained from standard tables (for $n = 4$), as the smallest passband ripple that will *just* meet the f_{PN}, f_{SN}, and $(A_{\min})_N$ requirements.

Summarizing, for filters in which the component changes are described by a Gaussian distribution, the nominal requirements are defined by:

- Passband edge frequency $f_{PN} = f_{PE}\left\{1 + \max\left[\mu\left(\frac{\Delta RC}{RC}\right) + 3\sigma\left(\frac{\Delta RC}{RC}\right)\right]\right\}$
- Stopband edge frequency $f_{SN} = f_{SE}\left\{1 + \min\left[\mu\left(\frac{\Delta RC}{RC}\right) - 3\sigma\left(\frac{\Delta RC}{RC}\right)\right]\right\}$
- Stopband loss $(A_{\min})_N = (A_{\min})_E + 2(A_{\max})_E$
- Passband ripple $(A_{\max})_N$ is minimized to just meet the f_{PN}, f_{SN}, and $(A_{\min})_N$ requirements without increasing the order.

The methods of this section are easily extended to high-pass, band-pass, and band-reject filters.

Note that deviations due to the op amp were not considered in the above discussions. The computation of this deviation is no easy task; however, in most filter designs (with pole frequencies below approximately 10 kHz) the deviations due to the op amp can be made much smaller than that due to the passive components (as shown in Examples 8.3, 9.3, and 10.2). For this reason, it is usually adequate to base the nominal requirements on the deviations due to the passive elements alone.

13.2.2 CHOICE OF COMPONENTS

Referring to Figure 13.2, after the nominal requirements are determined, the filter is synthesized to obtain the nominal element values, following the steps outlined in Section 13.1. The next step in the design is the choice of the components for building the filter. As mentioned previously, the objective is to pick the least expensive components that will allow the filter to meet the end-of-life requirements. This step is discussed in the following.

First, let us consider the factors that contribute to the cost of building a circuit. The cost related items can be grouped into two classes, namely, fixed costs and variable costs. In particular, filter testing, packaging, and the cost of the op amps are assumed to have fixed costs in that they do not vary significantly from filter to filter. These fixed cost items, by their very definition, need not be considered in the cost minimization algorithm. The variable cost items include the costs of the resistors and capacitors, and the tuning costs. The resistor and capacitor costs depend on their physical characteristics as defined by the temperature, humidity, and aging coefficients. In general, the cost increases as these coefficients decrease.* Furthermore, the cost will also depend on the manufacturing tolerances of the components. However, the gain deviations due to manufacturing tolerances can be nullified by tuning the filter. Therefore, if tuning is employed, components with wider manufacturing tolerances may be used; in other words, there is a trade-off between the component costs and the tuning costs.

The general problem of minimizing the cost of the filter, considering the component and tuning costs, is a rather complex one, requiring sophisticated computer aids. The problem has been researched by several authors and the interested reader is referred to [1, 7, 11, 14] for details. In the remainder of this section we will explain the design philosophy of cost minimization by considering a simplification of the general problem.

The first step in the cost minimization algorithm is to select a component type and tolerance for the filter realization. Let us hypothesize that three types of

* Since the resistors and capacitors always occur as an RC product in the transfer function (Equation 13.7), it is sufficient to consider the temperature, aging, and humidity coefficient of the RC product.

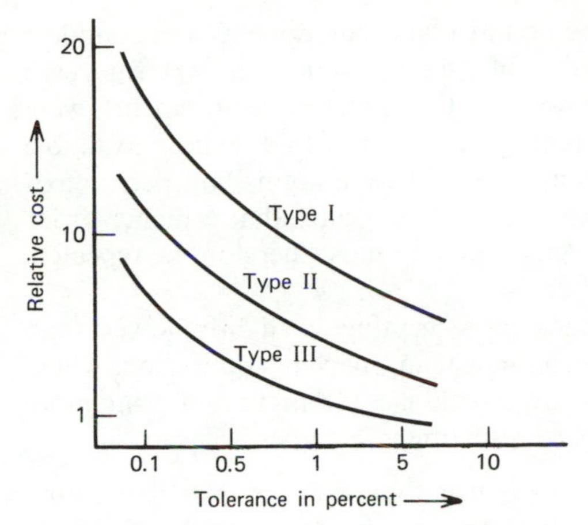

Figure 13.5 Relative cost versus manufacturing tolerances for three types of components.

components are available, whose relative costs are as indicated in Figure 13.5.* A reasonable first choice is the set of elements used for estimating the nominal filter requirements. With this component set, the filter is then analyzed, using Monte Carlo techniques, to determine the yield, which is the percentage of filters that pass the end-of-life specification. Suppose, for example, that the desired yield is 90 percent.† The analysis can give three possible results. First, the filter may *just* meet the end-of-life yield requirements. In this case any attempt at reducing the cost, by increasing the tolerances, will certainly result in more filters failing the end-of-life specifications. Therefore, this first choice of element tolerances corresponds to the desired cost optimized filter. A second possible result of the Monte Carlo analysis is that the filter meets the end-of-life specifications with a margin to spare. In this case the analysis is repeated iteratively with less expensive components, until the filter just meets the end-of-life yield requirement. Once again, this is the desired solution. The third possible result is that the yield is less than desired. In this case the most economical of the following alternatives is chosen:

(a) Use more expensive components. This can be done by decreasing the tolerances and/or by using a different type of component.
(b) Tune the filter, partially or completely, to reduce the effects of manufacturing tolerances.

* For simplicity, all the components in the circuit are assumed to have the same tolerance.

† The accuracy of the yield computation depends on the number of circuits considered in the Monte Carlo analysis. Typically, if 500 circuits are analyzed, the evaluated yield will be correct to within approximately ± 3 percent. However, to simplify the presentation, we assume the number of circuits to be large enough to establish the yield exactly.

(c) It is possible that even with the best available components and complete tuning the filter fails the end-of-life specification. This extreme case implies that the margins allowed in the nominal requirements were inadequate for a successful realization, even with the best available components and complete tuning. Since these margins had been maximized for the given order filter, this result indicates that a higher-order filter is needed. The entire design procedure must therefore be repeated, using the next higher-order filter.

(d) Finally, the designer may consider accepting a slightly lower than desired yield. This is the most economical approach in applications where the cost penalty of accepting a lower yield is less than that of using more expensive components, tuning, or resorting to a higher order.

The proper evaluation of these options is usually a very complex procedure. Therefore, in all but the simplest applications, one needs to use the computer aids referred to in the beginning of this section.

Example 13.1

An active filter is required to meet the low-pass filter requirements shown in Figure 13.6, from 0°C to 75°C for a 20-year life span. The characteristics of the resistors and capacitors are:

Resistors
- temperature coefficient (α_{TCR}) = 100 ± 10 ppm/°C
- aging in 20 years (α_{ACR}) = $\pm 0.5\%$

Capacitors
- temperature coefficient (α_{TCC}) = -120 ± 20 ppm/°C
- aging in 20 years (α_{ACC}) = $\pm 0.5\%$

The components are available with the following percent manufacturing tolerances (MT):

$$\pm 2, \quad \pm 1, \quad \pm 0.5, \quad \pm 0.25, \quad \text{and} \quad \pm 0.1$$

All the tolerance limits refer to the 3σ points of a Gaussian distribution. Assuming all the resistors and capacitors to have the same tolerances, and the op amp to be ideal, find the largest manufacturing tolerance for the components that allows 99.74 percent of the circuits to meet the end-of-life requirements.

Solution

The first step is to estimate the nominal requirements. The nominal passband edge frequency f_{PN} and the stopband edge frequency f_{PN} are evaluated using Equation 13.14 and 13.15, respectively. These equations require the mean and standard deviations of the components at the two temperature extremes of

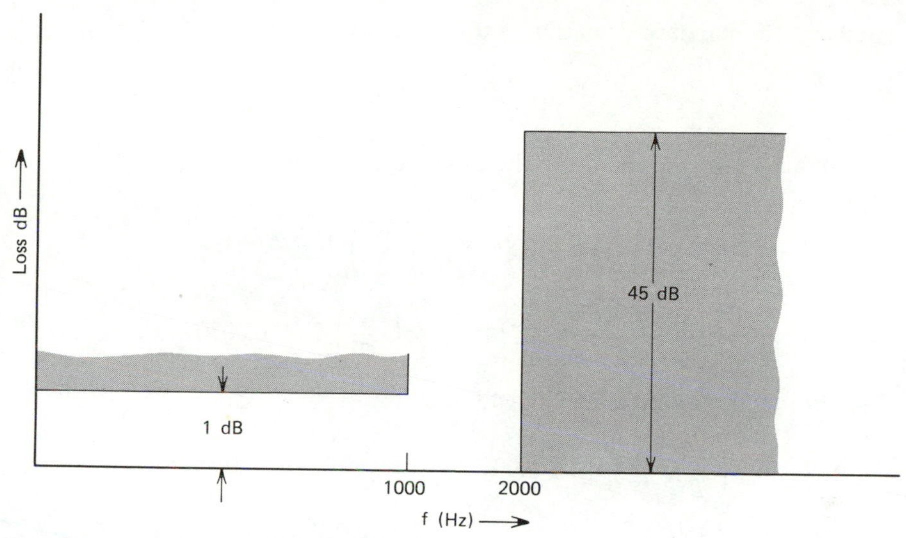

Figure 13.6 Requirements for Example 13.1.

0°C and 75°C.* As a first guess, assume ±0.5 percent resistors and capacitors will be used. The mean value for $\Delta R/R$, from Equation 5.55, is given by

$$\mu\left(\frac{\Delta R}{R}\right) = \mu(\alpha_{TCR})\Delta T = (10)^{-4}\Delta T$$

where ΔT is the deviation from room temperature. In this example, ΔT ranges from -25°C to $+50$°C. Thus

$$\mu\left(\frac{\Delta R}{R}\right)_{75^\circ\text{C}} = 0.005 \qquad \mu\left(\frac{\Delta R}{R}\right)_{0^\circ\text{C}} = -0.0025$$

Again from Equation 5.56, the standard deviation of $\Delta R/R$ is given by

$$3\sigma^2\left(\frac{\Delta R}{R}\right)_{75^\circ\text{C}} = [3\sigma(\alpha_{TCR})\Delta T]^2 + [3\sigma(\alpha_{ACR})]^2 + [3\sigma(MT)]^2$$

$$= [(10)^{-5}50]^2 + \left(\frac{0.5}{100}\right)^2 + \left(\frac{0.5}{100}\right)^2$$

$$= 5.025(10)^{-5}$$

So

$$3\sigma\left(\frac{\Delta R}{R}\right)_{75^\circ\text{C}} = 0.00709$$

* Since the aging coefficients have zero mean, we need only consider the two temperature extremes for computing the maximum and minimum values of $\Delta RC/RC$.

Similarly, the standard deviation at 0°C is seen to be

$$3\sigma\left(\frac{\Delta R}{R}\right)_{0^\circ\mathrm{C}} = 0.00706$$

Next, considering the change in the capacitors, we get:

$$\mu\left(\frac{\Delta C}{C}\right)_{75^\circ\mathrm{C}} = -0.006 \qquad \mu\left(\frac{\Delta C}{C}\right)_{0^\circ\mathrm{C}} = 0.003$$

$$3\sigma\left(\frac{\Delta C}{C}\right)_{75^\circ\mathrm{C}} = 0.00714 \qquad 3\sigma\left(\frac{\Delta C}{C}\right)_{0^\circ\mathrm{C}} = 0.00709$$

Substituting in Equation 13.5 and 13.6

$$\mu\left(\frac{\Delta RC}{RC}\right)_{75^\circ\mathrm{C}} = -0.001 \qquad \mu\left(\frac{\Delta RC}{RC}\right)_{0^\circ\mathrm{C}} = 0.0005$$

$$3\sigma\left(\frac{\Delta RC}{RC}\right)_{75^\circ\mathrm{C}} = 0.010062 \qquad 3\sigma\left(\frac{\Delta RC}{RC}\right)_{0^\circ\mathrm{C}} = 0.010006$$

From the above the minimum and maximum changes in $\Delta R/RC$ are easily identified as:

$$\min\left(\frac{\Delta RC}{RC}\right) = \mu\left(\frac{\Delta RC}{RC}\right)_{75^\circ\mathrm{C}} - 3\sigma\left(\frac{\Delta RC}{RC}\right)_{75^\circ\mathrm{C}} = -0.011$$

$$\max\left(\frac{\Delta RC}{RC}\right) = \mu\left(\frac{\Delta RC}{RC}\right)_{0^\circ\mathrm{C}} + 3\sigma\left(\frac{\Delta RC}{RC}\right)_{0^\circ\mathrm{C}} = 0.0105$$

Therefore, from Equation 13.14, the nominal passband frequency is

$$f_{PN} = 1000(1 + 0.0105) = 1010.5 \text{ Hz}$$

and the nominal stopband frequency, from Equation 13.15, is

$$f_{SN} = 2000(1 - 0.011) = 1978 \text{ Hz}$$

The nominal stopband attenuation, from Equation 13.16, is

$$(A_{\min})_N = 45 + 2(1) = 47 \text{ dB}$$

To evaluate the nominal passband ripple, we first find the order of the filter defined by the requirements

$$f_{PN} = 1010.5 \text{ Hz} \qquad f_{SN} = 1978 \text{ Hz} \qquad (A_{\min})_N = 47 \text{ dB} \qquad (A_{\max})_E = 1 \text{ dB}$$

From the standard tables [2], it can be verified that a fourth-order elliptic filter will meet these requirements. Moreover, the minimum passband ripple, for a fourth-order elliptic approximation that *just* meets the f_{PN}, f_{SN}, and $(A_{min})_N$ requirements, is 0.5 dB. Thus

$$(A_{max})_N = 0.5 \text{ dB}$$

The nominal requirements are therefore characterized by

$$f_{PN} = 1010.5 \text{ Hz} \qquad f_{SN} = 1978 \text{ Hz} \qquad (A_{max})_N = 0.5 \text{ dB} \qquad (A_{min})_N = 47 \text{ dB}$$

The fourth-order elliptic approximation satisfying these nominal requirements can be shown to be*

$$T(s) = 4.299(10)^{-3} \frac{[s^2 + 1.77565(10)^8][s^2 + 9.322148(10)^8]}{(s^2 + 1896.9s + 4.2789(10)^7)(s^2 + 5624.1s + 1.663268(10)^7)}$$

Following the guidelines established in Section 13.1, the pole closest to the passband edge is paired with the zero closest to the stopband edge. Also, to maximize the dynamic range, the ordering of the biquadratics and the distribution of the gain constant should be chosen so that the maximum voltage level at the output of both biquads is the same. This may be accomplished by using computer optimization algorithms or, if the programs are not available, by trial and error. The resulting biquadratic decomposition is given here:

$$T(s) = \frac{0.017842(s^2 + 9.332148(10)^8)}{s^2 + 5624.1s + 1.66328(10)^7} \cdot \frac{0.24097(s^2 + 1.77565(10)^8)}{s^2 + 1896.9s + 4.2789(10)^7}$$

This transfer function, sketched in Figure 13.7, is seen to meet the nominal requirements. Note that the maximum output voltage of the first biquad for a 1 volt input is 1 (Figure 13.8). This is the same as the maximum voltage for the second biquad, as observed from the overall filter response sketched in Figure 13.7.

One biquad circuit that gives reasonable element values and a low sensitivity is the Friend negative feedback circuit of Figure 9.11*b*. For this circuit, the nominal element values are obtained from Equation 9.60 and 9.61. One set of nominal element values (for $C_1 = C_2 = 0.01\ \mu\text{F}$, $R_A = 4\ \text{k}\Omega$, $R_B = \infty$) is given in Table 13.1.

The final step in the design process is the selection of the components. To start with, let us compute the yield using 0.5 percent components. A Monte Carlo analysis with these components, over the temperature range 0°C to 75°C and with 20 years of aging, results in the gain versus frequency response shown in Figure 13.9. The two curves shown are the $\pm 3\sigma$ boundaries, which enclose 99.74 percent of the networks. It is seen that the circuit with 0.5 percent components fails the passband requirement by approximately 0.2 dB. Therefore,

* This function was obtained from a computer program for elliptic approximations.

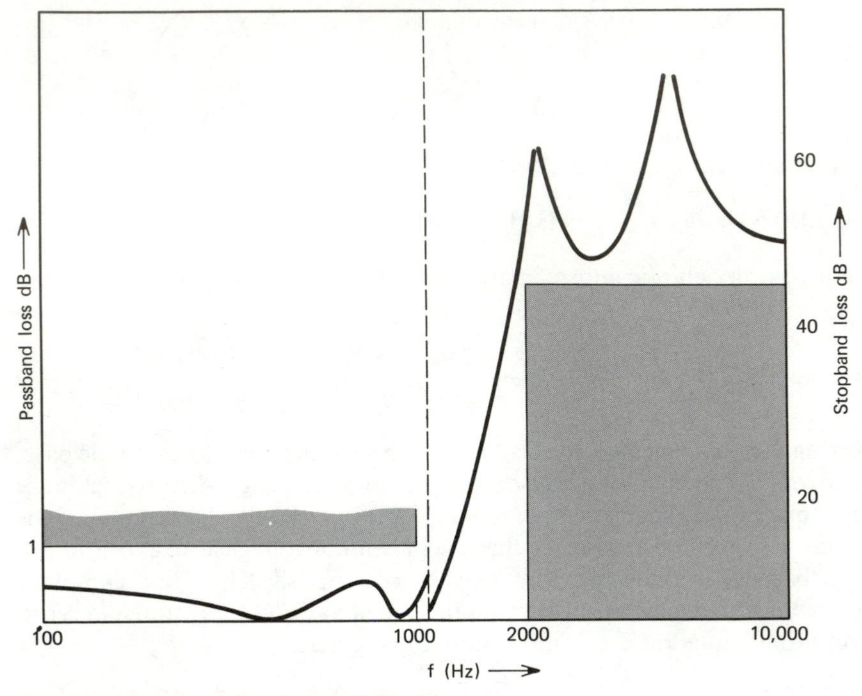

Figure 13.7 Response of nominal filter.

tighter tolerance elements are needed. A reasonable second choice is the 0.25 percent elements. The Monte Carlo analysis with these components results in the $\pm 3\sigma$ boundaries shown in Figure 13.10. In this case these boundaries are *just* within the required end-of-life requirements. It can therefore be concluded that the circuit with 0.25 percent tolerances satisfies the end-of-life requirements.

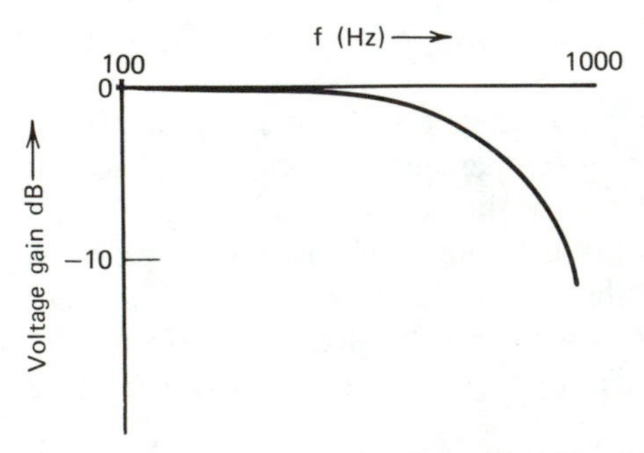

Figure 13.8 Voltage gain of the first biquadratic.

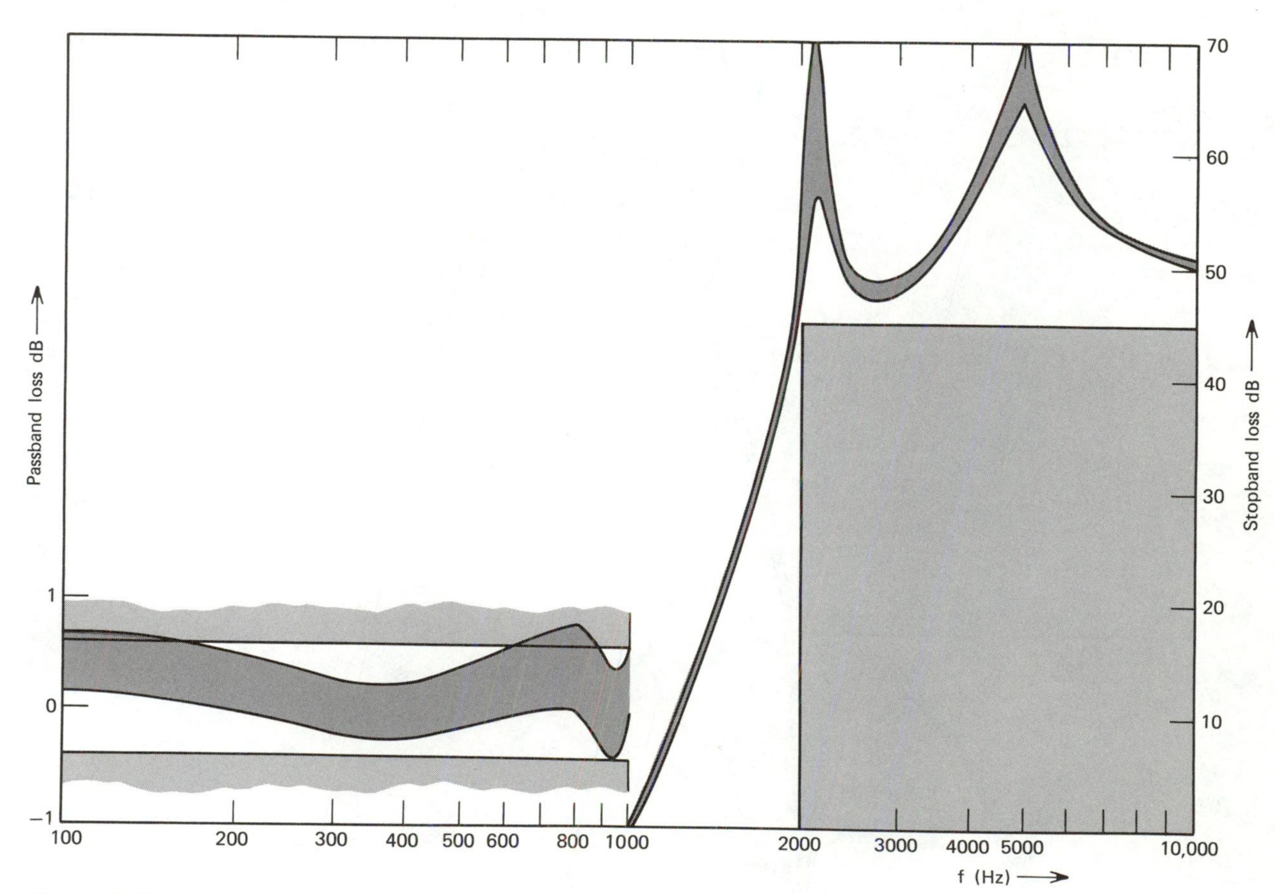

Figure 13.9 Monte Carlo analysis using ± 0.5 percent components.

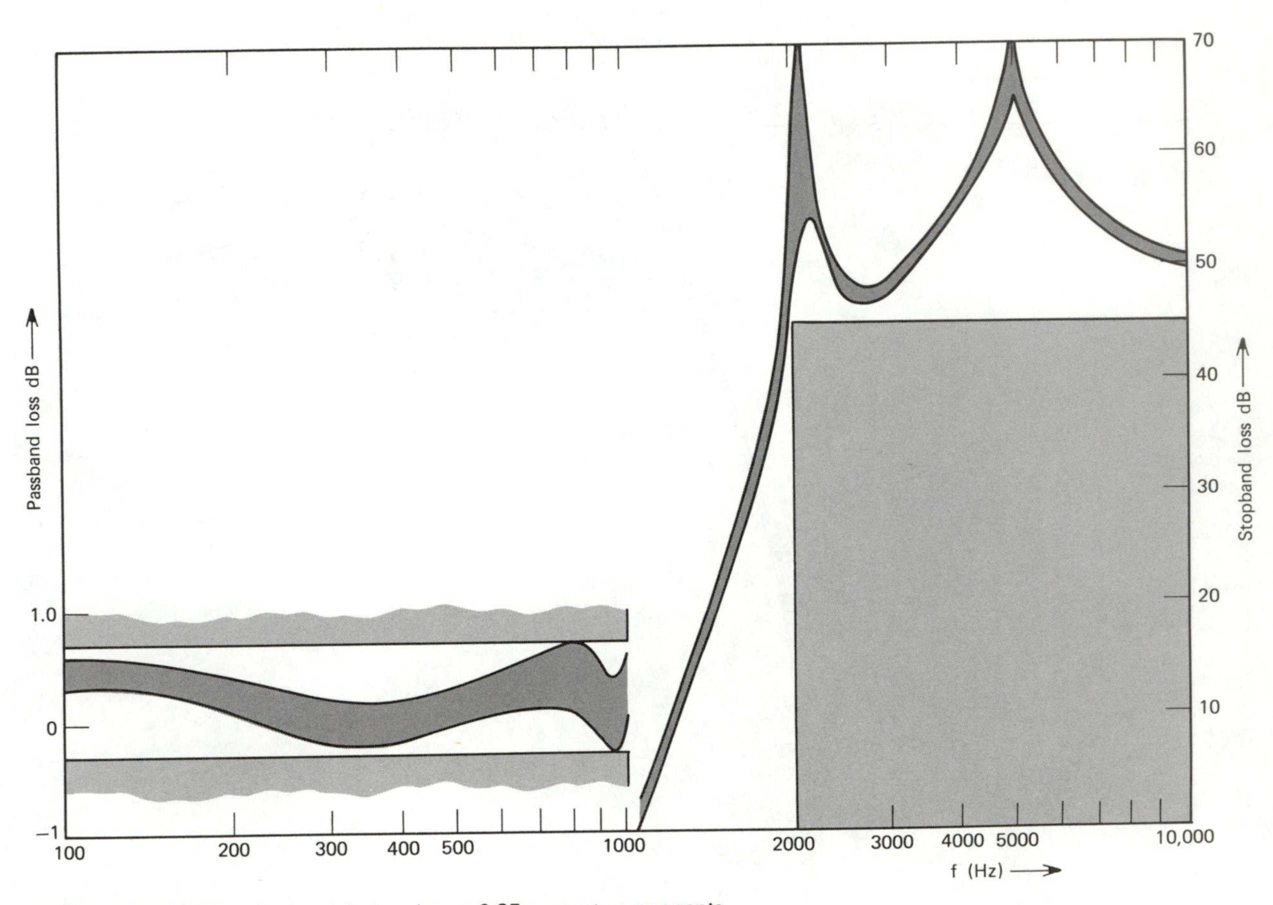

Figure 13.10 Monte Carlo analysis using ± 0.25 percent components.

Table 13.1 Element Values for the Nominal Filter in kΩ and μF

Element	Biquad 1	Biquad 2
R_2	35.6	105.0
R_4	17.45	7.83
R_5	540.0	3.09
R_6	∞	∞
R_7	0.646	33.5
R_B	∞	∞
R_C	224.0	16.6
R_D	4.07	5.27
C_1	0.01	0.01
C_2	0.01	0.01

Observations

1. The 0.5 percent components could only be used if the cost penalty of discarding some of the filters turned out to be less than the cost advantage of using the wider tolerance elements.
2. In the example, all the elements were assumed to have 0.25 percent tolerances. It is possible that the cost could be reduced by allowing some of the less critical resistors and capacitors to have a 0.5 percent (or even higher) tolerance. Further Monte Carlo analyses are needed to investigate this. ■

13.3 TECHNOLOGIES

In this section we describe some of the popular ways of manufacturing active filters. These include silicon **integrated circuits** used for manufacturing op amps, and the **discrete**, **thick-film**, and **thin-film** technologies for the manufacture of resistors and capacitors. To make an economic evaluation of these different methods, we must consider the costs of the components, the assembly and packaging costs, and the costs of tuning and testing the filter. In addition, the initial investment in the required equipment must be considered. Since the equipment is shared by all the circuits, the cost per circuit becomes a function of the total volume of production. Thus, as we will see, a technology that is used for high volumes of production is likely to be economically unsuitable for the production of small numbers of circuits.

In the following we restrict ourselves to a discussion of the main features of each technology, as related to active filter design. The interested reader is referred to [6] for a detailed description of the characteristics and processing steps of these different technologies.

13.3.1 INTEGRATED CIRCUIT OPERATIONAL AMPLIFIERS

Operational amplifiers are universally manufactured using silicon integrated circuits. The top view of an op amp integrated circuit chip (or IC chip, as it is called), is shown in Figure 13.11. In this technology the transistors are made by diffusing phosphorus and boron in a silicon wafer to form n and p type regions, respectively. These regions can also serve to provide resistors. Capacitors can be made by depositing a dielectric and a metal on the surface of the IC chip. The components are connected by gold or aluminum strips. The op amp shown in the figure consists of 35 transistors, 30 resistors, and 4 capacitors, all of which are contained in the IC chip, whose size is a miniscule $0.065'' \times 0.065''$ (i.e., $0.165 \text{ cm} \times 0.165 \text{ cm}$). The proximity of the devices makes their characteristics track very closely with changes in temperature, humidity, and aging, and this is used to advantage in the design of the op amp.

This technology requires much complex and expensive equipment, which can be economically justified for op amps, since the volume of production is very high. Besides offering the already mentioned size advantage, the integrated circuit technology results in an op amp of a high quality, which is also quite inexpensive. At the present time the resistors and capacitors for the active filter are not manufactured using this technology because the characteristics of these components (i.e., the manufacturing tolerance, temperature and aging coefficients) cannot be controlled well enough to meet the requirements of most active filters.

13.3.2 DISCRETE CIRCUITS

Discrete circuits consist of resistors, capacitors, and IC op amps hand or machine inserted onto a printed circuit board. A typical discrete active filter is shown in Figure 13.12. This technology offers a wide selection of resistor and capacitor types with different characteristics. In the circuit shown the resistors are made of a metal oxide on a glass rod; the capacitors consist of metallic plates or foils separated by polystyrene or dipped mica dielectric; and the op amp IC chips are packaged in a metal can. Discrete circuits require a minimum of engineering effort and equipment, and as such provide an easy way of producing small numbers of circuits at a low cost. Tuning may be performed by including potentiometers in the circuit. It is often more economical, however, to tune by means of a potentiometer and then replace the potentiometer by a discrete resistor of the nearest available value. Tuning, however, is expensive, so most manufacturers produce untuned and, therefore, medium accuracy discrete filters.

13.3.3 THICK-FILM CIRCUITS

Thick-film circuits consist of resistive, conductive, or dielectric inks fused onto the surface of a ceramic substrate. A typical thick-film circuit is shown in Figure 13.13. Here the op amp is an IC chip bonded onto the substrate.* In the thick-film technology, the inks are deposited by a squeegee driven across the ceramic surface, through a patterned silk or metallic screen. These inks are then fused in an oven to form the thick-film resistors, conductors, and capacitors. Typically, the layers are deposited to a thickness of approximately 10 μm to 50 μm [9]. The tuning of resistors can be achieved by laser trimming or by sandblasting. The processing steps in the thick-film technology are relatively simple and inexpensive, the circuit produced being small, light, and reliable. This technology proves to be quite economical for medium quality filters, when the number of filters produced is in the order of 10,000 to 100,000 per year.†

13.3.4 THIN-FILM CIRCUITS

Thin-film circuits consist of resistors and capacitors deposited on a ceramic substrate. The film thicknesses used are in the order of 0.005 μm to 1 μm, which is considerably less than in the thick-film process. This requires processing steps that must be kept under very tight control. Typically, the films may be deposited using evaporation, electroplating, or sputtering. To achieve the desired geometry, photolithographic and chemical etching techniques are used [9].

A thin-film active filter circuit used to realize biquad circuits is shown in Figure 13.14. This circuit uses two capacitors, ten resistors, and one IC op amp chip. In this technology, the resistors are typically composed of metal nitrides and the capacitors consist of tantalum and gold layers separated by a metal oxide dielectric. The technology allows the realization of high-quality resistors, which can be tuned very accurately by laser trimming techniques. However, there is a restriction on the range of resistor values that can be attained. At the present time special measures are needed for achieving resistors below 200 Ω and above 250 kΩ. Another limitation is that the quality of thin-film capacitors is below that available using discrete components, but they are adequate for

* In this filter circuit, the capacitors (not shown) are discrete elements external to the ceramic substrate. At the present time high quality thick-film capacitors are not available, so discrete capacitors are used instead.

† The number depends very much on the manufacturing setup.

most filter applications.* Thin-film technology is complex, requiring much engineering effort and equipment. However, the final result is a circuit that is very small and light, provides accurate filter characteristics, and is economically competitive when manufactured in large numbers (in the order of 100,000 or more per year).

13.4 CONCLUDING REMARKS

In this chapter we alluded to several computer programs that could be used to obtain a minimum cost filter design. Among these were the program for pairing, ordering and gain constant distribution, the tolerance assignment algorithms, and the Monte Carlo statistical analysis program. Often, however, the designer will not have access to such programs nor will he have the time to develop them. What most engineers in this situation do is to allow a wider margin in the nominal requirements, or else use narrower tolerance components than would be indicated by the computer optimization aids. As a result the design will usually be more expensive. However, the better the estimates of the engineer, the closer will his design be to optimal. The less computer aids the engineer has, the more he (she) must rely on his (her) engineering experience and judgment.

FURTHER READING

1. J. W. Bandler and P. C. Liu, "Automated network design with optimal tolerances," *IEEE Trans. Circuits and Systems*, *CAS-21*, No. 2, March 1974, pp. 219–222.
2. E. Christian and E. Eisenmann, *Filter Design Tables and Graphs*, Wiley, New York, 1966.
3. M. Fogiel, *Modern Microelectronics*, Research and Education Assoc., New York, 1972.
4. S. Halfin, "An optimization method for cascade filters," *Bell System Tech.*, *49*, No. 2, February 1970, pp. 185–190.
5. S. Halfin, "Simultaneous determination of ordering and amplifications of cascaded subsystems," *J. Optimization Theory and Appl.*, *6*, No. 5, 1970, pp. 356–363.
6. E. R. Hnatek, *A Users' Handbook of Integrated Circuits*, Wiley, New York, 1973.
7. B. Karafin, "The general component tolerance assignment problem in electrical networks," Ph.D. Thesis, Univ. of Pennsylvania, 1974.
8. E. Lueder, "A decomposition of a transfer function minimizing distortion and in-band losses," *Bell System Tech. J.*, *49*, No. 3, March 1970, pp. 455–469.
9. G. S. Moschytz, *Linear Integrated Networks Fundamentals*, Van Nostrand, New York, 1974, Chapter 6.

* The manufacturing tolerances of the capacitors can be compensated for by tuning the resistors to obtain the desired transfer function.

10. G. S. Moschytz, Linear Integrated Networks Design, Van Nostrand, New York, 1975, Chapter 1.
11. J. F. Pinel and K. A. Roberts, "Tolerance assignment in linear networks using non-linear programming," *IEEE Trans. Circuit Theory*, *CT-19*, No. 9, September 1972, pp. 475–479.
12. C. L. Semmelman, E. D. Walsh, and G. Daryanani, "Linear circuits and statistical design," *Bell System Tech. J. 50*, No. 4, April 1971, pp. 1149–1171.
13. L. Stern, *Fundamentals of Integrated Circuits*, Hayden, New York, 1968.
14. D. Sud and R. Spence, "Component tolerances assignment and design centering," *1974 European Conf. on Circuit Theory and Design*, July 1974, pp. 165–170.

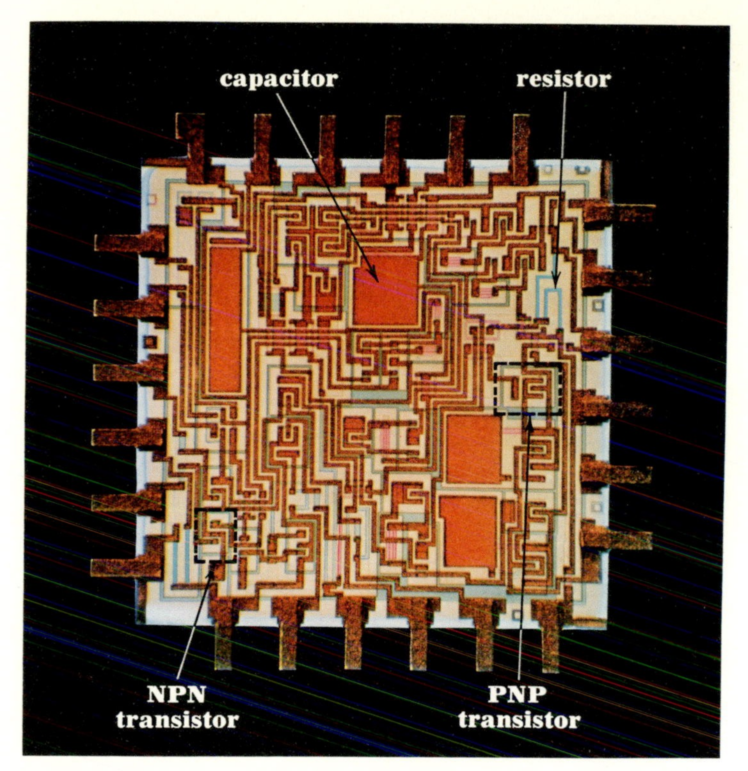

Figure 13.11 An integrated circuit op amp.

Figure 13.12 A discrete active filter circuit.

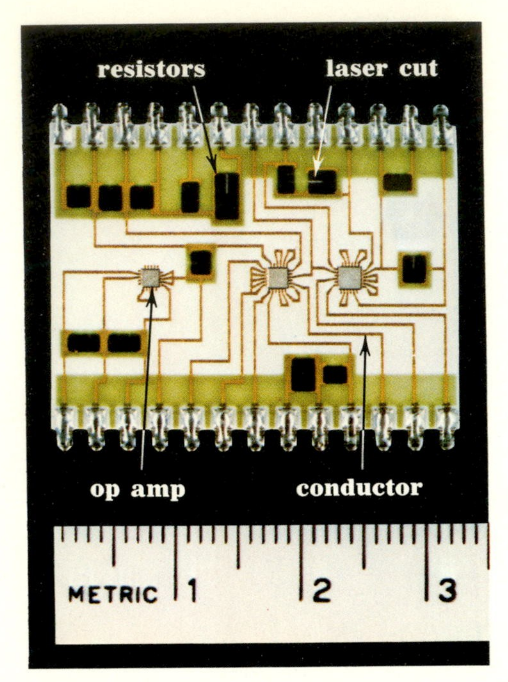

Figure 13.13 A thick-film circuit.

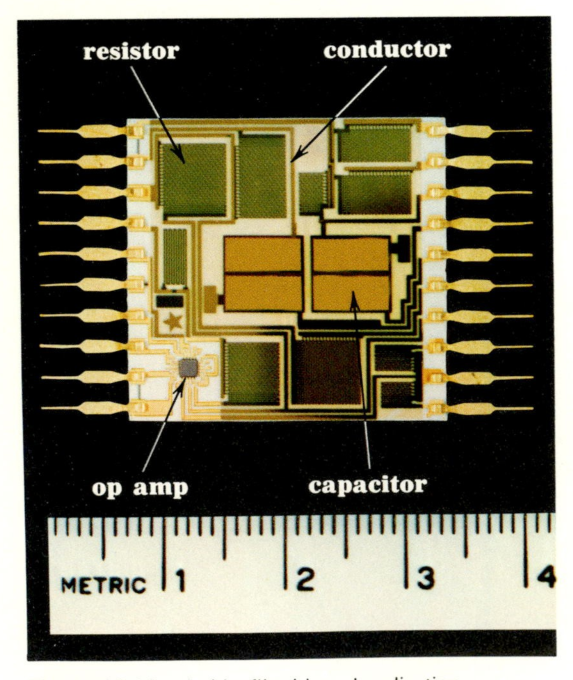

Figure 13.14 A thin-film biquad realization.

APPENDIX A

PARTIAL FRACTION EXPANSION

Consider the function

$$H(s) = \frac{n(s)}{d(s)} = \frac{n(s)}{(s - p_1)(s - p_2)\cdots(s - p_n)} \tag{A.1}$$

where all the poles, p_i, are simple and the degree of the numerator is less than that of the denominator.* Such a function can always be written in the form:

$$H(s) = \frac{K_1}{s - p_1} + \frac{K_2}{s - p_2} + \cdots + \frac{K_n}{s - p_n} \tag{A.2}$$

This is known as the *partial fraction expansion* of $H(s)$. In this expression, the constants K_i are referred to as the *residues* at the poles p_i. The residue K_i can be evaluated by multiplying both sides of Equation A.2 by $s - p_i$, and then letting $s = p_i$:

$$K_i = H(s)(s - p_i)|_{s = p_i} \tag{A.3}$$

* If the degree of $n(s)$ is equal to or greater than that of $d(s)$, then $d(s)$ is divided into $n(s)$ to yield

$$H(s) = q(s) + \frac{n'(s)}{d(s)}$$

where $n'(s)$ is of lower degree than $d(s)$. The partial fraction expansion is then performed on $n'(s)/d(s)$ [1].

To illustrate this, let us find the residues of the following function at its two poles:

$$H(s) = \frac{2(s+2)}{(s+1)(s+3)} = \frac{K_1}{s+1} + \frac{K_2}{s+3} \tag{A.4}$$

From Equation A.3

$$K_1 = H(s)(s+1)|_{s=-1} = 2\frac{(s+2)}{(s+3)}\bigg|_{s=-1} = 1$$

Similarly

$$K_2 = 2\frac{(s+2)}{(s+1)}\bigg|_{s=-3} = 1$$

Equation A.3 can also be used to obtain the partial fraction expansion of functions with complex conjugate poles. For example, consider the function

$$H(s) = \frac{1}{s(s^2+1)} = \frac{1}{s(s+j)(s-j)} \tag{A.5}$$

which has complex poles at $s = \pm j$. The partial fraction expansion of this function has the form

$$H(s) = \frac{K_1}{s} + \frac{K_2}{s+j} + \frac{K_3}{s-j}$$

where, from Equation A.3,

$$K_1 = \frac{1}{s^2+1}\bigg|_{s=0} = 1$$

$$K_2 = \frac{1}{s(s-j)}\bigg|_{s=-j} = -\frac{1}{2}$$

$$K_3 = \frac{1}{s(s+j)}\bigg|_{s=j} = -\frac{1}{2}$$

Thus

$$H(s) = \frac{1}{s} + \frac{-\frac{1}{2}}{s+j} + \frac{-\frac{1}{2}}{s-j} = \frac{1}{s} - \frac{s}{s^2+1} \tag{A.6}$$

If the function $H(s)$ contains a multiple pole of order n at p_1, the corresponding terms in the partial fraction expansion are [1]:

$$\frac{K_{11}}{(s-p_1)} + \frac{K_{12}}{(s-p_1)^2} + \cdots + \frac{K_{1n}}{(s-p_1)^n} \tag{A.7}$$

The constants K_{1i} in this equation are given by [1]:

$$K_{1i} = \frac{1}{(n-i)!}\frac{d^{n-i}}{ds^{n-i}} H(s)(s-p_1)^n \bigg|_{s=p_1} \tag{A.8}$$

For example, consider the partial fraction expansion of:

$$\begin{aligned} H(s) &= \frac{2}{(s+1)(s+2)^2} \\ &= \frac{K_1}{s+1} + \frac{K_{21}}{s+2} + \frac{K_{22}}{(s+2)^2} \end{aligned} \tag{A.9}$$

The residue K_1 is obtained using Equation A.3

$$K_1 = \frac{2}{(s+2)^2}\bigg|_{s=-1} = 2$$

The constants K_{21} and K_{22} are obtained from Equation A.8

$$\begin{aligned} K_{21} &= \frac{1}{(2-1)!}\frac{d}{ds}\left(\frac{2}{s+1}\right)\bigg|_{s=-2} \\ &= \frac{-2}{(s+1)^2}\bigg|_{s=-2} = -2 \end{aligned}$$

and

$$\begin{aligned} K_{22} &= \frac{1}{(2-2)!}\frac{d^0}{ds^0}\left(\frac{2}{s+1}\right)\bigg|_{s=-2} \\ &= \frac{2}{s+1}\bigg|_{s=-2} = -2 \end{aligned}$$

Thus

$$H(s) = \frac{2}{s+1} - \frac{2}{s+2} - \frac{2}{(s+2)^2}$$

FURTHER READING

1. F. F. Kuo, *Network Analysis and Synthesis*, Second Edition, Wiley, New York, 1966, Chapter 6.
2. M. E. Van Valkenburg, *Network Analysis*, Third Edition, Prentice-Hall, Englewood Cliffs, N.J., 1974, Chapter 7.

APPENDIX B

CHARACTERIZATION OF TWO-PORT NETWORKS

DEFINITIONS OF *z* AND *y* PARAMETERS

A general two-port network, shown in Figure B.1, can be characterized by the voltage-current pairs (V_1, I_1) and (V_2, I_2), at the two ports. These variables can be related as follows:

$$V_1 = z_{11}I_1 + z_{12}I_2 \tag{B.1}$$

$$V_2 = z_{21}I_1 + z_{22}I_2 \tag{B.2}$$

where z_{11}, z_{12}, z_{21}, and z_{22} are known as the z parameters of the network. They are defined by

$$z_{11} = \left.\frac{V_1}{I_1}\right|_{I_2=0} \qquad z_{12} = \left.\frac{V_1}{I_2}\right|_{I_1=0}$$
$$z_{21} = \left.\frac{V_2}{I_1}\right|_{I_2=0} \qquad z_{22} = \left.\frac{V_2}{I_2}\right|_{I_1=0} \tag{B.3}$$

In view of the above definitions, the z parameters are also referred to as the *open circuit* parameters.

The z parameters of a particular two-port network can be obtained by writing the node equations with the two ports driven by current sources I_1 and I_2. The voltages V_1 and V_2 can then be solved for from these node equations in the following form:

$$V_1 = \frac{\Delta_{11}}{\Delta}I_1 + \frac{\Delta_{21}}{\Delta}I_2 \tag{B.4}$$

$$V_2 = \frac{\Delta_{12}}{\Delta}I_1 + \frac{\Delta_{22}}{\Delta}I_2 \tag{B.5}$$

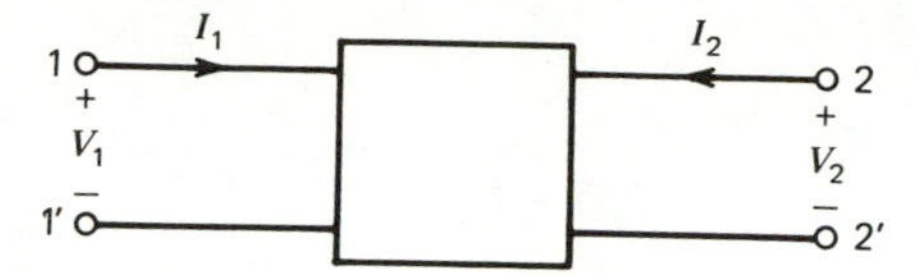

Figure B.1 A general two-port network.

where Δ is the nodal determinant and the Δ_{ij}'s are the ijth cofactors of the determinant. Comparing these equations with (B.1) and (B.2), we see that the z parameters can be identified in terms of the nodal determinant and cofactors by:

$$z_{11} = \frac{\Delta_{11}}{\Delta} \qquad z_{12} = \frac{\Delta_{21}}{\Delta} \tag{B.6}$$

$$z_{21} = \frac{\Delta_{12}}{\Delta} \qquad z_{22} = \frac{\Delta_{22}}{\Delta} \tag{B.7}$$

Observe that for passive *RLC* networks the nodal determinant is symmetrical, so $\Delta_{12} = \Delta_{21}$ and hence $z_{12} = z_{21}$.

An alternate way of relating the variables V_1, I_1, V_2, I_2 is in terms of the y parameters:

$$I_1 = y_{11}V_1 + y_{12}V_2 \tag{B.8}$$

$$I_2 = y_{21}V_1 + y_{22}V_2 \tag{B.9}$$

The y parameters of the two-port network are defined by

$$y_{11} = \left.\frac{I_1}{V_1}\right|_{V_2=0} \qquad y_{12} = \left.\frac{I_1}{V_2}\right|_{V_1=0}$$
$$y_{21} = \left.\frac{I_2}{V_1}\right|_{V_2=0} \qquad y_{22} = \left.\frac{I_2}{V_2}\right|_{V_1=0} \tag{B.10}$$

The y parameters are also known as the *short circuit* parameters.

These parameters can be obtained by writing the mesh equations with the two ports driven by voltage sources V_1 and V_2, in a manner analogous to that used for the z parameters.

DRIVING POINT AND TRANSFER FUNCTIONS

The driving point (dp) and transfer functions of a two port, defined in Section 2.1, can be expressed in terms of the z and y parameters. In particular, we are interested (Section 6.4) in the two-port configuration shown in Figure B.2, with port 2-2′ terminated by a resistance R_2. To find the input impedance of

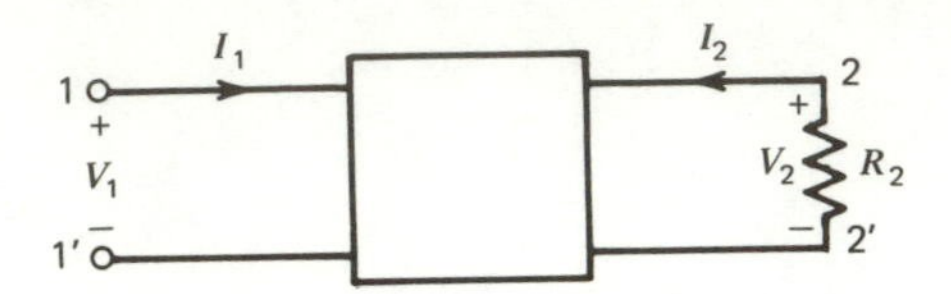

Figure B.2 A two-port network terminated by a resistor.

this network in terms of the z parameters, Equation B.1 and B.2 are combined with the following relationship, imposed by the load R_2 at port 2:

$$V_2 = -I_2 R_2 \tag{B.11}$$

This yields

$$V_1 = z_{11} I_1 + z_{12} I_2 \tag{B.12}$$

$$0 = z_{21} I_1 + (z_{22} + R_2) I_2 \tag{B.13}$$

Eliminating the variable I_2, the dp impedance at port 1-1′ is obtained:

$$Z_{IN} = \frac{V_1}{I_1} = \frac{z_{11}z_{22} - z_{12}z_{21} + z_{11}R_2}{z_{22} + R_2}$$

But

$$\Delta z = z_{11}z_{22} - z_{12}z_{21}$$

therefore

$$Z_{IN} = \frac{\Delta z/R_2 + z_{11}}{z_{22}/R_2 + 1} \tag{B.14}$$

In a similar way, substituting (B.11) in (B.8) and (B.9), we get

$$I_1 = y_{11} V_1 + y_{12} V_2 \tag{B.15}$$

$$0 = y_{21} V_1 + \left(y_{22} + \frac{1}{R_2}\right) V_2 \tag{B.16}$$

The dp admittance at port 1-1′ is obtained by eliminating the variable V_2:

$$Y_{IN} = \frac{I_1}{V_1} = \frac{y_{11} + \Delta_y R_2}{1 + y_{22} R_2} \tag{B.17}$$

where $\Delta y = y_{11}y_{22} - y_{12}y_{21}$

The voltage transfer function V_2/V_1 is also easily obtained from (B.16) in terms of the y parameters as:

$$\frac{V_2}{V_1} = -\frac{y_{21}}{1/R_2 + y_{22}} \tag{B.18}$$

APPENDIX C

MEAN AND STANDARD DEVIATION OF A RANDOM VARIABLE

In this appendix the reader is introduced to some elementary concepts from probability theory and statistics which are useful for a mathematical description of component deviations.

Consider, for example, a manufactured resistor whose nominal value is 100 Ω. Due to the production process, the values of the manufactured resistors will be spread about the 100 Ω nominal value. If this spread is given to be ± 5 percent, the measured values of the resistors will lie between 95 Ω and 105 Ω. A typical plot of the measured values for a large sample of resistors is shown in Figure C.1. In this plot, which is known as a *histogram*, the horizontal axis, representing resistance, is divided into 1 Ω bins. The vertical axis is the number of resistors $n(R_i)$ which have the value R_i to $R_i + 1$. For instance, there are 120 resistors in the range 96 Ω to 97 Ω. It should be apparent that the height of this histogram will depend on the total number of resistors measured N, and on the width of the bin ΔR. The shape of the plot can be made independent of N and ΔR by normalizing the vertical axis variable to

$$\frac{n(R_i)}{N\,\Delta R} \tag{C.1}$$

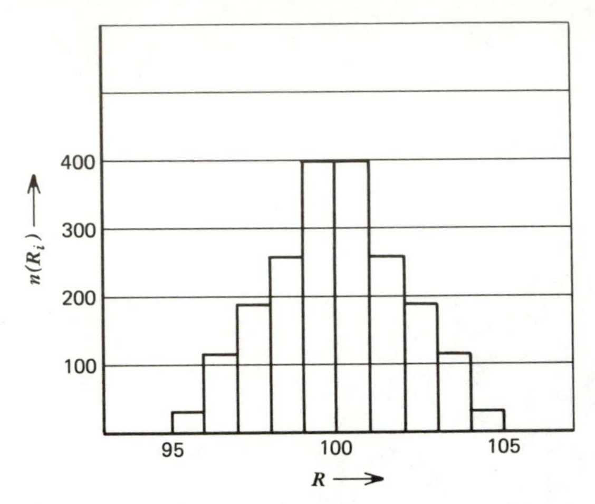

Figure C.1 Histogram for resistors.

as shown in Figure C.2. In this normalized histogram, the fraction of resistors in any interval R_i to $R_i + \Delta R$ is given by:

$$\frac{n(R_i)}{N\,\Delta R}\,\Delta R = \frac{n(R_i)}{N} \tag{C.2}$$

For instance, six percent of the resistors lie in the shaded bin 96 to 97 Ω. Moreover, the area under the whole histogram is given by:

$$\sum_{R_i=95}^{104} \frac{n(R_i)}{N\,\Delta R}\,\Delta R = \frac{1}{N} \sum_{R_i=95}^{104} n(R_i)$$

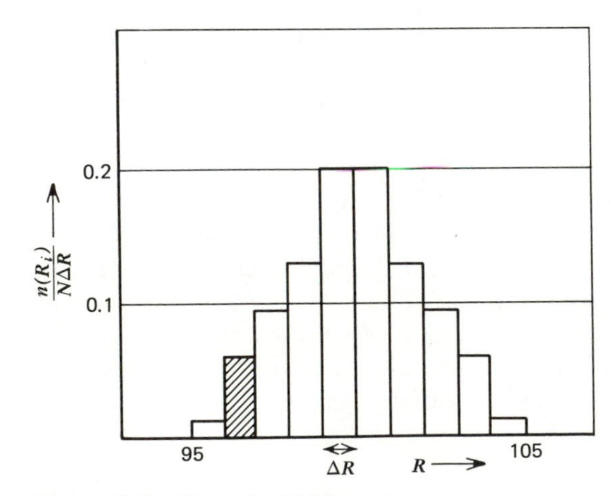

Figure C.2 Normalized histogram.

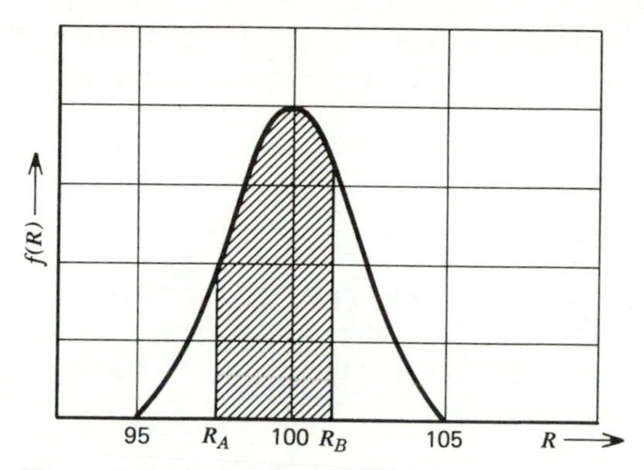

Figure C.3 Probability density function.

But the sum of the number of resistors in all the bins is the total number of resistor N. Thus the area under the normalized histogram is unity, that is

$$\frac{1}{N}\sum_i n(R_i) = 1 \tag{C.3}$$

Next, let us consider the limiting case as the sample size $N \to \infty$ and the width of the bin $\Delta R \to 0$. In the limit, the shape of the normalized histogram will approach a smooth curve as shown in Figure C.3. The variable defining the vertical axis is given by

$$f(R_i) = \lim_{\substack{N\to\infty \\ \Delta R\to 0}} \frac{n(R_i)}{N\,\Delta R} \tag{C.4}$$

Some interesting properties of this function, known as the *probability density function*, can be obtained as a limiting case of the discrete normalized function of Equation C.1.

1. The area under the probability density function is given by

$$\int_{-\infty}^{+\infty} f(R)dR = \lim_{N\to\infty} \sum_i \frac{n(R_i)}{N} \tag{C.5}$$

   From Equation C.3, this area is unity.
2. The fraction of the total number of resistors in the range R_A to R_B is given by

$$\sum_{\substack{R_i=R_A \\ \lim N\to\infty \\ \Delta R\to 0}}^{R_i=R_B-\Delta R} \frac{n(R_i)}{N} = \sum_{\substack{R_A \\ \lim \Delta R\to 0}}^{R_B-\Delta R} f(R_i)\Delta R = \int_{R_A}^{R_B} f(R)\,dR \tag{C.6}$$

Thus, the area under the curve from R_A to R_B is equal to the fraction of resistors in the range R_A to R_B. Stated differently, the probability of a resistor to have a value between R_A and R_B is

$$P(R_A < R < R_B) = \int_{R_A}^{R_B} f(R)\,dR \tag{C.7}$$

Observe that the probability of a resistor having a value between $-\infty$ and $+\infty$ is unity (being equal to the area under the probability density function from $-\infty$ to $+\infty$). As a special case of the above, the fraction of resistors with values greater than R_A, from Equation C.7, is

$$\int_{R_A}^{\infty} f(R)\,dR$$

Since probability density functions describe random phenomena, it is expected that the functions can have a large variety of shapes. Nevertheless, it is desirable to characterize these functions by a few parameters. One important parameter used to describe probability density functions is the *mean value.* In our example of manufactured resistors, if there are N resistors, the mean value is given by

$$\mu(R) = \frac{R_1 + R_2 + \cdots + R_i + \cdots + R_N}{N} \tag{C.8}$$

where R_i is the measured value of a resistor. If the resistors are put in bins, as in Figure C.1, the mean value can be expressed as

$$\mu(R) \cong \sum_i \frac{R_i n(R_i)}{N} \tag{C.9}$$

From Equation C.5, in the limit as $N \to \infty$, this expression generalizes to

$$\mu(R) = \int_{-\infty}^{+\infty} Rf(R)\,dR \tag{C.10}$$

The mean value of a symmetrical distribution, such as that shown in Figure C.3, is halfway between the end points of the resistor range. In the above example $\mu(R) = 100$.

Another important characteristic of the shape of the density function is its spread about the mean. The spread describes how far the measured value of resistors are from the mean. A parameter describing the spread is known as the *standard deviation* which, for N resistors, is defined as

$$\sigma(R) = \frac{\sqrt{[R_1 - \mu(R)]^2 + [R_2 - \mu(R)]^2 + \cdots + [R_i - \mu(R)]^2 + \cdots + [R_N - \mu(R)]^2}}{N}$$

Using the histogram notation

$$\sigma^2(R) \cong \sum_i \frac{[R_i - \mu(R)]^2 n(R_i)}{N}$$

In the limit as $N \to \infty$, using Equation C.5, this expression becomes

$$\sigma^2(R) = \int_{-\infty}^{+\infty} [R - \mu(R)]^2 f(R)\, dR \tag{C.11}$$

The two parameters μ and σ completely characterize most of the commonly used probability density functions.

So far, the mean μ, and standard deviation σ of one random variable has been considered. Let us next consider the algebraic sum y of a number of independent random variables* x_i:

$$y = ax_1 + bx_2 + cx_3 + \cdots$$

It can be shown that the mean and standard deviation of y are given by

$$\mu(y) = a\mu(x_1) + b\mu(x_2) + c\mu(x_3) + \cdots \tag{C.12}$$

$$\sigma^2(y) = a^2\sigma^2(x_1) + b^2\sigma^2(x_2) + c^2\sigma^2(x_3) + \cdots \tag{C.13}$$

These relationships are used in the study of deviations in the gain of a network due to the random variations in the circuit components.

Let us next consider some typical probability density functions. One function that occurs rather frequently is the Gaussian or normal probability function

$$f(x) = \frac{1}{\sigma\sqrt{2\pi}} \exp\left(-\frac{1}{2}\left[\frac{x - \mu}{\sigma}\right]^2\right) \tag{C.14}$$

where μ and σ are the mean and standard deviation of the probability density function $f(x)$. This function is sketched in Figure C.4. The area under this curve from $\mu - \sigma$ to $\mu + \sigma$ (found by integrating Equation C.14) can be shown to be 0.6826. This implies that 68.26 percent of the random variables lie within $\pm\sigma$ of the mean value. Similarly it can be shown that 95.44 percent of the random variables lie between $\pm 2\sigma$ of the mean, and 99.74 percent of the random variables lie between $\pm 3\sigma$ of the mean.

It can be also shown that the linear combination of a large number of independent random variables will tend to have a Gaussian distribution,† even when the individual variables themselves do not have Gaussian distributions. This interesting phenomenon explains why the Gaussian distribution occurs so frequently in nature.

* The random variables are assumed to be independent of one another, in the sense that the value any one random variable has no effect on the value of the other random variables.

† This law is known as the *law of large numbers*, or the *central limit theorem*.

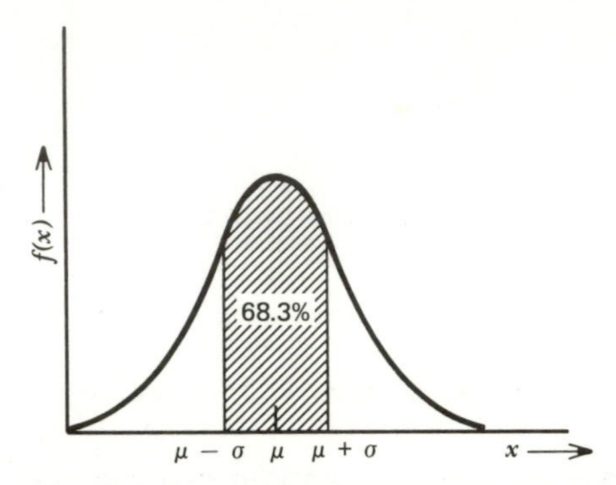

Figure C.4 Gaussian or normal density function.

Another common density function describing manufactured components is the uniform distribution, portrayed in Figure C.5. The range of possible values for the random variable extends from a to b. The flat density function implies that all the values in the range a to b occur with equal probability. For this distribution, from Equation C.10 and C.11

$$\mu = \frac{a + b}{2}$$

and

$$\sigma = \frac{(b - a)}{\sqrt{12}} \tag{C.15}$$

Manufacturers will often specify the nominal value of the resistor, its tolerance (i.e., spread in values at the time of manufacture), and the type of probability

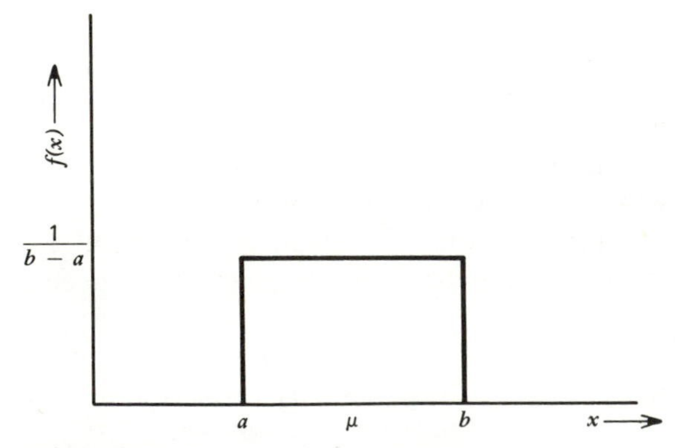

Figure C.5 Uniform density function.

density function. Typically, a resistance could be specified to have a nominal value of 100 Ω and a tolerance of ± 5 percent; where the ± 5 percent tolerance limits could refer to the 3σ points of a Gaussian distribution. From the above discussions, 99.74 percent of the resistors will have values between 95 Ω and 100 Ω. If a flat probability density function is specified, the ± 5 percent tolerance limits refer to the minimum and maximum resistor values (Figure C.5); in this case all the resistors will lie in the range 95 Ω to 105 Ω.

Although the examples in this appendix used resistor values as the random variable, the discussions are equally applicable to other components, to the response of a network, and indeed to any continuous random variable.

FURTHER READING

1. T. W. Anderson and S. L. Sclove, *Introductory Statistical Analysis*, Houghton Mifflin, Boston, Mass., 1974.
2. O. L. Davies and P. Goldsmith, *Statistical Methods in Research and Production*, Hafner, New York, 1972.
3. W. Feller, *An Introduction to Probability Theory and Its Applications, 1*, Second Edition, Wiley, New York, 1957.
4. T. C. Fry, *Probability and Its Engineering Uses*, Second Edition, Van Nostrand, Princeton, N.J., 1964.
5. I. Miller and J. E. Freund, *Probability and Statistics for Engineers*, Prentice-Hall, Englewood Cliffs, N.J., 1965.
6. P. L. Meyer, *Introductory Probability and Statistical Applications*, Addison-Wesley, Reading, Mass., 1965.

APPENDIX D

COMPUTER PROGRAMS

This appendix explains the use of the two programs, MAG and CHEB, referenced in the text. MAG (Section 2.7) computes the gain, phase, and delay of a product of biquadratics. CHEB (Section 4.8) computes the Chebyshev approximation function for low-pass, high-pass, band-pass, and band-reject filters.

The programs may be used in batch mode or in an interactive mode from a dial-up terminal. In the interactive mode the program asks for the required input (the answers may be given in any standard format). For the benefit of batch users, an example of the input cards follows each program listing.

The programs are written in standard ANSI FORTRAN IV which should be compatible with most FORTRAN compilers. The only nonstandard statements are the unformatted READs. Logical unit number 6 is used for both READ and WRITE.

Input Cards for MAG

Card 1: NSEC = the number of biquadratic sections

Card 2: Frequencies for computation
FS = start frequency Hz
FI = frequency increment Hz
FF = final frequency Hz

Card 3: Biquadratic parameters
M, C, D, N, A, B
Use one card for each biquadratic section

Card 4: Continue or end card.
1 = another run expected (data follows)
0 = terminate run

The input cards needed for Example 2.5 are (unformatted READ assumed):

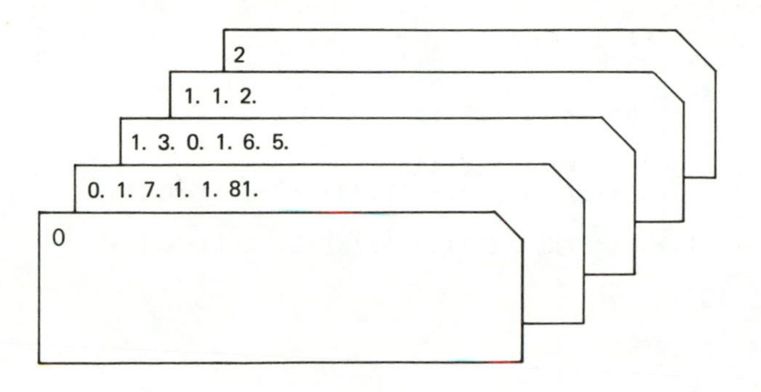

I. Program Listing for MAG

```
C   **   PROGRAM FOR COMPUTING THE GAIN (DB),
C   **   PHASE (DEG), DELAY (SECS)
C   **   OF A BIQUADRATIC IN THE FORM
C   **   (M*S**2 + C*S + D) / (N*S**2 + A*S + B)
C
C   **   NSEC = NO. OF BIQUADS
C   **   FS = START FREQUENCY (HZ)
C   **   FI = FREQUENCY INCREMENT (HZ)
C   **   FF = FINAL FREQUENCY (HZ)
C
         REAL M(10),C(10),D(10),N(10),A(10),B(10),RL,IMG
         COMPLEX S,NUMR,DEM
C   **   READ INPUT DATA
   10    WRITE (6,1010)
 1010    FORMAT (' ENTER NUMBER OF BIQUADRATIC SECTIONS')
         READ (6,*) NSEC
         WRITE (6,1012)
 1012    FORMAT (' ENTER START FREQ, FREQ INCREMENT, FINAL FREQ')
   20    READ (6,*) FS,FI,FF
         DO 100 I=1,NSEC
         WRITE (6,1014) I
 1014    FORMAT (' ENTER M, C, D, N, A, B FOR BIQUADRATIC ',I2)
  100    READ (6,*) M(I),C(I),D(I),N(I),A(I),B(I)
C   **   PRINT TABLE HEADING
         WRITE (6,1001)
 1001    FORMAT (1H ,4X,7HFREQ HZ,6X,7HGAIN DB,4X,9HPHASE DEG,
      +  3X,11HDELAY (SEC))
         FREQ = FS
C   **   CHECK IF ALL FREQS COMPLETE
  200    IF (FREQ.GT.FF) GO TO 400
C   **   INITIALIZE PARAMETERS
         W = 2.*3.141592654*FREQ
         S = CMPLX (0.,W)
         DB = 0.
         PHAS = 0.
         DLAY = 0.
C   **   COMPUTE MAG,PHASE,DEL FOR EACH FREQ
         DO 300 I=1,NSEC
         NUMR = M(I)*S**2+C(I)*S+D(I)
         DEM = N(I)*S**2+A(I)*S+B(I)
```

```
      X = CABS(NUMR/DEM)
      DB = DB+20.*ALOG10(X)
      RL = REAL(NUMR/DEM)
      IMG = AIMAG(NUMR/DEM)
      PHAS = PHAS+ATAN2(IMG,RL)
      X1 = D(I)-M(I)*W**2
      IF (ABS(X1).LT.0.00001) X1 = 10E-10
      X2 = B(I)-N(I)*W**2
      IF (ABS(X2).LT.0.00001) X2 = 10E-10
      D1 = (1./(1.+(C(I)*W)**2/X1**2))*(C(I)/X1+2.*C(I)*M(I)*
     + W**2/X1**2)
      D2 = (1./(1.+(A(I)*W)**2/X2**2))*(A(I)/X2+2.*A(I)*N(I)*
     + W**2/X2**2)
      DLAY = DLAY-D1+D2
  300 CONTINUE
      PHAS = PHAS*57.29577951
C  ** PRINT OUTPUT
      WRITE (6,1003) FREQ,DB,PHAS,DLAY
 1003 FORMAT (1H ,E12.5,3X,F10.3,3X,F9.3,4X,F10.6)
      FREQ = FREQ+FI
      GO TO 200
  400 WRITE (6,1004)
 1004 FORMAT (' ENTER 1 FOR MORE RUNS' )
      READ (6,1005) IJ
 1005 FORMAT (I1)
      IF (IJ.EQ.1) GO TO 10
      STOP
      END
```

Input Cards for CHEB

Card 1: Filter type
'LP' = low-pass; 'HP' = high-pass; 'BP' = band-pass; 'BR' = band-reject. Include the apostrophes.

Card 2: Filter requirements

(A) For LP and HP:
- FPASS = passband edge frequency Hz
- FSTOP = stopband edge frequency Hz
- AMAX = maximum loss in passband dB
- AMIN = minimum loss in stopband dB

(B) For BP and BR:
- FPASSL = lower passband frequency Hz
- FPASSH = upper passband frequency Hz
- FSTOPL = lower stopband frequency Hz
- FSTOPH = upper stopband frequency Hz
- AMAX = maximum loss in passband dB
- AMIN = minimum loss in stopband dB

Card 3: Frequencies for computing gain
- FS = start frequency Hz
- FI = frequency increment Hz
- FF = final frequency Hz

Card 4: Continue or end card
1 = another run expected (data follows)
0 = terminate run

The input cards needed for Example 4.11 are (unformatted READ assumed):

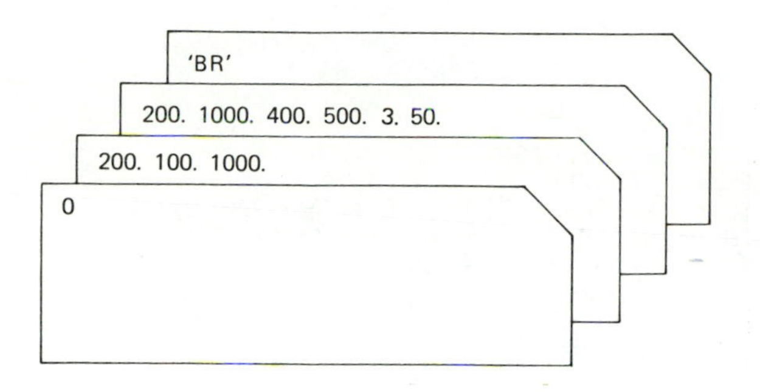

II. Program Listing for CHEB

```
C   ** PROGRAM FOR CHEBYSHEV APPROXIMATION
C   ** FOR LP HP BP BR FILTERS
       DIMENSION FTYPE(4),RE(10),IM(10),M(10),C(10),D(10),
      +N(10),A(10),B(10),W(20),B1(20)
       REAL IM,IX,N,PI,M
       COMPLEX S
       DATA FTYPE/'LP  ','HP  ','BP  ','BR  '/,PI/3.141592654/
C   ** PRINT 'LP' 'HP' 'BP' 'BR' TO IDENTIFY FILTER TYPE
     1 WRITE (6,1010)
  1010 FORMAT (' ENTER FILTER TYPE AS ',22H'LP', 'HP', 'BP', 'BR',
      +     ' (WITH QUOTES)')
       READ (6,*)FILT
       DO 10 I=1,4
       IFT=I
       IF (FILT.EQ.FTYPE(I))GO TO 15
    10 CONTINUE
       GO TO 900
    15 IF (IFT.GT.2) GO TO 20
C   ** READ FILTER REQUIREMENTS FOR LP & HP
C   ** FPASS=PASSBAND EDGE FREQ (HZ)
C   ** FSTOP=STOPBAND EDGE FREQ (HZ)
C   ** AMAX=PASSBAND RIPPLE (DB)
C   ** AMIN=STOPBAND LOSS AT FSTOP (DB)
     2 WRITE (6,1012)
  1012 FORMAT (' ENTER INPUT FOR LP OR HP'/9X,'FREQUENCIES (HZ):',
      +    '  FPASS, FSTOP'/9X,'LOSS (DB):  AMAX, AMIN')
       READ (6,*) FPASS,FSTOP,AMAX,AMIN
       GO TO 25
C   ** READ FILTER REQUIREMENTS FOR BP &BR
C   ** FPASSL=LOWER PASSBAND FREQ (HZ)
C   ** FPASSH=UPPER PASSBAND FREQ (HZ)
C   ** FSTOPL=LOWER STOPBAND EDGE FREQ (HZ)
C   ** FSTOPH=UPPER STOPBAND EDGE FREQ (HZ)
    20 WRITE (6,1014)
```

```
 1014 FORMAT (' ENTER INPUT FOR BP OR BR'/9X,'FREQUENCIES (HZ):',
     +     '  FPASSL, FPASSH, FSTOPL, FSTOPH'/9X,'LOSS (DB):',
     +     '  AMAX, AMIN')
      READ (6,*) FPASSL,FPASSH,FSTOPL,FSTOPH,AMAX,AMIN
C   ** READ FREQS FOR GAIN COMPUTATION
C   ** FS = START FREQUENCY (HZ)
C   ** FI = FREQUENCY INCREMENT (HZ)
C   ** FF = FINAL FREQUENCY (HZ)
   25 WRITE (6,1016)
 1016 FORMAT (' ENTER FREQUENCIES FOR GAIN COMPUTATION'/9X,
     +     'START FREQ, FREQ INCREMENT, FINAL FREQ')
      READ (6,*) FS,FI,FF
C   ** COMPUTE ORDER
   30 IF (IFT.EQ.1) FSTOPN=FSTOP/FPASS
      IF (IFT.EQ.2) FSTOPN=FPASS/FSTOP
      IF (IFT.EQ.3) FSTOPN=(FSTOPH-FSTOPL)/(FPASSH-FPASSL)
      IF (IFT.EQ.4) FSTOPN=(FPASSH-FPASSL)/(FSTOPH-FSTOPL)
      X1=(10.**(.1*AMIN)-1.)/(10.**(.1*AMAX)-1)
      X1=SQRT(X1)
      X1=ALOG(X1+SQRT(X1**2-1.))
      X2=FSTOPN
      X2=ALOG(X2+SQRT(X2**2-1.))
      ORDERN=X1/X2
      NO=ORDERN+.998
      INO=NO
C   ** COMPUTE LP ROOTS
      IF (IFT.EQ.3.OR.IFT.EQ.4)INO=NO*2
      N1=(NO+1.)/2.
      Y1=1./SQRT(10.**(.1*AMAX)-1.)
      Y1=ALOG(Y1+SQRT(Y1**2+1.))
      Y2=Y1/NO
      Y3=SINH(Y2)
      Y4=COSH(Y2)
      DO 40 J=1,N1
      Z=(PI/2.)*((2.*J-1.)/NO)
      RE(J)=Y3*SIN(Z)
   40 IM(J)=Y4*COS(Z)
C   ** BRANCH FOR HP BP BR POLES
      IF (IFT.EQ.2) GO TO 60
      IF (IFT.EQ.3) GO TO 100
      IF (IFT.EQ.4) GO TO 80
C   ** LP BIQUADRATIC FUNCTIONS
      DO 50 J=1,N1
      RE(J)=2.*PI*FPASS*RE(J)
      IM(J)=2.*PI*FPASS*IM(J)
      M(J)=0.
      C(J)=0.
      D(J)=RE(J)**2+IM(J)**2
      N(J)=1.
      A(J)=2.*RE(J)
   50 B(J)=D(J)
      IREM=MOD(NO,2)
      IF (IREM.EQ.0) GO TO 160
      D(N1)=RE(N1)
      N(N1)=0.
      A(N1)=1.
      B(N1)=RE(N1)
      GO TO 160
C   ** HP BIQUADRATIC FUNCTIONS
   60 DO 70 J=1,N1
      Y5=RE(J)**2+IM(J)**2
```

```
      RE(J)=(RE(J)/Y5)*2.*PI*FPASS
      IM(J)=(IM(J)/Y5)*2.*PI*FPASS
      M(J)=1.
      C(J)=0.
      D(J)=0.
      N(J)=1.
      A(J)=2.*RE(J)
   70 B(J)=RE(J)**2+IM(J)**2
      IREM=MOD(NO,2)
      IF (IREM.EQ.0) GO TO 160
      M(N1)=0.
      C(N1)=1.
      N(N1)=0.
      A(N1)=1.
      B(N1)=RE(N1)
      GO TO 160
C   ** BR BIQUADRATIC FUNCTIONS
   80 DO 90 J=1,N1
      Y9=RE(J)**2+IM(J)**2
      RE(J)=RE(J)/Y9
   90 IM(J)=IM(J)/Y9
C   ** BP & BR BIQUADRATIC FUNCTIONS
  100 IREM=MOD(NO,2)
      BW=FPASSH-FPASSL
      FC=SQRT(FPASSH*FPASSL)
      R=BW**2/(4.*FC**2)
      DO 120 J=1,N1
      IF (J.EQ.N1.AND.IREM.NE.0) GO TO 110
      RX=RE(J)**2
      IX=IM(J)**2
      U=.5+(RX+IX)*R/2.
      V=RX*R
      P=SQRT(ABS(U**2-V))
      XK=SQRT(ABS(U+P))
      XM=SQRT(ABS(U-P))
      W(J)=FC*(XK+SQRT(ABS(XK**2-1.)))
      W(NO+1-J)=FC**2/W(J)
      B1(J)=2.*XM*W(J)
      B1(NO+1-J)=2.*XM*W(NO+1-J)
      GO TO 120
  110 W(J)=FC
      B1(J)=RE(J)*BW
  120 CONTINUE
      DO 130 J=1,NO
      W(J)=W(J)*6.28318531
      B1(J)=B1(J)*6.28318531
      M(J)=0.
      C(J)=B1(J)
      D(J)=0.
      N(J)=1.
      A(J)=C(J)
      IF (IFT.EQ.4) C(J)=0.
      IF (IFT.EQ.4) D(J)=4*PI*PI*FPASSL*FPASSH
      IF (IFT.EQ.4) M(J)=1.
  130 B(J)=W(J)**2
      W0=2.*PI*FPASSL
      S=CMPLX(0.,W0)
      DBC=0.
      IF (IFT.EQ.4) GO TO 160
      DO 140 J=1,NO
      DBC=DBC+20.*ALOG10(CABS(M(J)*S**2+C(J)*S+D(J)))
```

```
  140 DBC=DBC-20.*ALOG10(CABS(N(J)*S**2+A(J)*S+B(J)))
      DO 150 J=1,NO
  150 C(J)=C(J)/(10.**((DBC+AMAX)/20./NO))
  160 NBQ=N1
      IF (IFT.EQ.3.OR.IFT.EQ.4) NBQ=NO
      NCH=NBQ/2
      NCH=NCH*2
      IF (NCH.EQ.NBQ) M(1)=M(1)/(10**(AMAX/20))
      IF (NCH.EQ.NBQ) D(1)=D(1)/(10**(AMAX/20))
C  ** PRINT APPROXIMATION FUNCTIONS IN BIQUADRATIC FORM
      WRITE (6,170)
  170 FORMAT (5X,'M',11X,'C',15X,'D',7X,'N',9X,'A',15X,'B')
      DO 190 J=1,NBQ
      WRITE (6,180)M(J),C(J),D(J),N(J),A(J),B(J)
  180 FORMAT (1X,F8.2,2X,F10.2,2X,F16.4,2X,F2.0,2X,F10.2,2X,F16.4)
  190 CONTINUE
C  ** COMPUTE GAIN AT GIVEN FREQUENCIES
      WRITE (6,200)
  200 FORMAT (3X,'FREQ(HZ)',3X,'GAIN(DB)')
      FREQ=FS
  300 IF (FREQ.GT.FF) GO TO 999
      WB=2.*PI*FREQ
      S=CMPLX(0.,WB)
      DB1=0.
      DO 210 J=1,NBQ
      DB1=DB1+20.*ALOG10(CABS(M(J)*S**2+C(J)*S+D(J)))
      DB1=DB1-20.*ALOG10(CABS(N(J)*S**2+A(J)*S+B(J)))
  210 CONTINUE
C  ** PRINT GAIN AT CRITICAL FREQUENCIES
      WRITE (6,220) FREQ,DB1
  220 FORMAT (3X,F8.1,3X,F8.3)
  230 CONTINUE
      FREQ=FREQ+FI
      GO TO 300
  900 WRITE (6,901)
  901 FORMAT (' INVALID FILTER TYPE')
  999 WRITE (6,933)
  933 FORMAT (' ENTER 1 FOR MORE RUNS')
      READ (6,934) II
  934 FORMAT (I1)
      IF (II.EQ.1) GO TO 1
      STOP
      END
```

ANSWERS TO SELECTED PROBLEMS

Chapter 1

1.1 $\dfrac{4}{2s^3 + 13s^2 + 22s + 4}$

1.7 4.7

1.9c $\left|\dfrac{V_O}{V_{IN}}\right|_{100\,\text{kHz}} = 1.961$

1.17 $\dfrac{-2s}{s^2 + 2s + 2}$

Chapter 2

2.9 6

2.18b $\dfrac{10.065s^2}{(s+2)^2(s+3)}$

2.22b 1.6 decades, 5.32 octaves

2.32b 50

Chapter 3

3.1 (b) high-pass, (d) high-pass-notch, (g) band-reject

3.5 0.316

3.9b 0.75, 0.335, 0.067

Chapter 4

4.2 41.2 dB

4.3 1872.8 Hz

4.5b 70.47 dB

4.8 $L_1 = 2\text{H}, C_1 = C_2 = 1\text{F}, K = 1/2$

4.10 6964.3 Hz

4.14 (a) n extrema

(b) min $= 1$, max $= \sqrt{1 + \varepsilon^2}$

(c) $\Omega_{\min} = \cos\left(\frac{k\pi}{2n}\right) \quad k = 1, 3, 5 \ldots$

$\Omega_{\max} = \cos\left(\frac{k\pi}{n}\right) \quad k = 0, 1, 2 \ldots$

4.18 49.6 dB

4.19 32.6 dB

4.23 (a) 36 dB
(b) 74 dB

4.24b 1270 rad/sec

4.31 12

Chapter 5

5.4 $S_{R_1}^{Q_p} = S_{C_1}^{Q_p} = -\frac{1}{2} + Q_p\sqrt{\frac{R_2C_2}{R_1C_1}}$

$S_{R_2}^{Q_p} = S_{C_2}^{Q_p} = -\frac{1}{2} + \left(1 - \frac{r_2}{r_1}\right)Q_P\sqrt{\frac{R_1C_1}{R_2C_2}}$

$S_{r_2}^{Q_p} = -S_{r_1}^{Q_p} = \frac{r_2}{r_1}Q_p\sqrt{\frac{R_1C_1}{R_2C_2}}$

5.10 $\frac{\Delta\omega_p}{\omega_p} = \pm 2\%; \quad \frac{\Delta Q_p}{Q_p} = \pm 3\%$

5.13a 0.174 dB, 0.087 dB, 0. dB, 0. dB

5.14 (a) $\mathscr{S}_{\omega_p}^G = -28.11$ dB, $\mathscr{S}_{Q_p}^G = 1.198$ dB
(b) $\mathscr{S}_{\omega_p}^G = -\mathscr{S}_{Q_p}^G = -8.686$ dB

5.16 At $\omega = 9$: $\mathscr{S}_{\omega_p}^G = -52.3$ dB, $\mathscr{S}_{Q_p}^G = 4.1$ dB, $\Delta G = -1.046$ dB

5.18 (a) $\pm 50\ \Omega$
(b) $175\ \Omega$
(c) $225\ \Omega$

5.20 $\mu\left(\frac{\Delta R}{R}\right) = 0.01, 3\sigma\left(\frac{\Delta R}{R}\right) = 0.0075$

5.21 $\mu\left(\frac{\Delta\omega_p}{\omega_p}\right) = -0.001, 3\sigma\left(\frac{\Delta\omega_p}{\omega_p}\right) = 0.0019$

$\mu\left(\frac{\Delta Q}{Q}\right) = 0., \quad 3\sigma\left(\frac{\Delta Q}{Q}\right) = 0.0019$

5.22 0.02 dB

5.25 1 percent: (a) -0.198 dB (b) -0.195 dB
10 percent: (a) -1.98 dB (b) -1.77 dB

Chapter 6

Abbreviations: ser ≡ series branch;
sh ≡ shunt branch;
anti ≡ series antiresonant branch;
(elements in henries, farads, and ohms).

6.3 $L_1(\text{ser}) = 1$, $C_2(\text{ser}) = \frac{1}{2}$, $L_3 = \frac{1}{2}$ and $C_4 = 1$ (anti)

6.5a $L_1(\text{ser}) = 1$, $C_2(\text{sh}) = \frac{1}{3}$, $L_3(\text{ser}) = \frac{9}{2}$, $C_4(\text{sh}) = \frac{1}{6}$

6.8a $C_1(\text{ser}) = \frac{2}{3}$, $R_2(\text{ser}) = 1$, $C_3 = 2$ and $R_4 = \frac{1}{4}$ (anti)

6.10a $R_1(\text{ser}) = \frac{1}{2}$, $C_2(\text{sh}) = \frac{1}{2}$, $R_3(\text{ser}) = 2$, $C_4(\text{sh}) = \frac{1}{2}$

6.13 $C_1 = \frac{3}{4}$, $L_1 = \frac{1}{2}$, $C_2 = 1$, $C_3 = \frac{1}{3}$, $L_3 = \frac{3}{10}$

6.14 $C_1(\text{sh}) = 1$, $L_2(\text{ser}) = \frac{1}{3}$, $C_3(\text{sh}) = \frac{9}{5}$,
$L_4(\text{ser}) = \frac{25}{24}$, $C_5(\text{sh}) = \frac{8}{15}$

6.18 $L_1 = \frac{3}{8}$, $C_2 = \frac{2}{7}$, $L_3 = \frac{7}{8}$, $L_4 = \frac{133}{80}$, $C_5 = \frac{8}{21}$, $L_6 = \frac{21}{40}$

6.23 $L_1(\text{ser}) = \frac{1}{2}$, $C_2(\text{sh}) = \frac{4}{3}$, $L_3(\text{ser}) = \frac{3}{2}$, $R_L(\text{sh}) = 1$

6.24a $L_1 = 0.00153$, $C_2 = 0.00158$, $L_3 = 0.00108$, $C_4 = 0.00038$, $R_L = 1$

6.27 $C_1(\text{ser}) = 0.00653$, $L_2(\text{sh}) = 0.00633$, $C_3(\text{ser}) = 0.00926$,
$L_4(\text{sh}) = 0.02631$, $R_L(\text{sh}) = 1$

6.30 (a) $K(s) = s^4$; $H(s) = (s^2 + 0.76537s + 1)(s^2 + 1.84776s + 1)$
(c) $K(s) = 0.9735s[s^2 + (0.866)^2]$;
$H(s) = 0.9735(s^2 + 0.76822s + 1.33863)(s + 0.76722)$

6.31 $R_s = 1$, $C_1(\text{sh}) = 1$, $L_2(\text{ser}) = 2$, $C_2(\text{sh}) = 1$, $R_L = 1$

6.34a $R_s = 1$, $C_1(\text{sh}) = 0.9484$, $L_2 = 1.0234$ and $C_2 = 0.1142$ (anti),
$C_3(\text{sh}) = 0.9484$, $R_L = 1$

Chapter 7

7.2 $$\frac{-Z_2/Z_1}{1 + \dfrac{1}{A}\left(1 + \dfrac{Z_2}{Z_1}\right)}$$

7.6a $$\frac{2s/R_1C_1 + 2/R_1R_2C_1C_2}{s^2 + s\left(\dfrac{1}{R_1C_1} + \dfrac{1}{R_2C_2} - \dfrac{1}{R_2C_1}\right) + \dfrac{1}{R_1R_2C_1C_2}}$$

7.9 Section 1: $R_1 = R_2 = \frac{1}{9}$,
Section 2: $R_1 = R_2 = \frac{1}{8}$, $K = 480$

7.13 $C_1 = C_2 = 0.1\ \mu\text{F}, R_1 = 1.56\ \text{k}\Omega, R_2 = 100\ \text{k}\Omega,$
$r_1 = \infty, r_2 = 0.$

7.20a $c_1 = a, c_2 = 0$

Chapter 8

8.1 Saraga design: $C_1 = 0.0866\ \mu\text{F}, C_2 = 0.01\ \mu\text{F}, R_1 = 40\ \text{k}\Omega,$
$R_2 = 116\ \text{k}\Omega, r_1 = 30\ \text{k}\Omega, r_2 = 10\ \text{k}\Omega$
(a) $C_1 = 0.216\ \mu\text{F}, C_2 = 0.025\ \mu\text{F}$
(b) See Figure 8.7: $R_4 = 53.3\ \text{k}\Omega, R_5 = 160\ \text{k}\Omega$

8.3 Section 1 Saraga design: $C_1 = 0.047\ \mu\text{F}, C_2 = 0.0159\ \mu\text{F},$
$R_1 = 5.48\ \text{k}\Omega, R_2 = 5.4\ \Omega, r_1 = 30\ \text{k}\Omega, r_2 = 10\ \text{k}\Omega$
Section 2 (Figure Pl.12): $R_1 = 15.96\ \text{k}\Omega, R_2 = 21.28\ \text{k}\Omega, C_1 = 0.0159\ \mu\text{F}$

8.6 (a) $R_1 = 1/\omega_p, R_2 = Q_p/\omega_p, C_2 = 1/Q_p$

(b) $S_{R_1}^{Q_p} = -S_{R_1}^{Q_p} = -\dfrac{1}{2} + Q_P;\ S_{C_1}^{Q_p} = -S_{C_2}^{Q_p} = \dfrac{1}{2} + Q_p;$

$S_{r_2}^{Q_p} = -S_{r_1}^{Q_p} = 1;\ S_{A_0\alpha}^{\omega_p} = -S_{A_0\alpha}^{Q_p} = \dfrac{2\omega_p}{A_0\alpha}$

(c) $3\sigma(\Delta G) = 1.21$ dB

8.10 $k \cong 1.4, 3\sigma(\Delta G) = 1.02$ dB

Answers to Problems 8.11 to 8.15 are based on Equation 8.34, and Problems 8.7 to 8.9.

8.11 $D1$: $3\sigma(\Delta G) = 14.3$ dB; $D3$: $3\sigma(\Delta G) = 4.28$ dB

8.12 $D1$: $3\sigma(\Delta G) = 4.43$ dB; $D3$: $3\sigma(\Delta G) = 1.065$ dB

8.14 (a) 1.4 percent (b) 1.44 kHz

8.16 *Hint*: $D1$ requires less elements than $D3$.

8.18 $K_{\max} = 75.78$

8.20 $R_1 = R_2 = R_3 = \sqrt{2}/\omega_p, C_1 = C_2 = 1, k = 5 - \sqrt{2}/Q_p$

8.24 BP, LP, and a HP (numerator of the form $s^2 + \alpha s$)

Chapter 9

9.1 $C_1 = C_2 = 0.01\ \mu\text{F}, R_1 = 1.953\ \text{k}\Omega, R_2 = 500\ \text{k}\Omega.$
To achieve $K = 400$, replace R_1 by $R_4 = 250\ \text{k}\Omega$ and $R_5 = 1.968\ \text{k}\Omega$

9.5 $R_2/R_1 = 84.86, R_B/R_A = 100, C_1 = C_2 = 0.01\ \mu\text{F}, R_B = 300\ \text{k}\Omega,$
$R_A = 3\ \text{k}\Omega, R_2 = 287.9\ \text{k}\Omega, R_4 = 252.2\ \text{k}\Omega, R_5 = 3.435\ \text{k}\Omega$

9.9 $k = 9$

9.12 $C_1 = C_2 = 0.01\ \mu\text{F}, R_1 = 6.42\ \text{k}\Omega, R_2 = 93\ \text{k}\Omega, R_4 = 13.26\ \text{k}\Omega,$
$R_5 = 12.46\ \text{k}\Omega, R_6 = 214\ \text{k}\Omega, R_7 = \infty, R_B = 10\ \text{k}\Omega, R_C = 2\ \text{k}\Omega,$
$R_D = 2\ \text{k}\Omega$

9.13 $R_2 = 2/a, R_1 = a/2b, K_1 = m + 2dR_1^2 - cR_1,$

$$R_3 = \frac{m - K_3}{R_1 b(d/b - m)}, R_D = 1, R_C = \frac{1}{m} - 1$$

Remaining elements are obtained from Equation 9.61 a and b

9.17 $C_1 = 0.1\ \mu\text{F}, C_2 = 0.01\ \mu\text{F}, R_1 = 98.18\ \text{k}\Omega, R_2 = 9.25\ \text{k}\Omega,$
$R_3 = 120.27\ \text{k}\Omega$

9.25 Saraga: $3\sigma(\Delta G) = 1.71$ dB (best choice);
Delyiannis: 2.32 dB; Designs A and C: 17.4 dB

9.27 *BP* and *HP*

Chapter 10

10.1a $C_1 = C_2 = 1, R_1 = 0.2, R_2 = R_3 = R_7 = R_{10} = 0.01581, R_4 = 0.02,$
$R_8 = 0.001, R_9 = 0.00158$. ($R_8$ is connected to $-V_{LP}$ node)

10.4 Section 1: $C_1 = C_2 = 0.0159\ \mu\text{F}, R_1 = R_4 = 18.6\ \text{k}\Omega,$
$R_2 = R_3 = R_6 = R_7 = R_8 = 9.33\ \text{k}\Omega, R_5 = 2.08\ \text{k}\Omega$
Section 2 (Figure P1.12): $R_1 = 14.4\ \text{k}\Omega, C_1 = 0.0159\ \mu\text{F},$
$R_2 = 64.8\ \text{k}\Omega$

10.7 (a) $3\sigma(\Delta G) = 0.33$ dB
(b) $3\sigma(\Delta G) = 0.53$ dB at $\omega = 275$ rad/sec

10.8 Delyiannis: $3\sigma(\Delta G) = 1.83$ dB; Saraga: 2.13 dB;
Three Amplifier Biquad: 2.14 dB

10.13 R_1: -2 percent, R_3: $+1.0025$ percent, R_4: $-.862$ percent

10.19 12.7 kHz

10.27 666 kHz

Chapter 11

11.9a $L_1 = 11.7$ mH, $C_1 = 2.99\ \mu$F, $L_2 = 8.16$ mH, $L_3 = 25.7$ mH

11.10 The gyrator-*RC* realization of the dual circuit is the least expensive. The costs are: gyrator 15.6; *FDNR* 14.4; gyrator (dual) 8.0; *FDNR* (dual) 24.0.

Chapter 12

12.1 $\omega_g \cong 9(10)^5, \omega_\phi \cong 1.6(10)^5, GM \cong -30$ dB, $PM \cong -40°$.
Unstable.

12.3 (a) $GM = 26$ dB, $PM \cong 30°$
(b) $GM = 40.8$ dB, $PM \cong 45°$

12.5 $\omega_g = 2\pi 2.2(10)^4, \omega_\phi = 2\pi 7(10)^3, GM = -20$ dB, $PM = -11°$

12.8b $A_0 = 10^5, \omega_1 = 6.41(10)^4, \omega_2 = 5.76(10)^6, \omega_3 = 4\pi(10)^6$

12.11 $\omega_g = 3.2(10)^7, \omega_\phi = 2(10)^7, GM = -10$ dB, $PM = -20°$

12.12 312.5 pF, $\omega_g = 6.5(10)^6$, $\omega_\phi = 2(10)^7$, $GM = 16$ dB, $PM = 45°$

12.14 -0.5 mV

12.15 9.901 kHz

12.17 $\omega_0' = 2\pi 2040$, $Q_0' = 9.82$

12.22 $3\sigma\left(\dfrac{\Delta\omega_0}{\omega_0}\right) = 0.015$, $3\sigma\left(\dfrac{\Delta Q_0}{Q_0}\right) = 0.0138$, $3\sigma(\Delta G) = 1.304$ dB

12.25b $R_L(\text{min}) = 357\ \Omega$

12.29 7.96 kHz

12.30 $A_1(\text{max}) = 4.77$ V, current limiting

12.32 284.5 mV

12.34b 25.9 kΩ

12.36 6.33 mV

INDEX